THE BIOLOGY OF DIATOMS

BOTANICAL MONOGRAPHS

BOTANICAL MONOGRAPHS · VOLUME 13

THE BIOLOGY OF DIATOMS

EDITED BY

DIETRICH WERNER

Dr. rer. nat.

Professor,
Botanisches Institut,
Universität Marburg-L

UNIVERSITY OF CALIFORNIA PRESS

BERKELEY AND LOS ANGELES

UNIVERSITY OF
CALIFORNIA PRESS
Berkeley and Los Angeles, California

ISBN: 0 520 03400 7
Library of Congress Catalog Card Number: 76–55574

Printed in Great Britain

CONTENTS

CONTRIBUTORS

W. MARSHALL DARLEY, Department of Botany, The University of Georgia, Athens, Georgia 30602, USA.

GERHARD DREBES, Biologische Anstalt Helgoland, Litoralstation List, D-2282 List (Sylt), Germany

ELEANOR L. DUKE, Department of Biological Sciences, The University of Texas at El Paso, Texas 79968, USA.

RICHARD W. EPPLEY, Institute of Marine Resources, University of California, San Diego, La Jolla, California 92093, USA.

ROBERT R.L. GUILLARD, Woods Hole Oceanographic Institution, Woods Hole, Massachusetts 02543, USA.

MARGARET A. HARPER, 169 Pembroke Road, Northland, Wellington 5, New Zealand

GRETHE RYTTER HASLE, Institutt for Marinbiologi og Limnologi avd. Marin Botanikk, Universitetet i Oslo, Postboks 1069, Blindern, Oslo 3, Norway

JOHAN A. HELLEBUST, Department of Botany, University of Toronto, Toronto, Ontario M5S 1 A 1, Canada

ERIK G. JØRGENSEN, Plantephysiologisk Institut, University of Kopenhagen, Øster Farimagsgade 2 A, Denmark

PETER KILHAM, Department of Ecology and Evolutionary Biology, Division of Biological Sciences, Natural Sciences Building, University of Michigan, Ann Arbor, Michigan 48104, USA.

JOYCE LEWIN, Department of Oceanography, University of Washington, Seattle, Washington 98105, USA.

C. DAVID MCINTIRE, Department of Botany and Plant Pathology, Oregon State University, Corvallis, Oregon 97331, USA.

WENDY M. MOORE, Department of Botany and Plant Pathology, Oregon State University, Corvallis, Oregon 973331, USA.

RUTH PATRICK, The Academy of Natural Sciences, Nineteenth and the Parkway, Philadelphia, Pennsylvania 19103, USA.

BERNHARD E.F. REIMANN, Department of Biology, New Mexico State University, Las Cruces, New Mexico 88003, USA.

DIETRICH WERNER, Botanisches Institut der Universität, Fachbereich Biologie, D-355 Marburg-L, auf den Lahnbergen, Germany

PREFACE

Our knowledge of the biology of a group of microorganisms is determined by various historical and scientific factors and influences. In the case of diatoms the early and intensive study of shell morphology and the taxonomy based upon that, contrasted markedly for a long time with the few scientists studying the physiology, the biochemistry and several ecological aspects. The breakthrough came with the successful cultivation of a number of species, which allowed diatoms to be used as laboratory organisms in the same way as many bacteria and other unicellular algae. This book therefore stresses these more recent developments and topics. The very recent and intensive use of the scanning electron microscope for taxonomical purposes may also lead to further progress in diatom systematics. But, as several expert scientists argued, a complete and revised taxonomy may not be available for several years.

For the completion of the book I have to thank particularly the contributors, who wrote their chapters in addition to the regular and doubtless formidable work load and Blackwell Scientific Publications for excellent cooperation. I also would like to thank Mrs. B. Kalkowski, Mrs. E. Stangier, Dr. J. Wilcockson and Mr. R. Roth from my laboratory for their help in preparing the book.

December 1976 *D. Werner*

CHAPTER 1

INTRODUCTION WITH A NOTE ON TAXONOMY

D. WERNER

Botanisches Institut der Universität,
Fachbereich Biologie, D-355 Marburg-L, Germany

THE USE OF ELECTRON MICROSCOPY IN MORPHOLOGICAL AND TAXONOMICAL DIATOM STUDIES

G. HASLE

1 CONTRIBUTIONS OF THE STUDY OF DIATOMS TO BIOLOGY

The total net primary production of plants on earth is estimated to be in the order of $1{\cdot}4 \times 10^{14}$ kg dry weight per year, of which at least 40% is produced by the marine phytoplankton species (Golley, 1972). This means, that about 2×10^{13} kg carbon is fixed in organic form per year by the various groups of marine phytoplankton species (Strickland, 1972). Due to their dominant role in the most productive areas of the oceans such as the upwelling areas and the continental shelves (see Chapter 12) we can estimate that diatoms contribute about 20 to 25% of the world net primary production. This figure puts the class of diatoms in at least the same position as regards world primary production as the Pinacean in the temperature and boreal forests and the Poaceae (grasses) in the savannas, grasslands and cultivated areas.

Diatoms played a key role in the origins of mariculture and sea farming. Allen & Nelson (1910) were the first to succeed in maintaining several species of marine diatoms in persistent culture and used them for rearing larvae of echinoderms, molluscs and copepods. *Phaeodactylum tricornutum* (the former *Nitzschia closterium* forma minutissima) (J. Lewin, 1958) was for many years the species most frequently cultured. During the next decades several other diatom species proved to be a much better food for larval growth and for bivalve molluscs and shrimps. These included *Chaetoceros calcitrans* (Imai & Hatanaka, 1950), *Thalassiosira pseudonana* (Guillard & Ryther, 1962). But mixtures of small flagellates such as *Isochrysis galbana, Monochrysis lutheri* and several centric diatoms such as *Skeletonema costatum* and *Chaetoceros calcitrans* (Walne, 1970) proved to be even better than feeding unialgal cultures.

Also, due to their ecological significance, diatoms were included in the first biological systems for evaluation of water quality and pollution (Kolkwitz & Marsson, 1908). Today, the number of zones used to classify the saprobicity levels varies considerably (Wuhrmann, 1951; Sladecek, 1961; Fjerdingstadt, 1971). The range of communities, typical for the various zones and the significance of diatom species as indicators in these communities (Patrick *et al.,* 1954; Fjerdingstadt, 1964; Cholnoky, 1968) is still rather controversial. The number of identified and accumulated pollutants in fresh and coastal waters is increasing so much faster than our knowledge of the influence of these substances on growth and reproduction of the indicator organisms, that the knowledge required before a biological indicator system of water quality can be operated is steadily increasing. The number of identified organic chemical substances in fresh and coastal waters is estimated as being greater than 100 000. The number of different polyarenes alone, identified with sophisticated gaschromatographic methods, is greater than 10 000 (Blumer, 1975).

A programme of a systematic study of how more than 10 000 species of diatoms take up these pollutants, at what rate and at what lowest possible concentration would involve more than a billion experiments. Combinations of several pollutants, influencing the common uptake and metabolism, would result in an almost countless number. Significant differences in the uptake of phenol (Ks and K_0) by more than a factor of ten were found for closely related *Nitzschia* species (Werner & Pawlitz, unpublished results) leaving less than 5×10^{-9} M phenol in the medium. This shows, that every species chosen as a biological indicator must be tested separately to evaluate its potential for uptake of pollutants. The uptake (or perhaps just binding) of inorganic or organic pollutants could be used as a physiological parameter for understanding the role of diatoms in the saprobic system.

No diatom species has yet gained a comparable broad use as a unicellular laboratory organism in physiology, biochemistry or genetics like some bacteria, unicellular green algae or yeasts. Considering the ubiquitous occurrence of diatoms and their impressive role in marine and fresh water ecology, this is perhaps somewhat surprising. Probably this was mainly due to the fact that the continuous cultivation and axenic culture of many diatom species has only been achieved during the last ten years, and that the sexual life cycle and the diminution of cell size complicated the microbiological handling of diatom species.

In the early stages of this research diatoms, with their shells and shell structure, were favourite objects of ultrastructure research and electron microscopy (Müller & Pasewaldt, 1942; Gerloff & Gölz, 1944). But this interest did not extend to the use of diatoms in other areas of ultrastructure research. In the comprehensive handbook of ultrastructure research (Lima-de Faria, 1969) diatoms remain completely unmentioned.

Mazia (1961) in a comprehensive review noted the fact that in many diatom species a normal mitotic cell division takes place while the cell size decreases continuously. He compares this with the division of cells with a subnormal size in some yeasts like *Schizosaccharomyces pombé* and ciliates. Robards (1970) includes in his book on plant cell ultrastructure only the valve structures and a few other points of interest in diatom cytology such as the quantitative studies on spindle formation and microtubule numbers in *Lithodesmium* (Manton *et al.,* 1969).

Diatoms were for a long time specially favoured organisms for permeability studies (Höfler, 1940). The recovery from plasmolysis was apparently so much faster in diatoms than in other plant cells, that an unusual permeability for the substances used was established. But Bogen & Follmann (1955) and Bogen (1956) showed and emphasized that inhibitors like dinitrophenol or azide reduce the deplasmolysis considerably, indicating, that processes of an active transport were involved in the deplasmolysis experiments. Also within the diatoms species used, *Caloneis obtusa* proved to be an exception of the generally high sugar permeability (Übeleis, 1957).

Diatoms were the most suitable and most often employed organisms for studying silicon metabolism (see Chapter 4), an element, which apart from oxygen, is as abundant in the earth's crust as all the other 90 elements together. Since silicon has been shown to be also an essential element for some higher plants (Werner, 1967) and for vertebrates (Carlisle, 1972), the general interest in studies on silicon metabolism in diatoms has increased significantly. In 1977 a Nobel symposium will be organized in Stockholm on *Silicon Biochemistry and Related Problems.*

2 SOME REMARKS ON THE HISTORY OF DIATOM BIOLOGY

In 1786 the term 'Diatomaceen' was used for the first time by O.F. Müller in his comprehensive treatise 'infusoria fluviatillia et marina'.

The term 'diatoms' was also employed by C.A. Agardh (1824), a few years after L. Nitzsch (1817) created for the same group of organisms the name 'bacillariae'. Possibly due to the similarity of 'bacillae' and 'bacteriae' the term 'diatoms' or 'diatomaceae' was favoured by several authors (J.W. Bailey, 1842; W. Smith, 1853; G.C. Wallich, 1860; A. Grunow, 1863; J.D. MacDonald, 1869) while others used both 'diatoms' and 'bacillariae' synonymously (F. Kützing, 1844; E. Pfitzer, 1871) for a long time.

The striking microscopic appearance of diatoms attracted the attention of many biologists and amateur scientists in the second half of the nineteenth century. In the volumes of the most respected botanical journals from this period, sometimes more than half of the articles on algae were concerned with diatom taxonomy, cytology or ecology. Famous controversies and discussions flared up on several aspects of diatom biology. The question whether the diatoms belong to the plant rather than to the animal kingdom was still discussed by W. Smith (1853). He confirmed the opinion that diatoms are unicellular plants, homologous to the Desmidiaceae and other unicellular protophyta. The movement, the unusual silica frustule (old name: silex) and the reproduction, were the main points which were investigated. Schmidt (1874) discussed the copulation of *Gomphonema mustela*. Kützing (1844) and Nägeli (1849) were the first to recognize that diatoms were single cells and a few years later several approaches for a classification of diatoms were available (L. Rabenhorst, 1853; W. Smith, 1853; A. Grunow, 1863). Problems of terminology in the description of diatom cells also arose at this time (v. Mohl, 1861). Kützing (1844) used the terms 'primary' and 'secondary' side for girdle band view and valvar view, Smith (1853) 'front view' and 'side view' and Ehrenberg (1854) 'ventral' and 'dorsal' side. A few years later, J.D. MacDonald (1869) and E. Pfitzer (1869) recognized the importance of the epitheca and hypotheca in the diminution of the cell size, leading to the 'MacDonald-Pfitzer-rule' to which the vast majority of all diatom species adhere.

A literature survey by De Toni in 1891, *Bibliotheca Diatomologica,* included already more than 1500 original communications. The *Treatise on Diatomaceae* by van Heurck (1896), reprinted in 1962, presented and summarized 192 genera with more than 900 species and varieties. Type collections of diatoms from the same time contained a similar number of preparations, 625 in the collection of J. Tempère and H. Peragallo, about 700 in the famous collection of H.L. Smith and 800 in the 'Typen-Platte' of J.D.

Möller. During this time, A. Grunow (1884, 1889) influenced and encouraged several other diatom taxonomists. These more comprehensive works used and continued the contributions of other very productive scientists like K.K. Greville (1865) and P.T. Cleve (1884). Other aspects of diatom biology received more attention a few years later. A vehement and sometimes personally insulting discussion between O. Bütschli (1892), R. Lauterborn (1894) and O. Müller (1897) concerning the mechanism of diatom movement arose. It seemed to be answered in the protoplasmic streaming theory for almost half a century before new ideas and observations questioned this theory. F. Schütt (1897) made use of more intracellular components for his classification than others before him. But progress in diatom cytology was slow; this was apparently caused partly by the difficulties in culturing and maintaining the collected species alive. In 1889 O. Müller still emphasized that auxospore formation was an event rarely observed in all genera of diatoms studied so far. At the beginning of the twentieth century 'microspore' formation was observed in *Coscinodiscus concinnus* and *Chaetoceros boreale* (Murray, 1896), in *Biddulphia mobiliensis* (Bergon, 1903) and in *Corethron valdiviae* (Karsten, 1904).

Also, during this time more comprehensive treatises on the morphology of diatoms (Karsten, 1896; Mereschkowsky, 1903) became available.

Detailed planktonological field studies and extensive cruises gave a boost to the knowledge of the ecological significance of diatoms. V. Hensen (1887) studied the planktonic diatoms in the Baltic Sea, P.T. Cleve (1878, 1884, 1900) the plankton from various areas including the Java Sea, the North Sea and the Arctic oceans, Schröder (1901) in the Gulf of Naples, C.H. Ostenfeld (1902) in the North Sea and from the Faëroe Islands, H. and M. Pergallo (1897–1903) from the waters around France, Mereschkowsky (1902) from the Black Sea and K. Okamura (1907) from the Japanese coast.

Gran (1908) reported on the diatoms of the northern plankton in the Atlantic Ocean in a comprehensive survey edited by K. Brandt and S. Apstein in Kiel. Gran described 43 genera with more than 300 species found in the plankton north of the 50th parallel including a few species which were found so far only north of the 40th parallel, but which he expected also to be found later north of the 50th parallel. A few genera, *Biddulphia, Chaetoceros, Coscinodiscus, Melosira, Nitzschia, Rhizosolenia* and *Thalassiosira* encompassed the majority of all species described. With regard to the distribution, Gran still supported the idea that most planktonic diatom species are restricted to a rather limited area and only a few are present in all oceans.

Today we know that a great number of planktonic diatom species are ubiquitous (K.E. Lohmann, 1960; Yamaji, 1972) as are other members of the phyto- and zooplankton. New methods (R. Lauterborn, 1896; H. Lohmann, 1911; W.E. Allen, 1922; H.W. Harvey, 1934) increased the number of identified species and connected diatom ecology to the progress of ecology in general. In more recent planktological and ecological works (C.S. Boyer,

1926; N.I. Hendey, 1937; E. Cupp, 1943; B. Cholnoky, 1968) physiological aspects gain increasing consideration (see Chapters 10, 11 and 12).

The contributions of numerous scientists to diatom biology in the last four decades are mentioned and considered in some detail in the chapters of this book. There are three names and their works, covering four decades or more, that are brought in the author's view into special prominence. F. Hustedt's (1930, 1966) work fundamentally influenced the taxonomy and systematics of diatoms during this time and L. Geitler (1932, 1973) and H.A. von Stosch (1942, 1973) both related their basic work on cytology and sexual reproduction of diatoms to the development of cytology and biology in general.

3 CLASSIFICATION OF DIATOMS

The original papers cited in the following chapters span at least five decades. Without going into great detail of the inherent advantages and the controversies surrounding the various different systems, just two of the most respected and used systems are summarized here and one recent revision of the Centrales.

The system of F. Hustedt (1930) keeps the main division between the Centrales and the Pennales at the level of orders and divides the genera into 15 families and, altogether, 30 subfamilies in the following scheme:

Division : Bacillariophyta
- **Class : Diatomatae**
  - **1. Order : Centrales**
    - **I. Suborder : Discineae**
      - **1. Family : Coscinodiscaceae**
        - (a) Subfamily : Melosiroideae (1)
        - (b) Subfamily : Sceletonemoideae (2)
        - (c) Subfamily : Coscinodiscoideae (3)
      - **2. Family : Actinodiscaceae**
        - (a) Subfamily : Stictodiscoideae (4)
        - (b) Subfamily : Actinoptychoideae (5)
        - (c) Subfamily : Asterolamproideae (6)
      - **3. Family : Eupodiscaceae**
        - (a) Subfamily : Pyrgodiscoideae (7)
        - (b) Subfamily : Aulacodiscoideae (8)
        - (c) Subfamily : Eupodiscoideae (9)
    - **II. Suborder : Soleniineae**
      - **4. Family : Soleniaceae**
        - (a) Subfamily : Lauderioideae (10)
        - (b) Subfamily : Rhizosolenioideae (11)

III. Suborder: Biddulphiineae
- **5. Family: Chaetoceraceae**
- **6. Family: Biddulphiaceae**
  - (a) Subfamily : Eucampioideae (12)
  - (b) Subfamily : Triceratioideae (13)
  - (c) Subfamily : Biddulphioideae (14)
  - (d) Subfamily : Isthmioideae (15)
  - (e) Subfamily : Hemiaulioideae (16)
- **7. Family: Anaulaceae**
- **8. Family: Euodiaceae**

2. Order: Pennales

IV. Suborder: Araphidineae
- **9. Family: Fragilariaceae**
  - (a) Subfamily : Tabellarioideae (17)
  - (b) Subfamily : Meridionoideae (18)
  - (c) Subfamily : Fragilarioideae (19)

V. Suborder: Raphidioidineae
- **10. Family: Eunotiaceae**
  - (a) Subfamily : Peronioideae (20)
  - (b) Subfamily : Eunotioideae (21)

VI. Suborder: Monoraphidineae
- **11. Family: Achnanthaceae**
  - (a) Subfamily : Achnanthoideae (22)
  - (b) Subfamily : Cocconeioideae (23)

VII. Suborder: Biraphidineae
- **12. Family: Naviculaceae**
  - (a) Subfamily : Naviculoideae (24)
  - (b) Subfamily : Amphiproroideae (25)
  - (c) Subfamily : Gomphocymbelloideae (26)
- **13. Family: Epithemiaceae**
  - (a) Subfamily : Epithemioideae (27)
  - (b) Subfamily : Rhopalodioideae (28)
- **14. Family: Nitzschiaceae**
  - (a) Subfamily : Nitzschioideae (29)
- **15. Family: Surirellaceae**
  - (a) Subfamily : Surirelloideae (30)

N.I. Hendey (1964) puts the whole class of the Bacillariophyceae in the order Bacillariales and divides this into 22 families in the following scheme:

Division: Chrysophyta
Class: Bacillariophyceae
Order: Bacillariales
Suborder: Coscinodiscineae
1. Family: Coscinodiscaceae

Genus:
Melosira Agardh
Paralia Heiberg

Groentvedia Hendey
Cyclotella Kützing
Stephanodiscus Ehrenberg
Coscinodiscus Ehrenberg
Planktoniella Schütt
Endictya Ehrenberg
Actinocyclus Ehrenberg
Roperia Grunow
Thalassiosira Cleve
Porosira Jörgensen
Coscinosira Gran
Hyalodiscus Ehrenberg
Podosira Ehrenberg
Druridgea Donkin
Skeletonema Greville
Stephanopyxis Ehrenberg
Cylindropyxis Hendey

2. Family : Hemidiscaceae
Hemidiscus Wallich

3. Family : Actinodiscaceae
Actinoptychus Ehrenberg
Asteromphalus Ehrenberg

Suborder : Aulacodiscineae
4. Family : Eupodiscaceae
Aulacodiscus Ehrenberg
Eupodiscus Bailey

Suborder : Auliscineae
5. Family : Auliscaceae
Auliscus Ehrenberg

Suborder : Biddulphineae
6. Family : Biddulphiaceae
Biddulphia Gray
Cerataulus Ehrenberg
Hemiaulus Ehrenberg
Eucampia Ehrenberg
Triceratium Ehrenberg
Trigonium Cleve
Isthmia Agardh
Lithodesmium Ehrenberg
Ditylum Bailey
Bellerochea Van Heurck
Streptotheca Shrubsole
Cerataulina Peragallo
Huttonia Grove & Sturt

7. Family : Anaulaceae
Anaulus Ehrenberg
Eunotogramma Weisse

8. Family : Chaetoceraceae
Chaetoceros Ehrenberg

Suborder: Rhizosoleniineae

9. Family: Bacteriastraceae

Bacteriastrum Shadbolt

10. Family: Leptocylindraceae

Leptocylindrus Cleve
Guinardia H. Peragallo
Bacteriosira Gran
Schroederella Pavillard
Dactyliosolen Castracane
Detonula Gran
Lauderia Cleve

11. Family: Corethronaceae

Corethron Castracane

12. Family: Rhizosoleniaceae

Rhizosolenia Brightwell

Suborder: Fragilariineae

13. Family: Fragilariaceae

Fragilaria Lyngbye
Rhaphoneis Ehrenberg
Dimeregramma Ralfs
Glyphodesmis Greville
Catenula Mereschkowsky
Campylosira Grunow
Auriculopsis Hendey
Asterionella Hassall
Opephora Petit
Trachysphenia Petit
Cymatosira Grunow
Striatella Agardh
Synedra Ehrenberg
Thalassiothrix Cleve & Grunow
Thalassionema (Grunow) Hustedt
Plagiogramma Greville
Licmophora Agardh
Podocystis Bailey ex Wm. Smith
Grammatophora Ehrenberg
Rhabdonema Kützing

Suborder: Eunotiineae

14. Family: Eunotiaceae

Suborder: Achnanthineae

15. Family: Achnanthaceae

Achnanthes Bory
Cocconeis Ehrenberg
Anorthoneis Grunow
Rhoicosphenia Grunow
Campyloneis Grunow

Suborder: Naviculineae

16. Family: Naviculaceae

Navicula Bory
Anomoeoneis Pfitzer

Stauroneis Ehrenberg
Cistula Cleve
Stenoneis Cleve
Diploneis Ehrenberg
Oestrupia Heiden
Caloneis Cleve
Pinnularia Ehrenberg
Scoliopleura Grunow
Scoliotropis Cleve
Pseudoamphiprora Cleve
Trachyneis Cleve
Mastogloia Thwaites ex Wm. Smith
Frustulia Agardh
Amphipleura Kützing
Brebissonia Grunow
Pleurosigma Wm. Smith
Gyrosigma Hassall
Rhoicosigma Grunow
Donkinia Ralfs
Toxonida Donkin
Amphiprora Ehrenberg
Tropidoneis Cleve

17. Family: Auriculaceae
Auricula Castracane

18. Family: Gomphonemaceae
Gomphonema Hustedt

19. Family: Cymbellaceae
Phaeodactylum Bohlin
Amphora Ehrenberg
Okedenia Eulenstein

20. Family: Epithemiaceae
Epithemia de Brébisson
Rhopalodia O. Müller

21. Family: Bacillariaceae
Bacillaria Gmelin
Cylindrotheca Rabenhorst
Nitzschia Hassall
Hantzschia Grunow

Suborder: Surirellineae
22. Family: Surirellaceae
Surirella Turpin
Campylodiscus Ehrenberg

A further development of the system of Hustedt, for the centric diatoms was proposed by R. Simonsen (1972 and personal communication), keeping, as did Hustedt, three suborders, but dividing the genera into 13 families instead of 8, according to the following scheme:

Order: Centrales

I. Suborder: Coscinodiscineae

1. **Family: Melosiraceae Kützing 1844**
   (*e.g. Endictya, Hyalodiscus, Melosira, Paralia, Pyxidicula, Stephanopyxis*)
2. **Family: Thalassiosiraceae Lebour 1930, sensu Hasle, 1973**
   (*e.g. Cyclotella, Detonula, Lauderia, Planktoniella, Skeletonema, Stephanodiscus, Thalassiosira*)
3. **Family: Coscinodiscaceae Kützing 1844**
   (*e.g. Coscinodiscus, Craspedodiscus, Fenestrella, Palmeria*)
4. **Family: Asterolampraceae H.L. Smith 1872**
   (*e.g. Asterolampra, Asteromphalus, Bergonia, Brightwellia, Rylandsia*)
5. **Family: Heliopeltaceae H.L. Smith 1872**
   (*e.g. Actinoptychus, Aulacodiscus, Glorioptychus*)
6. **Family: Stictodiscaceae Schütt 1896**
   (*e.g. Arachnoidiscus, Stictodiscus*)
7. **Family: Hemidiscaceae Hendey 1937, sensu Simonsen**
   (*e.g. Actinocyclus, Hemidiscus, Roperia*)

II. Suborder: Rhizosoleniineae

8. **Family: Pyxillaceae Schütt 1896**
   (Syn: Thaumatodiscaceae Cleve 1885)
   (*e.g. Gladius, Gyrodiscus, Pyrgupyxis, Pyxilla*)
9. **Family: Rhizosoleniaceae Petit 1889**
   (*e.g. Ditylum, Guinardia, Lithodesmium, Rhizosolenia*)
10. **Family: Chaetoceraceae H.L. Smith 1872**
    (*e.g. Attheya, Bacteriastrum, Chaetoceros*)

III. Suborder: Biddulphiineae

11. **Family: Hemiaulaceae Heiberg 1863**
    (*e.g. Cerataulina, Goniothecium, Hemiaulus, Trinacria*)
12. **Family: Biddulphiaceae Kützing 1844**
    (*e.g. Anaulus, Biddulphia, Eunotogramma, Terpsinoe, Trigonium*)
13. **Family: Eupodiscaceae Kützing 1849**
    (*e.g. Auliscus, Cerataulus, Eupodiscus, Odontella, Rutilaria, Triceratium*)

4 NOTES ON TAXONOMY AND SYSTEMATICS

The arrangement of this book is to stress several aspects of the biology of diatoms. This should help to bring diatom biology further out of the restriction as attractive shells and to taxonomic catalogues of the occurrence of diatoms at specific locations.

The species concept in diatoms is discussed in some detail in Chapter 12. Here only a few general remarks are given: The number of bona fide species is estimated to be about 12 000 (Hendey, 1964). According to VanLandingham (1967), of the 60 000 or so species and varieties in F.W. Mills' (1933–35) catalogue only about 10 000 were bona fide taxa and, since that time, about another 10 000 taxa have been proposed. The original purpose of

VanLandingham to compile a comprehensive revision of F.W. Mills' *An Index to the genera and species of the Diatomaceae and their synonyms* was changed to assembling a complete new catalogue. VanLandingham (1975) points out, that Mills and also Peragallo (1897–1903) missed so many entries in their references, that he had to re-check all the earlier ones. VanLandingham preserves the taxonomy of Hustedt and revises in critical cases other entries in favour of Hustedt's interpretation.

Since the genetics of diatoms is still almost completely unknown, the separation of 'species', 'varieties' and 'formae' by genetic crossing experiments is still impossible. So far, only in a few species can we handle experimentally the complete sexual life cycle (see Chapter 9), and even then only with considerable experience and skill. Further complications for genetic experiments of several diatom species may arise since the bigger forms may be polygenomic instead of diploid and thus resemble many protozoa in this aspect. The DNA content of all sizes of the vegetative cells of *Coscinodiscus asteromphalus* was determined with different methods to be $6\text{–}8 \times 10^{-5}$ μg DNA/cell (Werner, 1971). The volume of sperms *inside* spermatocytes of the same species where no DNA can be lost during extraction procedures was determined to be about 10 μm^3. This means, that a sperm, completely packed with DNA, could contain only $1\text{–}2 \times 10^{-5}$ μg DNA (Werner, unpublished results). These and other data suggest that many questions about developmental and biochemical genetics have to be answered, before routine crossing experiments can be used as methods for the discrimination of bona fide species and establishing a diatom genetics.

The systematic study of the variation of diatom shells under different cultural conditions and in different clones of the same species is just beginning, although promising and remarkable results have already been published (Holmes, 1967; Geissler, 1970). The extensive use of the scanning electron microscope (Hasle, 1968; Round, 1970) has provided so much new information about the fixed and variable details of the shells that Ross & Sims (1972) suggested a complete new terminology for the details of the frustules of centric diatoms. This was further developed by a group of scientists organized to propose a standardized terminology of diatom taxonomy (anonymous, 1975). This has already proved to be a very useful and effective contribution. Also the list of diatom collections (diatom shell collections) (Fryxell, 1975) was a necessary work. Some notes on the use of electron microscopy in morphological and taxonomical studies by Grethe Rytter Hasle, illustrated by eight figures, will conclude this chapter.

5 REFERENCES

AGARDH C.H. (1824) *Systema Algarum.* Lund, 312 pp. Editio Anastatica Amsterdamii. A. Ascher & Co. 1965.

ALLEN E.J. & NELSON E.W. (1910) On the artificial culture of marine plankton organisms. *J. mar. biol. Assoc. U.K.* **8,** 421–74.

ALLEN W.E. (1922) Quantitative studies on inshore marine diatoms and dinoflagellates of southern California in 1920. *Univ. Calif. Publs. Zool.* **22,** 369–78.

ANONYMOUS (1975) Proposals for a standardization of diatom terminology and diagnoses. *Beih. Nova Hedwigia* **53,** 323–54.

BAILEY J.W. (1842) A sketch of the Infusoria of the family Bacillaria, with some account of the most interesting species which have been found in a recent or fossil state in the United States. *Amer. J. Sci.* **42,** 88–105.

BERGON P. (1903) *Note sur un mode de sporulation observé chez le Biddulphia mobiliensis* Bailey. Société scientifique d'Arcachon, station biologique, traveaux des laboratoires, Paris.

BLUMER M. (1975) Organische Verbindungen in der Natur. Die Grenzen unseres Wissens. *Angew. Chemie* **87,** 527–34.

BOGEN H.J. (1956) Objekttypen der Permeabilität. In *Handbuch der Pflanzenphysiologie* Bd. II (ed. W. Ruhland) pp. 426–38. Springer: Berlin, Heidelberg, New York.

BOGEN H.J. & FOLLMANN G. (1955) Osmotische und nichtosmotische Stoffaufnahme bei Diatomeen (Untersuchungen über die spezifische Permeabilität der Diatomeen). *Planta (Berl.)* **45,** 125–46.

BOYER C.S. (1926–27) Synopsis of North American Diatomaceae. Part I, Coscinodiscatae, Rhizosoleniatae, Biddulphiatae, Fragillariatae. Part II, Naviculatae, Surirellatae. *Proc. Acad. nat. Sci. Philad.* **78,** suppl. 3–288 & **79,** suppl. 229–583.

BÜTSCHLI O. (1892) Mittheilung über die Bewegung der Diatomeen. *Heidelb. Nat. Med. Verh.* **4,** 580–6.

CARLISLE E. (1972) Silicon: an essential element for the chick. *Science* **178,** 619–21.

CHOLNOKY B.J. (1968) *Die Ökologie der Diatomeen in Binnengewässern.* 699 pp. J. Cramer, Lehre.

CLEVE P.T. (1878) Diatoms from the West Indian Archipelago. *Bihang till Kongliga Svenska–Vetenskaps Akademiens Handlingar* **5,** No. 8, 1–22.

CLEVE P.T. (1884) On the diatoms collected during the Arctic expedition of Sir George Nares. *Linn. Soc. Journ. (Bot.)* **20,** 313–7.

CLEVE P.T. (1900) The plankton of the North sea, the English Channel and the Skagerak in 1898. *Konigl. Svensk. Vet. Stockh. Ak. Handl.* **32,** No. 8, 53 pp.

CUPP E. (1943) *Marine plankton diatoms of the west coast of North America.* Univ. Calif. Press, Berkeley & London, 237 pp.

DETONI J.B. (1891–94) *Sylloge Algarum omnium hucusque cognitarum,* vol. II, *Bacillariae.* Sectio I–III. Typis Seminarii, Patavii, 1 556 pp.

EHRENBERG C.G. (1854) *Mikrogeologie. Das Erde und Felsen schaffende Wirken des unsichtbar kleinen selbständigen Lebens auf der Erde.* Leopold Voss, Leipzig, 374 pp.

FJERDINGSTADT E. (1964) Pollution of streams estimated by benthal phycomicroorganisms I. A saprobic system based on communities of organisms and ecological factors. *Int. Revue ges. Hydrobiol.* **49,** 63–131.

FJERDINGSTADT E. (1971) Microbial criteria of environment qualities. *Ann. Rev. Microbiol.* **25,** 563–82.

FRYXELL G.A. (1975) Diatom collections. *Beih. Nova Hedwigia* **53,** 355–65.

GEISSLER U. (1973) Die Schalenmerkmale der Diatomeen, Ursachen ihrer Variabilität und Bedeutung für die Taxonomie. *Beih. Nova Hedwigia* **31,** 511–35.

GEITLER L. (1932) Der Formwechsel der pennaten Diatomeen (Kieselalgen). *Arch. Protistenk.* **78,** 1–226.

GEITLER L. (1973) Auxosporenbildung und Systematik bei pennaten Diatomeen und die Cytologie von Cocconeis-Sippen. *Öst. Bot. Z.* **122,** 299–321.

GERLOFF J. & GÖLZ E. (1944) Über den Feinbau der Kieselschalen bei einigen Zentrischen Diatomeen. *Hedwigia* **81,** 283–97.

GOLLEY F.B. (1972) Energy flux in ecosystems. In *Ecosystem structure and function* (ed. J.A. Wiens) pp. 69–88. Oregon State University Press, Corvallis.

GRAN H.H. (1908) Diatomeen. In *Nordisches Plankton—Botanischer Teil* (ed. K. Brandt & C. Apstein) pp. XIX + 1–155. Lipsius & Tischler, Kiel & Leipzig.

GREVILLE R.K. (1865) Descriptions of new and rare diatoms. Ser. 14–17. *Microsc. Soc. Trans. Lond. N.S.* **13,** 1–10, 24–34, 43–75 & 97–105.

GRUNOW A. (1863) Über einige neue und ungenügend bekannte Arten und Gattungen von Diatomeen. *Verhandl. Zool.-Bot. Ges. Wien* **13,** 137–62.

GRUNOW A. (1884) Die Diatomeen von Franz Josefs Land. *Denkschriften der mathematisch naturwissenschaftlichen Classe der Kaiserlichen Akademie der Wissenschaften* **48,** 53–112.

GRUNOW A. (1889) Referat über A. Schmidt: *Atlas der Diatomaceenkunde* Heft 27–30. *Bot. Zbl.* **37,** 82.

GUILLARD R.R.L. & RYTHER J.H. (1962) Studies of marine planktonic diatoms I. *Cyclotella nana* Hustedt, and *Detonula converfacea* (Cleve) Gran. *Can. J. Microbiol.* **8,** 229–39.

HARVEY H.W. (1934) Measurement of phytoplankton population. *J. mar. biol. Ass. U.K.* **14,** 71–88.

HASLE G.R. (1968) The valve processes of the centric diatom genus Thalassiosira. *Norw. J. Bot.* **15,** 193–201.

HENDEY N.I. (1937) The plankton diatoms of the southern seas. *'Discovery' Rep.* **16,** 151–364.

HENDEY N.I. (1964) *An introductory account of the smaller algae of British coastal waters.* Part V: Bacillariophyceae (Diatoms). Fisheries Investigations ser. IV. 317 pp. HMSO, London

HENSEN V. (1887) *Über die Bestimmung des Planktons oder des im Meere treibenden Materials an Pflanzen und Thieren.* Deutsch. Meere Ber. 5. Bericht der Kommission zur wissenschaftlichen Untersuchung der deutschen Meere bei Kiel. für die Jahre 1882–1886. Kiel, 1–8, I–xix.

HÖFLER K. (1940) Aus der Protoplasmatik der Diatomeen. *Ber. dt. bot. Ges.* **58,** 97–120.

HOLMES R.W. (1967) Auxospore formation in two marine clones of the diatom genus *Coscinodiscus Am. J. Bot.* **54,** 163–8.

HUSTEDT F. (1930) Bacillariophyta (Diatomeae). In *Die Süßwasserflora Mitteleuropas* (ed. A. Pascher) 2. Aufl. 466 pp. G. Fischer, Jena.

HUSTEDT F. (1961–66) Die Kieselalgen Deutschlands. Österreichs und der Schweiz unter Berücksichtigung der übrigen Länder Europas sowie der angrenzenden Meeresgebiete. In *Kryptogamenflora von Deutschland, Österreich und der Schweiz* (ed. L. Rabenhorst) Bd. 7, Teil 3. 816 pp. Akademische Verlagsgesellschaft Geest & Portig, Leipzig.

IMAI T. & HATANAKA M. (1950) Studies on marine non colored flagellates, *Monas* sp., favorite food of larvae of various marine animals. Preliminary research on cultural requirements. *Sci. Res. Tohoku Univ.* Ser. 4, **18,** 304–15.

KARSTEN G. (1896) Untersuchungen über Diatomeen II. Synedra affinis Ktzg. *Flora* **83,** 33–53.

KARSTEN G. (1904) Die sogenannten 'Mikrosporen' der Plankton-diatomeen und ihre weitere Entwicklung, beobachtet an *Corethron valdiviae* n. sp. *Ber. dt. bot. Ges.* **22,** 544–54.

KOLKWITZ R. & MARSSON M. (1908) Ökologie der pflanzlichen Saprobien. *Ber. dt. bot. Ges.* **26,** 505–19.

KÜTZING F.T. (1844) *Die kieselschaligen Bacillarien oder Diatomeen.* 152 pp. Förstermann, Nordhausen.

LAUTERBORN R. (1894) Zur Fragenach der Ortsbewegung der Diatomeen. Bemerkungen zu der Abhandlung des Herrn O. Müller: Die Ortsbewegung der Bacillariaceen betreffend. *Ber. dt. bot. Ges.* **12,** 73–8.

LAUTERBORN R. (1896) Untersuchungen über Bau, Kernteilung und Bewegung der Diatomeen. 165 pp. Engelmann, Leipzig.

LEWIN J.C. (1958) The taxonomic position of *Phaeodactylum tricornutum. J. Gen. Microbiol.* **18,** 427–32.

LIMA-DE-FARIA A. (1969) (ed.) *Handbook of Molecular Cytology.* 1 508 pp. North Holland Publ., Amsterdam & London.

LOHMANN H. (1911) Über das Nannoplankton und die Zentrifugierung kleinster Wasserproben zur Gewinnung desselben im lebenden Zustand. *Int. Rev. ges. Hydrobiol.* **4,** 1–38.

LOHMANN K.E. (1960) The ubiquitous diatom, a brief survey of the present knowledge. *Amer. J. Sci.* **258,** 180–91.

MACDONALD J.D. (1869) On the structure of the diatomaceous frustule, and its genetic cycle. *Ann. Mag. Nat. Hist.*, Ser. 4, **3,** 1–8.

MANTON I., KOWALLIK K. & VON STOSCH H.A. (1969) Observations on the fine structure and development of the spindle at mitosis and meiosis in a marine centric diatom (*Lithodesmium undulatum*). I. Preliminary survey of mitosis in spermatogonia. *J. Microscopy* **89,** 295–320.

MAZIA D. (1961) Mitosis and the physiology of cell division. In *The Cell* (ed. J. Brachet & A.E. Mirsky) Vol. III, pp.77–412. Academic Press, New York and London.

MERESCHKOWSKY C. (1902) Note sur quelques diatomées de la Mer Noire. *J. Bot.* (Paris) **16,** 319–24, 358–60, 416–30.

MERESCHKOWSKY C. (1903) *Zur Morphologie der Diatomeen.* 427 pp. Kasan.

MILLS F.W. (1933–35) *An Index to the Genera and Species of the Diatomaceae and their Synonyms.* 1816–1932. 1 726 pp. Weldon & Wesley, London.

MOHL V. H. (1861) Ueber das Kieselskelett lebender Pflanzenzellen. *Botan. Zeitung* **19,** 209–15.

MÜLLER H.O. & PASEWALDT C.W. (1942) Der Feinbau der Testdiatomee Pleurosigma angulatum W.Sm. nach Beobachtungen und stereoskopischen Aufnahmen im Übermikroskop. *Naturwissenschaften* **30,** 55–60.

MÜLLER O. (1889) Auxosporen von *Terpsinoë musica* Ehr. *Ber. dt. bot. Ges.* **7,** 181–3.

MÜLLER O. (1897) Die Ortsbewegung der Bacillanaceen *V. Ber dt. bot. Ges.* **15,** 70–86.

MÜLLER O.F. (1786) *Diatomaceen.* (*Vibrio paxilifer, Vibrio bipunctatus, Vibrio tripunctatus, Gonium pulvinatum* etc.) Animalcua infusoria fluviatillia et marina quae detexit, systematice, descripsit et ad vivum delineare curavit. 376 pp. O.F. Müller. Havniae.

MURREY G. (1896) On the reproduction of some marine diatoms. *Proc. Roy. Soc. Edinb.* **21,** 207–19.

NÄGELI C. (1849) *Gattungen einzelliger Algen physiologisch u. systematisch bearbeitet.* 139 pp. Zürich. Schweizer Gesellschaft N. Denkschr. X.

NITZSCH L. (1817) Beitrag zur Infusorienkunde oder Naturbeschreibung der Zerkarien und Bazillarien. *Neue Schriften der naturforschenden Ges. Halle,* Bd. 3, Heft 1. 128 pp. Hindel's Verlag, Halle.

OKAMURA K. (1907) An annotated list of plankton microorganisms of the Japanese coast. *Annotnes zool. jap.* **6,** 125–52.

OSTENFELD C.H. (1902) Marine plankton diatoms. *Bot. Tidsskr.* **25,** 1–49.

PATRICK R., HOHN M. & WALLACE J.H. (1954) A new method for determining the pattern of the diatom flora. *Not. natn. Acad. Nat. Sci. Philad. No. 259,* 1–12.

PERAGALLO H. & M. (1897–1908) *Diatomées marines de France et des districts maritimes voisins.* 491 pp. M.J. Tempère, Grez-sur-Loing.

PERAGALLO M. (1897–1903) *Le catalogue général des Diatomées.* 973 pp. Clermont–Ferrand.

PFITZER E. (1869) Ueber den Bau und die Zelltheilung der Diatomeen. *Bot. Zeitung* **27,** 774–6.

PFITZER E. (1871) Untersuchungen über Bau und Entwicklung der Bacillariaceen (Diatomaceen). *Bot. Abh. aus dem Gebiet der Morphologie und Physiologie* **2,** 1–189.

RABENHORST L. (1853) *Die Süsswasser-Diatomaceen (Bacillarien)* für Freunde der Mikroskopie. 72 pp. Eduard Kummer, Leipzig.

ROBARDS A.W. (1970) *Electron microscopy and plant ultrastructure.* 297 pp. McGraw-Hill, Maidenhead.

ROSS R. & SIMS P.A. (1972) The fine structure of the frustule in centric diatoms: a suggested terminology. *Br. Phycol. J.* **7,** 139–63.

ROUND F.E. (1970) The genus *Hantzschia* with particular reference to *H. virgata v. intermedia* (Grun.) comb. nov. *Ann. Bot.* **34,** 75–91.

SCHMIDT A. (*et al.*) (1874–1959) *Atlas der Diatomaceenkunde.* 472 plates. R. Reisland, Aschersleben, Leipzig.

SCHRÖDER B. (1901) Das Phytoplankton des Golfs von Neapel. nebst vergleichenden Ausblicken auf das des Atlantischen Ozeans. *Neapel Zool. Stat. Mitth.* **14,** 1–38.

SCHÜTT F. (1896) Bacillariales (Diatomeae). In *Die natürlichen Pflanzenfamilien* (ed. A. Engler & K. Prantl.) Teil 1, Abt. 1b, pp. 31–153. Wilhelm Engelmann, Leipzig.

SIMONSEN R. (1972) Ideas for a more natural system of the centric diatoms. *Beih. Nova Hedwigia* **39,** 37–54.

SLADECEK V. (1961) Zur biologischen Gliederung der höheren Saprobitätstypen. *Arch. Hydrobiol.* **58,** 103–21.

SMITH W. (1853–56) *Synopsis of the British Diatomaceae.* Vol. 1–2. 107 pp. J. v. Voorst, London.

STOSCH V.H.A. (1942) Form und Formwechsel der Diatomee *Achnanthes longipes* in Abhängigkeit von der Ernährung. Mit besonderer Berücksichtigung der Spurenstoffe. *Ber. dt. bot. Ges.* **60,** 2–16.

STOSCH V.H.A., THEIL G. & KOWALLIK K. (1973) Entwicklungs-geschichtliche Untersuchungen an zentrischen Diatomeen. V. Bau und Lebenszyklus von *Chaetoceros didymum,* mit Beobachtungen über einige andere Arten der Gattung. *Helgol. Wiss. Meeresunters.* **25,** 384–445.

STRICKLAND J.D. (1972) Research on the marine planctonic food web at the institute of marine resources: a review of the past seven years of work. *Oceanogr. mar. biol. Ann. Rev.* (ed. H. Barnes) **10,** 349–414.

ÜBELEIS I. (1957) Osmotischer Wert, Zucker- und Harnstoffpermeabilität einiger Diatomeen. *Östr. Akad. Wiss.-Mathem.-Naturwiss. Klasse, Sitzungsber.* Abt. I, 166, 395–433.

VAN HEURCK V. (1896) *A treatise on the Diatomaceae.* 558 pp. Wesley & Sons, London.

VanLandingham v.S.L. (1967–75) *Catalogue of the fossil and recent genera and species of diatoms and their synonyms.* Vol. I–V. 2 963 pp. J. Cramer, Lehre.

Wallich G.C. (1860) On the development and structure of the diatom valve. *Transact. Microsc. Soc. Lond. n.s.* **8,** 129–45.

Walne P.R. (1970) Studies on the food value of nineteen genera of algae to juvenile bivalves of the genera Ostrea, Crasostrea, Mercenaria and Mytilus. *Fish. Invest. Ser.* **2,** 26, 1–62.

Werner D. (1967) Untersuchungen über die Rolle der Kieselsäure in der Entwicklung Höherer Pflanzen I. Analyse der Hemmung durch Germaniumsäure. *Planta (Berl.)* **76,** 25–36.

Werner D. (1971) Der Entwicklungszyklus mit Sexualphase bei der marinen Diatomee *Coscinodiscus asteromphalus.* II. Oberflächenabhängige Differenzierung während der vegetativen Zellverkleinerung. *Arch. Mikrobiol.* **80,** 115–33.

Wuhrmann K. (1951) Über die biologische Prüfung von Abwasserreinigungsanlagen. *Gesundheitsing,* **72,** 253–61.

Yamaji I. (1972) *Illustrations of the Marine Plankton of Japan.* 369 pp. Hoikusha Publ. Co., Osaka.

THE USE OF ELECTRON MICROSCOPY IN MORPHOLOGICAL AND TAXONOMICAL DIATOM STUDIES

GRETHE RYTTER HASLE

Institutt for Marinbiologi og Limnologi avd. Marin Botanikk, Universitetet i Oslo
Postboks 1069, Blindern, Oslo 3, Norway

Transmission as well as scanning electron microscopy are widely used to examine diatom morphology. Taxonomically important morphological features such as structure of valve processes and areolae have been revealed by these methods of investigation.

With the instruments and techniques in current use scanning electron microscopy is superior for the examination of:

a The whole diatom frustule, showing its architecture and gross morphology (Fig. 1.1).

b Diatom processes, their external (Fig. 1.2) and internal parts (Figs 1.3, 1.4, 1.5).

Transmission electron microscopy is preferable for the examination of:

a The smallest and most weakly silicified diatoms (Fig. 1.6).

b The most delicately structured morphological details of the diatom frustule, e.g. the areola velum (Figs 1.7, 1.8).

All micrographs were taken with instruments of the Electron Microscopical Unit for Biological Sciences, University of Oslo.

Fig 1.1. *Porosira glacialis* (Grun.) Jørg., whole frustule with one large labiate process and many small strutted processes, girdle composed of many bands. (× 1 200).

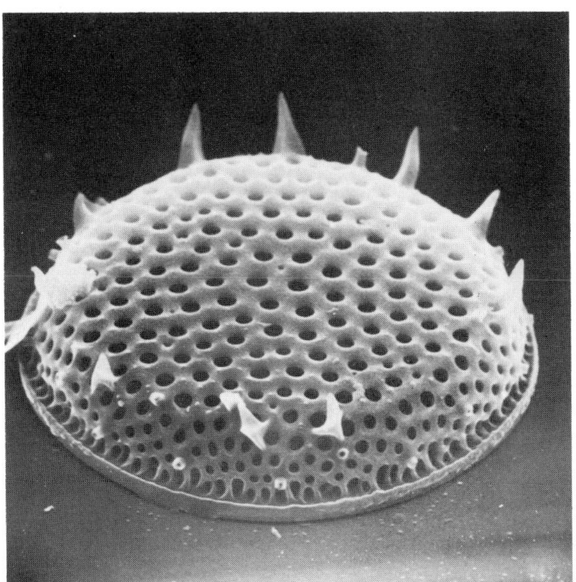

Fig. 1.2. *Thalassiosira eccentrica* (Ehrenb.) Cleve, valve in outside view, one labiate process, many small strutted processes, and coarse spines. (× 3 000).

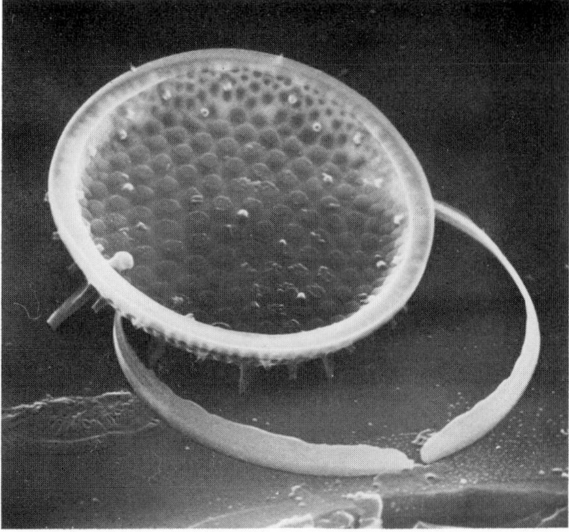

Fig. 1.3. *Thalassiosira eccentrica* (Ehrenb.) Cleve, valve in inside view, internal as well as external parts of labiate and strutted processes visible, areola velum on the internal valve surface. (× 2 400)

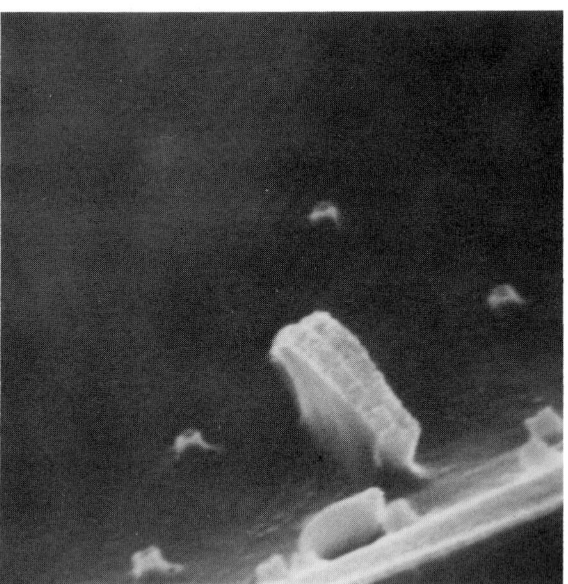

Fig. 1.4. *Porosira glacialis* (Grun.) Jørg., internal part of labiate process. (× 13 200).

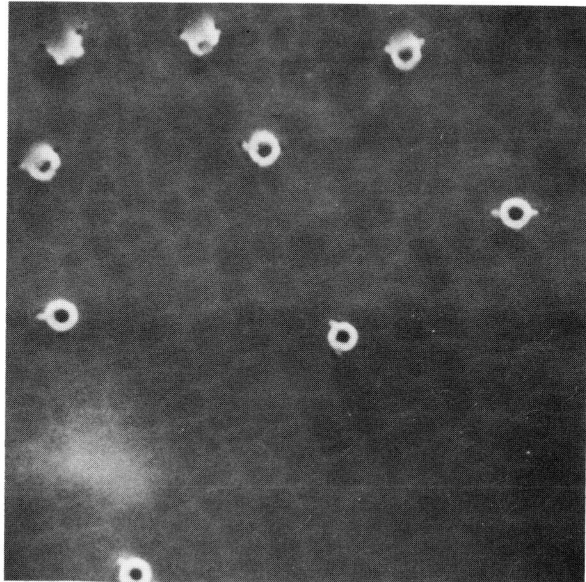

Fig. 1.5. *Porosira glacialis* (Grun.) Jørg., internal part of strutted processes. (× 12 600).

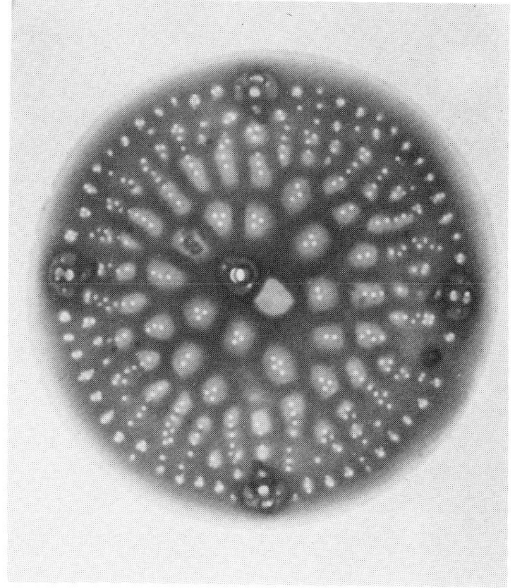

Fig. 1.6. *Thalassiosira profunda* (Hendey) Hasle, valve with diameter ca. 2·5 μm, small labiate and larger strutted processes, areolae of variable shape. (× 24 000).

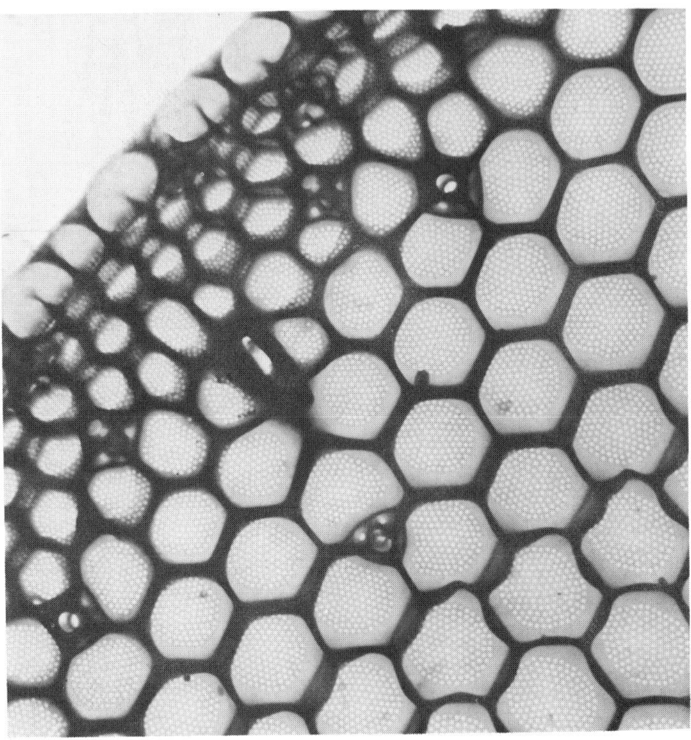

Fig. 1.7. *Thalassiosira eccentrica* (Ehrenb.) Cleve, part of valve, one large labiate process, several smaller strutted processes, and areolae with velum. (× 25 200).

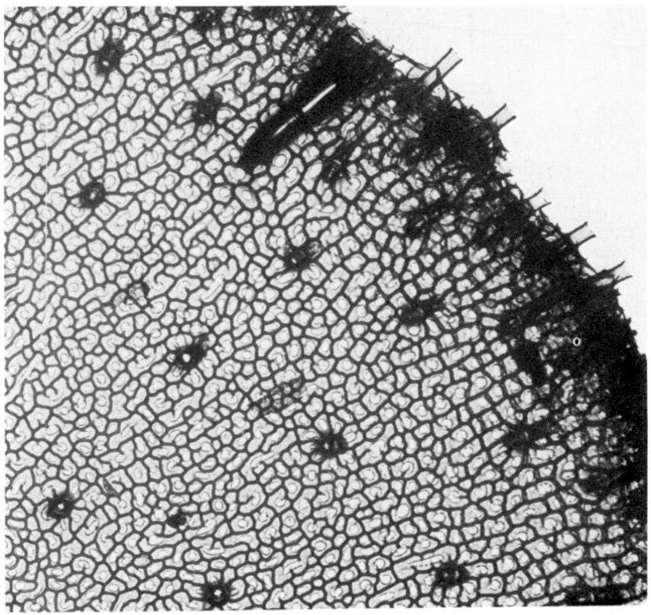

Fig. 1.8. *Porosira glacialis* (Grun.) Jørg., part of valve, one large labiate process, many small strutted processes, and areolae with velum. (× 9 900).

CHAPTER 2

THE GROWTH AND CULTURE OF DIATOMS

R.W. EPPLEY

Institute of Marine Resources,
University of California—San Diego,
P.O. Box 1529, La Jolla, Calif. 92093, USA

1 INTRODUCTION

The growing of algae means more than micro-horticulture under academic botanical auspices (Hutner, 1950).

The culture of diatoms goes back in time apparently to the late 1880s judging from references in the scientific literature. Allen & Nelson (1910) and

Pringsheim (1949) cite several predecessors, including Miguel, Houghton Gill, Richter, Lockwood, Karsten, Senft, and of course, Chodat and Beijerinck, for the period 1890–1905. The silicon requirement of diatoms was recognized early (Richter, 1905). Macronutrient requirements of unicellular algae, and their similarity to those of higher plants were determined by use of cultures by Molisch and Benecke (see Pringsheim, 1949). Allen & Nelson's (1910) interest in marine diatom cultures was as food organisms for marine invertebrate larvae. However, they also realized the utility of unialgal cultures for answering taxonomic and morphological questions and commented on life-cycles, again citing earlier workers, especially Miguel. Requirements of diatoms for specific organic substances in trace amounts (cf. Allen, 1914) and the nutritional needs of colourless forms (Richter, 1904) were already working concepts in these 'early' days.

The uses of algal cultures for providing chemically uniform experimental material and for studies of chemical composition seem to derive from later times. And, for example, the recognition that the soil extract of Erd-Schreiber (Schreiber, 1927) could be largely replaced by vitamin B_{12} (Hutner & Provasoli, 1953) has come about in the lifetime of most readers. Accounts by Bold (1942), Ketchum (1954), and Harvey (1957) give much of the historical flavour of culture and nutritional work, and Provasoli (1963) and Lewin & Guillard (1963) provide recent accounts.

The significance of cultures as experimental tools and the manifold insights their use has provided are difficult to comprehend these days as we take them for granted. For example, one could claim with justification that the whole concept of the eutrophication of natural waters depends upon culturing algae for its experimental verification. Diatom cultures have not figured as importantly in the derivation of biochemical pathways as have green algae, such as *Chlorella*. Nonetheless it is these organisms, more frequently than *Chlorella*, that are among the ecologically significant species in lakes and oceans. Environmental ecological studies will depend increasingly upon their use to gather autoecological data for individual important species; for species competition studies with mixed cultures; for aquaculture operations involving seminatural food chains; and for pollution effects studies. Very large outdoor cultures inoculated with the natural flora of lakes or oceans are coming into wider use for experimental ecological studies.

Before continuing I should state some disclaimers. The current compendium *Handbook of Phycological Methods. Culture Methods and Growth Measurements* (Stein, ed., 1973) is rich in practical methods and makes redundant any remarks on culture techniques, media, methods of measuring growth, materials to be used or avoided in designing culture apparatus, sterilization methods, and the like. The reader will also note that the present material is biased, my own experience having been limited to marine phytoplankton. Apologies are extended to those concerned with soil, freshwater, and marine benthic diatoms for the coarse treatment of their areas of interest.

The following section on the nutrition of diatoms will not attempt to be comprehensive, because of the massive literature on the subject and see Chapter 4. Lewin & Guillard reviewed the subject in 1963; the present writing will build upon their work.

2 NUTRITIONAL PECULIARITIES OF DIATOMS

2.1 *Silicic acid*

The important role of diatoms in the biological production of natural waters and the fact that diatoms require the element silicon, while most other algae do not, adds special significance to studies of the silicic acid nutrition of diatoms and the mechanism of diatom shell formation (Chapter 4). Silicic acid utilization by diatoms in culture can reduce the ambient Si concentration to less than 1 μg atom/litre (Jørgensen, 1953). The silica content of a diatom shell may vary over finite limits as a result of variations in growth rate (Lewin & Guillard, 1963), in ambient concentration (Braarud, 1948), or in the rate of Si supply to continuous cultures (Paasche, 1973a). Dissolution of silica from diatom shells is slow compared with remineralization rates of organic nitrogen and phosphorus (cf. Lewin, 1960; Antia *et al.* 1963). The dissolution rate of silica has ecological significance in terms of the buffering of natural waters (Garrels, 1965) and rates of nutrient supply to planktonic and benthic algae (Kilham, 1971).

Diatom cell division stops with the depletion of silicic acid from the culture medium (Lewin, 1955; Lewin & Chen, 1968; Werner, 1966). Synthesis of DNA also ceases (Darley & Volcani, 1969). Variation in the rate of cellular incorporation of silicic acid with silicic acid concentration in the medium follows saturation kinetics (Azam *et al.* 1973a) as does germanium incorporation in place of silica (Azam *et al.* 1973b). Germanium appears to act as a specific inhibitor of diatom growth (Lewin, 1966a; Werner, 1966) by interfering with silicic acid incorporation. However at very low concentrations germanium can be incorporated into diatom shells (Azam *et al.* 1973b). This has special significance in that Ge may serve as an isotopic tracer for Si, as well as an inhibitor of diatom growth, and the radioactive isotope ^{68}Ge is much more convenient in experimental work than that of silicon, ^{31}Si (Azam *et al.* 1973b).

2.2 *Boron*

By using synthetic media and plastic (polycarbonate or teflon) culture vessels Lewin (1966b; 1966c) demonstrated a boron requirement for 16 species of marine diatoms and 8 freshwater species (see also Werner, 1970). With

Cylindrotheca fusiformis growth rate was influenced by boron concentration in the range 0·05–0.5 mg B/litre but the final yield of cells was not influenced by boron concentration (Lewin, 1966b).

2.3 *Trace metals and chelation*

While research on macronutrients (K, Na, Mg, Ca) has been relatively inactive since Lewin & Guillard's review (1963), save for Hayward (1970), there has been much activity relative to trace metal nutrition, chelation, and heavy metals as environmental poisons. Harvey's early work on iron and manganese (Harvey, 1937; 1947; 1957) has new relevance to current interest in trace metals by virtue of two kinds of observations. (1) Certain 'enrichment experiments' in which nutrients are added to natural waters and a growth response, such as photosynthetic $^{14}CO_2$ assimilation, is measured have shown a marked growth response not to N, or P as expected, but to the addition of either iron, chelated iron, other di- or trivalent metals, or sometimes to chelators alone: the Sargasso Sea (Ryther & Guillard, 1959; Menzel & Ryther, 1961), the Indian Ocean (Tranter & Newell, 1963), and in certain lakes (Sakamoto, 1971). Johnston (1964) studied the growth of individual diatom species inoculated into filtered and sterilized seawater, taken from many areas of the sea around the British Isles, from a similar perspective, and see Provasoli (1963), for a review. (2) Deep ocean water and certain samples of freshly upwelled seawater off Peru (Barber *et al.* 1971) and in the Cromwell Current (Barber & Ryther, 1969) behave as culture media lacking in chelators. Planktonic diatoms grow in these waters only after a lag period of 2–4 days but the lag period is abolished by the addition of appropriate trace metal chelators.

Barber's hypotheses, now being tested with marine planktonic diatoms, are (1) growth-enhancing compounds are synthesized by phytoplankton and released into the surrounding seawater; (2) these compounds are functionally analogous to synthetic organic chelators; (3) the growth-enhancing compounds are organic (destroyed by UV-irradiation and removed by charcoal filtration) and are present in some seawaters and absent in others; and (4) the specific chemical benefit to phytoplankton involves altering the availability or the toxicity of trace metals (Barber, 1973).

Results of these 'enrichment experiments' and the water 'conditioning' effects of added chelators or products of plankton metabolism add ecological justification to the study of trace metals and chelation in the growth of diatoms. The water conditioning effect suggests that trace metal nutrition is complex and that trace metals, although present in the medium, must be in a proper chemical form for utilization. Similar conclusions were reached earlier from efforts to design chelated culture media (Provasoli *et al.* 1957). Similar considerations apply to heavy metal inhibition. That is, chemical speciation is important as well as total concentration of metal. Lewin & Chen (1971) found a reduced availability of iron for growth on storing seawater in containers. A soluble to particulate iron transformation accompanied the reduction in growth potential of the water.

2.4 *Roles of organic matter in diatom growth*

The water conditioning effects discussed in the previous section indicate one role of organic matter in the nutrition of diatoms. Others are the (1) direct utilization of organic substrates for energy and carbon skeletons, or for the nitrogen in urea (Grant *et al.* 1967; Carpenter *et al.* 1972; McCarthy, 1972) or in amino acids (Stephens & North, 1971), or for phosphorus (Kuenzler, 1966) in the organic molecule. (2) Specific vitamin requirements are common among diatoms for vitamin B_{12} (Provasoli, 1963; Guillard & Cassie, 1963; Guillard, 1968) and less common for thiamine (Lewin, 1965). These requirements have been of practical significance in bioassays of natural waters for vitamin B_{12} (Gold, 1964; Carlucci, 1973). (3) Extracellular organic products of certain species may inhibit growth of other species. For example *Olisthodiscus luteus* cultures contained a substance inhibitory to *Skeletonema costatum* (Pratt, 1966), and see Aubert *et al.* (1970). (4) Humic substances influenced the yield of cells in culture of *Thalassiosira nordenskioldii* (Prakash *et al.* 1973).

A large number of metabolic products have been identified in the media in which diatoms have grown. Chitan fibres (poly-*N*-acetylglucosamine) were produced by *Thalassiosira fluviatilis* (McLachlan *et al.* 1965). *Stephanodiscus hantzschii,* a freshwater diatom, released glycolate and other organics (Watt, 1969). Surface active substances were produced by *Cyclotella nana,* clone 3H, and several other species (Wilson & Collier, 1972). Extracellular polysaccharides were also noted in *Chaetoceros affinis* cultures (Myklestad & Haug, 1972; Myklestad *et al.* 1972). Hellebust (1965) identified a number of low molecular weight organics, many of them polyhydroxyl compounds, from marine diatom cultures.

A number of investigations have tested for interactions between species during growth in mixed cultures that might result from external metabolites (Talling, 1957; Fedorov & Kustenko, 1972). No interactions were noted between *Asterionella formosa* and *Fragilaria crotonensis* (Talling, 1957).

Release of vitamins into the culture medium may be fairly common among planktonic diatoms (Carlucci & Bowes, 1970a). In mixed cultures vitamin production by one species stimulated the growth of other species that require the vitamin (Carlucci & Bowes, 1970b).

3 SPECIES IN CULTURE

Success in achieving cultures of marine benthic and planktonic diatoms has been so widespread that it is easier to ask what ecologically significant species are not in culture rather than what species are. (These are mostly unialgal cultures, however, and more effort is usually required to obtain axenic cultures.) For example, Allen & Nelson in 1910 claimed 18 species in 'persistent' culture, i.e. surviving several transfers. One suspects culturing has not been so popular among limnologists as among oceanographers, judging from the relatively few species of freshwater diatoms in the University of Indiana collection (Starr, 1964), where only 27 species were listed. Many investigators of marine diatoms maintain small collections in their own laboratories in addition to the large collections. For example, Takano (1964) listed 44 species of benthic and planktonic diatoms cultured in his laboratory for at least 3 months. A similar number is apparently maintained at several locations, primarily as a convenient source of experimental material for the laboratory staff.

Lists of the species maintained at the large algal culture collections have been published from time to time: Cambridge collection (National Environment Research Council, 1971); Prague collection (Fott & Trucova, 1968). Tokyo collection (Watanabe & Hattori, 1966); Göttingen collection (Koch, 1964); Indiana University collection (Starr, 1964, 1971). Unpublished lists of cultures maintained at Plymouth, Woods Hole and La Jolla can be obtained by writing, respectively, Marine Biological Association of the UK, The Laboratory, Citadel Hill, Plymouth, PLI 2PB England; Culture Collection of Algae, Woods Hole Oceanographic Institution, Woods Hole, Massachusetts 02543, USA; Mr. J.B. Jordan, Institute of Marine Resources, University of California at San Diego, P.O. Box 109, La Jolla, California 92037, USA.

4 AUTECOLOGICAL STUDIES OF DIATOM GROWTH IN BATCH CULTURES

The term 'batch culture' refers to a culture of constant volume confined within a vessel and not subject to replacement of culture medium. That is, a flask of culture medium is inoculated with the test species and left be, save for aeration and the removal of samples for analysis. The time course of growth, measured for example as the increase in cell concentration (Fig. 2.1), shows a lag phase (no increase), an exponential phase (geometric increase in concentration with time where $dN/dt = N\mu$), a very brief period of linear growth (where $dN/dt =$ a constant), a stationary phase (where $dN/dt = 0$), and a death phase (declining cell concentration). Growth in the exponential phase can be described mathematically *as:*

$$\frac{dN}{dt} = \mu N \quad \text{or} \quad \mu = \frac{1}{t} \ln \left(\frac{N}{N_0}\right) \tag{1}$$

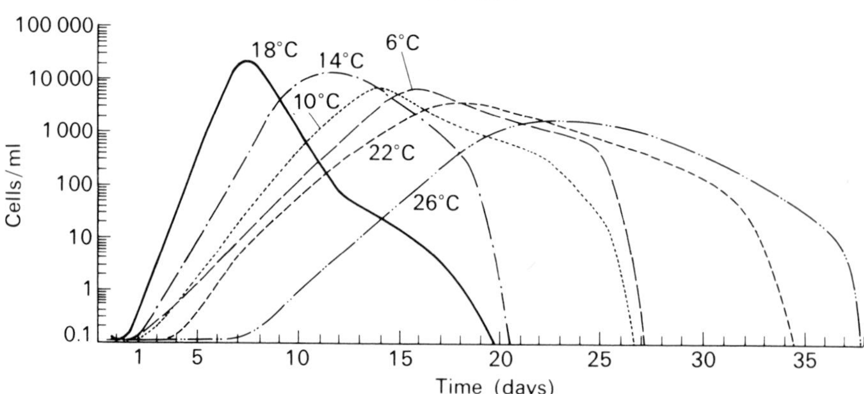

Fig. 2.1. Growth of *Chaetoceros socialis* in batch cultures held at different temperatures. Cell concentration (log scale) is shown as a function of time in days. The figure shows marked differences in the length of the lag phases, in the specific growth rate (from the slope of the exponential portion of the curves), in the duration of the stationary phase, and in the rate of decline (death phase) at different temperatures. From Kayser (1971).

or as:

$$N = N_0 2^{\mu t} \quad \text{or} \quad \mu = \frac{1}{t}\log_2\left(\frac{N}{N_0}\right) \tag{2}$$

where N is the cell concentration, N_0 the initial cell concentration, μ the specific growth rate and t is time. In equation (2) the number 2 applies only to species that increase in number by binary fission, as diatoms usually do in vegetative growth. Equation (1) gives μ in natural log units/day; equation (2) in cell divisions/day.

A more general expression for growth rate includes a rate constant for mortality (d). The ecologist's value, r, the instantaneous rate of net population growth, can be defined as $\mu - d$. *Then:*

$$\frac{dN}{dt} = rN = (\mu - d)N \tag{3}$$

Use of this formulation is probably unnecessary for batch culture work but it is useful for very dense or extremely nutrient limited cultures, as in dialysis cultures or in continuous cultures at low dilution rates, as well as with natural populations.

The mode of cell division of diatoms often results in a diminution in cell diameter. Round (1973) has recently reviewed the history of such observations and the present status. Although the valve diameter of a large diatom, such as *Coscinodiscus asteromphalus,* may decrease as much as 1·5 µm per cell division (Werner, 1971a), some species, and perhaps many species at certain cell diameters, do not show the size reduction (Round, 1973) and it now seems of little value for a general means of estimating growth rate.

The specific growth rate, μ, is a convenient parameter for characterizing the growth potential of a species, or clonal isolate, as a function of environmental variables and it provides a means of comparing one species with another. If measurements of cell concentration are made with an electronic particle counter, such as the Coulter® or Celloscope® counters, and samples are taken at appropriate time intervals, then the precision of estimating μ is probably about 10% of the mean rate. Haemacytometer or other microscopic methods would be less reproducible since the minimum counting error for any sample is of the order $\pm 10\%$ *vs.* about $\pm 1\%$ for the electronic counters. Detailed procedures for microscopic counting and estimating error are given by Guillard (1973). Use of electronic particle counters is described by Maloney *et al.* (1962) and Parsons (1973).

5 GROWTH RATES OF DIATOMS IN BATCH CULTURES *VS.* ENVIRONMENTAL VARIABLES

5.1 *Temperature*

Batch cultures have been used for many years to study autecological relationships between specific growth rate and temperature in diatoms (cf. Braarud, 1937; 1945). The upper and lower temperature limits and the optimum temperature, i.e. that resulting in the maximum growth rate, are reproducible parameters and can be used to characterize species or clones (Table 2.1). They can be expected to shift slightly when other factors are varied, for example salinity, but they are nevertheless very useful values. Temperature races, within marine species, have been defined (Braarud, 1961; Hulburt & Guillard, 1968) (Table 2.1). That is, clones of a species isolated from different areas, such as boreal *vs.* subtropical seas, display different optimum temperatures.

As a general trend, the higher the optimum temperature the higher will be the maximum specific growth rate (cf. Eppley, 1972). Diatoms isolated from Antarctic Sea ice showed optimum temperatures (T_{opt}) in batch cultures as low as 3–8°C with maximum growth rates about 1·0 doublings/day (Bunt, 1968) (Table 2.1). In contrast, a *Chaetoceros* species of the eastern tropical Pacific Ocean showed T_{opt} at about 32°C and μ_{max} about 4·0 doublings/day (Thomas, 1966). Temperate water species show intermediate values.

One of the puzzling results of μ *vs.* T measurements with cultures is that T_{opt} exceeds by several degrees the natural water temperatures where the species are found to be most abundant (Braarud, 1961; Smayda, 1969).

Table 2.1. Temperature and diatom growth: the optimum temperature (T_{opt}), the specific growth rate (divisions/day) at the optimum temperature, and the temperature range permitting growth. For similar tables elsewhere see Hoogenhout & Amesz (1965) and Canale & Vogel (1974).

Species	T_{opt}	μ	*Temperature range (°C)*	*Reference*
			Freshwater forms	
Asterionella formosa	17–20	2·1	<5 to >20	Talling (1955)
Asterionella formosa	20	2·2		Lund (1949)
Diatoma elongatum	20	1·2	<15 to >30	Müller (1972)
Fragilaria crotonensis	26	1·5	<15 to >30	Müller (1972)
Fragilaria capucina	20	1·3	<15 to >30	Müller (1972)
Nitzschia actinastroides	26	1·6	<15 to >30	Müller (1972)
Synedra acus var. *radians*	20	1·1	<15 to >30	Müller (1972)
			Antarctic ice forms	
Fragilaria sublinearis	5–6	0·91	−2 to 8–9	Bunt (1968)
Stauroneis membranacea	5	0·57	−2 to >12	Bunt (1968)
Synedra sp.	4–6	1·0	−2 to >7	Bunt (1968)
			Marine forms	
Asterionella japonica	20–25	1·2	<10 to 30	Kain & Fogg (1958)
Asterionella socialis	18	1·6	<5 to 24	Lewin & Mackas (1972)
Chaetoceros armatum	20	0·6	5 to >24	Lewin & Mackas (1972)
Chaetoceros gracilis	23–37	5–6	11 to 41	Thomas (1966)
Chaetoceros socialis	18	2·7	<4 to >26	Kayser (1971)
Detonula confervacea	10	1·3	<4 to >15	Guillard & Ryther (1962)
Detonula confervacea	12	1·5	<2 to 16	Smayda (1969)
Ditylum brightwellii	26	2·6	– to 32	Paasche (1968)
Phaeodactylum tricornutum	⩾25	2·4	<9 to >25	Spencer (1954)
Rhizosolenia fragilissima	18–25	1·2	9 to >30	Ignatiades & Smayda (1970)
Skeletonema costatum	16–26	2·4	6 to >28	Jitts *et al.* (1964)
Skeletonema costatum	20–30	4·3	<5 to >30	Curl & McLeod (1961)
Skeletonema costatum	>20	2·3	<7 to >20	Jørgensen (1968)
Skeletonema tropicum	(13–34)25	3·0	13 to 34	Hulburt & Guillard (1968)
Skeletonema tropicum	(17–31)24	1·6	17 to 31	Hulburt & Guillard (1968)
Thalassiosira pseudonana (5 clones)	20–25	1·8–2·8	<4 to >25	Guillard & Ryther (1962)
Thalassiosira nordenskioldii	11–14	1·6	<2 to 19	Jitts *et al.* (1964)

Further data can be found in Ukeles (1961) and Lanskaya (1961).

Temperature gradient incubators have been used to facilitate such studies with cultures. Designs are given for liquid cultures by Thomas *et al.* (1963) and by Jitts *et al.* (1964), and for solid medium by Jørgensen (1960). The latter two devices employ cross-gradients: a temperature gradient in one direction and a light gradient in the other. A general purpose cross-gradient device is described by van Baalen & Edwards (1973).

5.2 *Light*

Generalities on the growth response of diatoms to light are less easily discerned than with temperature. Doubtless this is because diatoms, like other unicellular algae, are very plastic in their photosynthetic response. Data for μ *vs.* light intensity, at constant temperature, have been published by several workers (Table 2.2). From intensive work with the green alga, *Chlorella* (cf. Sorokin & Krauss, 1962) one expects the growth response to light to vary with the temperature and this is apparent in Smayda's studies of *Detonula confervacea* (Table 2.2). Cells grown at a low temperature respond to lower intensities (i.e. saturating and supra-optimal inhibitory intensities are lower) than cells grown at a high temperature. Cell contents of chlorophyll, protein (Jørgensen, 1968) and carbon (Checkley cited in Eppley, 1972) are adjusted accordingly, as are pigment ratios and chromatophore size (Brown & Richardson, 1968).

The growth response to day length, as well as light intensity, has been studied but little in diatoms in spite of its potential ecological importance. Castenholz (1964) found maximum growth rates at intermediate day lengths (<24 hours light) for some species of benthic diatoms, and in continuous light for others. The coastal neritic diatom, *Ditylum brightwellii* falls in the former category (Fig. 2.2), *Nitzschia turgidula* in the latter (Paasche, 1968).

A recent report (Lipps, 1973) claims a red far-red effect on growth rates of some unicellular algae, including the diatom *Ditylum brightwellii*. Far-red light, given briefly (30 minutes) at the end of the light period of a light–dark illumination cycle, inhibited growth. The inhibition was relieved if the far-red exposure was followed by 30 minutes of red light. The ecological significance of such a presumptive phytochrome-type response may need consideration if the results are duplicated elsewhere.

5.3 *Salinity*

Growth responses to salinity have been studied in a number of estuarine and marine diatoms (Table 2.3). In general the former prove to be euryhaline, growing well over a broad salinity range, and the latter perhaps more stenohaline. Clones of *Thalassiosira pseudonana (Cyclotella nana)* with different salinity ranges and optima have been studied by Guillard & Ryther

Table 2.2. Variation in the specific growth rate (μ, in divisions/day) with light intensity; the saturating light intensity (I_{sat}), μ at saturation, the compensation light intensity (I_{comp}), and the experimental temperature. The authors give light intensity in different units to be noted by superscripts: a = lux of fluorescent light, b = foot candles of fluorescent light, c = cal.cm^{-2}.min^{-1} of tungsten light, d = cal.cm^{-2}.min^{-1} of fluorescent light. Very approximate conversion factors are 1 000 lux = 93 foot candles = 4–6 × 10^{-3} cal.cm^{-2}.min^{-1} according to Westlake (1965)

Species	μ	I_{sat}	I_{comp}	*Temperature (°C)*	*Reference*
				Freshwater form	
Stephanodiscus hantzschii	—	1 700 [a]	—	8	Swale (1963)
	0·67	1 700	—	14	Swale (1963)
	—	5 000	—	20	Swale (1963)
				Marine benthic forms	
Biddulphia aurita	1·5	0·03 [d]	0·004–0·006	11	Castenholz (1964)
Fragilaria striatula	1·7	0·04	0·008–0·01	11	Castenholz (1964)
Melosira moniliformis	0·75	0·02	0·001–0·002	11	Castenholz (1964)
				Marine planktonic forms	
Asterionella japonica	1·0	1 000–4 000 [a]	—	18	Kain & Fogg (1958)
Chaetoceros gracilis	1·5	600 [b]	10	23–26	Thomas (1966)
Detonula confervacea	1·2	200–600 [b]	—	2	Smayda (1969)
	1·5	600–1 200	—	7	Smayda (1969)
	1·5	1 200–1 800	—	12	Smayda (1969)
Ditylum brightwellii	2·1	⩾0·14 [c]	—	20	Paasche (1968)
Nitzschia turgidula	2·5	⩾0·03 [c]	—	20	Paasche (1968)
*Rhizosolenia fragilissima**	1·2	600 [b]	—	18–25	Ignatiades & Smayda (1970)
Skeletonema costatum	2·4	0·04–0·075 [c]	0·003–0·006	16–28	Jitts *et al.* (1964)
Skeletonema costatum	4·2	~0·10 [c]	0·006	20	McAllister, Shah & Strickland (1964)
Thalassiosira nordenskioldii	1·6	0·02–0·15 [c]	0·005–0·01	5–15	Jitts *et al.* (1964)

* I_{sat} varied with temperature and salinity in this detailed study.

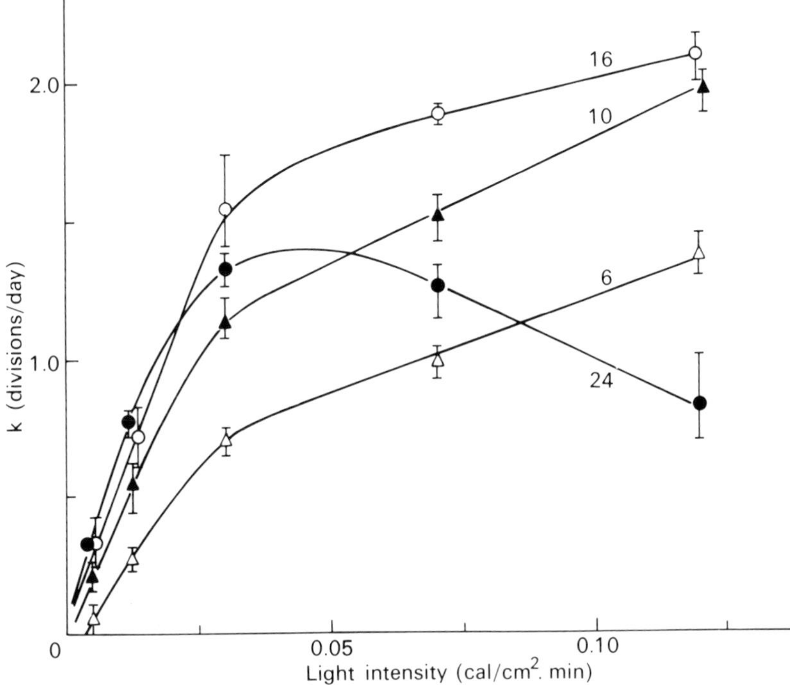

Fig. 2.2. The specific growth rate (k) of *Ditylum brightwellii* as a function of light intensity with different day lengths or hours of light per 24-hour cycle. Vertical bars indicate the range of observations. Temperature is 20°C. Growth rate is given as divisions/day. From Paasche (1968).

(1962). The oceanic isolate (clone 13-1) grew only at salinities greater than about 15‰ while the estuarine clones were euryhaline (Table 2.3).

5.4 *Nutrient concentration*

Earlier work indicated very low phosphate and/or nitrate requirements for the growth of marine diatoms (reviewed by Harvey, 1957; Ketchum, 1954; Lewin & Guillard, 1963). Much recent work has involved continuous cultures, to be discussed later. But studies of growth and of nutrient uptake kinetics in batch cultures are also current. Spencer (1954) anticipated that μ *vs.* concentration of N or P would be hyperbolic from the findings of Monod and Hinshelwood on bacterial growth. However, the N and P concentrations he used were too high to show this. Dugdale (1967) also suggested that phytoplankton growth rate *vs.* concentration of rate-limiting nutrient would be hyperbolic and would show an initial, almost linear increase of μ *vs.* nutrient concentration before

Table 2.3. Variation in specific growth rate (μ, in divisions/day) with salinity of the culture medium for some oceanic and estuarine diatoms: the optimum salinity in parts per thousand, μ at the salinity optimum, the range of salinities that allow growth, and the experimental temperature.

Species	*S*	μ	*Salinity range*	*Temperature (°C)*	*Reference*
Asterionella japonica	30	1·0	15 to >40	20	Kain & Fogg (1958)
Asterionella socialis	20–35	1·2	10 to 45	18	Lewin & Mackas (1972)
Chaetoceros armatum	25–40	0·7	<10 to >45	18	Lewin & Mackas (1972)
Detonula confervacea	~24	0·75	<8 to >32	10	Guillard & Ryther (1962)
Detonula confervacea	15–30	1·5	<5 to >35	7–12	Smayda (1969)
Rhizosolenia fragilissima	20–25	1·2	5 to >35	12–25	Ignatiades & Smayda (1970)
Thalassiosira pseudonana					
3 estuarine clones	0–35	3·0–3·4	0 to >34	20	Guillard & Ryther (1962)
estuarine clone 7–15	16–24	2·2	0 to >34	20	Guillard & Ryther (1962)
oceanic clone 13–1	24–32	2·8	<16 to 34	20	Guillard & Ryther (1962)

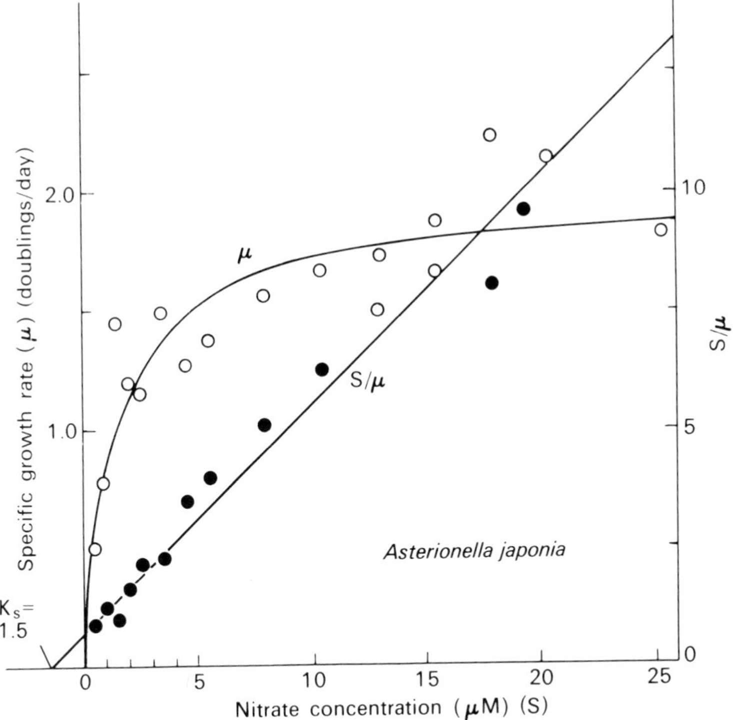

Fig. 2.3. The specific growth rate (μ) of *Asterionella japonica* as a function of the initial nitrate content of the culture medium (S). The straight line labelled S/μ represents a linear transformation of the Monod equation used to evaluate K_s and μ_m. From Eppley & Thomas (1969).

approaching a maximum, or saturation rate. The Langmuir adsorption isotherm (Hinshelwood, 1946) and the Michaelis-Menten equation of enzyme kinetics (Monod, 1950) are appropriate equations for such responses and are now in wide use. Replacing reaction rate with specific growth rate gives the Monod equation:

$$\mu = \frac{\mu_{\max} S}{K_s + S} \tag{4}$$

where μ_{max} is the maximum observed growth rate at nutrient saturation, S is the concentration of the limiting nutrient in the culture medium and K_s is the half-saturation constant (i.e. the value of S when $\mu = \mu_{max}/2$). When S is defined as the *initial* concentration in the culture and the value of μ is taken as the exponential growth rate observed later from growth measurements taken over time (Fig. 2.3), reasonable hyperbolic graphs of μ *vs.* S can be obtained

Table 2.4. Kinetic parameters for growth or nutrient uptake rate as a function of the concentration of limiting nutrient: μ_{max} in divisions/day, K_s in µg atoms/litre. Measurements were made in several ways: A = from the growth rates of batch cultures; B = from the growth of continuous cultures; C = from short-term measures of nutrient uptake. The experimental temperature is also given. Clonal designation in parenthesis. Superscripts: o = oceanic clone, e = estuarine clone.

Species	μ_{max}	K_s	*Method*	*Temperature (°C)*	*Reference*
			Phosphate		
Chaetoceros gracilis	>3·1	0·12	A	20	Thomas & Dodson (1968)
Nitzschia actinastroides	1·5	0·013	B	23	Müller (1972)
	1·4	1·0	C	23	Müller (1972)
Thalassiosira pseudonana	1·6	0·5	C	22	Fuhs *et al.* (1972)
	1·6	<0·03	B	22	Fuhs *et al.* (1972)
			Silicic acid		
Ditylum brightwellii	3.2	2·96	C	20	Paasche (1973b)
Licmophora sp.	1·4	2·58	C	20	Paasche (1973b)
Skeletonema costatum	2·4	0·80	C	20	Paasche (1973b)
Thalassiosira decipiens	1·3	3·37	C	20	Paasche (1973b)
Thalassiosira pseudonana	4·0	1·39	C	20	Paasche (1973b)
Thalassiosira pseudonana (3H)^{e}	3·6	0·98	A	20	Guillard, Kilham & Jackson (1973)
Thalassiosira pseudonana (13–1)^{o}	2·1	0·19	A	20	Guillard, Kilham & Jackson (1973)
			Vitamin B_{12}		
Skeletonema costatum	>1·4	0·32*	B	—	Droop (1970)

			Nitrate		
Asterionella japonica	1·9–2·0	1·2–1·5	A	21	Eppley & Thomas (1969)
	—	0·7–1·3	C	18	Eppley & Thomas (1969)
Chaetoceros gracilis	3·2	0·2	A	21	Eppley & Thomas (1969)
	—	0·1–0·3	C	18	Eppley & Thomas (1969)
Thalassiosira pseudonana (3H)[e]	—	1·9	C	20	Carpenter & Guillard (1971)
Thalassiosira pseudonana (7–15)[e]	—	1·2	C	20	Carpenter & Guillard (1971)
Thalassiosira pseudonana (13–1)[o]	—	0·38	C	20	Carpenter & Guillard (1971)
Thalassiosira pseudonana	3·1	0·05	B	25	Caperon & Meyer (1972b)
Thalassiosira pseudonana	—	0·35	C	25	Caperon & Meyer (1972b)
Comparison of marine diatoms of different cell size (nitrate)					
Chaetoceros gracilis (5 μm)**	—	0·1–0·3	C	18	Eppley, Rogers & McCarthy (1969)
Thalassiosira pseudonana (5 μm)	—	0·3–0·7	C	18	Eppley, Rogers & McCarthy (1969)
Skeletonema costatum (8 μm)	—	0·4–0·5	C	18	Eppley, Rogers & McCarthy (1969)
Asterionella japonica (10 μm)	—	0·7–1·3	C	18	Eppley, Rogers & McCarthy (1969)
Rhizosolenia stolterfothii (20 μm)	—	1·7	C	18	Eppley, Rogers & McCarthy (1969)
Leptocylindrus danicus (21 μm)	—	1·2–1·3	C	18	Eppley, Rogers & McCarthy (1969)
Ditylum brightwellii (30 μm)	—	0·6	C	18	Eppley, Rogers & McCarthy (1969)
Coscinodiscus lineatus (50 μm)	—	2·4–2·8	C	18	Eppley, Rogers & McCarthy (1969)
Rhizosolenia robusta (85 μm)	—	2·5–3·5	C	18	Eppley, Rogers & McCarthy (1969)
Coscinodiscus wailesii (210 μm)	—	2·1–5·1	C	18	Eppley, Rogers & McCarthy (1969)
Temperature effect (nitrate)					
Skeletonema costatum	—	~0·0	C	8	Eppley, Rogers & McCarthy (1969)
	—	0·4–0·5	C	18	Eppley, Rogers & McCarthy (1969)
	—	1·0	C	28	Eppley, Rogers & McCarthy (1969)

* Unit ng/litre vitamin B_{12}.
** Mean geometric cell diameter.

(Thomas & Dodson, 1968; Eppley & Thomas, 1969). Some estimates of K_s, obtained with batch and continuous cultures are given in Table 2.4 for diatoms with phosphate, nitrate, vitamin B_{12} or silicic acid as the rate-limiting nutrient.

Estimates of K_s derived from the growth of batch cultures seem to agree with K_s estimates from short-term uptake *vs.* concentration experiments (Eppley & Thomas, 1969). As will be seen (Table 2.4), these values are higher than those computed for growth in continuous, nutrient-limited cultures except as μ approaches μ_{max}. Nevertheless, there appears to be a relationship between observed K_s values and the nutrient concentration of the natural waters in which the species or clones are found (Dugdale, 1967; Eppley *et al.* 1969; Paasche, 1973b). Oceanic diatoms, for example, showed lower K_s values for nitrate (Carpenter & Guillard, 1971) and silicate (Guillard *et al.* 1973) than did estuarine clones of the same three species (Table 2.4). To date, corresponding comparisons of isolates from oligotrophic and eutrophic freshwaters have not, to my knowledge, included diatoms. Cell size and temperature also influence K_s (Table 2.4) as well as μ_{max}.

5.5 *Enrichment cultures*

The term 'enrichment cultures' refers to batch cultures with natural waters as the media and carried out in order to study the nutritional quality of the water (this usage differs from the microbiological sense where an enrichment culture is a means of inducing the selected growth of a particular organism, often as a step in its isolation into pure culture.) Samples of the natural water are supplemented with nutrients, either singly, or more usually by single omissions from complete medium (i.e. all minus *N,* all minus *P,* etc.). Either the naturally occurring plankton in the water serve as inoculum (Menzel *et al.* 1963) or the water is first sterilized, then inoculated with a single test species (Johnston, 1964; Smayda, 1970a, 1973). Growth is assessed as yield of cells after several days' growth, or better as the specific growth rate calculated from daily measurements of standing crop increase (cf. Erickson, 1972).

Uses of enrichment cultures and the interpretation of results fall outside the scope of this chapter and are too numerous to discuss in detail. Berland *et al.* (1972) have reviewed the experiments with sea water, and see also Chapters 10 and 12.

6 LARGE-SCALE CULTURES

Large-scale cultures of diatoms have been devised, in historical order, (1) to grow food for invertebrates (Allen & Nelson, 1910; Ansell *et al.* 1963, 1964; Gross, 1937; Raymont & Adams, 1958), (2) for ecological studies of natural waters, (3) a combination of (1) above, sometimes with nutrient removal from sewage effluent in order to discourage algal growth in receiving waters (Dunstan & Menzel, 1971; Dunstan & Tenore, 1972; Goldman *et al.* 1974).

The maximum cell production of a batch culture is achieved when the standing stock is a maximum and this occurs in the stationary phase of growth as μ approaches zero. However, in the long term, cell production is usually greater in continuous or semicontinuous cultures, i.e. with periodic dilution of the culture by harvesting and adding fresh medium (Ketchum & Redfield, 1938). Cultures for food production or nutrient removal usually use that stratagem (Dunstan & Menzel, 1971; Dunstan & Tenore, 1972; Ukeles, 1973).

Large-scale batch cultures are useful, however, in ecological studies, such as the giant plastic bags of Strickland and colleagues (Strickland, 1967; McAllister *et al.* 1961; Antia *et al.* 1963) and their 'deep tank' (Strickland *et al.* 1969). These cultures contained 70–100 m^3 of culture, either of natural phytoplankton diatoms (plastic bags) or of single species (deep tank). Even larger plastic bags are now in use in Scotland (Davies *et al.* 1973) and in the CEPEX programme of the International Decade of Ocean Exploration programme in Saanich Inlet near Victoria, B.C., Canada. Lund (1972) has isolated a portion of Blelham Tarn within a butyl rubber cylinder for ecological studies of the role of turbulence and mixing in diatom seasonal abundance. This device sets a record for volume of water enclosed: 16 000 m^3. In spirit, if not in scale, this is a culture, at least to the extent that a body of water is confined for the purpose of studying diatom growth in a semicontrolled manner. Other uses of large-scale cultures for lake studies are Goldman (1962); Bender & Jordan (1970); Edmondson *et al.* (1956); Kemmerer (1968); McLaren (1969); Schelske & Stoermer (1971); and Thomas (1961). In this context it should be noted that the spring bloom of temperate lakes (Pechlaner, 1970) and bays (Platt & Subba Rao, 1970) may behave kinetically as a batch culture. Exponential growth rates or Production/Biomass ratios for various lake diatom species have been calculated in some cases (Talling, 1955; Hughes & Lund, 1962; Happey, 1970; Pechlaner, 1970) and the rates are consistent with those observed in laboratory batch cultures.

Cultures of the scale of a few litres to a few hundred litres are commonly employed for chemical composition studies (cf. Antia & Kalmakoff, 1965). Large shipboard cultures (100–200 litres) have been useful for studies of natural phytoplankton growth and chemical composition (McAllister *et al.* 1960) and circadian periodicity in growth (Newhouse *et al.* 1967; Eppley *et al.* 1971a).

6.1 *Cell production rates*

The maximum expected production of diatoms in outdoor cultures illuminated by sunlight, and with production limited only by light, is significant for diatom production as food or for stripping nutrients from wastewaters. For *Chlorella,* Kok & Oorschot (1954) achieved production rates (dry matter) in laboratory cultures of up to 30 g.m^{-2}.day^{-1}. Production was an approximately linear function of total sunlight radiation up to about 500 cal.cm^{-2}.day^{-1}. Ryther (1959) computed somewhat lower net production rates *vs.* total sunlight radiation (see his Fig. 5.5).

Conversion factors between dry matter, cell carbon and calorific content are given by Platt & Irwin (1973) for mixed phytoplankton of a spring bloom, mostly diatoms (Platt & Subba Rao, 1970). Calories per mg dry weight varied from 2·1 to 3·7 but calories per mg carbon varied less, from 10·4 to 12·5. For *Chlorella* and other freshwater microalgae lacking mineralized cell walls the calorific content per mg dry weight is about 5·5 (Kok & Oorschot, 1954). Hence the *Chlorella* production rate cited above would be about 165 cal.m^{-2}.day^{-1} or about 15 g.C.m^{-2}.day^{-1} with total radiation 400–500 cal.cm^{-2}.day^{-1}. Recent measurements with mixed diatom cultures fertilized with secondary sewage effluent, and with a daily harvest rate of 50% of the culture, ranged from 1 to 5 g.C.m^{-2}.day^{-1} and averaged 2·6 over a 4-month period (Dunstan & Tenore, 1972).

Ketchum *et al.* (1949) pointed out that production of a culture is the product of the standing stock and the growth rate. These latter parameters tend to vary inversely and maximum production, on a continuous harvest basis, will occur at intermediate values such that their product is a maximum.

7 CONTINUOUS CULTURES

7.1 *Nutrient-limited cultures*

Continuous cultures are constant volume cultures provided with a continuous inflow of new medium and a corresponding outflow of culture (Fig. 2.4). There are two basic types. In the *chemostat* the flow rate of new medium is regulated. In the *turbidostat* the standing stock of cells is regulated, usually photometrically. The dilution rate, D, is defined as the flow rate divided by the volume of the culture and has dimensions time^{-1}. Net increase in the population can take place as long as the specific growth rate, μ, exceeds D. If D exceeds μ the culture washes out. The rate of change of the population is the difference between growth (μN) and washout (DN). At steady state the terms are equal and

$$\frac{dN}{dt} = \mu N - DN = 0 \qquad (5)$$

The rate of change of concentration of the limiting nutrient, S, is the difference between the input rate (DS_0), the washout rate (DS), and the nutrient

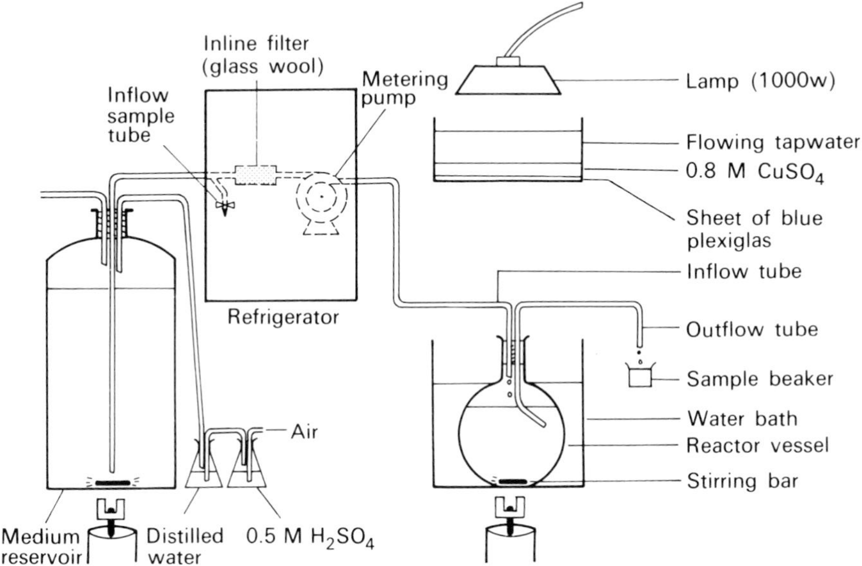

Fig. 2.4. Diagram of a chemostat culture device used for the study of growth of *Skeletonema costatum* under silicic acid limitation. The reactor vessel (culture vessel) is a 3-litre flat-bottomed boiling flask. From Davis, Harrison & Dugdale (1973).

utilization by the cells ($\mu N/Y$). At steady state ds/dt is zero and

$$\frac{ds}{dt} = DS_0 - DS - \frac{\mu N}{Y} = 0 \qquad (6)$$

where S_0 is the nutrient concentration in the incoming medium, S the concentration in the culture, and Y is the number of cells produced per mole or weight of the limiting nutrient. The dependence of μ on S is expected to follow the Monod equation, equation (1).

In a summary paper Droop (1970) concluded that the specific growth rate in a nutrient-limited continuous culture depended on the nutrient content of a cell, Q, rather than directly upon the ambient nutrient concentration in the medium

$$\mu = \mu_m(1 - K_Q/Q) \qquad (7)$$

where K_Q is the minimum nutrient content of a cell that permits growth. A number of other useful equations have been written for continuous cultures and the reader is referred to references listed in the following paragraph and see Tempest (1970) for a general review.

The chemical composition of diatoms and their physiological rate processes have been found to vary with μ in such cultures. For example, the cellular content of limiting nutrient is an approximately hyperbolic function of μ for vitamin B_{12} (Droop, 1970), phosphorus (Fuhs, 1969; Fuhs *et al.* 1972; Müller, 1972), nitrogen (Thomas & Dodson, 1972; Caperon & Meyer, 1972a,b; Eppley & Renger, 1974) and silicon (Paasche, 1973a,b,c; Davis *et al.* 1973), such that yield of cells per mole of nutrient consumed (Y) is not constant but varies with μ.

The Monod equation (equation 4) does not appear to hold rigorously in the studies cited above, where investigated. Droop (1970) and Caperon & Meyer (1972a,b) have written alternative equations for μ as a function of the cellular content of the limiting nutrient. Fuhs *et al.* (1972) and Caperon & Meyer (1972b) have computed how μ would vary with S, the latter being too low for measurement by current methods. Droop (1973) has also considered how the chemical composition of a cell would vary in the case of multiple nutrient limitations. Some estimates of μ_{max} and K_s for diatoms in continuous culture are provided in Table 2.4.

7.2 *Light-limited continuous cultures*

Continuous cultures have been used to study yield and photosynthetic parameters in cultures of *Chlorella* (cf. Myers & Graham, 1959) and other green algae, but not diatoms. Maximum photosynthetic rates, saturating and compensating light intensities can be readily studied as functions of ambient light in such cultures. Hence it is somewhat surprising that the light-limited culture of ecologically important diatoms has not been studied in view of the importance of the various illumination parameters (such as the compensation intensity for growth) and the difficulty of determining them with conventional methods. We still cite Jenkin (1937), for example, for compensating light intensities for diatom growth! Of course more recent estimates of I_{comp} have been made but still with batch cultures (Table 2.3).

7.3 *Other applications*

As suggested above, there are several additional and untried uses for continuous cultures. Riley (1966) has suggested that batch cultures may be analogues of growth in the spring bloom in temperate waters and that chemostats may be analogues of growth in tropical and subtropical oligotrophic waters where growth rate may be nutrient-limited and population size approximates a steady state. In both cases natural light is provided in light–dark illumination cycles. Continuous cultures illuminated as to provide light–dark cycles have been used to study circadian periodicity in cell division, photosynthesis, nutrient assimilation, etc. (Uno, 1971; Eppley *et al.* 1971b), which leads us to the next section.

8 SYNCHRONOUS CULTURES

8.1 *Vegetative life cycles*

When diatom cultures are illuminated under light–dark cycles, as for example in cultures kept facing a north window, cell division tends to be restricted to certain hours which may vary among species (Subrahmanyan, 1945). A period in darkness arrests cell development and results in the accumulation of physiologically young cells. On illumination a synchronous cell division takes place some hours later (Lewin *et al.* 1966; Darley, 1969). If the photoperiod and light intensity are suitably chosen essentially all the cells in such a culture will divide synchronously (von Denffer, 1949; Lewin *et al.* 1966; Paasche, 1968; Darley & Volcani, 1969). If the light intensity and duration are otherwise, then some fraction of the cells will divide in a restricted time period (Paasche, 1968). The time of cell division appears to vary from species to species, and also with environmental conditions (for example, compare Jørgensen (1966) and Eppley *et al.* (1971b), for differing results with *Skeletonema costatum*). Periodicity in photosynthetic capacity (Fig. 2.5) and in the various events in the cell cycle attends growth on light–dark cycles.

Adding silicic acid to silicon-starved diatom cultures also results in synchronous cell division (Lewin *et al.* 1966; and see Chapter 7). Darley & Volcani (1972) give details for achieving highly synchronous cultures combining light–dark cycles and Si deprivation. The sequence of events they noted in the cell cycle of *Cylindrotheca fusiformis* were: (1) DNA synthesis, mitosis and cytokinesis began in the 5th hour of the light period, (2) Si uptake and cell separation took place in the 7th or 8th hour (Darley & Volcani, 1969). In *Skeletonema costatum* grown on $L_{12}:D_{12}$ cycles cell division began in the 6th hour of the light period, chlorophyll *a* synthesis took place throughout the light period and photosynthetic capacity (i.e. photosynthetic rate at light saturation measured in a separate, constant light incubator) showed cyclic variation with a maximum midway in the light period and a minimum in the mid-dark period (Jørgensen, 1966). In *Ditylum brightwellii* grown on an $L_8:D_{16}$ cycle the sequence of events in cell development was as follows: daughter cells were formed late in the light period, in the dark they elongated and the numerous chromatophores began dividing. A minimum cell buoyancy was observed in the dark concurrent with cell expansion. Cell phosphorus increased in the dark period. In the following light period photosynthetic rate increased to a maximum, and synthesis of photosynthetic pigments, proteins and carbohydrates took place (Eppley, Holmes & Paasche, 1967).

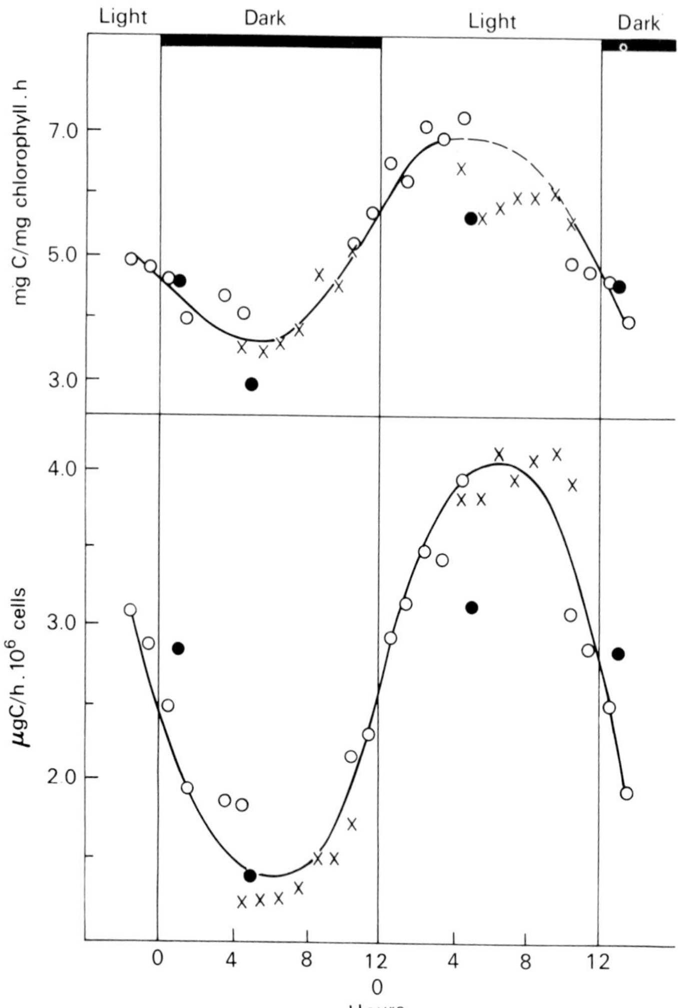

Fig. 2.5. The variation in the rate of light saturated photosynthesis in a synchronous culture of *Skeletonema costatum* during a 24-hour period. The upper curve shows the rate as mg carbon assimilated per mg chlorophyll per hour. The lower curve shows the rate as μg carbon assimilated per 10^6 cells per hour. Light intensity is 18 klux. From Jørgensen (1966).

Nutrient assimilation also shows periodicity in batch cultures grown with light–dark cycles. The enzyme nitrate reductase shows highest activity in the light in *Skeletonema costatum* (Packard *et al.* 1971). Periodicity in nitrate, ammonium and phosphate uptake, with *N* and *P* uptake out of phase, was noted in a mixed species culture of diatoms (Eppley *et al.* 1971a).

8.2 *Resting spores*

By experimentally manipulating the cell size (von Stosch, 1965) and adjusting the temperature and illumination of cultures it has been possible to induce resting spores at will in *Stephanopyxis turris* and *S. palmeriana* (Drebes, 1964, 1966). With the latter, *P*- or *P* and *N*- deprivation also promoted resting spore formation. Kashkin (1964) and Drebes (1966) found that the resting spores germinated in complete culture media in the light, whether collected in sediment samples or taken from experimental cultures.

9 DIALYSIS CULTURES

Dialysis cultures provide some of the characteristics of both batch cultures and continuous cultures. Reduced to its basic components a dialysis culture consists of two chambers, separated by a semipermeable membrane. The organisms are retained in one of the chambers and fresh medium in the other. Or, alternatively, different species of organisms are placed in the chambers as might be done for studies of species interaction mediated by diffusible metabolic products. The various uses and mathematical theory behind these uses in microbiology were reviewed by Schultz & Gerhardt (1969). Only recently have such cultures come into use with diatoms (Jensen *et al.* 1972; Jensen & Rystad, 1973; Prakash *et al.* 1973). In all three of these reports two types of experimental apparatus has been used: (1) a two-chambered glass vessel with the chambers separated by a membrane filter, and (2) bags of dialysis tubing suspended in a reservoir. The latter have provided the most extensive use to date, including use with axenic cultures (Jensen *et al.* 1972).

Growth curves for these dialysis cultures show the usual lag and exponential phases followed by rather prolonged linear growth periods ($dN/dt =$ a constant) before maximum cell concentrations were reached in the stationary phase. In all three reports the maximum cell concentrations attained in the dialysis cultures exceeded those produced in batch cultures. To date a dozen or so marine diatoms have been grown in dialysis cultures and it appears that there will be no particular restrictions as to the use of other species.

Jensen & Rystad (1973) report the imaginative use of dialysis cultures of *Thalassiosira pseudonana* clone 3H, *Skeletonema costatum,* and *Phaeodactylum tricornutum* to measure the capacity of natural seawater for supporting phytoplankton growth. They observed a marked dependence of the rate of cell production in linear growth with ambient nitrate concentration of the waters (3 and 30 m water from Trondheimsfjiord) and a similar dependence of final cell concentration. For example (Fig. 2.6) linear regressions of cell production rate on ambient nitrate concentration were:

Ph. tricornutum: 10^6 cell.ml^{-1}.day^{-1} = 0·38 (µg at NO_3/litre) + 0·57
Th. pseudonana: 10^6 cell.ml^{-1}.day^{-1} = 0·16 (µg at NO_3/litre) + 0·22
Sk. costatum: 10^6 cell.ml^{-1}.day^{-1} = 0·17 (µg at NO_3/litre) + 0·25

The linear growth apparently results from a constant diffusion rate of the limiting nutrient across the dialysis membrane.

Potential uses of dialysis cultures noted by Jensen *et al.* (1972) include (1) to monitor natural water quality and its capacity to support phytoplankton growth; (2) to use phytoplankton cells for accumulation of trace substances

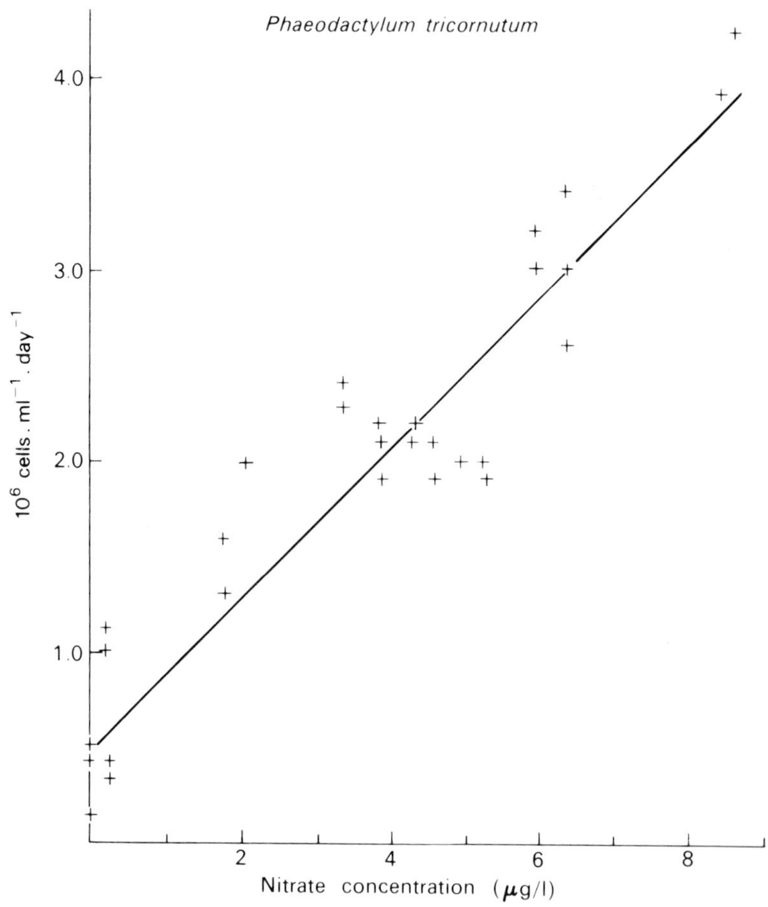

Fig. 2.6. Rate of cell production of *Phaeodactylum tricornutum* (10^6 cells. ml^{-1}. day^{-1}) growing in dialysis bags exposed to natural seawater. The rate is shown as a function of the ambient concentration of nitrate in the seawater in µg/litre. Temperature about 10°C. From Jensen & Rystad (1973).

from large volumes of water; (3) since phytoplankton in dialysis bags are not available to animals in the outside water for food, the production of phytoplankton can be measured in the absence of grazing (grazing effects on production of phytoplankton might be judged by the difference in plant production inside and outside of the dialysis bags); (4) dialysis cultures may be the method of choice for studies of exocrine production and species interaction (allelochemistry) involving diffusible, external metabolites.

10 SOME GENERALITIES ON SPECIFIC GROWTH RATE

10.1 *Cell size*

When cultures of various diatom species were grown in batch cultures under the same conditions (media, illumination, temperature) the specific growth rates varied with the mean cell size of the species. Larger cells grew more slowly than small-celled species of both marine pennate (Williams, 1964) and centric diatoms (Eppley & Sloan, 1966; Werner, 1970). However, this trend of declining μ with increased cell volume was not very evident in comparisons within species where isolates with different valve diameter showed only slight or no differences in μ (Werner, 1971a; Paasche, 1973c). McMahon (1973) suggested that metabolic rates among organisms would be expected to vary with body weight to the 0·75 power

$$\frac{dW}{dt} = a \cdot W^{0\cdot 75}$$

Therefore, the specific growth rate might be expected to vary with body weight to the –0·25 power

$$\mu = \frac{1}{W}\frac{dW}{dt} = \frac{a \cdot W^{0\cdot 75}}{W} = a \cdot W^{-0\cdot 25} \quad (9)$$

However, regression analysis of μ *vs.* cell volume in place of weight, using data of Williams (1964) and Eppley & Sloan (1966) for 20°C, gave a mean coefficient of –0·070 instead of the expected –0·25 (R.W. Eppley & E. Stewart, unpublished). But the scatter of experimental points is too great to permit any confidence in this value.

Other parameters displaying systematic variation of cell size in diatoms are carbon content (Mullin *et al.* 1966; Strathman, 1967), DNA content (Holm-Hansen, 1969), sinking rate of marine species (Fig. 2.7) (reviewed by Smayda, 1970b), and half-saturation concentrations for nutrient uptake (Eppley *et al.* 1969). While these generalities are interesting and useful, it is dangerous to expect them to hold in every case. Particularly interesting in this regard are the

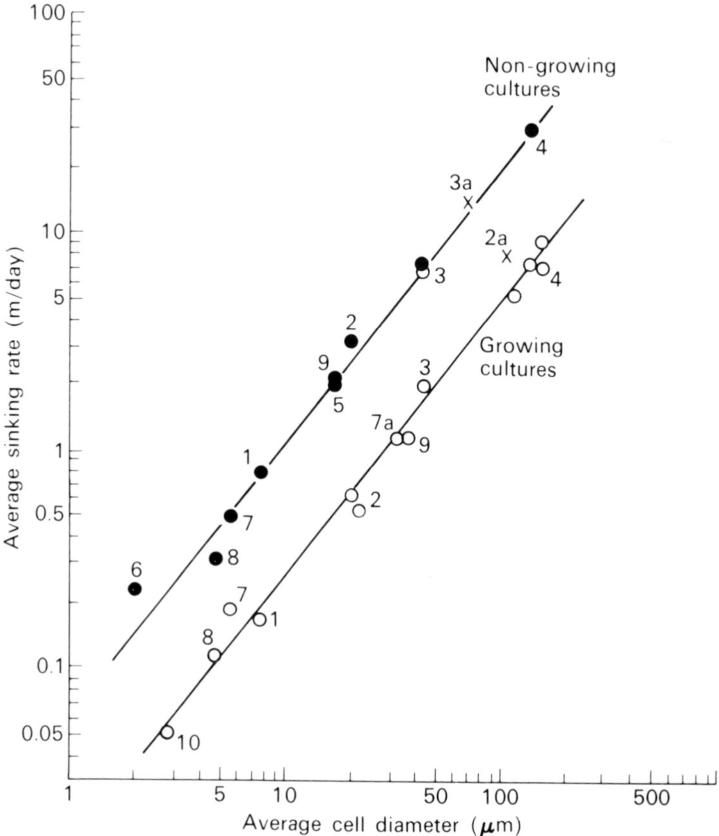

Fig. 2.7. The sinking rate of some unicellular marine diatoms in seawater at 20°C. Sinking rate varied directly with the average cell diameter. Rates for cells taken in the exponential phase of growth were lower than those for the same organisms in the stationary phase (non-growing). The numbers refer to species: (1) *Thalassiosira pseudonana,* (2) *Ditylum brightwellii,* (3) *Coscinodiscus* sp., (4) *C. wailesii,* (5) *Rhizosolenia stolterfothii,* (6) *R. hebetata* f. *semispina,* (7) same after auxospore formation, (8) *Thalassiosira nana,* (9) *Stephanopyxis turris,* (10) *Phaeodactylum tricornutum.* From Eppley, Holmes & Strickland (1967).

differences in chemical composition and sensitivity to growth inhibitors of different size classes of *Coscinodiscus asteromphalus* (Werner, 1971a,b) and the wide variations in μ observed in different laboratories for species such as *Ditylum brightwellii* (Paasche, 1973c) and *Skeletonema costatum* (Davis *et al.* 1973).

10.2 *Temperature*

Each species or clone appears to show a maximum specific growth rate, probably genetically determined, at its temperature optimum and with optimal light, salinity, pH and nutrients provided. A graph of experimentally measured values of μ *vs*. temperature (with data for more than 100 species of unicellular algae) suggested that a smooth curve of maximum expected μ *vs*. temperature could be drawn with all values falling below the line (Eppley, 1972). There is always an exception to a rule and new data for *Skeletonema costatum* (Davis *et al.* 1973) fall above the line, perhaps requiring some revision. Nevertheless, an approximate equation for the line of maximum expected μ *vs*. T is

$$\log_{10} \mu = 0{\cdot}0275T - 0{\cdot}070 \qquad (10)$$

10.3 *Light*

Similar 'models' could no doubt be undertaken for μ *vs*. illumination (incident irradiance, or better, light absorbed by the photosynthetic pigments) with proper provisos for temperature and past history of the culture. However, the product law (response = intensity × duration) does not appear to hold for photosynthetic growth. In particular, some species of diatoms display maximum μ values with 14–18 hour day lengths rather than in continuous light (Castenholz, 1964; Paasche, 1968). A more elaborate model based upon treating the light absorbed by chlorophyll *a* as S in a Monod-type equation is subject to the same difficulties (Eppley & Sloan, 1966). Nevertheless, marine planktonic diatoms do appear to use light more effectively for growth than do unicellular green algae under comparable conditions (Jitts *et al.* 1964; Dunstan, 1973).

11 EPILOGUE

The technique of culturing diatoms has been with us more than seven decades. Yet new techniques, uses and insights keep appearing. For example, the use of planktonic diatoms as bioassay organisms in pollution studies is justified wherever these organisms are ecologically significant and particularly in lakes, estuaries and both the coastal and open ocean. Lund *et al.* (1971), Kayser (1971, 1973) and Erickson (1972) have taken steps in this direction for freshwater and marine forms, respectively. (*Asterionella formosa, Chaetoceros socialis* and *Thalassiosira pseudonana* were proposed test species.) It is interesting and significant that marine diatoms appear to be more sensitive to growth rate inhibition by chlorinated hydrocarbons (PCBs and DDT) than other unicellular algae tested (Mosser *et al.* 1972a,b) and an oceanic clone of

Thalassiosira pseudonana was more sensitive than estuarine clones of the species (Fisher *et al.* 1973).

A fundamental stumbling block in studies of planktonic diatom growth in nature is our appalling inability to measure standing stocks and growth rates, either of each species or of an assemblage, by some convenient and widely available method. A rapid, automated means of species identification and enumeration would broaden horizons and move this field ahead more quickly perhaps than any of us can imagine.

12 ADDENDUM

The literature review for this report was completed in June, 1974. Meanwhile, the use of diatom cultures as experimental tools continues apace and many new papers have appeared. These recent references, and those missed earlier which are now to be cited, will be mentioned in the same order of topics as the main body of the text using the numerical abbreviations for the chapter subheadings.

2.1 A mass-spectrometer method for measuring silicic acid incorporation by diatoms has now been published (Goering, J.J., Nelson, D.M. & Carter, J.A. (1973) Silicic acid uptake by natural populations of marine phytoplankton. *Deep-Sea Res.* **20,** 777–89). Azam has shown, using diatom cultures, that a radioisotope of germanium, ^{68}Ge, is a suitable tracer for silicic acid (Azam, F. (1974) Silicic acid uptake in diatoms studied with [^{68}Ge] germanic acid as a tracer. *Planta (Berl.)* **121,** 205–12; Azam F. & Volcani, B.E. (1974) Role of silicon in diatom metabolism. VI. Active transport of germanic acid in the heterotrophic diatom *Nitzschia alba*. Arch. Mikrobiol. **92,** 11–20). These new methods should facilitate studies of silicic acid utilization by diatoms and are especially needed for measurements with natural populations.

2.3 The element molybdenum is known to play a special role in nitrogen metabolism and to be a constituent of the enzyme nitrate reductase in certain microorganisms. Effects of molybdenum deficiency have now been studied in *Navicula pelliculosa,* a freshwater diatom (Wallen, D.G. & Carter, L.D. (1975) Molybdenum dependence, nitrate uptake and photosynthesis of freshwater plankton algae. *J. Phycol.* **11,** 345–49). Molybdenum deficiency resulted in lower chlorophyll *a* content and reduced rates of photosynthesis and nitrate uptake. At the same time these authors measured nitrate uptake rates *vs.* concentration of nitrate and noted that the half-saturation contents for the two freshwater species investigated were considerably higher than those of marine species. Further, the value for the freshwater diatom was lower by an order of magnitude than that of the green alga they studied, *Chlamydomonas reinhardtii* (reinhardii).

5.1–5.3 Salinity and temperature limits for growth have been determined for several species since Tables 1 and 3 were compiled. These include *Bellerochea polymorpha, B. spinifera, Fragillaria pinnata, F. rotundissima, Navicula pelliculosa* and *Synedra fragilaroides* (Hargraves, P.E. & Guillard, R.R.L. (1974) Structural and physiological observations on some small marine diatoms. *Phycologia* **13,** 163–72), and a detailed study of *Thalassiosira rotula* (Schöne, H.K. (1974) Experimentelle Untersuchungen zur Ökologie der marinen Kieselalge *Thalassiosira rotula*. II. Der Einfluss des Salzgehaltes. *Mar. Biol.* **27,** 287–98). In the latter case higher growth rates and a broader salinity optimum were noted when illumination was provided in light dark cycles than in continuous light (see 10.3 for another example of reduced growth rate in continuous light).

5.4 The question of how growth is related to ambient nutrient concentration in diatoms was illuminated by the report of an ATP-ase, presumably localized at the cell surface, which is activated by nitrate. Comparison of the K_m for nitrate activation of the ATP-ase activity and the half-saturation constants for nitrate uptake by intact cells indicated that the two values were very similar and lead to the suggestion that the ATP-ase may be involved in nitrate transport (Falkowski, P.G. (1975) Nitrate uptake in marine phytoplankton: (nitrate, chloride)—activated adenosine triphosphate from *Skeletonema costatum* (Bacillariophyceae). *J. Phycol.* **11,** 323–26; Falkowski, P.G. (1975) Nitrate uptake by marine phytoplankton: Comparison of half-saturation constants for seven species. *Limnol. Oceanogr.* **20,** 412–17).

An interesting approach to the use of batch cultures to study the relation between specific growth rate and ambient nutrient concentration has appeared (Klaveness, D. & Guillard, R.R.L. (1975) The requirement for silicon of *Synura petersenii* (Chrysophyceae). *J. Phycol.* **11,** 349–55). The authors added very small inocula to a series of flasks containing a range of silicic acid concentrations and were able to measure specific growth rates before the ambient silicic acid levels were significantly reduced by growth. In effect their growth conditions were steady-state in that both μ and Si concentration (the growth rate-limiting nutrient) were constant over the measurement period. It is this steady-state condition that makes continuous cultures so attractive. The K_s they report for silicic acid-limited growth was 0·23 μM and μmax was 1·1 divisions.day^{-1} (see Table 2.4 for values for other species). One wonders if Si-limited chemostat growth would also yield this K_s value, or a value 1/10 of this as in previous comparisons between batch and continuous cultures.

7.1 Use of continuous cultures to investigate ecological problems continues, particularly for the identification of growth-rate limiting nutrients (Peterson, B.J., Barlow, J.P. & Savage, A.E. (1974) The physiological state with respect to phosphorus of Cayuga Lake phytoplankton. *Limnol. Oceanogr.* **19,** 396–408) and phytoplankton metabolism (Bienfang, P.K. (1975) Steady state analysis of nitrate-ammonium assimilation by phytoplankton. *Limnol. Oceanogr.* **20,** 402–11).

8 A recent study of diatom growth in outdoor tanks has significance for several sections of this report, i.e. large scale cultures, continuous cultures, synchronous cultures. But the dominant effects resulted from the natural daylight cycles of illumination and the variation in nitrate assimilation patterns with growth rate under these conditions. At low growth rate (dilution rate) the diatom crops continually kept the medium depleted of nitrate, but at high dilution rates the ambient nitrate level displayed a circadian oscillation, being low in the light and higher at night. Such complexity of interaction was not previously suspected but has considerable importance for the allocation of limiting nutrients in certain natural situations as well as for the use of mass cultures for tertiary waste treatment (Malone, T.C., Garside, C., Haines, K.C. & Roels, O.A. (1975) Nitrate uptake and growth of *Chaetoceros sp.* in large outdoor continuous cultures. *Limnol. Oceanogr.* **20,** 9–19).

8.1 Synchronous cultures as models of nature receive some support from the observation of partial synchrony in cell division in a natural diatom population (Smayda, T.J. (1975) Phased cell division in natural populations of the marine diatom *Ditylum brightwellii,* and the potential significance of diel phytoplankton behaviour in the sea. *Deep-Sea Res.* **22,** 151–65).

10.1 A summary of published data on the variation in growth rate, sinking rate, respiration and on other physiological processes has appeared (Laws, E.A. (1975) The importance of respiratory losses in controlling the size distribution of marine phytoplankton. *Ecol.* **56,** 519–26). More data on the covariation of growth rate and cell size within a species are also available (Findlay, I.W.O. (1972) Effects of external factors and cell size on the cell division rates of a marine diatom, *Coscinodiscus pavillardii* Forti. *Int. Rev. ges. Hydrobiol.* **57,** 523–33).

11 Diatom cultures are becoming more and more useful in pollution studies as they seem relatively more sensitive than other algal groups to many pollutants. Further examples are given for petroleum products and copper (Pulich, W.M. jr., Winters, K. & Van Baalen, C. (1974) The effects of No. 2 fuel oil and two crude oils on the growth and photosynthesis of microalgae. *Mar. Biol.* **28,** 87–94; Saward, D., Stirling, A. & Topping, G. (1975) Experimental studies on the effects of copper on a marine food chain. *Mar. Biol.* **29,** 351–61).

ACKNOWLEDGEMENTS

I thank Ms Marian Tate for typing the manuscript and my colleagues at the Food Chain Research Group for many helpful and stimulating discussions on how best to gain insights on the behaviour of marine planktonic diatoms from their study in laboratory cultures.

Supported by US National Science Foundation Grant No. GA-31167X2 and US Atomic Energy Commission Contract No. AT(11-1)GEN 10, P.A. 20.

13 REFERENCES

ALLEN E.J. (1914) On the culture of the plankton diatom *Thalassiosira gravida* Cleve in artificial sea water. *J. mar. biol. Ass. U.K.* **10,** 417–39.

ALLEN E.J. & NELSON E.W. (1910) On the artificial culture of marine plankton organisms. *J. mar. biol. Ass. U.K.* **8,** 421–74.

ANSELL A., RAYMONT J.E.G., LANDER K.F., CROWLEY E. & SCHACKLEY P. (1963) Studies on the mass culture of *Phaeodactylum* II. The growth of *Phaeodactylum* and other species in outdoor tanks. *Limnol. Oceanogr.* **8,** 184–204.

ANSELL A., COUGHLAN D.L., LANDER K.F. & LOOSEMORE F.A. (1964) Studies on the mass culture of *Phaeodactylum* IV. Production and nutrient utilization in outdoor mass cultures. *Limnol. Oceanogr.* **9,** 334–42.

ANTIA N.J. & KALMAKOFF J. (1965) Growth rate and cell yields from axenic mass culture of fourteen species of marine phytoplankton. *Fish. Res. Bd. Canada mss Rep. Ser. No. 203.*

ANTIA N.J., MCALLISTER C.D., PARSONS T.R., STEPHENS K. & STRICKLAND J.D.H. (1963) Further measurements of primary production using a large-volume plastic sphere. *Limnol. Oceanogr.* **8,** 166–83.

AUBERT M., PESANDO D. & PARCEMIN J.M. (1970) Médiateur chimique et relations inter-espèces mise en evidence d'un inhibiteur de synthèse metabolique d'une diatomée produit par un péridinéen (étude 'in vitro'). *Rev. Intern. Oceanogr. Med.* **17,** 5–21.

AZAM F., HEMMINGSEN B.B. & VOLCANI B.E. (1973a) Transport and metabolism of silicic acid in a marine heterotrophic diatom *Nitzschia alba. Bact. Proc.* p. 51.

AZAM F., HEMMINGSEN B.B. & VOLCANI B.E. (1973b) Germanium incorporation into the silica of diatom cell walls. *Arch. Mikrobiol.* **92,** 11–20.

BARBER R.T. (1973) Organic ligands and phytoplankton growth in nutrient-rich seawater. In *Trace Metals and Metal-Organic Interactions in Natural Waters* (ed. P.C. Singer) pp. 321–38. Ann Arbor Science Publishers, Ann Arbor, Michigan.

BARBER R.T. & RYTHER J.H. (1969) Organic chelators: factors affecting primary production in the Cromwell Current upwelling. *J. exp. mar. Biol. Ecol.* **3,** 191–99.

BARBER R.T., DUGDALE R.C., MACISAAC J.J. & SMITH R.L. (1971) Variation in phytoplankton growth associated with the source and conditioning of upwelled waters. *Invest. Pesq.* **35,** 171–93.

BENDER M.E. & JORDAN R.A. (1970) Plastic enclosures versus open lake productivity measurements. *Trans. Am. Fish Soc.* **99,** 607–10.

BERLAND B.R., BONIN D.J., MAESTRINI S.Y. & POINTIER J.P. (1972) Etude de la fertilité des eaux marines au noyen de tests biologiques effectués avec des cultures d'algues. I. Comparaison des méthodes d'estimation. *Int. Rev. ges. Hydrobiol.* **56,** 933–44.

BOLD H.C. (1942) The cultivation of algae. *Bot. Rev.* **8,** 69–138.

BRAARUD T. (1937) A quantitative method for the experimental study of plankton diatoms. *J. Cons. perm. int. Explor. Mer.* **12,** 321–32.

BRAARUD T. (1945) Experimental studies on marine plankton diatoms. Avhandl. Norske Videnskaps—Akad. Oslo. I. *Mat. Naturv. Klasse No. 10.* 15 pp.

BRAARUD T. (1948) On variations in form of *Skeletonema costatum* and their bearing on the supply of silica in cultures of diatoms. *Nyt Mag. Naturvid.* **86,** 31–44.

BRAARUD T. (1961) Cultivation of marine organisms as a means of understanding environmental influences on populations. In *Oceanography* (ed. Mary Sears) pp. 271–98. AAAS, Washington, D.C.

BROWN T.E. & RICHARDSON F.L. (1968) The effect of growth environment of the physiology of algae: light intensity. *J. Phycol.* **4,** 38–54.

BUNT J. (1968) Some characteristics of microalgae isolated from Antarctic sea ice. *Antarctic Res. Ser. Vol. 11.* Biology of the Antarctic Seas III, 1–14.

CANALE R.P. & VOGEL A.H. (1974) Effects of temperature on phytoplankton growth. *J. Envir. Eng. Div. (Amer. Soc. Civil Engr.)* **100,** 231–41.

CAPERON J. & MEYER J. (1972a) Nitrogen-limited growth of marine phytoplankton I. Changes in population characteristics with steady-state growth rate. *Deep-Sea Res.* **19,** 601–18.

CAPERON J. & MEYER J. (1972b) Nitrogen-limited growth of marine phytoplankton II. Uptake kinetics and their role in nutrient limited growth of phytoplankton. *Deep-Sea Res.* **19,** 619–32.

CARLUCCI A.F. (1973) Bioassay: cyanocobalamin. *Handbook of Phycological Methods.* (ed. J. Stein) pp. 387–94. Cambridge Univ. Press. London & New York.

CARLUCCI A.F. & BOWES P.M. (1970a) Production of vitamin B_{12}, thiamine and biotin by phytoplankton. *J. Phycol.* **6,** 351–57.

CARLUCCI A.F. & BOWES P.M. (1970b) Vitamin production and utilization by phytoplankton in mixed culture. *J. Phycol.* **6,** 393–400.

CARPENTER E.J. & GUILLARD R.R.L. (1971) Intraspecific differences in nitrate half-saturation constants for three species of marine phytoplankton. *Ecology* **52,** 183–85.

CARPENTER E.J., REMSEN C.C. & WATSON S.W. (1972) Utilization of urea by some marine phytoplankters. *Limnol. Oceanogr.* **17,** 265–69.

CASTENHOLZ R.W. (1964) The effect of daylength and light intensity on the growth of littoral marine diatoms in culture. *Physiologia Pl.* **17,** 951–63.

CURL JR., H. & MCLEOD G.C. (1961) The physiological ecology of a marine diatom *Skeletonema costatum* (Grev.) Cleve. *J. mar. Res.* **19,** 70–88.

DARLEY W.M. (1969) Silicon and the division cycle of the diatoms *Navicula pelliculosa* and *Cylindrotheca fusiformis. Proceedings of the North American Paleontological Convention Part G.* pp. 994–1 009.

DARLEY W.M. & VOLCANI B.E. (1969) Role of silicon in diatom metabolism. A silicon requirement for deoxyribonucleic acid synthesis in the diatom *Cylindrotheca fusiformis* Reimann and Lewin. *Expl. Cell Res.* **58,** 334–42.

DARLEY W.M. & VOLCANI B.E. (1972) Synchronized cultures: diatoms. In *Methods in Enzymol.* (ed. A. San Pietro) **23A,** 85–96.

DAVIES J.M., GAMBLE J.C. & STEELE J.H. (1973) Preliminary studies with a large plastic enclosure. In *Estuarine Research* (ed. L.E. Cronin) pp. 251–264. Academic Press, New York.

DAVIS C.O., HARRISON P.J. & DUGDALE R.C. (1973) Continuous culture of marine diatoms under silicate limitation I. Synchronized life cycle of *Skeletonema costatum. J. Phycol.* **9,** 175–80.

DENFFER V.D. (1949) Die planktische Massenkultur pennater *Grunddiatomeen. Arch. Mikrobiol.* **14,** 159–202.

DREBES G. (1964) Uber den Lebenszyklus der marinen Planktondiatomee *Stephanopyxis turris* (Centrales) und seine Steuerung im Experiment. *Helgoländer wiss. Meeresunters* **10,** 152–3.

DREBES G. (1966) On the life history of the marine plankton diatom *Stephanopyxis palmeriana. Helgoländer wiss. Meeresunters.* **13,** 101–14.

DROOP M.R. (1970) Viatmin B_{12} and marine ecology V. Continuous cultures as an approach to nutritional kinetics. *Helgoländer wiss. Meeresunters* **20,** 629–36.

DROOP M.R. (1973) Some thoughts on nutrient limitation in algae. *J. Phycol.* **9,** 264–72.

DUGDALE R.C. (1967) Nutrient limitation in the sea: dynamics, identification and significance. *Limnol. Oceanogr.* **12,** 685–95.

DUNSTAN W.M. (1973) A comparison of the photosynthesis–light intensity relationship in phylogenetically different marine microalgae. *J. exp. mar. Biol. Ecol.* **13,** 181–87.

DUNSTAN W.M. & MENZEL D.W. (1971) Continuous cultures of natural populations of phytoplankton in dilute, treated sewage effluent. *Limnol. Oceanogr.* **16,** 623–32.

DUNSTAN W.M. & TENORE K.R. (1972) Intensive outdoor culture of marine phytoplankton enriched with treated sewage effluent. *Aquaculture* **1,** 181–92.

EDMONDSON W.T., ANDERSON G.C. & PETERSON D.F. (1956) Artificial eutrophication of Lake Washington. *Limnol. Oceanogr.* **1,** 47–54.

EPPLEY R.W. (1972) Temperature and phytoplankton growth in the sea. *Fish. Bull. (U.S.)* **70,** 1 063–85.

EPPLEY R.W. & RENGER E.R. (1974) Nitrogen assimilation of an oceanic diatom in nitrogen-limited continuous culture. *J. Phycol.* **10,** 15–23.

EPPLEY R.W. & SLOAN P.R. (1966) Growth rates of marine phytoplankton: correlation with light absorption by cell chlorophyll *a*. *Physiologia Pl.* **19,** 47–59.

EPPLEY R.W. & THOMAS W.H. (1969) Comparison of half-saturation constants for growth and nitrate uptake of marine phytoplankton. *J. Phycol.* **5,** 375–79.

EPPLEY R.W., HOLMES R.W. & PAASCHE E. (1967) Periodicity in cell division and physiological behaviour of *Ditylum brightwellii,* a marine planktonic diatom, during growth in light–dark cycles. *Arch. Mikrobiol.* **56,** 305–23.

EPPLEY R.W., HOLMES R.W. & STRICKLAND J.D.H. (1967) Sinking rates of marine phytoplankton measured with a fluorometer. *J. exp. mar. Biol. Ecol.* **1,** 191–208.

EPPLEY R.W., ROGERS J.N. & MCCARTHY J.J. (1969) Half-saturation 'constants' for uptake of nitrate and ammonium by marine phytoplankton. *Limnol. Oceanogr.* **14,** 912–20.

EPPLEY R.W., CARLUCCI A.F., HOLM-HANSEN O., KIEFER D., MCCARTHY J.J., VENRICK E. & WILLIAMS P.M. (1971a) Phytoplankton growth and composition in shipboard cultures supplied with nitrate, ammonium or urea as the nitrogen source. *Limnol. Oceanogr.* **16,** 741–51.

EPPLEY R.W., ROGERS J.N., MCCARTHY J.J. & SOURNIA A. (1971b) Light/dark periodicity in nitrogen assimilation of the marine phytoplankters *Skeletonema costatum* and *Coccolithus huxleyi* in N-limited chemostat culture. *J. Phycol.* **7,** 150–4.

ERICKSON S.J. (1972) The toxicity of copper to *Thalassiosira pseudonana* in unenriched inshore sea water. *J. Phycol.* **8,** 318–22.

FEDOROV V.D. & KUSTENKO N.G. (1972) Interregulation of marine planktonic diatoms in mono- and mixed cultures. *Oceanologia* **12,** 111–22.

FISHER N.S., GRAHAM L.B., CARPENTER E.J. & WURSTER C.F. (1973) Geographic differences in phytoplankton sensitivity to PCB's. *Nature (London)* **241,** 548–9.

FOTT B. & TRUCOVA E. (1968) List of species in the culture collection of algae at the Department of Botany, Charles University of Prague. *Acta Univ. Carol. Biol.* **1967,** 197–221.

FUHS G.W. (1969) Phosphorus content and rate of growth in the diatoms *Cyclotella nana* and *Thalassiosira fluviatilis*. *J. Phycol.* **5,** 312–1.

FUHS, G.W., DEMMERLE S.D., CANELLI E. & CHEN M. (1972) Characterization of phosphorus-limited plankton algae. *Limnol. Oceanogr. Spec. Sympos.* Vol. I, 113–32.

GARRELS R.M. (1965) Silica: role in the buffering of natural waters. *Science* **148,** 69–70.

GOLD K. (1964) A microbiological assay for vitamin B_{12} in sea water using radiocarbon. *Limnol. Oceanogr.* **9,** 343–47.

GOLDMAN C.R. (1962) A method of studying nutrient limiting factors *in situ* in water columns isolated by polyethylene film. *Limnol. Oceanogr.* **7,** 99–101.

GOLDMAN J.R., TENORE K.R., RYTHER J.H. & CORWIN N. (1974) Inorganic nitrogen removal in a combined tertiary treatment-marine aquaculture system-J. removal efficiencies. *Wat. Res.* **8,** 45–54.

GRANT B.R., MADGWICK J. & DALPONT G. (1967) Growth of *Cylindrotheca closterium* var californica (Mereschk) Reimann & Lewin on nitrate, ammonia and urea. *Aust. J. mar. Freshwat. Res.* **18,** 129–36.

GROSS F. (1937) Notes on the culture of some marine plankton organisms. *J. mar. biol. Ass. U.K.* **31,** 753–68.

GUILLARD R.R.L. (1968) B_{12} specificity of marine centric diatoms. *J. Phycol.* **4,** 59–64.

GUILLARD R.R.L. (1973) Division rates. In *Handbook of Phycological Methods* (ed. J. Stein) pp. 289–311. Cambridge Univ. Press, London & New York.

GUILLARD R.R.L. & CASSIE V. (1963) Minimum cyanocobalamin requirements of some marine centric diatoms. *Limnol. Oceanogr.* **8,** 161–5.

GUILLARD R.R.L. & RYTHER J.H. (1962) Studies of marine planktonic diatoms. I. *Cyclotella nana* Hustedt and *Detonula confervacea* (Cleve) Gran. *Can. J. Microbiol.* **8,** 229–39.

GUILLARD R.R.L., KILHAM P. & JACKSON T.A. (1973) Kinetics of silicon-limited growth in the marine diatom *Thalassiosira pseudonana* Hasle and Heimdal (=*Cyclotella nana* Hustedt). *J. Phycol.* **9,** 233–7.

HAPPEY C.M. (1970) The effects of stratification on phytoplanktonic diatoms in a small body of water. *J. Ecol.* **58,** 635–51.

HARVEY H.W. (1937) The supply of iron to diatoms. *J. mar. biol. Assoc. U.K.* **22,** 205–19.

HARVEY H.W. (1947) Manganese and the growth of phytoplankton. *J. mar. biol. Assoc. U.K.* **26,** 562–79.

HARVEY H.W. (1957) *The Chemistry and Fertility of Sea Waters.* 234 pp. Cambridge.

HAYWARD V. (1970) Studies on the growth of *Phaeodactylum tricornutum.* VI. The relationship to sodium, potassium, calcium and magnesium. *J. mar. biol. Assoc. U.K.* **50,** 293–9.

HELLEBUST J.A. (1965) Excretion of some organic compounds by marine phytoplankton. *Limnol. Oceanogr.* **10,** 192–206.

HINSHELWOOD C.N. (1946) *The Chemical Kinetics of the Bacterial Cell.* Oxford Univ. Press, London.

HOLM-HANSEN O. (1969) Algae: amounts of DNA and organic carbon in single cells. *Science* **163,** 87–8.

HOOGENHOUT H. & AMESZ J. (1965) Growth rates of photosynthetic microorganisms in laboratory cultures. *Arch. Mikrobiol.* **50,** 10–24.

HUGHES J.C. & LUND J.W.G. (1962) The rate of growth of *Asterionella formosa* Hass. in relation to its ecology. *Arch. Mikrobiol.* **42,** 117–29.

HULBURT E.M. & GUILLARD R.R.L. (1968) The relationship of the distribution of the diatom *Skeletonema tropicum* to temperature. *Ecology* **49,** 337–9.

HUTNER S.H. (1950) Introduction: the role of algal cultures in research. In *The Culturing of Algae* (eds. J. Brunel, G.W. Prescott & L.H. Tiffany) pp. 3–10. C.F. Kettering Foundation. Antioch Press, Ohio.

HUTNER S.H. & PROVASOLI L. (1953) A pigmented marine diatom requiring vitamin B_{12} and uracil. *Phycol. News Bull.* **18,** 7–8.

IGNATIADES L. & SMAYDA T.J. (1970) Autecological studies of the marine diatom *Rhizosolenia fragilissima* Bergen. I. The influence of light, temperature and salinity. *J. Phycol.* **6,** 332–9.

JENKIN P.M. (1937) Oxygen production by the diatom *Coscinodiscus excentricus* Ehr. in relation to submarine illumination in the English Channel. *J. mar. biol. Assoc. U.K.* **22,** 301–43.

JENSEN A. & RYSTAD B. (1973) Semi-continuous monitoring of the capacity of sea water for supporting growth of phytoplankton. *J. expl. mar. Biol. Ecol.* **11,** 275–85.

JENSEN A., RYSTAD B. & SKOKLUND L. (1972) The use of dialysis culture in phytoplankton studies. *J. expl. mar. Biol. Ecol.* **8,** 241–8.

JITTS H.R., MCALLISTER C.D., STEPHENS K. & STRICKLAND J.D.H. (1964) The cell division rates of some marine phytoplankters as a function of light and temperature. *J. Fish. Res. Bd Canada* **21,** 139–57.

JOHNSTON R. (1964) Seawater, the natural medium of phytoplankton. II. Trace metals and chelation, and general discussion. *J. mar. biol. Assoc. U.K.* **44,** 104–16.

JØRGENSEN E.G. (1953) Silicate assimilation by diatoms. *Physiologia Pl.* **6,** 301–15.

JØRGENSEN E.G. (1960) *Carnegie Inst. Wash. Yearbook* **59,** 348–9.

JØRGENSEN E.G. (1966) Photosynthetic activity during the life cycle of synchronous *Skeletonema* cells. *Physiologia Pl.* **19,** 789–99.

JØRGENSEN E.G. (1968) The adaptation of plankton algae. II. Aspects of the temperature adaptation of *Skeletonema costatum. Physiologia Pl.* **21,** 423–7.

KAIN J.M. & FOGG G.E. (1958) Studies on the growth of marine phytoplankton. I. *Asterionella japonica* Gran. *J. mar. biol. Assoc. U.K.* **37,** 397–413.

KASHKIN N.I. (1964) On the water deposits of phytoplankton in the sublittoral regularity of the distribution of oceanic plankton. *Trudy Okeonol. Inst. Akad Nauk, U.S.S.R.* **65,** 49–57. (*Abst. Deep-Sea Res.* 13,333.)

KAYSER H. (1971) Produktivitätsmessungen an Phytoplanktonorganismen aus Küstengewässern als Standardmethode für einen Abwassertest. *Thalassia Jugoslav.* **7,** 139–50.

KAYSER H. (1973) Über den Einfluss von Rotschlamm auf die Kultur einiger mariner Planktonalgen. *Helgoländer wiss. Meeresunters.* **25,** 357–83.

KEMMERER A.J. (1968) A method to determine fertilization requirements of a small fishing lake. *Trans. Am. Fish. Soc.* **97,** 425–8.

KETCHUM B.H. (1954) Mineral nutrition of phytoplankton. *Ann. Rev. Pl. Physiol.* **5,** 55–74.

KETCHUM B.H. & REDFIELD A.C. (1938) A method for maintaining a continuous supply of marine diatoms by culture. *Biol. Bull.* **75,** 165–9.

KETCHUM B.H., LILLICK L. & REDFIELD A.C. (1949). The growth and optimum yield of algae in mass culture. *J. cell comp. Physiol.* **33,** 267–79.

KILHAM P. (1971) A hypothesis concerning silica and freshwater planktonic diatoms. *Limnol. Oceanogr.* **16,** 10–18.

KOCH W. (1964) Verzeichnis der Sammlung von Algenkulturen am Pflanzenphysiologischen Institut der Universität Göttingen. *Arch. Mikrobiol.* **47,** 402–32.

KOK B. & VAN OORSCHOT J.C.P. (1954) Improved yields in algal mass cultures. *Acta bot. neerl.* **3,** 533–46.

KUENZLER E.J. (1966) Glucose-6-phosphate utilization by marine algae. *J. Phycol.* **1,** 156–64.

LANSKAYA L.A. (1961) Fission rate of plankton algae of the Black Sea in cultures. Chap. 13 in *Marine Microbiology* (ed. C.H. Oppenheimer) pp. 127–32.

LEWIN J.C. (1955) Silicon metabolism in diatoms. *J. gen. Physiol.* **39,** 1–10.

LEWIN J.C. (1960) The dissolution of silica from diatom walls. *Geochim. Cosmochim. Acta* **21,** 182–98.

LEWIN J.C. (1965) The thiamine requirement of a marine diatom. *Phycologia* **4,** 141–4.

LEWIN J.C. (1966a) Silicon metabolism in diatoms. V. Germanium dioxide, a specific inhibitor of diatom growth. *Phycologia* **6,** 1–12.

LEWIN J.C. (1966b) Physiological studies of the boron requirement of the diatom *Cylindrotheca fusiformis*. *J. Expl. Bot.* **17(52),** 473–9.

LEWIN J.C. (1966c) Boron as a growth requirement for diatoms. *J. Phycol.* **12,** 160–3.

LEWIN J.C. & CHEN C.H. (1968) Silicon metabolism in diatoms. VI. Silicic acid uptake by a colorless marine diatom, *Nitzschia alba*. *J. Phycol.* **4,** 161–6.

LEWIN J.C. & CHEN C.H. (1971) Available iron: a limiting factor for marine phytoplankton. *Limnol. Oceanogr.* **16,** 670–5.

LEWIN J.C. & GUILLARD R.R.L. (1963) Diatoms. *Ann. Rev. of Microbiol.* **17,** 373–408.

LEWIN J.C. & MACKAS D. (1972) Blooms of surf-zone diatoms along the coast of the Olympic Peninsula, Washington. I. Physiological investigations of *Chaetoceros armatum* and *Asterionella socialis* in laboratory cultures. *Mar. Biol.* **16,** 171–81.

LEWIN J.C., REIMANN B.E., BUSBY W.F. & VOLCANI B.E. (1966) Silica shell formation in synchronously dividing diatoms. In *Cell Synchrony-studies in biosynthetic regulation* (eds. I.L. Cameron and G.M. Padilla) pp. 169–88. Academic Press, New York.

LIPPS M.J. (1973) The determination of the far-red effect in marine phytoplankton. *J. Phycol.* **9,** 237–42.

LUND J.W.G. (1949) Studies on *Asterionella*. I. The origin and nature of the cells producing seasonal maxima. *J. Ecol.* **37,** 389–419.

LUND J.W.G. (1972) Preliminary observations on the use of large experimental tubes in lakes. *Verh. int. Verein. theor. angew. Limnol.* **18,** 71–7.

LUND J.W.G., JAWORSKI G.H.M. & BUCKA H. (1971) A technique for bioassay of freshwater, with special reference to algal ecology. *Acta Hydrobiol.* **13,** 235–49.

MCALLISTER C.D., PARSONS T.R. & STRICKLAND J.D.H. (1960) Primary productivity at Station 'P' in the north-east Pacific Ocean. *J. Cons. perm. int. Explor. Mer.* **25,** 240–59.

MCALLISTER C.D., PARSONS T.R., STEPHENS K. & STRICKLAND J.D.H. (1961) Measurements of primary production in coastal sea water using a large volume plastic sphere. *Limnol. Oceanogr.* **6,** 237–58.

MCALLISTER C.D., SHAH N. & STRICKLAND J.D.H. (1964) Marine phytoplankton photosynthesis as a function of light intensity; a comparison of methods. *J. Fish. Res. Bd. Canada* **21,** 159–81.

MCCARTHY J.J. (1972) The uptake of urea by marine phytoplankton. *J. Phycol.* **8,** 216–22.

MCLACHLAN J., MCINNES A.G. & FALK M. (1965) Studies on the chitan (chitan: poly-*N*-acetylglucoseamine) fibers of the diatom *Thalassiosira fluviatilis* Hustedt. I. Production and isolation of chitan fibers. *Can. J. Bot.* **43,** 707–13.

MCLAREN I.A. (1969) Population and production ecology of zooplankton in Ogac Lake. *J. Fish Res. Bd Can.* **26,** 1 485–559.

MCMAHON T. (1973) Size and shape in biology. *Science* **179,** 1 201–4.

MALONEY T.E., DONOVAN E.J. & ROBINSON E.L. (1962) Determination of numbers and sizes of algal cells with an electronic particle counter. *Phycologia* **2,** 1–8.

MENZEL D.W. & RYTHER J.H. (1961) Nutrients limiting the production of phytoplankton in the Sargasso Sea, with special reference to iron. *Deep-Sea Res.* **7,** 276–81.

MENZEL D.W., HULBURT E.M. & RYTHER J.H. (1963) The effects of enriching Sargasso Sea water on the production and species composition of phytoplankton. *Deep-Sea Res.* **10,** 209–19.

MONOD J. (1950) La technique de culture continue; theorie et applications. *Annls. Inst. Pasteur, Paris* **79,** 390–410.

MOSSER J.L., FISHER N.S. & WURSTER C.F. (1972a) Polychlorinated biphenyls and DDT alter species composition in mixed cultures of algae. *Science* **176,** 533–5.

MOSSER J.L., FISHER N.S., TANG T.C. & WURSTER C.F. (1972b) Polychlorinated biphenyls; toxicity to certain phytoplankters. *Science* **175,** 191–2.

MÜLLER H. (1972) Growth and phosphate requirements of *Nitzschia actinastroides* (Lemm) v. Goor in batch and chemostat culture under phosphate limitation. *Arch. Hydrobiol. Suppl.* **38, 4,** 399–484.

MULLIN M.M., SLOAN P.R. & EPPLEY R.W. (1966) Relationship between carbon content, cell volume and area in phytoplankton. *Limnol. Oceanogr.* **11,** 307–11.

MYERS J. & GRAHAM J.R. (1959) On the mass culture of algae. II. Yield as a function of cell concentration under continuous sunlight irradiance. *Pl. Physiol.* **34,** 345–52.

MYKLESTAD S. & HAUG A. (1972) Production of carbohydrates by the marine diatom *Chaetoceros affinis* var. willei (Gran) Hustedt. I. Effect of the concentration of nutrients in the culture medium. *J. expl. mar. Biol. Ecol.* **9,** 125–36.

MYKLESTAD S., HAUG A. & LARSEN B. (1972) Production of carbohydrates by the marine diatom *Chaetoceros affinis* var. willei (Gran) Hustedt. II. Preliminary investigation of the extracellular polysaccharide. *J. expl. mar. Biol. Ecol.* **9,** 137–44.

NATIONAL ENVIRONMENT RESEARCH COUNCIL (1971) *Culture collection of algae and protozoa—list of strains* pp. 1–73. Culture Centre of Algae and Protozoa, Cambridge.

NEWHOUSE J., DOTY M.S. & TSUDA R.T. (1967) Some diurnal features of a neritic surface plankton population. *Limnol. Oceanogr.* **12,** 207–12.

PAASCHE E. (1968) Marine plankton algae grown with light–dark cycles. II. *Ditylum brightwellii* and *Nitzschia turgidula. Physiologia Pl.* **21,** 66–77.

PAASCHE E. (1973a) Silicon and the ecology of marine plankton diatoms. I. *Thalassiosira pseudonana (Cyclotella nana)* grown in a chemostat with silicate as the limiting nutrient. *Mar. Biol.* **19,** 117–26.

PAASCHE E. (1973b) Silicon and the ecology of marine plankton diatoms. II. Silicate-uptake kinetics in five diatom species. *Mar. Biol.* **19,** 262–9.

PAASCHE E. (1973c) The influence of cell size on growth rate, silica content, and some other properties of four marine diatom species. *Norw. J. Bot.* **20,** 151–62.

PACKARD T.T., BLASCO D., MACISAAC J.J. & DUGDALE R.C. (1971) Variations of nitrate reductase activity in marine phytoplankton. *Inv. Pesq.* **35,** 209–19.

PARSONS T.R. (1973) Coulter counter for phytoplankton. In *Handbook of Phycological Methods* (ed. J. Stein) pp. 345–58. Cambridge Univ. Press, London & New York.

PECHLANER R. (1970) The phytoplankton spring outburst and its conditions in Lake Erken. *Limnol. Oceanogr.* **15,** 113–30.

PLATT T. & IRWIN B. (1973) Calorific content of phytoplankton. *Limnol. Oceanogr.* **18,** 306–10.

PLATT T. & SUBBA RAO D.V. (1970) Primary production measurements on a natural plankton bloom. *J. Fish. Res. Bd Can.* **27,** 887–99.

Prakash A., Rashed M.A., Jensen A. & Subba Rao D.V. (1973) Influence of humic substances on the growth of marine phytoplankton: diatoms. *Limnol. Oceanogr.* **18,** 516–24.

Pratt D.M. (1966) Competition between *Skeletonema costatum* and *Olisthodiscus luteus* in Narraganset Bay and in culture. *Limnol. Oceanogr.* **11,** 447–55.

Pringsheim E.G. (1949) *Pure cultures of algae.* Cambridge Univ. Press, London & New York, 119 pp.

Provasoli L. (1963) Organic regulation of phytoplankton fertility. In *The Sea* (ed. M.N. Hill) **2,** 165–219.

Provasoli L., McLaughlin J.J.A. & Droop M.R. (1957) The development of artificial media for marine algae. *Arch. Mikrobiol.* **25,** 392–428.

Raymont J.E.G. & Adams M.N.E. (1958) Studies on the mass culture of *Phaeodactylum. Limnol. Oceanogr.* **3,** 119–36.

Richter O. (1905) Über Reinkulturen von Diatomeen und die Notwendigkeit von Kieselsäure für die Diatomee *Nitzschia palea* (Kütz) W. Sm. *Verh. Ges. dt. Naturf. u. Ärzte,* 76. Versammlung zu Breslau. Verlag F.C.W. Vogel, Leipzig, 249–50.

Riley G.A. (1966) In *Marine Biology, II* (ed. C.H. Oppenheimer). 369 pp. AIBS, Washington, D.C.

Round F.E. (1973) The problem of reduction of cell size during diatom cell division. *Nova Hedwigia* **23,** 291–303.

Ryther J.H. (1959) Potential productivity of the sea. *Science* **130,** 602–8.

Ryther J.H. & Guillard R.R.L. (1959) Enrichment experiments as a means of studying nutrients limiting phytoplankton production. *Deep-Sea Res.* **6,** 65–9.

Sakamoto M. (1971) Chemical factors involved in the control of phytoplankton production in the Experimental Lakes Area, northwestern Ontario. *J. Fish Res. Bd Can.* **28,** 203–13.

Schelske C.L. & Stoermer E.F. (1971) Eutrophication, silica depletion and predicted changes in algal quality in Lake Michigan. *Science* **173,** 423–4.

Schöne H.K. (1972) Experimentelle Untersuchungen zur Ökologie der marinen Kieselalge *Thallassiosira rotula.* I. Temperatur und Licht. *Mar. Biol.* **13,** 284–91.

Schreiber E. (1927) Die Reinkultur von marinem Phytoplankton. *Wiss. Meeresuntersuch. Abt. Helgoland, N.F.* **16,** 1–34.

Schultz J.S. & Gerhardt P. (1969) Dialysis culture of microorganisms: design, theory and results. *Bact. Rev.* **33,** 1–47.

Smayda T.J. (1969) Experimental observations on the influence of temperature, light and salinity on cell division of the marine diatom *Detonula confervacea* (Cleve) Gran. *J. Phycol.* **5,** 150–7.

Smayda T.J. (1970a) Growth potential bioassay of water masses using diatom cultures: Phosphorescent Bay (Puerto Rico) and Caribbean waters. *Helgoländer wiss. Meeresunters.* **20,** 172–94.

Smayda T.J. (1970b) The suspension and sinking of phytoplankton in the sea. *Oceanogr. Mar. Biol. Ann. Rev.* **8,** 353–414.

Smayda T.J. (1973) The growth of *Skeletonema costatum* during a winter–spring bloom in Narragansett Bay, Rhode Island. *Norw. J. Bot.* **20,** 219–47.

Sorokin C. & Krauss R.W. (1962) Effects of temperature and illuminance on *Chlorella* growth uncoupled from cell division. *Pl. Physiol.* **37,** 37–42.

Spencer C.P. (1954) Studies on the culture of a marine diatom. *J. mar. biol. Assoc. U.K.* **33,** 265–90.

Starr R.C. (1964) The culture collections of algae at Indiana University. *Am. J. Bot.* **51,** 1013–44.

Starr R.C. (1971) The culture collection of algae at Indiana University—additions to the collection July 1966–July 1971. *J. Phycol.* **7,** 350–62.

STEIN J.R. ed. (1973) *Handbook of Phycological Methods.* Culture methods and growth measurements. 448 pp. Cambridge Univ. Press, Cambridge.

STEPHENS G.C. & NORTH B.B. (1971) Extrusion of carbon accompanying uptake of amino acids by marine phytoplankters. *Limnol. Oceanogr.* **16,** 752–7.

STOSCH V.H.A. (1965) Manipulierung der Zellgrösse von Diatomeen im Experiment. *Phycologia* **5,** 21–4.

STRATHMANN R.R. (1967) Estimating the organic carbon content of phytoplankton from cell volume or plasma volume. *Limnol. Oceanogr.* **12,** 411–8.

STRICKLAND, J.D.H. (1967) Between Beakers and Bays. *New Scient.* p. 276–8.

STRICKLAND J.D.H., HOLM-HANSEN O., EPPLEY R.W. & LINN R.J. (1969) The use of a deep tank in phytoplankton ecology. I. Studies of the growth and composition of phytoplankton crops at low nutrient levels. *Limnol. Oceanogr.* **14,** 23–34.

SUBRAHMANYAN R. (1945) On the cell division and mitosis in some south Indian diatoms. *Proc. Indian Acad. Sci.* **22,** 331–54.

SWALE E.M.F. (1963) Notes on *Stephanodiscus hantzschii* Grun. in culture. *Arch. Mikrobiol.* **45,** 210–6.

TAKANO H. (1964) Diatom culture in artificial sea water. II. Cultures without using soil extract. *Bull. Tokai reg. Fish. Res. Lab.* **38,** 45–57.

TALLING J.F. (1955) The relative growth rates of three plankton diatoms in relation to underwater radiation and temperature. *Ann. Bot. N.S.* **19,** 329–41.

TALLING J.F. (1957) The growth of two plankton diatoms in mixed cultures. *Physiologia Pl.* **10,** 215–23.

TEMPEST D.W. (1970) The continuous culture of microorganisms. In *Methods in Microbiology* (eds. J.R. Norris & D.W. Rebbons), Academic Press, New York.

THOMAS E.A. (1961) Vergleiche über die Plankton Produktion in Flaschen und im Plankton-Test-Lot nach Thomas. *Verh. Ver. Limnol.* **14,** 140–6.

THOMAS W.H. (1966) Effects of temperature and illuminance on cell division rates of three species of tropical oceanic phytoplankton. *J. Phycol.* **2,** 17–22.

THOMAS W.H. & DODSON A.N. (1968) Effects of phosphate concentration on cell division rates and yield of a tropical oceanic diatom. *Biol. Bull.* **134,** 199–208.

THOMAS W.H. & DODSON A.N. (1972) On nitrogen deficiency in tropical Pacific oceanic phytoplankton. II. Photosynthetic and cellular characteristics of a chemostat-grown diatom. *Limnol. Oceanogr.* **17,** 515–23.

THOMAS W.H., SCROTTEN H.L. & BRADSHAW J.S. (1963) Thermal gradient incubators for small aquatic organisms. *Limnol. Oceanogr.* **8,** 357–61.

TRANTER D.J. & NEWELL B.S. (1963) Enrichment experiments in the Indian Ocean. *Deep-Sea Res.* **10,** 1–9.

UKELES R. (1961) The effect of temperature on the growth and survival of several marine algal species. *Biol. Bull.* **120,** 255–64.

UKELES R. (1973) Continuous culture—a method for the production of unicellular algal foods. In *Handbook of Phycological Methods* (ed. J. Stein) pp. 233–54. Cambridge Univ. Press, London & New York.

UNO S. (1971) Turbidometric continuous culture of phytoplankton. Construction of the apparatus and experiments on the daily periodicity in photosynthetic activity of *Phaeodactylum tricornutum* and *Skeletonema costatum. Bull. Plankton Soc. Japan* **18,** 14–27.

VAN BAALEN C. & EDWARDS P. (1973) Light-temperature gradient plate. In *Handbook of Phycological Methods* (ed. J. Stein) pp. 267–73. Cambridge Univ. Press, London & New York.

WATANABE A. & HATTORI A. (1966) Cultures and collections of algae. Proc. U.S.-Japan Conf., Hakone, Japan. *Jap. Soc. Pl. Physiol.* 100 pp.

WATT W.D. (1969) Extracellular release of organic matter from two freshwater diatoms. *Ann. Bot.* **33,** 427–37.

WERNER D. (1966) Die Kieselsäure im Stoffwechsel von *Cyclotella cryptica* Reimann, Lewin and Guillard. *Arch. Mikrobiol.* **55,** 278–308.

WERNER D. (1970) Productivity studies on diatom cultures. *Helgoländer wiss. Meeresunters.* **20,** 97–103.

WERNER D. (1971a) Der Entwicklungscyclus mit Sexualphase bei der marinen Diatomee *Coscinodiscus asteromphalus.* I. Kultur und Synchronisation von Entwicklungsstadien. *Arch. Mikrobiol.* **80,** 43–9.

WERNER D. (1971b) Der Entwicklungscyclus mit Sexualphase bei der marinen Diatomee *Coscinodiscus asteromphalus.* II. Öberflachenabhängige Differenzierung während der vegetativen Zellverkleinerung. *Arch. Mikrobiol.* **80,** 115–33.

WESTLAKE D.F. (1965) Some problems in the measurement of radiation under water: a review. *Photochem. Photobiol.* **4,** 849–68.

WILLIAMS R.B. (1964) Division rates of salt marsh diatoms in relation to salinity of cell size. *Ecology* **45,** 877–80.

WILSON W.B. & COLLIER A. (1972) The production of surface-active material by marine phytoplankton cultures. *J. mar. Res.* **30,** 15–26.

CHAPTER 3

THE ULTRASTRUCTURE OF THE DIATOM CELL

ELEANOR L. DUKE

Department of Biological Sciences,
The University of Texas at El Paso,
El Paso, Texas 79968, USA

amd

B.E.F. REIMANN

Department of Biology
New Mexico State University
Las Cruces, New Mexico 88003, USA

1 INTRODUCTION

The Bacillariophyceae or diatoms represent among the Chrysophyta the protococcal type. As such their vegetative cells are encased by a rigid shell. In diatoms this shell, always composed of at least two structural siliceous elements, the valva (valve) and the pleura (girdle), is so regularly reproduced that for many years it has been the main criterion by which diatoms were systematically classified. Of course, it was also known relatively early that diatoms possess chromatophores (chloroplasts), cell nuclei and other microscopic structures commonly present in unicellular algae. In comparison with the knowledge accumulated on the subject of shell morphology their existence as a living cellular entity remained neglected for a long time. More than 100 years ago this fact had been acknowledged by Pfitzer (1871) who compared in his *Untersuchungen über Bau und Entwicklung der Bacillariaceen* (Investigations about structure and development of the Bacillariaceae) the direction which the research on diatoms had taken with the progress in research on other shell-bearing biological objects: 'Es ist bekannt, dass das Sammeln und Unterscheiden der Gehäuse von Muscheln und Schnecken schon auf eine hohe Stufe der Ausbildung gelangt war, ehe man daran ging auch das in der Schale steckende Tier einer genaueren Beobachtung zu unterziehen'. (It is known that the collecting and the classification of the shells of mussels and snails had already attained a high level of development before one proceeded to subject also the animal inside the shell to a more intensive investigation.)

Today, even after the introduction of the techniques for both transmission and scanning electron microscopy, diatom research still focuses to some extent on the shell morphology. Of course, there have been and still are exceptions. One of the earliest was the attempt by Pfitzer (1882) to classify diatoms according to the number of chromatophores. Subsequently, there have been many other cytological investigations published which are concerned with various morphological aspects of diatom biology. And during the last decade, first information on the ultrastructural construction of the diatom cell became available. This will be the primary subject of this chapter.

The assumed title of an article as e.g. 'The Ultrastructure of the Diatom . . .' today still refers almost inevitably to the morphology of the shell structure. From the estate of this attitude the technical term 'frustule (-la)' is inherited. A translation of this Latin word into English means 'a little bit' or 'a little piece', a perfectly meaningless description which has been indiscriminatingly applied to the entire cell, to the separated cell wall, or to the silica shell after removal of the protoplast. Yet, diatoms are not different from other plant cells in having a cell wall-bound protoplast. As a consequence the following nomenclature will be applied throughout this chapter: The term 'cell' will be used for the entire diatom i.e. the cell part which is surrounded by the plasmalemma (the protoplast) and the cell wall surrounding it. With the term 'protoplast' the plasmalemma-bound cell parts are meant without the cell wall. The cell wall which surrounds the protoplast is composed of various amounts of organic constituents, the 'organic' cell wall part (sometimes also referred to as being of a pectic nature), and the silica shell or the 'inorganic' cell wall part.

The overwhelming attitude favouring shell morphology, however, was not the only reason for the relatively late start of the exploration of the ultrastructural features of the diatom cell. There were a number of technical obstacles which had to be mastered before the quality of the ultrathin sections reached a level equivalent to that obtained with other biological objects. Daniels & Hayes (1960) have to be credited with the first account of the ultrastructure of some unidentified pennate freshwater diatoms found growing epiphytically on *Nitella*. Although this publication demonstrates many of the ultrastructural details which are now common knowledge for many of the phycological objects (e.g. mitochondria, lamellae of chloroplasts, etc.), it also has still, without diminishing its value, an air of an incidental finding. More often, the investigator had the discouraging experience of observing the explosion of one diatom cell after another during the polymerization of the methacrylate monomer as the result of his efforts.

Today, we are relatively well-informed of the basic ultrastructural cytology of the vegetative cells of a number of diatom species, among them *Achnanthes brevipes* (Drum & Pankratz, 1964a; Holdsworth, 1968), *A. longipes, A. minutissima* (Drum *et al.* 1966), *Amphipleura pellucida* (Stoermer & Pankratz, 1964; Stoermer *et al.* 1965; Drum *et al.* 1966), *A. rutilans* (Drum, 1969a), *Bacillaria paxillifer* (Drum *et al.* 1966), *B.* sp. (Drum & Pankratz, 1966), *Biddulphia pulchella* (Ueda, 1961), *Cocconeis diminuta* (Taylor, 1972), *Cyclotella meneghiniana* (Drum & Pankratz, 1964a), *Cylindrotheca fusiformis* (Reimann *et al.* 1965; Lewin *et al.* 1966), *Cymbella affinis* (Drum, 1963; Drum & Pankratz, 1964a; Drum *et al.* 1966), *Gomphonema parvulum* (Drum & Pankratz, 1964a, 1964b; Drum *et al.* 1966; Dawson, 1973a,b), *Hantzschia* sp. (Lauritis *et al.* 1967), *Mastogloia grevillei* (Stoermer *et al.* 1964; Drum *et al.* 1966), *Melosira granulata* (Drum *et al.* 1966), *M. varians* (Drum *et al.* 1966; Crawford, 1971, 1973), *Navicula gracilis* (Drum & Hopkins, 1966), *N. pelliculosa* (Reimann *et al.* 1966; Drum *et al.* 1966; Coombs & Volcani, 1968; Coombs *et al.* 1968a,b), *N. tripunctata* (Drum *et al.* 1966), *Neldium affine* (Drum *et al.* 1966), *Nitzschia alba* (Azam *et al.* 1973; Lauritis *et al.* 1967), *N. angularis* (Gibbs, 1962a), *N. palea* (Drum, 1962, 1963; Drum *et al.* 1966), *N.* sp. (Ueda, 1961; Drum & Pankratz, 1964a), *Pinnularia noblilis* (Drum, 1966; Drum & Hopkins, 1966; Drum *et al.* 1966), *Phaeodactylum tricornutum* (Reimann & Volcani, 1968), *Pleurosigma angulatum* (Drum *et al.* 1966), *Rhopalodia gibba* (Drum & Pankratz, 1965; Drum *et al.* 1966), *R. gibberula* (Drum & Pankratz, 1965), *Stephanodiscus niagarae* (Drum, Pankratz *et al.* 1966), *Streptotheca thamesis* (Esser, 1968), *Subsilicea fragilarioides* (von Stosch & Reimann, 1970), *Surirella ovalis* (Drum & Pankratz, 1963, 1964a; Drum *et al.* 1966) and *Thalassiosira fluviatilis* (Dweltz & Colvin, 1968).

In addition to this information on the ultrastructural organization of vegetative cells, the first reports of electron microscope investigations of gametogenesis and gametes have recently become available (*Lithodesmium undulatum,* Manton & von Stosch, 1966; Manton *et al.* 1969a,b, 1970a,b; and *Biddulphia levis,* Heath & Darley, 1972).

The significance of these investigations is that they provide a base line for studies of a more experimental nature, such as the metabolic role of silicic acid in biological systems and its associated morphological reaction norm. The uniformity with which the valve pattern of the vegetative diatom cell is repeated after every cell division challenges one to search for the biogenetic reasons for this consistency. In this respect, diatoms seem to offer most suitable test objects for ultrastructural and molecular genetic investigations.

2 TECHNICAL PROBLEMS IN ULTRASTRUCTURAL INVESTIGATION OF DIATOMS

The problems encountered in the preparation of diatoms for ultrathin sectioning are threefold:

1 The fixan has to be fast enough in penetrating the cell to prevent the formation of postmortem artifacts.
2 The embedding medium should be able to accommodate both the silica shell and the other cell organelles.
3 The sectioning presents a number of problems such as cell orientation and precautions required relating to the hardness of the silica shell.

A primary requirement for a successful fixation is the knowledge of the conditions under which the respective diatom grows. Both osmolality and ionic strength are of importance, and yet, these factors are only rarely taken into account (see Maser *et al.* 1967).

Initially, the fixation was accomplished employing the conventional buffered osmium tetroxide solutions. Daniels & Hayes (1960) used 1% osmium tetroxide in veronal buffer at pH 7·4. The research group in Ames, Iowa (Stoermer *et al.* 1964) used 1% osmium tetroxide but adjusted the pH to 6·1 as in Kellenberger, Ryter & Sechaud's fixan for bacteria (1958). Reimann *et al.* (1966) employed 1% osmium tetroxide but used a buffer diluted to the approximate osmolality of the culture medium for the diatoms and also mentioned the importance of the ionic product for the fixan. They also employed, for special purposes, 0·6% potassium permanganate buffered and adapted in the same fashion. Many modifications of this basic scheme have been reported. Dweltz & Colvin (1968) added one volume of culture medium to 9 volumes of 4% glutaraldehyde or 1·2% potasium permanganate or 1% osmium tetroxide. Better general cell preservation was obtained with aldehyde fixans. Stoermer & Pankratz (1964) used for the fixation of freshwater diatoms 3% glutaraldehyde buffered with phosphate buffer to pH 7·3 as proposed by Sabatini *et al.* (1963) for other objects.

It appears that there are various possibilities for the fixation of freshwater diatoms. The exposure to the fixan is usually longer than that used for other cell suspensions, e.g. the cells are exposed overnight to the fixan. In many instances the temperature is kept at $\pm 0°$ to $+4°C$ (ice water or refrigerator). Perhaps the relatively rigid cell wall is accountable for the slow penetration of the fixan, for the slow exchange of water and dehydrant during dehydration, and for the slow penetration of the embedding resin.

As a general recommendation, particularly in cases where the relation between deposited silicic acid and the adjacent cell structures are to be studied, it is advisable to keep the osmotic pressure of the fixan 50 to 100 mOsm/kg higher than the environmental medium to ensure an active penetration into the cell but not to cause plasmoptysis. The most suitable pH seems to be slightly above 7·0 for aldehyde fixation while the pH for the postfixation with osmium tetroxide is not critical. However, it should be in the same range as that used during the aldehyde fixation.

The fixation for marine or brackwater diatoms follows the same principle as the fixation for freshwater species. Here, it is even more important that the high osmotic pressure and the ionic strength of the marine environment be taken into account for the formula of the fixan. The osmotic pressure of the enriched artificial seawater as used, e.g. for the culture experiments with *Cylindrotheca fusiformis* (Lewin *et al.* 1966) is about 780 mOsm/kg. The osmolality of 3·5% glutaraldehyde buffered with phosphate buffer (Sabatini *et al.* 1963) as it is conventionally used for human and animal tissue, ranges between 650 and 660 mOsm/kg and is therefore too low for the fixation of marine diatoms. This is probably the reason why fixation with glutaraldehyde concentrations as high as 4% is successful (Dweltz & Colvin, 1968; Holdsworth, 1968; Manton *et al.* 1969a,b) although it may be more desirable to keep the concentration of the aldehyde lower. The natural capacity of seawater to act as a buffer was found to be not quite adequate for most purposes, although the fixation of a chain-forming diatom, *Subsilicea fragilarioides* (von Stosch & Reimann, 1970) which has extremely thin silica shells, with chrom-osmium tetroxide in seawater neutralized with potassium hydroxide showed acceptable results. It is also interesting to note that formaldehyde has rarely been applied for the fixation of diatoms (Reimann, 1961) although its more rapid penetration into tissue blocks is well-known.

In our experiments we are using glutaraldehyde at an approximate concentration of 3% in phosphate buffer (pH 7·2 to 7·3) with an osmotic pressure adjusted to about 960 mOsm/kg with NaCl crystals. The conductivity of this fixan is 46 Siemens/cm (the seawater nutrient is approximately 30 Siemens/cm). A corresponding adaptation is made for both buffer and postfixation with osmium tetroxide while the osmolarity of the buffer is kept at the level of the marine nutrient.

For the special purpose of demonstrating the silicalemma (the biological membrane which envelopes the silica shell during its appositional expansion), a special fixan was developed (Reimann *et al.* 1965). The prefixation uses potassium permanganate, buffered and adjusted in osmolarity to the nutrient, for 45 minutes to 1 hour at ±0°C and is followed by a main fixation in 1% osmium tetroxide overnight.

The choice of the embedding medium is of equal importance. Today, most of the epoxy resins will not cause particular problems for adequate preservation of the general cell structure of diatoms. The primary requirements which all of the commonly used epoxy resins meet is their low percentage of volume decrease during polymerization. Most of these media do not shrink more than 2%. Araldite (Ueda, 1961), Epon 812 (Drum and Pankratz, 1963) and Maraglas (Reimann, 1964) are among the embedding media which have been successfully employed. The use of methacrylate (Daniels & Hayes, 1960; Ueda, 1961; Drum, 1962, 1963) and polyester embedding media has only historical significance. For faster and more thorough penetration the low viscosity embedding resin described by Spurr (1969) and applied by Dawson (1973a) may be indicated.

The actual sectioning does not provide any particular difficulties provided that diamond knives are used. Glass knives will be able to make only a few suitable cuts and become dull. Diamond knives can last several years as long as they are properly cared for.

It is one of the characters of the Bacillariophyceae that the arrangement of their shell elements takes place in precise order. This character, already recognized by very early investigators using light microscopy, has led to a nomenclature concerned with the axes and planes of symmetry of the cell and its shell components specifically applicable to diatoms. This nomenclature is old and well established and there is no need to change it. Although the cellular components such as chromatophores, mitochondria or the endoplasmic reticulum do not follow necessarily the symmetrical arrangement of the shell, the application of the nomenclature pertaining to any plane through the cell permits a very precise determination of every particular location or orientation of a cellular entity within the cell. Sections through cells oriented in one of the planes of symmetry permit a much easier interpretation of the location of cellular entities. In diatoms of a strictly centric, i.e. radially symmetrical construction, there are two directions in which sections should traverse the cell; first in levels parallel to the valvar (paravalvar) orientation and second in levels parallel to the pervalvar plane (parapervalvar orientation). Since centric diatoms deviate readily from the basic radially symmetric concept to become bilateral (triangular or irregular), the use of the terms apical and transapical (paraapical and paratransapical) as used in pennate diatoms seems to be appropriate. In sections of randomly oriented pennate diatoms the paratransapical orientation is most frequently found, as e.g. also cross-sections in bacteria.

In most sections the silicic acid of the mature valve will be more or less shattered. If the silica of a growing valve is to be investigated, particularly in reference to its adjacent cell structures, some precautions have to be taken to assure a successful interpretation of the morphological results: (1) The sections have to be extremely thin. A low clearance angle of the diamond knife and selected cutting speeds assure section thicknesses below 30 nm. (2) The plane of the section has to traverse the siliceous structures in a perpendicular direction. If this prerequisite is not taken into account already a deviation of 5° from the necessary perpendicular direction will obscure membranous elements adjacent to the siliceous material (Reimann *et al.* 1966). An acceleration potential has to be found which prevents the occurrence of an excess of inelastically deflected electrons in the proximity of the siliceous structures. Higher acceleration potentials reduce the amount of inelastic deflection within the dense silicic acid only to a limited degree, even if the thickness of the section is below the conventional thickness of about 30 nm. The thinner the section, the more problems due to insufficient contrast occur and in these instances unintended inelastic electron scattering becomes apparent (Fig. 20, Reimann *et al.* 1966).

3 ULTRASTRUCTURE OF THE CELL

The ultrastructural concept of the diatom cell follows that of most other photosynthetically active plant cells. The cell is characterized by the possession of a nuclear region, bordered by a nuclear envelope, which projects tubules of endoplasmic reticulum throughout the cell cytoplasm. There are dictyosomes, mitochondria and chromatophores (chloroplasts), and as in most other algal cells, large vacuoles. The cell body is surrounded by a plasmalemma. The only entity which is truly specific for diatoms is the cell wall, a laminated component of a siliceous core enveloped by layers of organic matter. This cell organization is described in the following part in detail with special emphasis on morphological appearances specific for diatoms.

3.1 *Nucleus*

According to our present knowledge the actual interphase cell nucleus is a zone occupied by stretched desoxyribonucleic acid strands, ribonucleic acid and protein. This is true for the cells of higher plants and animals, for protista and algae, and consequently also for diatoms. The ultrastructure of the interphase nucleus however, does not reveal a complex arrangement as may be expected from the biochemical composition. The nuclear area in diatoms exhibits a relatively homogeneous fine granular appearance with little indication of chromatin aggregates. Occasionally, perinuclear chromatin can be observed. However, its appearance seems to be the exception. Relatively larger amounts of chromatin can be seen in the picture of male gametes of *Lithodesmium undulatum* (Manton & von Stosch, 1966). Nucleoli have been seen in many diatom species. In some instances the nucleoli are of rather large sizes, as in *Melosira varians* (Crawford, 1973). In many diatoms the nucleolus is of a coarse granular texture differentiating itself from the rest of the nucleus only by its grain size and electron opacity. Drum & Pankratz (1964a) show pictures of a quite vacuolated nucleolus surrounded by a rather coarse nucleoplasm in *Surirella ovalis*. The authors also report the presence of up to eight nucleoli in the rather large H-shaped nucleus in this species. This, however, appears to be an exception. More often a single nuclear zone will be the host of only one nucleolus. It is probably possible to mistake isolated larger chromatin aggregates for nucleoli.

It is common practice to describe the cell nucleus as being surrounded by the nuclear envelope. It seems to us however, that the nuclear envelope is more a part of the endoplasmic reticulum rather than a part of the nucleus itself. This appears to be a rather intriguing thought. It implies that the entity which is associated in our present thoughts with the idea of a controlling centre, is in reality very dependent on the rest of the structural elements of the cell. Its shape is dictated by the nuclear envelope in coordination with other structural elements of the cell. As soon as chromosomal separation begins, the nucleus becomes dependent upon the microtubular system of the cell. Both the membranous as well as the tubular systems are described below.

The area occupied by the nuclear zone varies from species to species. In smaller diatom cells, the nucleus occupies a relatively large space, for instance, in *Navicula pelliculosa* (Reimann *et al.* 1966). In large cells, especially those which exhibit large vacuoles, the nuclear area is proportionally small, as in *Melosira varians* (Crawford, 1973), *Mastogloia grevillei* (Stoermer *et al.* 1964) and *Surirella ovalis* (Drum & Pankratz, 1964a). The shape of the nuclear area as seen in ultrathin sections varies from circular, implying a spherical shape as in *Cylindrotheca fusiformis* (Reimann *et al.* 1965) and *Phaeodactylum tricornutum* (Reimann & Volcani, 1968) to a lenticular shape as in *Melosira varians* (Crawford, 1973) to H-shaped in *Surirella ovalis* (Drum & Pankratz, 1964a). In some instances the nuclear zone extends a blunt conical projection into the cytoplasm. The tip of the cone has a small shallow sella. This is the site where microtubules originate in a zone of a specific electron opacity, e.g. in *Melosira varians* (Crawford, 1973; see also below).

The location of the cell nucleus is generally polar. This can be easily recognized in centric diatoms as in *Melosira varians* (Crawford, 1973) or *Stephanopyxis turris* (Reimann, 1960). Indeed, it is an indication that mitosis is to be expected in the near future if the nucleus migrates into the centre of the cell (Reimann, 1960). After cell division the two nuclei are located relatively near each other in the two adjacent hypothecal regions of the daughter cells. Only shortly before the new cell division, the nuclei spiral into the centre of the cell again. The location of the nucleus is more difficult to discern in the pennate diatoms with long apices. Here the nucleus seems to be located primarily in the centre of the cell, whereas it is in reality in the hypothecal portion of the protoplast. A central location of the nucleus of the vegetative cell of *Lithodesmium undulatum* has been reported by Manton *et al.* (1969a).

Under certain conditions, the nuclear area can become very small. Lewin *et al.* (1966) report that the nuclear region in *Cylindrotheca fusiformis* becomes extremely small if the cells are grown under conditions of silica starvation. The nuclear zone is almost entirely traversed by the membranes of the endoplasmic reticulum. Under these conditions the cells seem to be deprived of the ability to undergo mitosis and cell division.

Multinuclear cells have been reported in *Navicula pelliculosa* after treatment with colchicine (Coombs *et al.* 1968a).

3.2 *The membranous system*

The choice of the term 'membranous system' is based on the fact that a number of membranes within the cell and on the cell border are morphologically identical. They are, in fact, so uniform in appearance and show so little deviation in their dimensions that Robertson (1964) coined the name 'unit membrane'. Diatoms particularly show, however, that unit membranes are not uniform at all in their physiological functions. It lies in the nature of all entity-limiting layer systems that they can enclose structures and spaces of various dimensions. Membranes can form a sheet-like structure, e.g. the nuclear envelope, the central parts of dictyosomes or the cell border; or tubular structures, e.g. the endoplasmic reticulum; or small spherical structures, e.g. vesicles. In the discussion of the membranous system, membranes of obviously different dimensions, such as the thylakoid membranes of the chromatophores and the mitochondrial membranes are excluded. Thus, under the term 'membranous system' the following membranes are considered:

1 The nuclear envelope with the particular extensions surrounding the chromatophores.
2 The surface covering membranes, i.e. the membranes of the tonoplast and the plasmalemma.
3 The endoplasmic reticulum, the Golgi apparatus, the silica shell producing silicalemma and related structures.

This classification cannot be, of course, a rigid one since there are transitions between forms of membranes. In addition there are so-called vesicular complexes which Drum described first (1963), the enveloping membranes of which seem to have the same dimensions as the other membranes which are discussed here.

(a) *Nuclear envelope*

The nuclear envelope in diatoms represents one of the larger sheet-like formations of the membranous system which borders the nuclear zone as in most other animal and plant cells. As most other membrane formations, perhaps with the exception only of the plasmalemma and the tonoplast, the nuclear envelope is a large flat vesicle with its shape dictated by the nuclear and cytoplasmic areas. We will use the term enchylema for the content of the vesicle. This term was originally used for the part of the cytoplasm which is present in the liquid state (Wilson, 1896). It can quite well also be used for the space which it occupies (perinuclear space; Watson, 1955). The tubular structures of the endoplasmic reticulum, the dictyosomes and their vacuoles indeed suggest the presence of fluid matter.

The emphasis of such a membrane-bound space avoids the often used term 'double membrane' which had led to difficulties in the description of the morphology of the nuclear envelope. If the membrane is considered to be a bimolecular leaflet, then obviously, it consists basically of two layers, thus it can be called a double (layer) membrane. On the other hand, the two adjacent membranes of the nuclear envelope are never, for any large extent, so closely attached to one another as to form a double membrane comparable to the layers of the myelin sheath. In order to differentiate between the two layers of the nuclear envelope we will use the terms perinuclear membrane and pericytoplasmic membrane.

The perinuclear membrane of the nuclear envelope is usually smooth surfaced. Chromatin or nucleoprotein abuts directly its outside surface. The enchylema is electron translucent. The space which it occupies, in cross-section through the nuclear envelope, has variations between 15–50 nm.

The continuity of both the perinuclear and the pericytoplasmic membranes is interrupted by nuclear pores. They are constructed by the merging of the two membranes of the envelope to form a circular to near pentagonal or hexagonal opening approximately 80–100 nm in diameter (Drum & Pankratz, 1964a). This opening is bridged by a diaphragm, a thin lamella with an occasional nodular annulus in the centre, approximately 19 nm in diameter (*Streptotheca thamesis*; Esser, 1968). Eight additional spherical granules, 10 nm in diameter, lining the edge of the pore have been described by Crawford (1973) in *Melosira varians*. The granules are similar in size and arrangement with those described by Fabergé (1973) on isolated nuclear pores from oocytes of *Taricha granulosa*.

Esser (1968) estimates that in the nuclear envelope of *Streptotheca thamesis* there are 6000 to 7000 pores per nucleus. The pericytoplasmic membrane of the nuclear envelope occasionally bears the character of rough-surfaced endoplasmic reticulum. A number of scattered solitary ribosomes abut, in these cases, the membrane adjacent to the cytoplasm.

The pericytoplasmic membrane is of importance as it can extend beyond the nuclear region to form a vesicular cover of the chromatophore. We propose to call this enveloping vesicle simply (in accordance with Gibbs, 1962b, 'chloroplast envelope') the chromatophore envelope (= 'periplastidäre Cisterne'; Falk & Kleinig, 1968; = 'chloroplast endoplasmic reticulum: CER', Dawson, 1973a). In its arrangement it is very similar to the nuclear envelope, different however, by the absence of pores. A chromatophore envelope is present regardless of whether the chromatophore is located adjacent or remote from the nucleus. Only in species in which the nucleus and chromatophore are adjacent to one another, the nuclear envelope participates in the formation of the chloroplast envelope. A structure of particular interest is a tubular-reticular system which branches off from the inner membrane of the chromatophore envelope, occupying the space between chromatophore membrane and chromatophore envelope. In *Gomphonema parvulum,* it has been demonstrated that this reticulum (periplastidal network, Dawson, 1973a; 'periplastidäres Reticulum', in *Tribonema,* Falk & Kleinig, 1968) is located in the proximity of the pyrenoid of the chromatophore. If one concurs with the assumption of Bouck (1965) that the periplastidal reticulum as a surface extension of the chromatophore envelope could be capable of taking up the soluble products of photosynthesis, then an uninterrupted channel system has been established connecting the enchylema of the periplastidal reticulum with the chromatophore envelope and thus the nuclear envelope and finally the endoplasmic reticulum. The periplastidal reticulum is not only present in the proximity of the pyrenoid. It is indeed more often located beneath the chromatophore envelope near the nuclear envelope junction, e.g. *Nitzschia palea* (Drum, 1963).

The pericytoplasmic side of the chromatophore envelope exhibits small areas of ribosomal attachment as well as direct connections to the endoplasmic reticulum. The outer, pericytoplasmic part of the chromatophore envelope is completed by a corresponding inner ('perichromatophore') envelope part. At the zones where the chromatophore abuts the nucleus directly, this inner part of the chromatophore envelope becomes simultaneously the outer (pericytoplasmic) membrane of the nuclear envelope.

The importance of a differentiation between perinuclear and pericytoplasmic parts of the nuclear envelope is elucidated by experiments reported by Esser (1968). In *Streptotheca thamesis* RNAse leaves the two membranes intact; pepsin destroys both of them with the exception of the nuclear pores; DNAse leaves the pericytoplasmic membrane part intact, but destroys the perinuclear portion of the nuclear envelope. Whether similar reactions will be found for the chromatophore envelope remains to be explored.

Little is known of the ultrastructural changes of the nuclear envelope during mitosis in the vegetative cells of diatoms. In the development of spermatogonia of *Lithodesmium undulatum* its fate is to a certain degree indicated by Manton *et al.* (1969a). However, there are certain differences in the mode of the cell division of vegetative cells and of spermatogonia particularly in reference to the silica shell formation which does not occur in spermatogonia. Thus, it remains to be seen whether the events during the mitosis of vegetative cells, as far as it concerns the resolution and reformation of the nuclear envelope are identical in the different developmental stages. Manton *et al.* (1969a) report a breaking up of the nuclear envelope at the end of prophase to form large vacuoles near the poles of the spindles and chains of spherical membrane-bound cavities at other cell localities during metaphase. A similar appearance is present during the metaphase of the first meiotic division (Manton *et al.* 1969b). The nuclear envelope ceases to exist except for a small residual disc in the proximity of the zone of the peripheral spindle attachment. Of interest are a number of small vesicles, circular in cross-section, between the attachment zone and the envelope fragments, from where they seem to originate. Their function is unknown.

During or at the end of telophase, and during the ongoing cytokinesis the reorganization of the nuclear envelope takes place. This process first becomes evident by an assembly of still unorganized vacuoles or vesicles at the periphery of the nuclear zone (Manton *et al.* 1970a). The final structure of the nuclear envelope seems to be completed by a confluence of these vesicles.

In the enchylema of the fully developed spermatocytes in *Biddulphia levis,* Heath & Darley (1972) have reported small tubular structures which are circular in cross-section. They have a diameter of 11 nm, thereby falling in the same size range as tonofilaments. However, the morphological appearance of these tubules is very much like that of the flimmer hairs in the flagella in these gametes.

The character of the nuclear envelope as a membranous entity coherent with the endoplasmic reticulum is enhanced by a few additional findings. During the conversion of the spermatogonia of *Lithodesmium undulatum* into male gametes a swelling of the spermatogonia is reported which commences after the last vegetative cell division and stops before the first meiotic event. This swelling is concurrent with a distension of the nuclear envelope, a process during which large vacuoles are formed abutting the nuclear zone. A similar swelling, yet not to the extent exhibited by the nuclear envelope can be seen in the adjacent endoplasmic reticulum (Manton *et al.* 1969b). This simultaneous distension of the nuclear envelope and the endoplasmic reticulum supports also the concept of a liquid phase, an enchylema, kept within the membranous ducts. Under silicon starvation, one of the striking phenomena is the damage to the nuclear envelope. Lewin *et al.* (1966) report a traversing of many tubular membrane-bound structures through the nuclear zone and at least partial loss of the original envelope in *Cylindrotheca fusiformis*.

(b) *Surface covering membranes: tonoplast and plasmalemma*

The plasmalemma and the tonoplast are the two other larger sheet-shaped parts of the membranous system, disregarding at this point the smaller, flattened vesicular parts of the dictyosomes. Whereas the morphological appearance of the tonoplast follows that of the vacuome of most other plant cells, the plasmalemma can exhibit peculiarities unique to diatoms. However, the uniqueness is not equally distributed over all the species. It will depend upon the morphological type of a particular species to what degree the unique features of the plasmalemma are developed. Despite this, it has to be kept in mind that the basic configuration of the plasmalemma remains that of a unit membrane. The peculiarities are threefold:

1 The plasmalemma is always located directly adjacent to the cell wall, the shape of which is greatly determined by the silica shell. This proximity, however, is only rarely seen in electron micrographs. Only in cells which have recently formed the valvar part of the cell wall, a direct abutting of the plasmalemma to the cell wall can be observed, e.g. in *Gomphonema parvulum* illustrated in Fig. 22 by Dawson (1973b). The plasmalemma of the epithecal valva in the same picture is disjointed from the cell wall. The shape of the surface which it forms is in many ways a replica of the inner surface of the cell wall. The feature specific for diatoms lies in the fact that larger parts of the plasmalemma are covered at many places by heavy silica shell parts. Whether the cell wall inhibits, fosters or remains inert in the exchange of nutrients or metabolic products remains unclear.

2 As indicated above, the plasmalemma follows directly adjacent to the inner outline of the cell wall. This can be a very simple straight line, e.g. in *Melosira varians* (Crawford, 1971, 1973) or in *Navicula pelliculosa* (Reimann *et al.* 1966). In more complicated cell walls the plasmalemma will follow all the indentations, projections, septa, and internal spaces formed by the siliceous cell wall parts, e.g. *Mastogloia grevillei* (Stoermer *et al.* 1964). By this, an enormous increase of the surface of the protoplast can be obtained, e.g. in species of *Rhabdonema,* while on the other hand a relatively simple general outline of the cell is obtained. *A priori,* there does not appear to be a relationship between the porosity of the siliceous cell wall parts and the increase in cell surface.

The fact that parts of the cell wall extend down into the protoplast, e.g. the fibulae in the genus *Cylindrotheca* (Reimann & Lewin, 1964; Reimann *et al.* 1965) may account for the strong connection between two otherwise relatively simple and loosely arranged cell wall thecae.

3 After the siliceous parts of a new valve have been completed they are transported to the cell surface to form, together with additional organic material, the final new cell wall. The plasmalemma in diatoms is able to permit the passage of large silica parts. This is known to happen in *Cylindrotheca fusiformis* (Reimann, 1964; Lewin *et al.* 1966). In this process the parts of the raphae, which are formed beneath the plasmalemma within the cytoplasm, move upwards to the membrane, through it, and come to rest directly above it. The fibulae become surrounded by newly formed membranous material and remain embedded in the cytoplasm.

In other species the silica parts are too large to speak of a true passage. Here, the primary plasmalemma is abandoned after completion of the siliceous cell wall parts and reformed as a new secondary cell border abutting the completed cell wall (*Gomphonema parvulum,* Dawson 1973b; *Navicula pelliculosa,* Reimann *et al.* 1966). The exact biophysical, biochemical and morphological processes leading to the movement of the silica parts through the plasmalemma are unknown.

Direct connections between plasmalemma and endoplasmic reticulum have not been reported. A final subject of discussion in which the plasmalemma may play one of the key functions is the movement of the pennate diatoms which possess a raphe (see Chapter 8). The movement depends on attachment to a solid surface and occurs in the same direction in which the raphe, a slit in the cell wall, is oriented. *Nitzschia linearis* moves in a straight line, *Cymatopleura elliptica* in a somewhat circular direction, and species of *Cylindrotheca* rotate around their apical axes during the movement. The motion can be quite suddenly reversed and may be so forceful in certain species that a cell can selfdestruct by its forward movement when it encounters an obstacle which it cannot move out of its way. The forces and structures which are capable of producing such powerful movement have not yet been discovered. The plasmalemma can only very rarely be seen to penetrate the raphe fissure. The cytoplasm adjacent to the plasmalemma does not show microfibrils or microtubules, structural elements believed to be associated with cell movement. Gordon & Drum (1970) postulate that the movement is caused by a dual action of the capillarity of the raphe fissure and a fluid medium within the raphe. The fluid is excreted through the plasmalemma into the raphe at a site located at the forward-moving end of the cell. Due to a change in the physicochemical nature of the fluid a more solid matter is extruded at the other end of the raphe causing a continuous consumption and consequently a continuous flow along the raphe. But this concept would require morphological evidence such as excretion vesicles penetrating through the plasmalemma or similar structures, none of which have been yet discovered. Lauritis *et al.* (1968) have illustrated the projection of fibrillar elements through the raphe fissure of *Nitzschia alba.* Locomotion based on such fibrillar structures appears *a priori* more probable.

(c) *Reticular membranous system*

The protoplast of the diatom cell is traversed by irregularly distributed elements of the endoplasmic reticulum, i.e. tubular structures with and without adhering ribosomes and the spherical elements disjoined from the tubular system, generally called cisternae of the endoplasmic reticulum. In this respect diatoms do not differ from other plant cells. In most publications the presence of an endoplasmic reticulum is mentioned but little attention is paid to its various elements. A particular subject may be the connection of the reticular portion which is 'smooth'-surfaced with the dictyosomal elements of the Golgi apparatus. Many of the early observations do not differentiate between the 'rough' or 'smooth'-surfaced tubular parts and the structural elements of the dictyosomes.

The Golgi system in diatoms has played a significant role although early investigators did not associate the structures they observed with the structure named after Camillo Golgi. Such structures were first reported by Pfitzer (1871) in *Pinnularia viridis*. Subsequent investigators found so-called double rods or double plates in many other diatoms. Drum (1966) finally confirmed their dictyosomal nature and their paired arrangement in *Pinnularia nobilis*.

Up to 20 dictyosomes are reported in the plane of a section of *Achnanthes longipes* (Drum *et al.* 1966) and there may be more if the entire perinuclear area of this diatom is taken into account. In most other diatoms there are fewer dictyosomes, usually located also in the proximity of the cell nucleus. In evaluating these structures one must be aware that the ultrathin section often reveals less than 1% of the total cell, thus the number of dictyosomes may increase if the entire cell is considered. There are up to 12 cisternae discernible in certain diatoms, e.g. in *Pinnularia nobilis* (Drum & Hopkins, 1966) but usually this number is by far not so high and small dictyosomes may exhibit only four cisternae, e.g. in *Phaeodactylum tricornutum* (Reimann & Volcani, 1968) and in *Nitzschia alba* (Lauritis *et al.* 1968). There does not seem to be a relationship between cell size and number of cisternae or number of dictyosomes.

One particular process in which dictyosomes or their lacunae seem to play a key role is during meiosis where they invade the region of the spindle, apparently initiating the formation of the new nuclear envelope (*Lithodesmium undulatum*; Manton *et al.* 1970a,b). The same authors report the presence of dictyosomes adjacent to a structure termed 'spindle precursor' which is located at the periphery of the nucleus (1969b, 1970b). Dawson (1973a) reports of at least five different kinds of vesicles which are released by the dictyosomes in *Gomphonema parvulum*: (1) 'small electron-lucent vesicles' with apparently no known function; (2) 'granular vesicles with distinct contents' which are believed to function as 'plasmalemmasomes' because they occur close to the plasmalemma; (3) granular vesicles without distinct contents, frequently seen at the 'edge of the plasmalemma'; (4) large electron-lucent vesicles which are supposed to fuse into the silicalemma; and (5) multivesicular bodies. It is conceivable that vesicle types 2 and 3 which show the greatest amount of activity during cell wall formation may supply the material for the inner organic portion of the cell wall. Whether the type 4 vesicles indeed form the silicalemma or transport material into the silica deposition vesicle remains to be further investigated. It is well-known that membranes can extend *ad libitum,* thus a contribution of membranous material is not really necessary. The idea of a silica transport in these vessels cannot presently be proven as long as there is no indication of any silica compounds in these vesicles. The opinion that the dictyosomes are produced from the pericytoplasmic part of the nuclear envelope as postulated by Crawford (1973) requires also further investigation. Dictyosomes are structures which originate from smooth-surfaced endoplasmic reticulum. The pericytoplasmic membrane of the nuclear envelope shows quite often attachments of ribosomes, thus this membrane has more the character of a rough-surfaced membrane. This membrane quite frequently produces small vesicles. Whether these vesicles form the cisternae by confluence on one side of the dictyosome is a questionable assertion since it is known that all the lacunae of the dictyosomes have the ability to separate vesicles often with quite different contents.

In 1942 Chadefaud described in several pennate diatoms delicate, spherical colourless structures which are slightly light-refringent and always located in the perinuclear cytoplasm. With neutral red they react faintly orthochromatic (as opposed to food vacuoles in *Phaeophyceae* which are strongly metachromatic with neutral red). This reaction corresponds to similar structures reported in *Phaeophyceae* by Crato (1892) who had named them 'physodes'. Chadefaud (1942) assigned the name 'corps (or granule) physoides' for structures which he observed in diatoms. These particles have been associated by Drum & Pankratz (1964a) with a so-called membrane-limited vesicular complex ('single-membrane-limited vesicular complex'). Such a complex consists of a membrane which forms a larger vesicle of irregular outline. The space encompassed by the vesicle contains numerous small cisternae with or without electron opaque contents or tubular structures, and in addition, widely distributed randomly scattered small granular material. Such complexes have been described in *Achnanthes minutissima* by Drum & Pankratz (1964a), in *Amphipleura pellucida* by Stoermer *et al.* (1965), in *Gomphonema parvulum* by Drum & Pankratz (quoted in 1964b) and by Dawson (1973a), in *Mastogloia grevillei* by Drum & Pankratz (quoted in 1964b), in *Melosira varians* by Crawford (1973), in *Nitzschia varians* by Crawford (1973), in *Nitzschia palea* by Drum & Pankratz (quoted in 1964b) and in *Surirella ovalis* by Drum & Pankratz (1964a) and by Drum *et al.* (1966). In *Cylindrotheca fusiformis* the vesicular

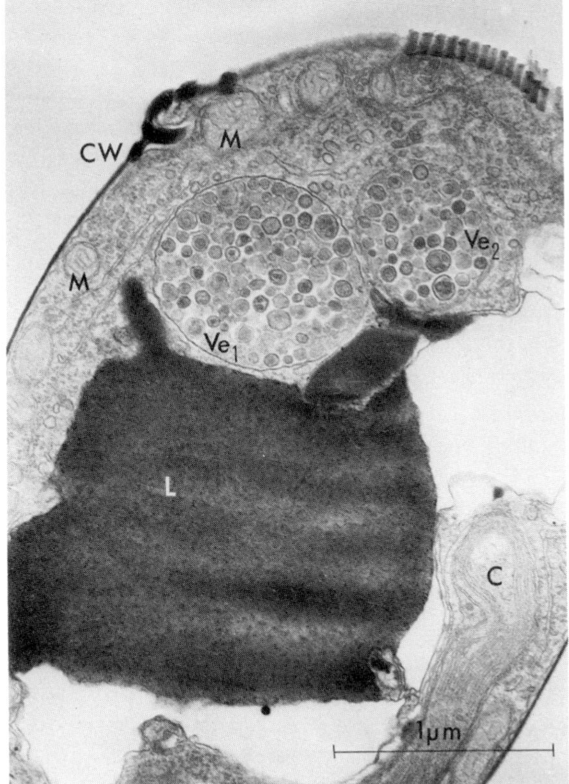

Fig. 3.1. Near paratransapical section through a cell of *Cylindrotheca fusiformis* showing two vesicular complexes (Ve_1 and Ve_2) adjacent to each other with a variety of internal vesicles of different sizes, contents, and development. The vacuole exhibits a large electron-opaque body (L). CW, cell wall, C, chromatophore, M, mitochondria.

complexes are particularly developed and can occur in pairs in various parts of the cell (Fig. 3.1.). Similar structures have been reported in related algal groups, e.g. *Dictyocha fibula* (van Valkenburg, 1971).

The appearance of vesicular complexes, however, is not limited to plant cells (noted by Drum & Pankratz, 1964a). They have also been reported in stages of differentiation of fetal rat duodenum (Behnke, 1963) and are known to be ubiquitous in many tissue cells. Dawson (*Gomphonema parvulum*, 1973a) has shown that vesicular complexes occur near a dictyosome in association with other vacuolar elements which leads to the belief that they have their origin in the Golgi complex. Earlier authors (Crato, 1892; Chadefaud, 1942) have indicated that there may be a relation between physodes and granule physoides and the production of fucosan. However, presently there is nothing known about the function of vesicular complexes.

An unanswered question is whether or not dictyosomes play a particular role in the process of silica shell formation. Earlier investigations by Manton & Leedale (1961) on *Paraphysomonas*, a heterokont flagellate member of the *Chrysophyceae*, thus a close relative of diatoms, report that silica scales are formed in vesicles of or deriving from dictyosomes. Based on this observation, Reimann (1964) discussed a similar origin of the vesicles within which diatoms deposit and solidify silicic acid. The object of this investigation was *Cylindrotheca fusiformis*, a diatom which exhibits in the nuclear plane of a paratransapical section dictyosomes in the proximity of the silica deposition vesicle. However, this location of the dictyosome is also in the perinuclear location, a place in which dictyosomes are generally known to occur in diatoms. Considering the fact that most dictyosomes in diatoms are in a perinuclear location, the proximity of the dictyosomes to the site of silica deposition in *Cylindrotheca fusiformis* is only a secondary feature of this particular species and thus the importance of the Golgi apparatus for the silica deposition may not be as significant as initially assumed.

Some doubt about the dictyosome as the origin of the silica deposition vesicle is shed in the report of the silica shell formation of *Navicula pelliculosa* (Reimann *et al.* 1966). The authors state that the origin of the silica deposition vesicles is still unclear. Further reasons for this uncertainty is the extent of this vesicle; it extends over the entire length of the newly forming cells, even those with far drawn out apices, i.e. quite remote from any dictyosomes which are generally located in the proximity of the nucleus. If the formation of a girdle band is seen in one of the rostra of a recently divided cell, the deposition vesicle is clearly seen within the cytoplasm. But there are no dictyosomal structures in the neighbourhood of this vesicle (Fig. 3.2). The opinion that dictyosomes contribute to both a newly formed plasmalemma as well as the membranes surrounding the intracellularly formed silica shell part was considered by Stoermer *et al.* (1965) in *Amphipleura pellucida*. The same basic concept was considered by Dawson (1973a,b) in studies of *Gomphonema parvulum* and by Lauritis *et al.* (1968) in *Nitzschia alba*. It is known in diatoms that the cell separation after mitosis is concluded by an invagination of the cytoplasm, e.g. in *Melosira varians*, and *Stephanopyxis turris* (Reimann, 1960) which is always bordered by a coherent plasmalemma. Thus, the initial plasmalemma is not formed by a confluence of vesicles as in higher plants. The schematic drawings referring to such a vesicle confluence to form the initial plasmalemma may be incorrect (Stoermer *et al.* 1965). Whether or not the final plasmalemma is formed in such a way will be a subject for further investigation.

Reimann *et al.* (1966) in studies of *Navicula pelliculosa* and Dawson (1973b) with *Gomphonema parvulum* consider that the parietal silicalemma of

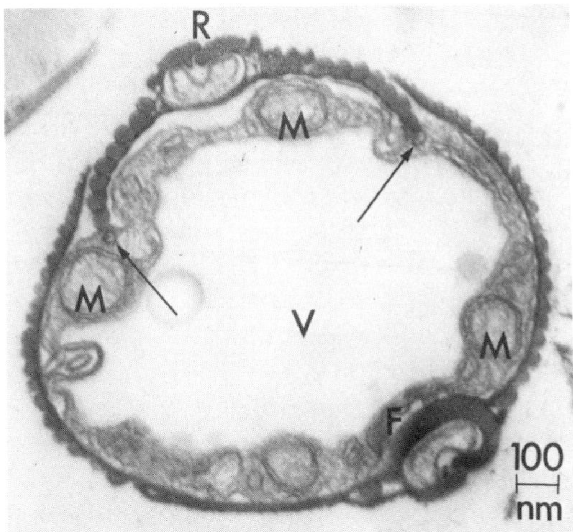

Fig. 3.2. Paratransapical section through one of the rostra of *Cylindrotheca fusiformis*. The hypotheca is still formed in the girdle region within the cytoplasm (arrows). Adjacent structures are mitochondria (M) and vesicular parts of the membranous system. F, fibula of the raphe system, R, raphe fissure, V, vacuome.

the deposition vesicle under change of function becomes the final plasmalemma.

The membranes of the system involved with the production of the morphological entity most characteristic for diatoms, the silica shell, necessarily deserved particular discussion. The silica shell, a structural entity of almost pure silicic acid, is so regularly repeated during numerous cell generations that if the term 'replication' has any place in biological morphology it could be applied here. To a great extent, the taxonomy of the Bacillariophyceae is based on this fact. There are relatively few factors which vary to a certain degree, such as length and width of the valve. The general layout of the shell construction, particularly at the ultrastructural level changes little. The observer of a sample of shells deriving from a clonal culture can hardly avoid the impression of evidence for a high degree of genetic influence. Despite this, little is known about the genetic control of the shell production. Every part of deposited silicic acid seen at the end of cell separation (and subsequently) appears within the cytoplasm in the proximity of the cell surface and always surrounded by a membrane (*Gomphonema parvulum*, Drum & Pankratz, 1964b; Dawson, 1973b; *Cylindrotheca fusiformis*, Reimann, 1964; Reimann *et al.* 1965; *Mastogloia grevillei*, Stoermer *et al.* 1964; *Amphipleura pellucida*, Stoermer *et al.* 1965; *Navicula pelliculosa* Reimann *et al.* 1966; *Nitzschia alba*, Lauritis *et al.* 1968). This membrane, because of its functional uniqueness was given the name silicalemma (Reimann *et al.* 1966). It is, in dimension and appearance, identical with the membranes existing in other parts of the membranous system. When it becomes discernible due to the electron-opaque deposits of polycondensed silicic acid it can be seen that it forms a vesicle which already has the approximate shape of the valve or girdle which is eventually formed, without having the dimension and thickness of the completed silica parts. The silicalemma is able to grow and extend with the progressing deposition of silicic acid.

It is one of the particular features of the silicalemma that it possesses sites where silica is preferentially deposited within the vesicle which it forms and other sites where no silica deposition takes place. In *Navicula pelliculosa* the areas of preferential silica deposition are those adjacent to the raphe fissure. The membrane which envelopes these two areas is continuous with the two adhering membranes which bridge the fissure as such, a portion which does not show silica deposition (Reimann *et al.* 1966). Thus, the membrane of a site exhibiting a particularly high activity of silica deposition extends into a membrane part which is morphologically identical but is totally inactive relative to silica deposition.

The space surrounded by a silicalemma seems always to encompass one individual, finally coherent siliceous structure, e.g. a valva, an intercalary band or a girdle band. In *Cylindrotheca fusiformis* the silicalemma surrounds the two silica bands which form the raphe between them and the fibulae which maintain the distance between the two bands. All other silica bands are enveloped by individual membranes (Fig. 3.2; see also Reimann, 1964; Reimann *et al.* 1965; Lewin *et al.* 1966).

Of considerable importance is the relationship between silicalemma and deposited silicic acid. Observing special precautions relative to factors of fixan, osmolality and ionic product it is impossible to detect a space between the membrane part facing the inside of the vesicle and the polycondensed siliceous structure itself. If it is assumed that the resolving power which electron microscopy provides for ultrathin sections is limited to about 1·5 nm, this value must at least be considered even if no visible space is present between polymer silicic acid and silicalemma. If such a morphologically unaccounted space were present, it could only amount to 4 or 5 molecular layers of un-polycondensed silicic acid between the membrane and the site of progressing polycondensation. The size of this possibly unaccounted space would not change a concept of a membrane involvement in the process of silicic acid condensation. Accepting the result that no space at all is present between the membranes, the influence of the silicalemma towards the silicic acid condensation becomes an even more acceptable thought. If one considers also that an amount of less than 10 ppm of silicic acid (Reimann, unpublished) can still be used by certain diatoms to form complete shells, i.e. *Cylindrotheca fusiformis,* the role of an active silicic acid pump mechanism for the silicalemma becomes a highly probable assumption (Reimann, 1964; Reimann *et al.* 1965; Lewin *et al.* 1966).

Consolidating all of these factors, one comes to the astonishing result that in the silicalemma a membrane is present about which all functions (at least grossly) which it performs are known. It pumps silicic acid actively and selectively, apparently under strict genetic control. It permits water to pass in the direction opposite to the passage of silicic acid, and in doing so it may remove water as it is liberated by the condensation process which transfers soluble silicic acid into the solid preopaline state of the diatom shell (see Chapter 4, 3.2 and 3.3). At the time of silica shell formation this membrane appears to have no other functions.

3.3 *Mitochondria*

The mitochondria in diatoms represent solely the tubular type, i.e. the projections which originate from the inner mitochondrial membrane and extend into the interior of this organelle are generally of tubular construction. The shape of the mitochondria varies considerably from a near spherical outline to oval, bacillar, or in some instances (*Cylindrotheca fusiformis*) long tubular formations. Similarly, the number of mitochondria in sections through diatoms differs from cell to cell and from species to species.

Although the major constructive element of a mitochondrion is a membrane, this membrane differs from other membranes in the cytoplasm. Measurements of mitochondrial membranes in *Cylindrotheca fusiformis, Phaeodactylum tricornutum* and *Navicula pelliculosa* reveal an average thickness of 5 nm (range 4–6 nm) as compared to 7·5 nm (range 7–9 nm) of the average thickness for other cytoplasmic membranes. A rough survey of the size range of the mitochondria in various species is compiled in Table 3.1 using mainly published illustrations, values given in descriptions and compiled values from pictures of *Cylindrotheca fusiformis* and *Navicula pelliculosa* in our possession. It must be noted that measurements of this kind are affected with various errors, one of which is a possibly incorrect indication of the magnification which cannot be eliminated by the reader. In addition, ultrathin sections rarely permit the observation of the full length of rod-shaped objects which are randomly distributed within a space through which unoriented sections are made. Similar conditions exist in contorted tubular objects as they are found in the mitochondrial tubules.

Mitochondria with an electron-opaque as well as with a translucent intertubular (intercristal) space have been illustrated; this appearance seems greatly to depend upon the fixation and the secondary electron contrast. Intertubular electron-dense granular inclusions have been portrayed in many diatom mitochondria, however they have not been the subject of any specific discussion.

Table 3.1. Mitochondrial size range in diatoms (estimates from ultrathin sections)

	Mitochondria			Tubular structures				
Diatom species	*Maximum length (μm)*	*Diameter Average (μm)*	*Diameter Range (μm)*	*Length (μm)*	*Diameter Average (nm)*	*Diameter Range (nm)*	*Comments*	*References*
Achnanthes minutissima	1·4	0·5			32	17–47	Irregular in outline	Drum & Pankratz (1964a)
Amphipleura pellucida	1·0	0·4						Stoermer *et al.* (1964)
Amphipleura pellucida	2·5	0·4	0·2–0·5	0·2	55	32–65	85 profiles per mitochondrial section	Stoermer *et al.* (1965)
Biddulphia levis (male gametes)	0·8	0·7		0·14	53	38–61		Heath & Darley (1972)
Cocconeis diminuta	0·6	0·5	0·4–0·6				'Tubular cristae'	Taylor (1972)
Cylindrotheca fusiformis	20·0	0·3	0·2–0·4	0·22	32	24–51	251 mitochondria measured; longitudinal shape	Reimann (compiled data)
Cymbella affine	0·6			0·07	53	32–69	Irregular-oval	Drum & Pankratz (1964a)
Gomphonema parvulum	0·6	0·5	0·4–0·6	0·05	45	32–55	Oval	Drum & Pankratz (1964a)

Gomphonema parvulum	1·9	0·2	0·1–0·4	0·05	30	20–53	Oval with spherical inclusions	Dawson (1973a)
Melosira varians	3·1	0·3	0·1–0·8	0·12	43	31–53	Bacillar-oval	Crawford (1973)
Navicula gracilis	2·0	0·6	0·5–0·7	0·21	56	33–110	Circular to oval	Drum & Hopkins (1966)
Navicula pelliculosa	1·3	0·5	0·1–1·0	0·21	30	10–69		Coombs *et al.* (1968b)
Navicula pelliculosa	1·4	0·5	0·1–0·9	0·19	32	10–70	Irregular, oval to circular	Reimann (compiled data)
Nitzschia alba	1·0	0·5					Author's estimate	Lauritis *et al.* (1968)
Nitzschia palea	0·5	0·2	0·1–0·4	0·11	34	20–41		Drum (1963)
Phaeodactylum tricornutum	0·7	0·3	0·2–0·6	0·13	38	33–50		Reimann & Volcani (1968)
Pinnularia nobilis	1·5	0·6	0·4–0·7		78	50–100	Long-oval to oval	Drum (1966)
Pinnularia nobilis		0·6		0·30	59	45–71		Drum & Pankratz (1966)
Rhopalodia gibba	8·0		0·2–0·3				Author's measurements	Drum & Pankratz (1965)
Thalassiosira fluviatilis	5·0	1·0					Author's measurements	Dweltz & Colvin (1968)

3.4 *Chromatophores*

The chlorophyll containing photosynthetically active organelles in diatoms are variously designated by three different names; Crawford (1973) and Gibbs (1962a) use the term *plastid*, Heath & Darley (1972), Holdsworth (1968), Taylor (1972) and Ueda (1961) employ the term *chloroplast* while Daniels & Hayes (1960), Dawson (1973a,b) and Reimann *et al.* (1966 and later), have used the term *chromatophore*. Others have employed various terms in their publications, e.g. Drum *et al.* (1966). The term chromatophore was first used by Schmitz (1883) in the report of his painstaking investigations on the presence of these organelles in algae. The term may have validity to be retained in favour of the term chloroplast since the latter refers only to the green colour due to the presence of chlorophyll and carotene. In diatoms and all *Chrysophyta* this organelle usually exhibits various hues of brown resulting from the presence of fucoxanthin. The term plastid, derived during early light microscopic investigation (Schimper, 1883) encompasses actually all pigment containing, well-circumscribed particles in the cytoplasm of the plant cell as chloroplasts or chromatoplasts according to their respective colour or even as leucoplasts in the absence of a pigment. Recently, the term chromatophore has been applied to the pigment-carrying membranous structures in *Cyanophyta* (see e.g. Scagel *et al.* 1965). For those structures however, a much earlier term chromatoplasma existed and there seems no need to change it. These are a few of the reasons why we retain the name chromatophores.

The general outline of chromatophores in diatoms is a subject which has been studied by many investigators particularly in the period prior to the introduction of the electron microscope as a scientific tool. Therefore, we will not make any particular comments on this topic. The ultrastructural arrangement of the diatom chromatophore, however, was one of the earlier electron microscopic studies. Besides the initially reported article by Daniels & Hayes (1960), the investigations by Ueda (1961) and by Gibbs (1962a) on the relation between various algal chromatophores merit a special citation.

Ueda (1961) discussed the number of lamellae in 9 taxa of algae and distinguished 2 and 4 lamellar systems and also stacked lamellae as they are found in the chloroplast grana of higher plants. Diatoms and other *Chrysophyta,* as well as *Phaeophyta, Pyrophyta,* and *Euglenophyta* exhibit 4 'lamellae'. The lamellae constitute the subunits of the larger sheet-like, disc-shaped structures which are comparable with the grana in chloroplasts of higher plants. Ueda measured 40 to 60 nm for the diameter of the disc. He furthermore mentions the presence or absence of pyrenoids; in the two diatom species *Biddulphia pulchella* and an unidentified species of the genus *Nitzschia* he reported the presence of pyrenoids. The diatom *Nitzschia angularis* exhibits, according to Gibbs (1962a), bands of 3 to 4 appressed lamellae for which the author also uses the term 'appressed discs'. The lamellae measure approximately 12–13 nm in diameter, thus a total of 4 appressed discs would amount to about 52 nm, a size range which falls well within the measurements reported by Ueda (1961). The comparison of the values reported by these two authors is difficult insofar as the terms 'lamella' and 'stacked' or 'appressed' discs are applied to different entities. Therefore, we use for the smallest structural unit the term 'membrane'. This membrane, with a width of approximately 4–5 nm (Gibbs, 1962a) is thinner than membranes of the membranous system. The membrane forms a flat, disc-shaped vesicle or sac which we call, according to Menke (1961), the thylakoid. In cross-section this thylakoid is approximately 12–15 nm thick, a value which varies considerably with different preparation techniques. Several coherent thylakoids form a disc, a structure morphologically homologous to the granum in the chloroplasts of higher plants. The site where two thylakoids adhere shows a somewhat thicker membrane which at higher magnification resolves into three layers of higher electron opacity, two outer layers of equal size and an inner one of greater thickness and usually higher electron opacity, but which is not twice the thickness of the outer layers. The entire entity measures between 6·5–7·5 nm (Gibbs, 1962a). The original idea (Ueda, 1961) that there are only 4 lamellae per disc must be modified according to later reports. In *Amphipleura rutilans* (Drum, 1969a) only 3 thylakoids can be discerned composing a disc, *Navicula pelliculosa* exhibits commonly 4 to 6 but can vary between 3 and 7 thylakoids and *Nitzschia palea* can have discs composed of 9 thylakoids (Drum, 1963).

In paratransapical sections the discs are in the simplest case arranged in parallel strips which follow the outline of the cell border, i.e. their ends are usually curved to a certain degree. In paraapical section they run in an apical direction. This general pattern is followed in centric diatoms with small lenticular chromatophores, as e.g. *Melosira varians* (Crawford, 1973) and *Lithodesmium undulatum* (Manton & von Stosch, 1966) as well as in many pennate diatoms, e.g. the araphid *Subsilicea fragilarioides* (von Stosch & Reimann, 1970), the monoraphid *Cocconeis diminuta* (Taylor, 1972) and *Achnanthes minutissima* (Drum & Pankratz, 1964a; Drum *et al.* 1966), the biraphid, naviculoid *Amphipleura rutilans* (Drum, 1969) and *A. pellucida* (Stoermer & Pankratz, 1964; Stoermer *et al.* 1965) and *Navicula pelliculosa* (Reimann *et al.* 1966), *Gomphonema parvulum* (Drum & Pankratz, 1964a,b), *Rhopalodia gibba* and *R. gibberula* (Drum & Pankratz, 1965) among the *Epithemiaceae,* and *Nitzschia palea* (Drum, 1962 & 1963) of the *Nitzschiaceae* as well as in the fusiform unsilicified type of *Phaeodactylum tricornutum* (Reimann & Volcani, 1968). This general pattern can be interrupted by the space occupied by pyrenoids. The discs are usually displaced by pyrenoids or interrupted and parts of the thylakoids can invade or traverse the pyrenoid. The only disc which apparently follows an independent pattern is the one which abuts the chromatophore membrane. Dawson (1973a) therefore calls this disc the 'girdle lamella'. This can be a closed, ring-shaped structure, or it can be interrupted once or in various places. Deviation of the general pattern is shown in short thylakoids which lie in an acute angle to the apical axis, e.g. in *Navicula gracilis* (Drum & Hopkins, 1966) or in *Cymbella affinis* (Drum, 1963; Drum & Pankratz, 1964a). Less regularly distributed discs are found in *Surirella ovalis* (Drum & Pankratz, 1963). When observed in a paraapical plane, the discs are either long and run parallel to one another, or are interrupted many times. Furthermore, they can form wave-like or curled patterns (Drum & Pankratz, 1964a). *Cylindrotheca fusiformis* possesses two chromatophores which spiral along the somewhat twisted thecae. The discs in this species are in part concentrically arranged. A special pattern is formed by the H-shaped chromatophore of *Mastogloia grevillei* (Stoermer *et al.* 1964), in which the discs follow the outline of the H, as seen in paratransapical sections. In not fully developed chromatophores the discs abut the pyrenoid in a near right angle, and pursue a somewhat slanted path to the periphery of the chromatophore. In apical section the appearance of a fully developed chromatophore approximates that of *Cymbella affinis* with the discs arranged in a diagonal direction.

There is nothing yet known of the ultrastructure of the dividing chromatophore in diatoms. Manton & von Stosch (1966) report that chromatophores become smaller during gametogenesis in *Lithodesmium undulatum*.

Under silica starvation the chromatophore in *Cylindrotheca fusiformis* tends to respond with a considerable increase in the number and extent of the discs while the number of thylakoids remain unchanged.

The discs can incorporate electron-dense particles which are a little smaller or larger than the total diameter of the disc. Their electron opacity is less than that of fat droplets. Between discs many granular structures are usually found which exhibit a morphology closely resembling ribosomes. They seem to be slightly smaller than ribosomes, however. In addition, there are electron-opaque, usually circular structures which are considered to be oil droplets. The entire chromatophore is surrounded by a chromatophore membrane (Dawson, 1973a; Crawford, 1973) which is characteristically a pair of membranes of apparently similar dimensions as the mitochondrial membranes.

All diatom chromatophores exhibit one or more pyrenoids and these structures have initially been one of the primary subjects in the description of the ultrastructure of these algae (Drum & Pankratz, 1964a). Pyrenoids are basically of lenticular or fusiform outline but there are many variations of this concept. The pyrenoid of a *Nitzschia* sp. described by Drum & Pankratz (1964a) is plane-convex, that of *Amphipleura pellucida* (Stoermer *et al.* 1965) is circular in paratransapical view, and *Mastogloia grevillei* exhibits pyrenoids of an almost rectangular outline. Most pyrenoids exhibit membranous structures which traverse the pyrenoid matrix in various directions. It will depend upon placing oriented planes of sections through at least the more elaborate forms in order to understand fully the various directions in which the membranes cross the pyrenoids. The crossing membranes originate from the discs and are consequently extended thylakoids. But usually only one or two thylakoids of one disc project into the pyrenoid. Border-forming membranes can be present but are not always discernible. Indeed, the pyrenoid border against the chromatophore matrix can form quite elaborate structures, e.g. in *Mastogloia grevillei* (Stoermer *et al.* 1964). In this diatom marginal parts of the pyrenoid form a row of vesicles. These vesicles seem to have direct contact with the thylakoids on one side and connect inside the pyrenoid with an indistinct and only marginally visible lamellar system.

The internal structure of the pyrenoid body, often called the matrix, has been described as fine to coarse granular and of higher electron density than the chromatophore matrix. In early pictures, e.g. in the profile of the pyrenoid of *Cymbella affinis* (Drum, 1963) the matrix appears to consist of an electron-dense material within which numerous electron-lucent, spot-like structures are distributed and which are tightly packed but not particularly oriented, giving the matrix the appearance of a granularity. Subsequently it was found that under certain but not yet understood conditions the matrix can exhibit a distinct lamellation. This was found in *Cocconeis diminuta* (Taylor, 1972), and in *Melosira varians* (Crawford, 1973; who also described this occurrence in *M. nummuloides*). Lauritis *et al.* (1968) depict the matrix of a pyrenoid as a series of parallel dense lines within a granular background. The authors call this a paracrystalline organization. The object of their observations was *Navicula pelliculosa* grown in a culture treated with 2 000 μg/ml colchicine. Under colchicine-free culture conditions, the lamination of the pyrenoid of this species is not apparent.

The most far reaching contribution has been made by Holdsworth (1968) who demonstrated a true crystalline lattice appearance of the pyrenoid matrix in *Achnanthes brevipes*. The quality of the preparation and the resolution of the obtained micrographs permit the reconstruction of a three-dimensional model of the morphological building blocks of the matrix. Accordingly, the building blocks are spherical in shape with a diameter of 5 nm which are arranged either in linear rows or in a hexagonal pattern. In the linear arrangement, the building blocks are either tightly packed or connected by short bridges. Similar bridges connect the building blocks in a hexagonal or crisscrossed pattern. Both patterns occur in the same pyrenoid together with areas which do not show this organization. The author assumes that both patterns are caused by the same basic molecular arrangement which is sectioned at different angles. The author suggests that the crystal is mainly composed of proteins and a small amount of RNA. The granular, non-crystalline appearance of some of the parts of the matrices are supposedly composed of the same globular subunits but in no organized arrangement.

A logical interpretation of the function of the pyrenoid has yet to be

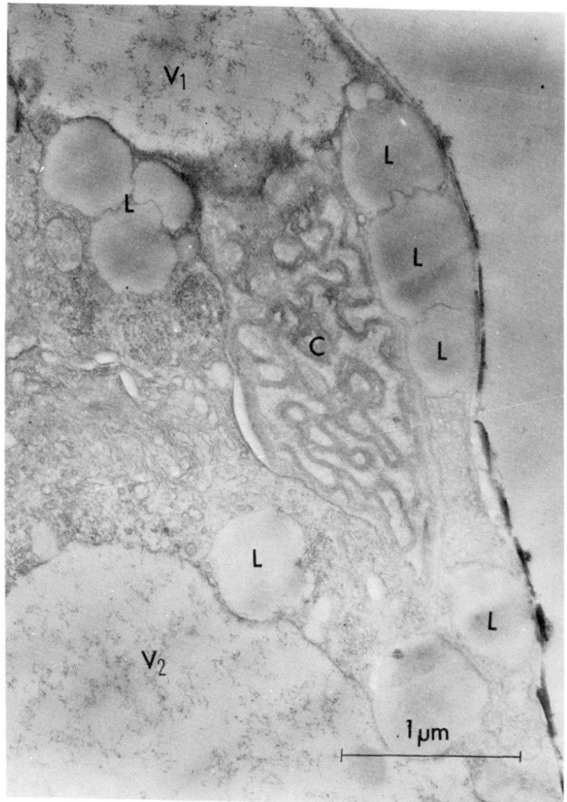

Fig. 3.3. Paraapical section through a cell of *Cylindrotheca fusiformis* derived from a silicic acid free culture. The cell which was still able to form a greatly reduced silica shell shows a number of large secretory vesicles with a homogeneous slightly electron-dense content (L) directly abutting the chromatophore (C) of which only a small part is visible. Note the curled arrangement of the thylakoid discs. Some of the vesicles are moved to the tonoplast of the vacuome (V_1 and V_2). Note the outward bulging of the cell wall due to the formation of large secreted vesicles.

discovered. The accumulation of fat droplets at the periphery of pyrenoids presents no conclusive evidence for their involvement in the production of metabolic products, particularly since this has only been observed in a limited number of cases, and electron-dense secretions directly into the cytoplasm can be seen in *Cylindrotheca fusiformis* particularly in silicic acid depleted cultures (Fig. 3.3).

In the colourless diatom *Nitzschia alba,* Lauritis *et al.* (1968) describe a vesicular structure for which they choose the term proplastid. There are no chromatophores of the above described arrangement present in this diatom. The proplastids are large vesicles limited by two independent parallel-running membranes, and approximately 0·5–1·0 μm in size. Although this structure is located close to the cell nucleus, there is no membranous connection between the nuclear envelope and the vesicle: it lacks any chromatophore envelope. Its matrix is described as homogeneous with occasional (tubular? or lamellar?) structures in the lumen of the vesicle. There is nothing known of the size of the membranes involved in forming the vesicle. This would permit one to determine whether the membranes belong to the membranous system or whether they have different dimensions as e.g. mitochondria or chromatophores. It is quite conceivable that these structures represent greatly reduced chromatophores which have ceased their original function. This may explain why a direct connection with the enchylema of the nuclear envelope or with the tubular system is unnecessary and therefore absent. The term proplastid is perhaps not a suitable choice: obviously, these structures do not produce new chromatophores. According to the general appearance the structures may represent a pyrenoid as the only residual part of a previous chromatophore, the membrane of which is the only other constructive element which remains.

3.5 *Ribosomal fraction*

Ribosomes are present in all diatoms as solitary entities or as polysomes or abutting parts of the membranous system. That they are involved in protein synthesis, as in any other organism, can be deduced from the fact that they are present in larger amounts in cells which are in a phase of extensive growth, whereas the number of ribosomes is reduced in cells with little or no essential growth. In *Cylindrotheca fusiformis*, in the cells resulting from recent cell division, two daughter cells are visible which are strikingly different in appearance immediately after the separation of the protoplasts. One of the cells, occupying the location encircled by the parental hypotheca exhibits a much denser cytoplasm than its adjacent sister cell in the space of the epitheca (Lewin *et al.* 1966). The density of the cytoplasm is a result of the large amount of ribosomes.

3.6 *Microtubular system*

The recognition that there exists a tubular system in every cell, different and independent from the tubular structures which can be formed by the membranous system is quite a recent discovery (Burnside, 1975). The appearance of tubular structures was at first thought to be a specific structure confined to the mitotic spindle apparatus and to the cilia and flagella of many algae and protista, and of certain cells of more complex organisms (Slauterback, 1963). As a result, little attention was paid to the presence of microtubules in ultrathin sections of diatoms.

Microtubules are understood as cylindrical structures which are long, straight and relatively rigid. They are composed of 13 protofilament subunits of tubulin, 5 nm thick, forming the cylinder wall. The outer diameter ranges from 18 to 25 nm. In ultrathin sections the tubular structure is not immediately apparent. For good preservation, particularly of the intracytoplasmic microtubules, the aldehyde fixation is a prerequisite (Ledbetter & Porter, 1963). In longitudinal view, microtubules are seen as two narrow parallel lines of the approximate diameter mentioned above. In cross-section, a narrow circular structure is discernible, usually with an apparently empty core, about 15 nm in diameter.

In diatoms microtubules have been observed in three different locations of the cell: (1) in the flagella of the spermatocytes (of two species); (2) in the cytoplasm; and (3) in the nuclear area in which the microtubules form the spindle fibres during mitosis and meiosis.

One of the surprising discoveries was made by Manton & von Stosch (1966), and Manton *et al.* (1970b) who reported that the microtubular arrangement in the male gamete (spermatocyte) of *Lithodesmium undulatum* deviates from the 9/9 + 2 norm by the absence of the two central microtubules in the axoneme of the flagellum. The same arrangement was found in the male gametes of *Biddulphia levis* by Heath & Darley (1972). Exceptions to the prevailing pattern of microtubules in flagella have been previously reported, e.g. in the sperm flagella of certain insects (Phillips, 1969). The significance of such changes has to be discussed from two angles. One aspect is the flagellar motion, and Heath & Darley (1972) have touched the subject in their discussion. So far no observations revealing a different movement of flagella without the central microtubules have been made. Secondly, the change may have phylogenetic and taxonomic significance. Motile diatom gametes may always lack the two central microtubules. This, however, should be the subject of further investigations.

The axoneme, upon entering the cytoplasm of the spermatocyte in which it is inserted, changes into the structure commonly called the kinetosome (or basal body). The kinetosomes in the two species in which spermatocytes have been observed are relatively short. Heath & Darley (1972) report that cross-sections through the kinetosome of *Biddulphia levis* indicate the usual cartwheel configuration at the proximal end of the structure. However, the microtubular triplets seen in most sections through the middle of the kinetosomes in other flagellated cells are replaced by two attached microtubules in an arrangement similar to that in the axoneme. The end of the kinetosome is frequently seen in the proximity of intercytoplasmic microtubules. The flagella in both *Lithodesmium undulatum* and *Biddulphia levis* possess flimmer hairs on their periphery. The flimmer hairs appear as hollow tubular structures with a diameter of about 11 nm. They are inserted into the flagellum in two rows through ill-defined, tapering, electron-dense spots beneath the plasmalemma of the flagellum (Heath & Darley, 1972). According to their size the flimmer hairs are morphologically in the same size range as the tonofilaments. Tonofilaments, however, are not tubular structures.

The presence of cytoplasmic microtubules was described relatively early by Drum & Pankratz (1963, 1964a) in *Surirella ovalis*. Although other publications soon followed reporting the presence of cytoplasmic microtubules (Stoermer *et al.* 1965; *Amphipleura pellucida*), the early description by Drum & Pankratz (1963) bears a particular significance. The subject of this article is an illustration of a structure in *Surirella ovalis* which the authors call 'centrosome'. A similar structure was observed by Crawford (1973) in *Melosira varians* for which he designates the term 'polar body'. Despite the fact that there are morphological differences between these two structural entities, they have in common the feature that in either, microtubules radiate toward or away from them. Particularly in the illustration by Drum & Pankratz (1963), one cannot avoid the impression that the depicted structure is a centre of microtubular generation. The individual microtubules continue for a short distance into the otherwise coarse granular matrix of the centrosome. This is, of course, no proof for the presence of a centre for microtubular generation; the centrosome could as well be a point of attachment for microtubules.

The polar body described by Crawford (1973) in *Melosira varians* consists of three regions which are different in the attraction of the electron-contrast yielding media (cells fixed with sodium cacodylate buffered glutaraldehyde and contrasted with osmium tetroxide). The zone of highest electron opacity is, relative to the nuclear envelope, distally located. It consists of a sheet-like ill-defined layer placed in a direction parallel to the nuclear envelope. Directly beneath this structure microtubules can be discerned in both cross-section and in longitudinally sectioned planes. Another membranous, tubular or rod-like structure with broad, electron-dense margins is located between the microtubule-containing zone and the nuclear envelope. The nuclear body forms out of its generally lenticular shape, a wide, cone-like projection toward the polar body with a small indentation on the tip of the cone. A cone-like projection can also be deduced from the illustrations by Drum & Pankratz (1963) in *Surirella ovalis*. (A similar cone-shaped projection can also be observed in *Cocconeis diminuta*, by Taylor, 1972; microtubules or a centrosome, however, have not been observed in this species.) Microtubules radiate from the centre part of the polar body and from the directly adjacent cytoplasm either into other parts of the cytoplasm or in groups of three or four in a groove-like indentation formed by the nuclear zone along the side of the cone. They split into single tubules near the foot of the cone where they terminate. A similar arrangement of microtubules has been observed in *Amphipleura pellucida* by Stoermer *et al.* (1965).

Centres of microtubular assemblage are doubtless the 'spindle precursors' described by Manton *et al.* (1969a,b, 1970a,b) in *Lithodesmium undulatum*. The spindle precursor is a relatively poorly outlined structure which initially forms a nearly rectangular plate about 500 nm long, 600 nm wide and 170 nm thick. It is not a homogeneous structure, but it is composed of two marginally located, distinctly electron-dense, rod-like zones separated by parallel bands and lines of alternating higher and lower electron opacity, the middle part apparently consisting of a row of globular subunits. The delineation of these structures is better defined on their lateral borders than on their ends, the electron opacity of which fades away into the background of the cytoplasm and does not have a true border visible. This plate-like structure initially abuts the nuclear envelope by an additional zone of fine granular electron opacity. Microtubules are first visible in this zone under a simultaneous stretching of the outer electron-dense, rod-like zones in a lateral direction.

The way in which the microtubules are physically formed remains unclear. If, by the term 'precursor', an entity is meant which heralds the coming of the spindle at the beginning of mitosis or meiosis, then this term may be acceptable. However, a precursor implies (in the context of a 'structure-and-function' complex) more an entity which, by modification, forms another entity. Is the spindle precursor a material similar to that which forms the microtubular wall, e.g. unoriented aggregates of tubulin? This, however, cannot be morphologically proven so far and one may wonder whether the spindle precursor is not an entity for which a term deriving from the family of 'centrosome/centriole' should be applied (Bernhard & Harven, 1960). Whether the abbreviation MTOC ('microtubule organizing centre' as in Helper & Palevitz, 1974) is a good choice as a term for these structures could also be questioned, although its descriptive nature comes perhaps somewhat closer to the true functions of the spindle precursor.

About the function of microtubules, there are many comprehensive discussions (Olmsted & Borisy, 1973; Helper & Palevitz, 1974; Burnside, 1975). Microtubules are involved in the chromosome movement during cell division, in the motility of flagellate cells, and they have also been associated with the intracellular organization and movement of material, with the maintenance as well as with the development of the cell shape. In diatoms, it would be conceivable to connect microtubular function with the silica shell formation as well as with the gliding motion which some pennate species exhibit. Investigations directly aimed toward this microtubular function are lacking. Studies by Coombs *et al.* (1968a), using colchicine to influence the cell wall formation of *Navicula pelliculosa* have failed to demonstrate a direct morphological relation between microtubules and the silicalemma. Otherwise, there seems to occur a distinct disorientation of the general cell layout of this species, and subsequently of course, deviations of the general shell morphology are the consequence. Of interest in this respect may be the fact that a clone of *Navicula pelliculosa*, obtained after colchicine treatment which the authors term 'polyploid' (Coombs *et al.* 1968b) shows distinctly larger shells but no change in the general pattern and arrangement of the shell. This may indicate that in diatoms the pattern of the silica shell is not controlled by microtubular action and that the general shape of the shell follows the general shape of the protoplast which may be controlled by microtubular action.

Drum & Hopkins (1966) have demonstrated fibrillar bundles oriented below the raphe system in apical direction in *Navicula gracilis*. The size of the fibrillar bundle, as described by the author, varies between 150 and 250 nm. The individual fibre as measured from their illustration (Fig. 11) is approximately 18 nm in diameter, the wall thickness an estimated 6 nm. This range appears to be exactly fitting the sizes reported for microtubules although the authors doubt that the structures are indeed microtubules. Location and size of these structures, however, provide so far the best clue for an explanation for the method which causes the sliding motion of these algae (see Chapter 8). This concept may be also supported by the protrusion of fibrils 14 nm in diameter through the raphe fissure in *Nitzschia alba* (Lauritis *et al.* 1968). Another possibility would be a finding of microfibres, structures resembling morphologically and chemically actin filaments with a size range between 4 to 7 nm. However, in the absence of such a finding other modes for locomotion have to be taken into account. The possible function, the biochemical composition, and thus, the biological significance of the tubular structures found forming the flimmer hairs on the flagella of the diatom spermatocytes also await an explanation and interpretation.

3.7 *Storage and secretory products*

In diatoms under certain nutritional conditions, accumulations of lipid droplets can be observed by light microscopy. The nature of these lipids can be verified by various staining methods employing colour reactions with Nile blue or Sudan stains (Crawford, 1973; Dawson, 1973a). In ultrathin sections the droplets appear as either electron-opaque or as hyaline planes of lower density, usually with a somewhat increased density in the marginal parts. They are commonly enclosed by a membrane. Such lipid or oil droplets, or oil bodies, as they are termed by the various authors, have been observed in diatom species in which investigations of the cell contents have been made. Stoermer & Pankratz (1964) report extremely large oil bodies, occupying the entire vacuole in *Amphipleura pellucida*, they are present as smaller dense bodies in the vacuole and cytoplasm of the spermatocytes of *Biddulphia levis* (Heath & Darley, 1972), in *Cylindrotheca fusiformis* (Lewin *et al.* 1966) particularly under silica starvation. Small and homogeneously electron-dense bodies are deposited in the vacuole along with very large and more homogeneously hyaline bodies which the authors assume to be chrysolaminaran. *Gomphonema parvulum* has been especially intensively investigated by Drum & Pankratz (1964a) and by Dawson (1973a,b). Dawson (1973a) has particularly verified by a combination of conventional methods for ultrathin sectioning, with freeze etching, and by employing Nile blue and Sudan black B for light microscopic investigation, the lipid nature of these granules. Similar granules have been observed in *Mastogloia grevillei* (Stoermer *et al.* 1964) in which several irregularly delineated oil bodies in the vacuome can become 6 μm in diameter. They are also present in *Melosira varians* (Crawford, 1973), *Nitzschia alba* (Lauritis *et al.* 1968), *N. palea* (Drum, 1963) and also as particularly large bodies in *Rhopalodia gibba* and *R. gibberula* (Drum & Pankratz, 1965).

With the aid of a toluidine blue staining reaction the presence of polyphosphates was established (Dawson, 1973a; Crawford, 1973). In ultrathin sections the polyphosphate granules exhibit a high electron opacity paired with a distinct coarse porosity of its matrix, e.g. in *Amphipleura pellucida* (Stoermer & Pankratz, 1964), *Melosira varians* (Crawford, 1973), or in *Gomphonema parvulum* (Dawson, 1973a). Dawson, in the same publication finds a morphological difference in the plane of lipid and polyphosphate bodies as exposed by the freeze-etching technique; whereas the lipid plane is basically flat and without a particular structure, the fractioned plane of polyphosphate is marked with numerous minute indentations (or projections).

Of interest is the fact that in morphological investigations one of the major storage products in diatoms, chrysolaminaran is rarely mentioned. Dawson (1973a) mentions that in her investigations it was morphologically not demonstrated by either light nor by electron microscopy. In the absence of investigations employing cytochemical reactions for the identification of chrysolaminaran the assumed presence of this important storage product in *Cylindrotheca fusiformis* under silica starvation (Lewin *et al.* 1966) and in *Thalassiosira fluviatilis* (Dweltz & Colvin, 1968) has to be regarded with some reservation. On the other hand, chrysolaminaran (leucosin) can occur in such abundance in diatoms that it changes the refractive index of the cell (von Stosch, 1951). The membrane-bound large hyaline bodies in the vacuole of *Cylindrotheca fusiformis* which the authors assume to be chrysolaminaran (Lewin *et al.* 1966) are initially directly abutting the chromatophore (Fig. 3.3). They are excreted through the chromatophore membrane into the chromatophore envelope which provides their membrane. The quite voluminous structures are formed at any side of the chromatophore surface, even into the relatively narrow space between chromatophore and plasmalemma. The consequence is an irregular wave-shaped outline of the border of the cell which is already very enlarged as a result of the silica starvation. The hyaline bodies are transported through the cytoplasm into the vacuome of the cell. There are no structures inside the chromatophore which would indicate a specific aggregation of similar corresponding material.

Other electron-dense material is described as released by the dictyosome, e.g. in *Amphipleura pellucida* (Stoermer *et al.* 1965). The function of the Golgi complex as a secretory cell system, particularly involved in the production of certain enzymes is today a generally accepted conception. In the absence of exact knowledge of the nature of the material produced in the Golgi lacunae and in Golgi vesicles, it is perhaps premature to discuss their probable function within the cell.

3.8 *Cell wall*

There are two types of cell walls which have to be taken into consideration and depending on how rigid a cell wall taxonomy is applied, perhaps a third type. The first cell wall type is found in almost every vegetative diatom cell. It is composed of the silica shell and organic components surrounding the shell.

The second cell wall type is the zygote membrane which is formed specifically to envelop the fertilized egg cell or the fused isogametes. It consists mainly of organic material, but can include siliceous parts, e.g. *Melosira varians* (Reimann, 1960) and in *Rhabdonema adriaticum* (von Stosch, 1958, 1962), in which two zygote membranes are formed.

The third structural element which could be mentioned in this category is the perizonium which is developed during the auxospore formation by pennate diatoms (von Stosch, 1962).

One can argue whether the cell wall of the vegetative cell should have been discussed in the previous section under secretory products. Indeed, the silica shell is secreted and deposited into a vesicle before it is moved to the cell surface. Similar statements could be given for the organic component, however, with some reservation. If the formation of the final cell wall is observed as it occurs in most diatoms, the cytoplasmic part which had comprised the area between cell surface and silicalemma usually remains attached to the silica parts. It forms a generally very compressed thin layer of organic material. The silicalemma which initially faced the cell interior becomes, after disjoining the silica shell, the new plasmalemma. In the course of the further development, new organic material is deposited against the inner surface of the shell, forming in this way the inner part of the shell envelope, e.g. in *Amphipleura pellucida* (Stoermer *et al.* 1965), in *Gomphonema parvulum* (Dawson, 1973b), *Navicula pelliculosa* (Reimann *et al.* 1966), and in *Nitzschia alba* (Lauritis *et al.* 1968). Various amounts of organic material of different origin, such as membranous parts, ribosomes and other cytoplasmic components which were trapped between the plasmalemma and the extending silicalemma constitute the outer organic layer of the cell wall. In comparison, the material of the inner layer of the shell could be quite similar in various species, although its amounts vary from species to species. The inner organic cell wall part of *Subsilicea fragilarioides* (von Stosch & Reimann, 1970) is in comparison with other species, enormously thick and laminated whereas the outer layer is extremely thin. It may be of significance that this species was the object used in the initial exploration of ruthenium red as a biological stain for both light and electron microscopy (Reimann, 1961). Indeed, the cell wall of *Subsilicea* stains deep red and is quite electron-opaque when treated with ruthenium red. The presence of polyglucuronic acids or its derivatives which react specifically with ruthenium red have to be assumed. The cell wall in *Subsilicea fragilarioides* is morphologically not only characterized by the large amount of organic material comprising the inner cell wall part but also in its differentiated electron contrast. Pleural, mantle and valve parts of the inner organic layer exhibit a narrow electron-lucent lamina. This is followed towards the shell layer by a broad laminated electron dense layer which tapers off toward the ends of the girdle and the middle of the valve plane. The dense layer is followed again by a more lucent lamina upon which the silica is located. One of the additional differences in *Subsilicea fragilarioides* is a splitting of this upper translucent lamina at the transition between valve mantle and valve plane forming a triangular lacuna filled with an electron-opaque fibrillar substance. This lacuna runs parallel along the edges of the valvar plane and is only interrupted at the sites of the poles of the cell.

Another diatom with a weakly silicified cell wall is *Cylindrotheca fusiformis* (Reimann & Lewin, 1964; Reimann, 1964; Reimann *et al.* 1966; Lewin *et al.* 1966; Lewin & Reimann, 1969). The species possesses a wide valvar area lacking silica shell parts with the exception of the somewhat centrally located raphe system. The extent of the two lateral unsilicified valve planes corresponds with the cell diameter. This cell wall part is produced subsequent to the appearance of the silica parts of the valve and the two adjacent silica strips above the surface of the protoplast. At this time the two plasmalemma layers of the just-separated daughter cells are still running closely parallel to one another across the middle of the former parent cell. In the narrow space between the two plasmalemmae originate the new silicified cell walls by a condensation of material which is secreted through the cell membrane. The so-formed organic layer is distinctly differentiated into at least three layers and joins on both margins the somewhat thinner layers which carry the silica parts of the valve and the girdle. After the cells have separated, the flexibility of the cell wall of this species permits the newly-formed thecae to protrude outward to form the final circular cross-section of the transapical plane. The end parts of the girdle are formed as all other silica parts of the shell within the cytoplasm of the protoplast (Fig. 3.2). The organic material exhibits, in the light microscope, a positive reaction with ruthenium red.

The ability of the organic part of the cell wall to stain specifically with ruthenium red has quite practical implications. In cases of forensic investigations, involved in the determination of the possibility of death by drowning, the finding of diatom cells or cell walls in the alveolar spaces of the lung which have entered these spaces by aspiration is important. Using the staining method with ruthenium red, it is possible to differentiate between the cell walls of recently alive diatoms and the shells of fossil diatoms which have derived from the airborne dust of diatomaceous earth used as packing material and for other industrial purposes. It is almost impossible to destroy the organic skin by biological means (enzymes).

In *Pinnularia nobilis,* Drum & Hopkins (1966) report a feather-like formation of the outer layer of organic material covering the silica shell in this particular species. The role which the organic skin may play in providing protection for the deposited silica shell parts from dissolution has been discussed (Reimann *et al.* 1966; see also Lewin, 1961). In this context it may be of interest to note that infrared spectra of lyophilized diatom shells derived from cultures which are prepared by employing enzymes for removal of the protoplast and after removal of the organic cell wall parts with hypochloric acid do not show patterns identical with spectra of diatomaceous earth, or of natural opal, or of silicic acid which was artificially polymerized in various ways (Duke & Reimann, unpublished).

About the ultrastructure of the zygote membranes in diatoms there is only one observation reported by Reimann (1960) of the siliceous fine structure of the zygote membrane in *Melosira varians* of which the author referred to as perizonium in this publication. In pennate diatoms we believe today that the zygote membranes do not contain silica deposits with few exceptions as e.g. in *Rhabdonema adriaticum* (von Stosch, 1962).

A special position among diatoms relative to the construction of the cell wall is occupied by *Phaeodactylum tricornutum* (Lewin, 1958). One type of vegetative cell of this species is fusiform or, more rarely triradiate in outline and lacks any silica in the cell wall. The other type has an oval outline and one of its thecae possesses a silica valve similar to those of the genus *Cymbella* (Lewin *et al.* 1958). This part of the cell is furnished with a raphe and if the cell is attached with its valvar side to a substrate, it is able to perform the same sliding motion as are most of the raphe-possessing diatoms. The theca opposite to the shell carrying the valve does not possess such a silica part and is not capable of locomotion. If the univalvar oval cell form is already a significant deviation from the general characteristics of diatoms, the unsilicified cell wall of the fusiform type is biologically a puzzle. The presence of one form or the other depends upon the nature of the environment; on a solid substrate the motile form becomes predominant after a prolonged period. If the substrate is lacking the majority of cells are of the fusiform type. Whether the phenotypical changes from one form into the other accompanies a change of the nuclear phase is unknown. Is the fusiform cell a zygote membrane indicating a suppressed sexual event? The fact that this cell wall consists of two thecae speaks against such a proposal. Yet, this organic cell wall is very different from the cell wall of other diatoms (Reimann & Volcani, 1968). A base layer is composed of three laminae: an electron-dense lamina 3 nm thick which faces the plasmalemma, a lamina of lesser electron opacity 4–6 nm thick and a third more electron-opaque lamina which is crowned by an arrangement of ridges 7 nm high and 10 nm apart, and of various lengths. The ridges run parallel to each other in sets of 2 to 5 or more and are surrounded by groups of similar ridges which run in various other directions. In this way a pattern is produced in which sets of ridges are interrupted by other sets of ridges running in a different direction. There is no preference for a particular angle between sets of ridges in the production of this pattern. Nothing is known of the mode by which this cell wall is produced.

Perizonia, according to the definition of von Stosch (1962) are siliceous wall structures, limited to pennate diatoms. Nothing is known of their formation or composition.

3.9 *Extracellular, structure-forming secretions*

Under this title three subjects are discussed which are quite different relative to their structure and composition: slime capsules and pseudothalli, stalks, and spines or setae of non-siliceous material. The term extracellular is used to designate structures which are produced by the cell and which occur outside the cell wall.

Among those diatoms which have been investigated by electron microscopical sectioning techniques, there is one, *Navicula pelliculosa,* which regularly produces large and substantial slime capsules. If the cells are carefully treated during fixation and dehydration, this capsule can be seen in the sections as an irregularly granular matter in which the many centres of higher electron opacity are connected by narrow fibrillar strands (Reimann *et al.* 1966). This structural pattern is quite likely a product of both fixation and dehydration. The slime substance is in its natural state gelatinous and coagulates during the fixation. According to Lewin (1955) the chief component of the capsule material is glucuronic acid. The capsule itself occurs as a poly-glucuronide and shows a specific staining reaction with ruthenium red. The secretion of gelatinous material can lead to the formation of a sheath which develops into either a colony within which the cells are quite irregularly located or to the formation of plant-like structures often referred to as pseudothalli. Stoermer *et al.* (1964), describe the accumulation of secretion products in so-called locules in the shell of *Mastogloia grevillei* from where it is apparently channelled through locular tubes to the outside of the cell. The locules are part of the intercalary bands. Drum (1969a) describes the construction of a pseudothallus of the diatom *Amphipleura rutilans*. The pseudothallus is composed of tube-like structures which branch in various ways forming tufts 2–5 cm in height. In cross-section the tube wall is between 0·8–1·6 μm thick. Its structural subunits are filaments (tubules) deposited in the form of a network. The lumen of the tube is divided in compartments of various sizes occupied by one or more diatom cells. The compartmental wall material is morphologically the same as that of the wall of the tube. The thread-like compartmental wall material is described by the author as a polysaccharide, and is 2·0–2·5 nm thick. The compartmental walls are between 50 and 100 nm thick. Although the pseudothallus wall yields a distinct staining reaction with ruthenium red, its chemical composition is not a pectin, but according to R. Lewin (1958) composed of a carbohydrate fraction with xylose and mannose as main contributors with traces of rhamnose as well as of two unidentified components, and of a protein fraction with 12 ninhydrin-positive substances. Drum (1969a) discusses the protective action of the tubes against the osmotic pressure of varying salinity values.

Another entity which reacts with ruthenium red with an intensive red colour is the stalk of many littoral diatoms. Hufford & Collins (1972), employing scanning electron microscopy, have demonstrated that the stalk is secreted through the terminal pores in *Cymbella cistula*. In *Gomphonema olivaceum*, the fine structure of such a stalk was investigated by Drum (1969b). The stalk, composed of sulphated polyxylose-galactan, consists of a more rigid wall which is irregularly and deeply invaginated and is composed of tightly packed fibrils. The interior of the stalk is filled with a loosely arranged felt-like network of similar fibrils.

One of the surprising discoveries was the ability of diatoms to secrete a chitin-like substance to which the term 'chitan' has been assigned (McLachlan *et al.* 1965). It was well-known by most investigators who used centric discoid plankton diatoms that some of them are able to produce long filamentous projections which originate from the periphery of the valve disc and from the centre of the valvar plane. These filaments are somewhat rigid, nearly invisible and only well observed in a near darkfield. McLachlan *et al.* (1965) were able to isolate these filaments in *Thalassiosira fluviatilis* and to make a preliminary assessment of their composition. It was found that the filaments after hydrolysis are entirely composed of glucosamine. Later investigations have confirmed this result and found this also to be true in *Cyclotella cryptica* (Blackwell *et al.* 1967). Dweltz *et al.* (1968) give the following morphological assessment: the chitan thread is about 60–80 μm long and at their root, i.e. the seta or pore in the valve where they penetrate the cell wall, 0·1–0·2 μm thick. The thread consists of a number of filamentous subunits which are oval to ribbon shaped in cross-section, 200 nm wide and 50 nm thick. In these units, which the authors call segments, one can distinguish a central part of higher and two marginal parts of lower electron opacity which are distinct from one another by two small grooves on either side of the more electron-dense central part. The threads as well as the individual ribbons taper off in size the longer they become by feathering off of the marginal filament and by a decrease in the width of the filament towards the end of the thread. Based upon X-ray diffraction, the authors propose a three-dimensional chemical structural formula for the chitan.

3.10 *Symbiotic, intracellular organisms in diatoms*

Drum & Pankratz (1965) report an unusual cytoplasmic inclusion in both *Rhopalodia gibba* and *R. gibberula.* The inclusions are ovoid in shape, 4 to 6 μm wide and 5 to 7 μm long. The diatom cell separates these objects from its cytoplasm by membrane-covered plasma strands and one can therefore argue whether or not the location of the objects is intravacuolar or intracytoplasmic. These cytoplasmic envelopes project into the vacuolar space. The inclusions resemble the basic constructional concept of Cyanophyta. They have an irregularly waved surface and a structural border which closely resemble both the cell wall and the plasmalemma. Drum & Pankratz (1965) describe this border as a five-layered wall. The sequence of the morphological layering from the outside to the interior of the cell is as follows: the outer layer, 12 nm thick can be occasionally resolved into 2 individual layers each 5 nm thick. This outer layer is followed by a layer of lesser density, 6 nm thick, a further layer, 4 nm thick of higher density, and a layer concluding the wall on the inside, which is 6 nm thick and again of lesser electron opacity. The cell body is finally surrounded by a cell membrane, 6 nm thick, a size range which corresponds with a size of the cell membrane of most bacterial or blue-green algae. There are membranous thylakoid-like vesicles projecting from the cell periphery in a perpendicular direction for various distances into the cytoplasm in a fashion as it is observed in the chromatoplasma of the blue-green algae. The picture, however, could also resemble a very inflated mitochondrion. The matrix of this cellular body is very fine granular with occasional aggregations of spherical particles in a size range of 20 nm. A few of the membranous projections cross the body entirely, others form an irregular arrangement of more inflated vesicles within its centre.

The structural appearance of this body is similar to that exhibited by blue-green algae. The reduction of the photosynthetic lamellae of the chromatoplasma may indicate a cell with an insignificant photosynthesis. The consequence would be that the cell depends upon its host cell. Thus, the inclusion body might quite well be a phytoparasite. The authors (Drum & Pankratz, 1965) discuss the possibility that these cells could be involved in nitrogen fixation, a process which would give them more the character of a symbiotic organism.

4 CONCLUSION AND OUTLOOK

We believe our survey to show that significant advances in our knowledge of the organization of the diatom cell has been made during the last 20 years. This progress is chiefly a result of the application of ultrathin sectioning as a method for ultrastructural investigations. Our knowledge has shifted to structural dimensions about 100 times smaller than those in which we used to think in the past. This process of exploring and learning has explained the nature of some of the structures previously considered to be solely confined to diatoms, e.g. the mysterious perinuclear double plates revealed to be dictyosomes of the Golgi complex. The same process has also lead to exciting and unexpected discoveries. The absence of the two microtubules in the axoneme of the flagella in spermatocytes in diatoms belongs in this category. The finding that the silica shell is produced within the protoplast of the diatom cell is another of these events. However, there are also numerous new questions which have been raised during this process which are questions confined to diatoms and questions of general cytological interest. There is the problem of the specific biochemical action of the Golgi complex, and its morphological manifestations which has to be explored and the functional nature of the spindle precursor, and the morphological events which permit the movement of some diatoms cells. These are again just a few obvious questions which await answers.

In one particular respect, however, diatoms may provide the key to the answer of a very fundamental question: in the silicalemma a membrane is available which could serve as a morphological and biochemical model for the exploration of the structure and function complex which we today call the membrane or cytomembrane (see Chapter 4).

5 REFERENCES

AZAM F., HEMMINGSEN B. & VOLCANI B.E. (1973) Germanium incorporated into the silica of diatom cell walls. *Arch. Mikrobiol.* **92,** 11–20.

BEHNKE O. (1963) Demonstration of acid phosphatase-containing granules and cytoplasmic bodies in the epithelium of foetal rate duodenum during certain stages of differentiation. *J. Cell Biol.* **18,** 251–65.

BERNHARD W. & HARVEN E. DE (1960) L'ultrastructure du centriole et d'autres éléments de l'appareil achromatique. In *Fourth International Conference on Electron Microscopy.* pp. 217–27. Springer-Verlag.

BLACKWELL J., PARKER K.D. & RUDALL K.M. (1967) Chitin fibers of the diatom *Thalassiosira fluviatilis* and *Cyclotella cryptica. J. molec. Biol.* **28,** 383–5.

BOUCK J.B. (1965) Fine structure and organelle association in brown algae. *J. Cell Biol.* **26,** 523–37.

BURNSIDE B. (1975) The form and arrangement of microtubules: an historical, primarily morphological, review. In *The Biology of Cytoplasmic Microtubules* (David Soifer, editor). *Ann. N.Y. Acad. Sci.* **253,** 14–26.

CHADEFAUD M. (1942). Existence de corps paranucléaries physoïdes chez les diatomées pennatées. *Trav. Algol.* **1,** 1–15.

COOMBS J. & VOLCANI B.E. (1968) Studies on the biochemistry and fine structure of silica-shell formation in diatoms. Chemical changes in the wall of *Navicula pelliculosa* during its formation. *Planta* (Berl.) **82,** 280–92.

COOMBS J., LAURITIS J.A., DARLEY W.M. & VOLCANI B.E. (1968a) Studies on the biochemistry and fine structure of silica shell formation in diatoms. V. Effects of colchicine on wall formation in *Navicula pelliculosa* (Bréb.) Hilse. *Z. Pflanzenphysiol.* Bd. 59. S. 124–52.

COOMBS J., LAURITIS J.A., DARLEY W.M. & VOLCANI B.E. (1968b) VI. Fine structure of colchicine-induced polyploids of *Navicula pelliculosa* (Bréb). Hilse. *Pflanzenphysiol.* **59**(3), 274–84.

CRATO E. (1892) Die Physode, ein Organ des Zellenleibes. *Ber. dt. bot. Ges.* **10,** 295.

CRAWFORD R.M. (1971) The fine structure of the frustule of *Melosira varians* C.A. Agardh. *Br. phycol. J.* **6**(2), 175–86.

CRAWFORD R.M. (1973) The protoplasmic ultrastructure of the vegetative cell of *Melosira varians* C.A. Agardh. *J. Phycol.* **9,** 50–61.

DANIELS S.C., JR. & HAYES T.L. (1960) The internal structure of a diatom. *Mikroskopie* **15**(1/2), 5–7.

DAWSON P.A. (1973a) Observations on the structure of some forms of *Gomphonema parvulum* Kütz. II. The internal organization. *J. Phycol.* **9,** 165–75.

DAWSON P.A. (1973b) Observations on the structure of some forms of *Gomphonema parvulum* Kütz. III. Frustule formation. *J. Phycol.* **9,** 353–65.

DRUM R.W. (1962) The non-siliceous fine structure of a diatom. In *Fifth International Congress for Electron Microscopy.* Edit. Sydney S. Breese, Jr. Vol. 2. UU-14. Academic Press.

DRUM R.W. (1963) The cytoplasmic fine structure of the diatom, *Nitzschia palea. J. Cell Biol.* **18**(2), 429–40.

DRUM R.W. (1966) Electron microscopy of paired Golgi structures in the diatom *Pinnularia nobilis. J. Ultrastruct. Res.* **15,** 100–7.

DRUM R.W. (1969a) Light and electron microscope observations on the tube-dwelling diatom *Amphipleura rutilans* (Trentepohl) Cleve. *J. Phycol.* **5**(1), 21–6.

DRUM R.W. (1969b) Electron microscope observations of diatoms. *Öst. bot. Z.* **116,** 321–30.

DRUM R.W. & HOPKINS J.T. (1966) Diatom locomotion: an explanation. *Protoplasma* **62**(1), 1–33.

DRUM R.W. & PANKRATZ H.S. (1963) Fine structure of a diatom centrosome. *Science* **142**(3588), 61–3.

DRUM R.W. & PANKRATZ H.S. (1964a) Pyrenoids, raphes, and other fine structure in diatoms. *Amer. J. Bot.* **51**(4), 405–18.

DRUM R.W. & PANKRATZ H.S. (1964b) Post mitotic fine structure of *Gomphonema parvulum. J. Ultrastruct. Res.* **10,** 217–23.

DRUM R.W. & PANKRATZ H.S. (1965) Fine structure of an unusual cytoplasmic inclusion in the diatom genus, *Rhopalodia. Protoplasma* **60**(1), 141–9.

DRUM R.W. & PANKRATZ H.S. (1966) Locomotion and raphe structure of the diatom *Bacillaria. Nova Hedwigia* **10**(3/4), 315–7.

DRUM R.W., PANKRATZ H.S. & STOERMER E.F. (1966) *Diatomeenschalen im elektronenmikroskopischen Bild*. Teil VI. J.-G. Helmcke-W. Krieger. Verlag von J. Cramer. 3301 Lehre.

DWELTZ N.E. & COLVIN J.R. (1968) The structure of the diatom *Thalassiosira fluviatilis*. *Can. J. Microbiol.* **14,** 1049–52.

DWELTZ H.E., COLVIN J.R. & MCINNES A.G. (1968) Studies on chitan (B-(1-4)-linked 2-acetoamido-2-desoxy-D-glucan) fibers of the diatom *Thalassiosira fluviatilis* Hus. III. The structure of chitan from X-ray diffraction and electron microscope observation. *Can. J. Chem.* **46,** 1513–21.

ESSER K. (1968) Elektronenmikroskopische-cytochemische Untersuchungen an der Kernmembran von *Streptotheca thamesis*. *Experientia* **24**(1), 61–2.

FABERGÉ A.C. (1973) Direct demonstration of eight-fold symmetry in nuclear pores. *Z. Zellforsch.* **136,** 183–90.

FALK H. & KLEINIG H. (1968) Feinbau und Carotinoide von *Tribonema* (Xanthophyceae). *Arch. Mikrobiol.* **61,** 347–62.

GIBBS S.P. (1962a) The ultrastructure of the chloroplasts of algae. *J. Ultrastruct. Res.* **7,** 418–35.

GIBBS S.P. (1962b) Nuclear envelope-chloroplast relationships in algae. *J. Cell Biol.* **14**(3), 433–44.

GORDON R. & DRUM R.W. (1970) A capillarity mechanism for diatom gliding locomotion. *Proc. Natl. Acad. Sci. USA* **67,** 338–44.

HEATH I.B. & DARLEY W.M. (1972) Observations on the ultrastructure of the male gametes of *Biddulphia levis* Ehr. *J. Phycol.* **8,** 51–9.

HELPER P.K. & PALEVITZ B.A. (1974) Microtubules and microfilaments. *Ann. Rev. Pl. Physiol.* **25,** 309–62.

HOLDSWORTH R.H. (1968) The presence of a crystalline matrix in pyrenoids of the diatom, *Achnanthes brevipes*. *J. Cell Biol.* **37,** 831–7.

HUFFORD T.L. & COLLINS G.B. (1972) The stalk of the diatom *Cymbella cistula* SEM observation. *J. Phycol.* **8,** 208–10.

KELLENBERGER E., RYTER A. & SECHAUD J. (1958) Electron microscope study of DNA-containing plasmas. II. Vegetative and mature phage DNA as compared with normal bacterial nucleoids in different physiological states. *J. Biophys. Biochem. Cyt.* **4,** 671–8.

LAURITIS J.A., HEMMINGSEN B.B. & VOLCANI B.E. (1967) Propagation of *Hantzschia sp.* Grunow daughter cells by *Nitzschia alba* Lewin & Lewin. *J. Phycol.* **3,** 236–7.

LAURITIS J.A., COOMBS J. & VOLCANI B.E. (1968) Studies on the biochemistry and fine structure of silica shell formation in diatoms. IV. Fine structure of the apochlorotic diatom *Nitzschia alba* Lewin & Lewin. *Arch. Mikrobiol.* **62,** 1–16.

LEDBETTER M.C. & PORTER K.R. (1963) A 'microtubule' in plant cell fine structure. *J. Cell Biol.* **19,** 239–50.

LEWIN J.C. (1955) The capsule of the diatom *Navicula pelliculosa*. *J. Gen. Microbiol.* **13,** 162–9.

LEWIN J.C. (1958) The taxonomic position of *Phaeodactylum tricornutum*. *J. Gen. Microbiol.* **18,** 427–32.

LEWIN J.C. (1961) The dissolution of silica from diatom walls. *Geochim. Cosmochim. Acta* **21,** 182–98.

LEWIN J.C., LEWIN R.A. & PHILPOTT D.C. (1958) Observations on *Phaeodactylum tricornutum*. *J. Gen. Microbiol.* **18,** 418–26.

LEWIN J.C. & REIMANN B.E. (1969) Silicon and plant growth. *Ann. Rev. Pl. Physiol.* **20,** 289–304.

LEWIN J.C., REIMANN B.E., BUSBY W.F. & VOLCANI B.E. (1966) Silica shell formation in synchronously dividing diatoms. In *Cell Synchrony* (eds I.L. Cameron & G.M. Padilla) pp. 169–88. Academic Press, Inc., N.Y.

LEWIN R.A. (1958) The mucilage tubes of *Amphipleura rutilans*. *Limnol. Oceanogr.* **3,** 111–3.

MANTON I. & LEEDALE G.F. (1961) Observations on the fine structure of *Paraphysomonas vestita,* with special reference to the Golgi apparatus and the origin of scales. *Phycologia* **1,** 37–57.

MANTON I. & STOSCH v.H.A. (1966) Observations on the fine structure of the male gamete of the marine centric diatom *Lithodesmium undulatum*. *J. R. Microsc. Soc.* **85**(2), 119–34.

MANTON I., KOWALLIK K. & STOSCH v.H.A. (1969a) Observations on the fine structure and development of the spindle at mitosis and meiosis in a marine centric diatom (*Lithodesmium undulatum*). I. Preliminary survey of mitosis in spermatogonia. *J. Microsc.* **89,** 295–320.

MANTON I., KOWALLIK K. & STOSCH v.H.A. (1969b) Observations on the fine structure and development of the spindle at mitosis and meiosis in a marine centric diatom (*Lithodesmium undulatum*). II. The early meiotic stages in male gametogenesis. *J. Cell Sci.* **5,** 271–98.

MANTON I., KOWALLIK K. & STOSCH v.H.A. (1970a) Observations on the fine structure and development of the spindle at mitosis and meiosis in a marine centric diatom (*Lithodesmium undulatum*). III. The later stages of meiosis I in male gametogenesis. *J. Cell Sci.* **6,** 131–57.

MANTON I., KOWALLIK K. & STOSCH v.H.A. (1970b) Observations on the fine structure and development of the spindle at mitosis and meiosis in a marine centric diatom (*Lithodesmium undulatum*). IV. The second meiotic division and conclusion. *J. Cell Sci.* **7,** 407–43.

MCLACHLAN J., MCINNIS A.G. & FALK M. (1965) Studies on the chitan (chitin: poly-*N*-acetylglucosamine) fibers of the diatom *Thalassiosira fluviatilis* Hustedt. I. Production and isolation of chitan fibers. *Can. J. Bot.* **43,** 707–13.

MASER M.D., POWELL T.E., III & PHILPOTT C.W. (1967) Relationships among pH, osmolality, and concentrations of fixative solutions. *Stain Technology* **42**(4), 175–82.

MENKE W. (1961) Über die Chloroplasten von *Anthoceros punctatus*. *Z. Naturforsch.* **16b,** 334–6.

OLMSTED J.B. & BORISY G.G. (1973) Microtubules. *Ann. Rev. Biochem.* **42,** 507–40.

PFITZER E. (1871) Untersuchungen über Bau und Entwicklung der Bacillariaceen. In *Hanstein's Bot. Abh a.d. Geb. d. Morphologie und Physiologie.* Vol. 1.

PFITZER E. (1882) Die Bacillariaceen (Diatomaceen). In *Schenk's Handb. d. Bot.* **2,** 443. Breslau.

PHILLIPS D.M. (1969) Exceptions to the prevailing pattern of tubules (9 + 9 + 2) in the sperm flagella of certain insect species. *J. Cell Biol.* **40,** 28–43.

REIMANN B. (1960) Bildung, Bau und Zusammenhang der Bacillariophyceenschalen. *Nova Hedwigia* **2,** 349–73.

REIMANN B. (1961) Zur Verwendbarkeit von Rutheniumrot als elektronenmikroskopisches Kontrastierungsmittel. *Mikroskopie* **16,** 224–6.

REIMANN B.E.F. (1964) Deposition of silica inside a diatom cell. *Expl. Cell Res.* **34,** 605–8.

REIMANN B.E. & LEWIN J.C. (1964) The diatom genus *Cylindrotheca* Rabenhorst. *J. R. microsc. Soc.* **83**(3), 283–96.

REIMANN B.E.F., LEWIN J.C. & VOLCANI B.E. (1965) Studies on the biochemistry and fine structure of silica shell formation in diatoms. I. The structure of the cell wall of *Cylindrotheca fusiformis* Reimann & Lewin. *J. Cell Biol.* **24**(1), 39–55.

REIMANN B.E.F., LEWIN J.C. & VOLCANI B.E. (1966) Studies on the biochemistry and fine structure of silica shell formation in diatoms. II. The structure of the cell wall of *Navicula pelliculosa* (Bréb.) Hilse. *J. Phycol.* **2**(2), 74–84.

REIMANN B.E.F. & VOLCANI B.E. (1968) Studies on the biochemistry and fine structure of silica shell formation in diatoms. III. The structure of the cell wall of *Phaeodactylum tricornutum* Bohlin. *J. Ultrastruct. Res.* **21**, 182–93.

ROBERTSON J.D. (1964) Unit membranes. A review with recent studies of experimental alterations and a new subunit structure in synaptic membranes. pp. 1–82. In *Soc. Study Develop. Growth Symp. 22, Cellular Membranes in Development*. Academic Press, N.Y.

SABATINI D.D., BENSCH K. & BARRNETT R.J. (1963) Cytochemistry and electron microscopy. The preservation of cellular ultrastructure and enzymatic activity by aldehyde fixation. *J. Cell Biol.* **17**, 19–58.

SCAGEL R.F., BANDONI R.J., ROUSE G.E., SCHOFIELD W.B., STEIN J.R. & TAYLOR T.M.C. (1965) *An Evolutionary Survey of the Plant Kingdom*. Wadsworth Publishing Co., Inc. Belmont, California.

SCHIMPER A.F.W. (1883) Über die Entwicklung der Chlorophyllkörner und Farbkörper. *Bot. Ztg* **41**, 105–12, 121–31, 137–46, 153–62.

SCHMITZ C.J.F. (1883) Die Chromatophoren der Algen. Vergleichende Untersuchungen über Bau und Entwicklung der Chlorophyllkörper und analogen Farbstoffkörper der Algen. *Nat. Hist. Ver. Verhandl.* (Bonn) **40**, 1–180.

SLAUTTERBACK D.B. (1963) Cytoplasmic microtubules. I. Hydra. *J. Cell Biol.* **18**, 367–88.

SPURR A.R. (1969) A low-viscosity epoxy resin embedding medium for electron microscopy. *J. Ultrastruct. Res.* **26**, 31–43.

STOERMER E.F. & PANKRATZ H.S. (1964) Fine structure of the diatom *Amphipleura pellucida*. I. Wall structure. *Am. J. Bot.* **51**(9), 986–90.

STOERMER E.F., PANKRATZ H.S. & DRUM R.W. (1964) The fine structure of *Mastogloia grevillei* Wm. Smith. *Protoplasma* **59**(1), 1–13.

STOERMER E.F., PANKRATZ H.S. & BOWEN C.C. (1965) Fine structure of the diatom *Amphipleura pellucida*. II. Cytoplasmic fine structure and frustule formation. *Am. J. Bot.* **52**(10), 1067–78.

STOSCH v.H.A. (1951) Über das Leukosin, den Reservestoff der Chrysophyten. *Naturwissenschaften* **38**, 192–3.

STOSCH v.H.A. (1958) Kann die oogame Araphidee *Rhabdonema adriaticum* als Bindeglied zwischen den beiden grossen Diatomeengruppen angesehen werden? *Ber. dt. bot. Ges.* **71**, 241–9.

STOSCH v.H.A. (1962) Ueber das Perizonium der Diatomeen. *Vorträge aus dem Gesamtgebiet der Botanik, Neue Folge.* **1**, 43–52.

STOSCH v.H.A. & REIMANN B.E.F. (1970) *Subsilicea fragilarioides* gen. et spec. nov., Eine Diatomee (Fragilariaceae) mit vorwiegend organischer Membran. *Nova Hedwigia.* **31**, 1–36.

TAYLOR D.L. (1972) Ultrastructure of *Cocconeis diminuta* Pantocsek. *Arch. Mikrobiol.* **81**, 136–45.

UEDA K. (1961) Structure of plant cells with special reference to lower plants. VI. Structure of chloroplasts in algae. *Cytologia* **26**, 344–58.

VAN VALKENBURG S.D. (1971) Observations on the fine structure of *Dictyocha fibula* Ehrenberg. II. The protoplast. *J. Phycol.* **7**, 118–32.

WATSON M.L. (1955) The nuclear envelope. Its structure and relation to cytoplasmic membranes. *J. Biophys. Biochem. Cytol.* **1**, 257–80.

WILSON E.B. (1896) *The cell in development and inheritance.* Columbia University Biological Series. IV. The MacMillan Co. New York.

CHAPTER 4

SILICATE METABOLISM

D. WERNER

Botanisches Institut der Universität,
Fachbereich Biologie,
D-355 Marburg-L, Germany

1 GENERAL ASPECTS: SILICON AS AN ESSENTIAL ELEMENT

Since the diatoms have contributed most by far to our present knowledge of silicon metabolism, this chapter will include also some more general and methodological aspects.

Silicon occupies 25·5% (by weight) of the earth crust, together with the 49·4% of oxygen, this implies that the silicon–oxygen bond of silica minerals is by far the most abundant chemical bond in the earth crust. In seawater on the other hand, the predominant habitat of diatoms, silicon is, with about 0·0003 atom % a rather scarce element, less abundant than fluorine or boron (Pytkowicz & Kester, 1971; Kester *et al.* 1967). In organisms the silicon content varies between 5 atom % in diatoms (from dry weight) to less than 0·001% in some vertebrates.

From the 25 elements today considered as essential for members of various classes of organisms (Frieden, 1972), silicon has attracted special attention due to its position under carbon in the periodic system of elements. As it could be concluded from the atomic properties (Table 4.1), the elements reveal more essential differences than similarities. Special points of difference would include the particular strength of the C–H bond compared to the Si–H bond, the marked stability of the Si–O bond compared to the C–O bond, the much higher stability of the C–C bond than the Si–Si bond, the molecular status of the oxides, SiO_2 as a polymer with an indefinite number of different structures, and CO_2 as a gaseous monomer. So far, no bonds, other than Si–O, have been demonstrated in biological systems. Thus Si metabolism is so far a silicate metabolism. In 1906 Richter established that silicon is an essential element for the development of diatoms. For several higher plants, by direct or indirect methods, this also was established for monocotyledons (*Oryza sativa, Hordeum vulgare*) by Ishibashi (1936) and for the dicotyledons *Nicotiana tabacum* and *Lemna minor* (Wagner, 1940; Werner, 1967b) and for *Equisetum arvense* (Chen & Lewin, 1969). For the chick Si as an essential element was proven in sophisticated dietary experiments (Carlisle, 1972), avoiding the numerous contaminating sources of the ubiquitous silica.

Valuable ideas about the interaction of silicates with biopolymers came from studies on prebiotic synthesis of biopolymers on silica templates. Clay can function as a catalyst, for example, in oxidizing colourless aromatic amines to blue or black pigments (Weiss & Hoffmann, 1951). Polypepetides have been synthesized on the surface of kaolin as a catalyst (Wacker, 1958). The most advanced studies so far have been conducted by Katchalsky (1973) and Paecht-Horowitz (1970, 1973). These authors obtained heterogeneous polymerization of polyalanines in the presence of montmorillonite (220–480 mg/litre). Within a few hours, they had synthesized in an aqueous

Table 4.1. Comparison of atomic properties of carbon, silicon and germanium.

	Carbon	*Silicon*	*Germanium*
Bond energies (kcal/mole)	C–C : 85 C–O : 85 C–H : 99 C–Cl : 81	Si–Si : 53 Si–O : 108 Si–H : 77 Si–Cl : 91 Si–C : 63	Ge–Ge : 62 Ge–O : 90 Ge–H : 69 Ge–Cl : 81 Ge–C : 58
Covalent radius Ionic radius Bond length *	0·077 nm C^{4+} : 0·015 nm C–C : 0·154 nm C=C : 0·133 nm C–O : 0·143 nm C=O : 0·122 nm	0·117 nm Si^{4+} : 0·041 nm Si–Si : 0·234 nm Si–O : 0·164 nm	0·122 nm Ge^{4+} : 0·053 nm Ge–Ge : 0·245 nm Ge–O : 0·165 nm GeO_6 : 0·190 nm (octahedra)
Electron configuration and atomic number in brackets	$1s^2$ $2s^2$ (6) $2p^2$	$1s^2$ $2s^2$ (14) $2p^6$ $3s^2$ $3p^2$	$1s^2$ $2s^2$ (32) $2p^6$ $3s^2$ $3p^6$ $3d^{10}$ $4s^2$ $4p^2$
Reaction of halides with H_2O	$CCl_4 + H_2O$: ——	$SiCl_4 + 2H_2O \rightarrow SiO_2 + 4HCl$	$GeCl_4 + 2H_2O \rightarrow GeO_2 + 4HCl$
Reaction with O_2	$C + O_2 \rightarrow CO_2$ 194 kcal	$Si + O_2 \rightarrow SiO_2$ 191 kcal	$Ge + O_2 \rightarrow GeO_2$ 132 kcal (hexag.) 138 kcal (tetrag.)
Polarity of hydrogen compounds	$\overset{\delta^-}{C}-\overset{\delta^+}{H}$	$\overset{\delta^+}{Si}-\overset{\delta^-}{H}$	$\overset{\delta^\pm}{Ge}-\overset{\delta^\pm}{H}$
Electronegativities (Allred-Rochow scale with H = 2·10)	2·50	1·74	2·02
Boiling point of hydrides	CH_4 : −162°C C_3H_8 : −42°C	SiH_4 : −112°C Si_3H_8 : −53°C	GeH_4 : −88°C Ge_3H_8 : 111°C

References: Fritz, 1963; Heslop & Robinson, 1967; Glockling, 1969. Noll, 1956; Cottrell, 1958.
* Bond lengths vary considerably in different compounds.

system at pH 8·5 chains of up to 50 amino acids using alanine adenylate as a precursor. The molecular weights obtained showed a discontinuous spectrum with four or more maxima. In the same systems, according to Calvin (1974), polynucleotides are also found. Assuming similar systems did in fact play a role in prebiotic evolution, the idea arises, that the functions of silicates in living organisms today might be a further development of those early catalytic functions.

2 METHODS FOR THE STUDY OF SILICATE METABOLISM IN DIATOMS

2.1 *Spectroscopical methods and tracer studies ^{31}Si, ^{32}Si, ^{68}Ge, ^{71}Ge*

The classical method for determination of silicon compounds is the hydrolysis of the polymer silica to monosilicic acid and its reaction with molybdate followed by a reduction to a blue coloured complex (Jolles & Neurath, 1898; King, 1939; Stegemann & Fitzek, 1954; Mullin & Riley, 1955; Engel &

Table 4.2. Methods for determination of silicon and germanium.

Method	*Reactive compound*	*Quantities determined*	*Reference*
Molybdate blue reaction	$Si(OH)_4$	1–100 μg SiO_2	Engel & Holzapfel (1960)
Atomic absorption spectroscopy	Si and $(SiO_2)_n$ with $n < 10^5$	10^{-2}–1 μg SiO_2	Werner (1970a)
IR-spectroscopy	opals		Langer & Flörke (1974)
Mass spectroscopy	^{29}Si	10^{-6}–10^{-4} μg SiO_2	Goering *et al.* (1973)
Liquid scintillation counting (LSC)	^{31}Si	10^{-2}–1 μg SiO_2	Coombs & Volcani (1968)
LSC	$^{32}Si \rightarrow {}^{32}P$	10^{-1}–10 μg SiO_2	Werner *et al.* (1975)
LSC	^{71}Ge	10^{-4}–10^{-2} μg GeO_2	Werner & Petersen (1973)
LSC	$^{68}Ge \rightarrow {}^{68}Ga$	10^{-7}–10^{-5} μg GeO_2	Azam *et al.* (1973)

Holzapfel, 1960; Tůma, 1962; Heinen, 1963; Strickland & Parsons, 1968). As other new methods for determination of Si, atomic absorption spectroscopy (Werner, 1970a), mass spectroscopy (Goering *et al.* 1973) and the use of radioactive tracers (Coombs & Volcani, 1968; Werner & Petersen, 1973; Azam *et al.* 1973; Werner *et al.* 1975) have been used. The range of sensitivity for these different methods is compared in Table 4.2. In the standard colourimetric assay, the interference by phosphate is overcome by the addition of fluoride (Engel & Holzapfel, 1960). A method for the simultaneous determination of germanium and silicon in the presence of arsenic and phosphorus was communicated by Sorrentino & Paul (1970).

The sensitivity of atomic absorption spectroscopy (AAS) of silicon has made significant advances in the last few years. The measurement at 251·6 nm using a N_2O/C_2H_2 flame is not interfered with by a 100-fold concentration of Ge and is superior by a factor of 100 in sensitivity to the molybdate blue reaction. Another advantage of atomic absorption spectroscopy is the possibility to determine also the Si content of silicate particles up to 20–50 nm diameter. Particles with up to 6×10^5 Si atoms can be measured without the need to hydrolyse (Werner, 1970a).

The use of radioactive tracers has further increased the analytical sensitivity by several orders of magnitude with all the advantages of this method of simplicity and precision. However, the use of some of the available radioactive isotopes of silicon and germanium gives rise to difficulties. First of all, the radiochemical qualities have to be considered. Figure 4.1 shows the isotopes of silicon and germanium with the stable isotopes (black), the positron decay (left side) and the negatron decay (right side). The four radioactive isotopes used so far in studies on silicate metabolism of diatoms are marked with an arrow pointing to the generated daughter nuclides. ^{31}Si has the disadvantage of a half-life of only 2·6 hours making this a useful tracer only for short term experiments. Additionally a generator for producing the isotope has to be available nearby. ^{32}Si decays to ^{32}P as the daughter nuclide and after 14 days the ^{32}P in the sample reaches 50% of the activity of ^{32}Si. Only a separation from the daughter nuclide for example by incorporation of silicic acid into diatom shells (Werner *et al.* 1975) and the application of two counting techniques (Cerenkov-counting, liquid scintillation counting) which differentiate between the isotopes can overcome this problem. With the germanium isotopes, ^{71}Ge raises no problems since it decays to the stable ^{71}Ga. The half-life of 11 days is quite reasonable and comparable to the widely used ^{32}P (14 days) and ^{131}I (8 days). ^{68}Ge on the other hand decays to ^{68}Ga, a nuclide with a half-life of only 68 minutes. This means, in all experiments lasting several hours or days, a high percentage of the measured radioactivity comes from ^{68}Ga and not from ^{68}Ge. To avoid any interference, all samples containing $^{68}Ge/^{68}Ga$ have to be measured not earlier than 12 hours after the end of the biological uptake or incorporation experiment. Then more than

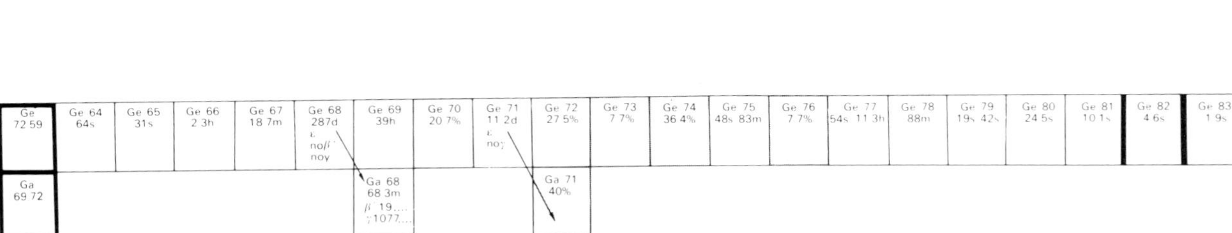

Fig. 4.1. Nuclides of silicon and germanium, with indicated daughter nuclides (arrows) of phosphorus (for ^{31}Si and ^{32}Si) and gallium (for ^{68}Ge and ^{71}Ge). For the stable isotopes (black) the percentages of occurrence are given, for the radioactive isotopes the half-life (in ms, s, m, h, d or a) and the type and energy of radiation (, β^+, β^- and γ). β^- energies in MeV, γ energies in keV. Data from Seelmann-Eggebert *et al.* (1974).

99·9% of the ^{68}Ga taken up or incorporated has decayed, and virtually all the measured ^{68}Ga activity derives from ^{68}Ge.

At least the same sensitivity as with the tracer methods is available from mass spectroscopy, which can register, with modern equipment, 100 particles per second. Considering a loss factor of 10^3 for ionization and of 10^3 for transmission, this means that, for a measurement of 100 seconds, 10^{10} Si atoms are needed. This gives a sensitivity range of 10^{-4} to 10^{-6} μg SiO_2 or GeO_2. With this sensitivity the incorporated GeO_2 in a single cell of *Coscinodiscus asteromphalus* (2×10^{-4} μg GeO_2) (Werner & Petersen, 1973) or the total SiO_2 of a single cell of *Cyclotella cryptica* ($1{\cdot}6$–$2{\cdot}5 \times 10^{-5}$ μg SiO_2) (Werner, 1966) could be determined.

2.2 *Contamination in culture*

In studies of silicate metabolism in diatoms and other organisms the controlled supply or the complete removal of silicates from the culture system is taken for granted. The ubiquitous occurrence of silica renders the complete elimination of SiO_2 from other nutrients, the water, the walls of the culture vessels and from the gases used more difficult than for any other nutrient. An error, often not realized, is the assumption that the use of plastic containers and equipment eliminates the danger of contamination from the culture vessels. In fact dust contamination in production can add significant amounts of SiO_2 to the plastic vessels. Also ultrapure chemicals, like 'K_2CO_3–suprapure–Merck No. 4926', can contain up to 5 μg SiO_2/g, a concentration added per litre of medium in the same order of magnitude (10^{-7} M) as that of many trace elements added to media (personal communication from Merck Company).

3 UPTAKE OF $Si(OH)_4$

3.1 *Concentration of silicic acid in natural environments*

Due to its role as a nutrient for various plankton species, higher plants and animals, the silicate concentration in aquatic habitats varies considerably. Krey (1942) determined values in the Kieler Fjord between 0·0 and 17 μg Si/litre. Armstrong (1951) found a concentration of 11 μg Si/litre in seawater near Plymouth. Jørgensen's measurements (1955) gave concentrations between 30 and 40 ± 5 μg Si/litre in eutrophic Danish lakes during maximum growth of diatoms. Tessenow (1966) studied in a comprehensive and detailed work, the silicon cycle in several lakes in Holstein. He found average concentrations of 4–6 mg Si/litre in the interstitial soil waters, 11–16 mg Si/litre in spring waters and 0·02–12 mg Si/litre in the waters leaving the lakes. In spring, after blooms of diatoms, concentrations of 20–50 µg Si/litre were left behind, unrelated to the concentrations before the growth period.

3.2 *Energetics of uptake*

The requirement for energy in the uptake of $Si(OH)_4$ was demonstrated by J. Lewin (1955b) in *Navicula pelliculosa*. 2,4-dinitrophenol (DNP) inhibited the uptake in concentrations (2×10^{-5} M) which stimulated O_2 consumption by uncoupling oxidative phosphorylation. Other inhibitors such as cyanide and azide inhibited respiration and $Si(OH)_4$ uptake at the same concentration. Attempts to evaluate the energy requirements for $Si(OH)_4$ uptake were performed by Werner (1965) in *Cyclotella cryptica*. In a N-free medium (to prevent net protein and nucleic acid synthesis) cells could take up, in the dark, about 12 µg SiO_2 per µg carbohydrate or 36 µmoles $Si(OH)_4$ by the consumption of 1 µmole glucose (from the stored chrysolaminaran). The addition of 1×10^{-4} M DNP also inhibited this uptake completely. Thus the molar ratio of ATP/$Si(OH)_4$ used for the uptake could not exceed 1. Since more than 95% of this silicic acid is accumulated in a polycondensed form in the cells, this would imply that besides the translocation process of the uptake no further energy is needed for the polycondensation itself. (The term polymerization should be avoided since H_2O is a product of the reaction). 10^{-4} M DNP and 10^{-3} M azide also inhibited the $Si(OH)_4$ uptake in barley seedlings and reduced the silicic content in the shoots to 1·4 and 0·6 mg SiO_2/g dry weight, respectively, compared to 5·3 mg in the controls (Barber & Shone, 1966).

3.3 *Mechanism and kinetics of uptake*

Uptake of silicic acid in *Navicula pelliculosa* was completely inhibited in cells washed repeatedly with distilled water. Addition of reduced sulphur compounds such as glutathione or thiosulphate to those cells restored this function (J. Lewin, 1954). More recent studies on uptake of silicic and germanic acids gave further understanding of the mechanism. Using the colourless heterotrophic diatom *Nitzschia alba*, J. Lewin & Chen (1968) demonstrated that a heat stable factor could be washed from the cells reducing their capacity to take up silicic acid. This factor was not identified, but the addition of aspartic or glutamic acids (5×10^{-4} M) promoted a full recovery of the function. Several other compounds such as glutamine, asparagine, histidine and tyrosine proved to be ineffective, some others including pyruvate and 2-oxoglutarate were partially active. These findings are in good agreement with studies on the regulation of metabolism by silicic acid (see page 126). The same species was used by Azam *et al.* 1974) in an excellent work on the active transport of silicic acid using $^{31}Si(OH)_4$. The Michaelis constant K_m was determined to be $4 \cdot 5 \times 10^{-6}$ M, V_{max} as 3·35 μmoles $Si(OH)_4$.g weight^{-1}.min^{-1} at 30°C and the inhibition by germanic acid to have K_i $2 \cdot 2 \times 10^{-6}$ M. The uptake proceeds with the same rate between pH 7·0 and pH 9·0, whereas, for example, glucose uptake had a sharp optimum pH. From these data it still remains open whether the undissociated form $Si(OH)_4$ or the dissociated forms, $H_2SiO_4^{2-}$ and $H_3SiO_4^{-}$, are directly involved in the uptake process. The authors have suggested that experiments with very low substrate concentrations such that the non-metabolized form falls below the Si_0 value (the concentration, not used by the cells) should be performed.

Uptake and incorporation of germanic acid as a model for silicic acid uptake was studied by Azam *et al.* (1973) and Azam (1974), using ^{68}Ge as a tracer. The main results obtained were in good agreement with experiments from Werner & Petersen (1973) with ^{71}Ge.

Silicate uptake kinetics in five different diatom species were compared by Paasche (1973a,b,c) in a series of excellent and definitive studies. The Si_0 and K_v values ($K_v = K_m - SiO_0$) found for *Skeletonema costatum* were $0 \cdot 32 \times 10^{-6}$ M and $0 \cdot 80 \times 10^{-6}$ M, for *Thalassiosira pseudonana* $0 \cdot 67 \times 10^{-6}$ M and $1 \cdot 39 \times 10^{-6}$ M, for *Licmophora* sp. $1 \cdot 30 \times 10^{-6}$ M and $2 \cdot 58 \times 10^{-6}$ M, for *Ditylum brightwellii* $0 \cdot 38 \times 10^{-6}$ M and $2 \cdot 96 \times 10^{-6}$ M and for *Thalassiosira decipiens* $1 \cdot 33 \times 10^{-6}$ M and $3 \cdot 37 \times 10^{-6}$ M. With the exception of *Ditylum brightwellii* the ratio of the values is always between 2 and 3. However, the author himself expresses some doubts as to whether the Si_0 values are valid (Paasche, 1973a). He discusses several reasons such as interference of released 'binding factors', which could complex reactive $Si(OH)_4$ or the interference of reacting but non-silicic substances. The use of other analytical methods (see Table 4.2) might overcome these problems. Paasche (1973b,c) also emphasizes the discrepancies between the measured V_{max} of $Si(OH)_4$ uptake in pg Si.cell^{-1}.h^{-1} compared to the growth rate and the silicate content of the cells. While the values for *Ditylum brightwellii* ($V_{max} = 26 \cdot 6$ pg Si.cell^{-1}.h^{-1}; generation time 7·5 hours; silicate content 200–900 pg Si/cell) were in good agreement, the V_{max} for the other species was in general too low.

The importance of the Si_0 and the K_m values for the succession of phytoplankton species was already emphasized and discussed earlier by Dugdale (1967) and Kilham (1971). Further determinations of these values, also using tracer methodology or mass spectroscopy for a larger range of the species in plankton communities of major ecological areas, should be considered very desirable work (see Guillard *et al.* 1973).

3.4 *Uptake in the light and in the dark*

The uptake of silicic acid in the light and in the dark by autotrophic diatoms is, at least in some species, occurring by different mechanisms. At concentrations below 100 μg Si/l a natural plankton population, mainly *Asterionella formosa* and *Fragillaria crotonensis,* could take up further silicate only during the dark period (Tessenow, 1966). Also the specificity towards $Si(OH)_4$ and $Ge(OH)_4$ is significantly different in the light and in the dark. Growing cells of *Cyclotella cryptica* take up $Si(OH)_4$ more than three times faster in the light (10·000 lux) than in the dark (Werner, 1966) but $^{71}Ge(OH)_4$ at the same rate under both conditions (Werner & Petersen, 1973). But cells without a normal pool of chrysolaminaran (depleted by a 24 hour dark period) take up $Ge(OH)_4$ in the light three times faster than cells with a regular carbohydrate reserve. These experiments indicate that the specificity of $Si(OH)_4$ uptake is influenced by factors which are acting in a different way under strong light and in the dark. But the mechanism in detail remains open to further study.

3.5 *Uptake studies in chemostat cultures and simulation of natural conditions*

In silicic acid limited cultures under steady state conditions a concentration of 10^{-5}M $Si(OH)_4$ supported the vegetative growth of 10^8 cells/l of *Skeletonema costatum* with a dilution rate of 0·04/hour (Davis *et al.* 1973). An almost complete transition of a $Si(OH)_4$ limited chemostat culture to sexual reproduction in *Skeletonema costatum* was achieved by the same authors using a stationary phase inoculum instead of a continuous culture sample. During a 4-day period no vegetative growth of the cells was noted, $Si(OH)_4$ uptake dropped to zero and almost all cells were involved in sexual reproduction. Harrison (1974) determined the mean uptake ratio (N:Si:P) under $Si(OH)_4$ limitation in *Skeletonema costatum* to be 10:2:1. He noticed also an interesting phase with a high $Si(OH)_4$ uptake rate for about 3–4 hours (phase V_s) with values of 0·1 to $0{\cdot}13 \times 10^{-7}$ μg atom Si.cell^{-1}.hour^{-1}, which is about 4 times higher as Paasche's V_{max} for the same species. This phase was followed by a rapid decrease in the uptake rate of silicic acid (phase V_e). In comparison, phosphate was assimilated during both phases (V_s and V_e) with very similar rates. This apparent surplus uptake of silica will be discussed in more detail in part 4.1.

4 SILICA CONTENT OF DIATOMS

4.1 *Variation in the silica content of diatoms*

The silica content of living diatom cells can be caused to vary by at least five different factors.

1 Limitation in the silica supply in the medium.
2 Variation of other culture conditions (e.g. light, temperature, pH, nutrient concentration, density of the culture).

3 The variation of the valve diameter and thereby the cell surface (MacDonald–Pfitzer rule).
4 The stage in the mitotic life cycle.
5 Variation in the formation of special silica containing structures e.g. spines and processes.

Some of the available data are summarized in Table 4.3. Einsele & Grim (1938) compared in a comprehensive study the Si content of 15 diatom species from plankton samples. They calculated additionally the surface area of the cells, not considering the complex topography of the cell surface, the SiO_2 weight per μm^2, the thickness of the frustules from these data, the volume

Table 4.3. Silicon-content of diatoms.

Species *c: from cultures* *p: from plankton samples*	*pg Si/cell*	*Size (μm)*	*Reference*
Asterionella formosa (p)	45–200	45–65 × 2–3	Einsele & Grim (1938)
Fragillaria crotonensis (p)	70–200	80 × 3	Einsele & Grim (1938)
Tabellaria fenestrata (p)	170–350	57 × 6	Einsele & Grim (1938)
Synedra acus f. angustissima (p)	450–550	420 × 3·5	Einsele & Grim (1938)
Melosira italica (p)	120–180	7 × 30	Einsele & Grim (1938)
Cyclotella glomerata (p)	100	7 × 5	Einsele & Grim (1938)
Cyclotella compta (p)	900	18 × 9	Einsele & Grim (1938)
Cyclotella bodanica (p)	15 000	50 × 20	Einsele & Grim (1938)
Asterionella formosa (p)	120–140	46–130 × 1–2	Lund (1950)
Nitzschia palea (c)	190	20–65 × 2·5–5	Jörgensen (1953)
Nitzschia palea (c) (Si-limited)	50	20–65 × 2·5–5	Jörgensen (1953)
Bacillaria paradoxa (c)	250	60–150 × 4–8	Jörgensen (1953)
Bacillaria paradoxa (c) (Si-limited)	50	60–150 × 4–8	Jörgensen (1953)
Cyclotella cryptica (c) continuous light culture	9·8	9 × 12	Werner (1966) and Werner & Pirson (1967)
Cyclotella cryptica (c) 3 hour dark	7·9	9 × 8	Werner (1966) and Werner & Pirson (1967)
Cyclotella cryptica (c) 12 hour dark	11·9	9 × 10	Werner (1966) and Werner & Pirson (1967)
Cyclotella cryptica (c) Si-limited	9·8	9 × 14	Werner (1966) and Werner & Pirson (1967)
Navicula pelliculosa (c)	1·3–1·4	8–11 × 4–5	Busby & Lewin (1967)
Nitzschia alba (c)	9		Lewin & Chen (1968)
Skeletonema costatum (c)	1–2	5 × 4–12	Paasche (1973c)
Thalassiosira pseudonana (c)	0·6–1·5		Paasche (1973c)
Ditylum brightwellii (c)	200–900	13–52 × 80–130	Paasche (1973c)
Thalassisora decipiens (c)	90–330	17–28 × 8–18	Paasche (1973c)
Skeletonema costatum (c) (Si-limited)	0·56		Harrison (1974)

of the cells or the percentage of Si per cell volume. Especially the last parameter with values between 0·25 and 1·1 pg Si/μm^3 cell volume raises some doubts whether the SiO_2 determination from these plankton samples might be considerably too high. As an example, the Si content of *Cyclotella glomerata* with a size of 7 × 5 μm was found to be about 100 pg Si/cell. In pure culture the value for *Cyclotella cryptica* with a bigger cell size was determined to be between 7 and 11 pg Si/cell (Werner, 1966). In *Cyclotella cryptica* this value corresponds with a high value of 20–30% SiO_2 per dry weight. Also other data from logarithmically growing culture cells suggest that the first estimations from plankton samples might be much too high. More data obtained by modern methods are required to decide if there is any real difference in the SiO_2 content in cultured cells and in cells from plankton samples.

The Si content in *Ditylum brightwellii* changes during the vegetative cell diminution from 900 pg/cell in cells with a 52 μm valve-diameter to 200 pg in cells with a 13 μm valve diameter (Paasche, 1973b), which is in good agreement with the concomitant reduction in cell surface.

The Si content throughout the mitotic life cycle can vary only by a factor of two on a per cell basis with a maximum just before the separation of the daughter cells. By use of inhibitors of cell division such as colchicine, the silica content in *Cyclotella cryptica* per cell was increased by a factor of three (Badour, 1968).

Under $Si(OH)_4$ limitation Si content per cell decreases in some species like *Skeletonema costatum* and *Thalassiosira pseudonana* (Harrison, 1974; Paasche, 1973a). This reduction is clearly made possible by the fact that these species still divide even after cessation of $Si(OH)_4$ uptake and by developing very thin shells (Braarud, 1948). Other species, e.g. *Cyclotella cryptica,* will not divide at all when Si-deficient (Werner, 1966). Exponentially growing cells of *Cyclotella cryptica* were able to take up an excess of silicic acid in the dark 5 hours after cell division had ceased. These cells could use this excess of stored silica to maintain protein, DNA and chlorophyll syntheses in a $Si(OH)_4$-free medium much longer than cells without this excess silica (Werner & Pirson, 1967). It was demonstrated that this capacity was only to a small extent a consequence of the different 'cell ages' compared in these experiments. This was shown not only by inhibitor experiments but also by calculations of the surface area of cells at different stages of the mitotic life cycle and by direct measurements of the decreasing SiO_2 content of synchronized dividing cells in the light (Werner & Pirson, 1967).

4.2 *Dissolution of silica from diatom shells*

The huge deposits of infusorial or diatomaceous earth in various areas give evidence of the stability of many diatom shells under certain conditions

(Lisitzin, 1971). About 30 million square kilometres of the ocean floor are covered by diatom oozes, with a maximum occurrence in southern parts of the Indian and the Pacific Oceans and in the upwelling areas, such as near the coast of Peru (Fairbridge, 1966).

Experimental studies on the dissolution of silica from diatom shells or cells should take into account the following factors:

1 The size of the species.
2 The diameter of the valve in the sexual life cycle.
3 The stage in the mitotic life cycle.
4 The growth rate and phase of the cells.
5 The average thickness of the cell wall.
6 The average specific surface area (surface per g) of the shells.
7 The temperature, at which the cells have been grown.
8 The light conditions (continuous light, light-dark regime).
9 The $Si(OH)_4$ concentration, at which the cells have grown (Si-limited, Si-surplus).
10 The concentration of the other nutrients including the gases.
11 The relative proportions of the different parts of the shells (valves, girdle bands, copulae, spines).
12 The 'cleaning' methods of the frustules to separate the organic material (acid, heat, enzymes).
13 The dissolution conditions (pH, concentration and type of the buffer used, the 'substrate' (shells or cells) concentration, the product ($Si(OH)_4$, SiO_xH_x) concentration, temperature, duration, atmospheric pressure).
14 The presence or the addition of protective agents.
15 The geological age of fossil material.

Jørgensen (1955) demonstrated, using methyl alcohol-killed cells, that at pH 10·0 in a carbonate buffer, the silicate of *Thalassiosira nana* cells was completely dissolved within 40 days, whereas that of *Nitschia linearis* dissolved to only about 20%. After each determination he transferred the cells to a new container to avoid any influence of the product ($Si(OH)_4$) concentration on the process. For *Nitzschia linearis* the rate of dissolution falls off after a certain time, but was linear throughout for *Thalassiosira nana*. He concluded that, in the former, some parts of the shells hydrolyze more rapidly than others. J. Lewin (1961), using cells of *Navicula pelliculosa* from a batch culture at the end of the growth period, obtained several additional results. All treatments that killed the cells, such as the addition of organic solvents, protein-denaturating agents or heat, increased the rate of dissolution of silica compared to living cells. However it still remained at only about 10% of the rate at which $Si(OH)_4$ was solubilized from acid-cleaned shells (pH 9–9·3 in 0·01 M Tris buffer at 19°C). Addition to acid-cleaned shells of 0·0037 M Fe^{2+} or Al^{3+} reduced the subsequent solubilization of $Si(OH)_4$ under these conditions by more than 90%. A treatment of heat-killed plankton diatoms (mainly *Thalassionema nitzschioides* and *Coscinodiscus* sp.) with 0·01 M EDTA or with 0·01 M DHEG increased the dissolution. Both types of experiment indicate that ions, possibly Fe^{2+} or Al^{3+} (or $Al(OH)_3$), can protect parts of diatom shells and that the removal of these compounds by acid cleaning is one of the reasons for the greatly enhanced dissolution of $Si(OH)_4$ from cleaned shells.

Fossil diatomite was studied by Cooper (1952) and Krauskopf (1965), and was shown to dissolve very slowly. J. Lewin (1961) explained this by a reduction of the specific surface area from 89 m^2/g of walls of cultured cells of *Coscinodiscus asteromphalus* to about 22 m^2/g in diatomaceous earth (Teichner & Pernoux, 1951; Iler, 1955). Kamatani (1971) re-examined the dissolution of diatomaceous earth in comparison to cultured cells. The most striking difference he obtained was in neutral 2% solutions of $CaCl_2$, $MgCl_2$ and Na_2SO_4 (at 100°C for 60 minutes); 96 to 99% of the silicate from the shells of *Thalassiosira decipiens* and of *Nitzschia closterium* were dissolved, but only 2 to 4% of two different kinds of diatomaceous earth from Japan. The first sample came from an upper Pliocene location and was predominantly *Stephanodiscus niagarae* while in the other, from a Pleistocene–Pliocene location, the much smaller *Cyclotella comuta* was predominant. By the same treatment sponge spicules were dissolved but only to an extent of less than 1%, whereas a silica gel preparation (60–100 μm) dissolved to 97 to 99%. In all cases, 2% NaCl was less effective in dissolution than $CaCl_2$, $MgCl_2$ and Na_2SO_4. Perhaps a comparison of the salts on an equimolar basis would have been better, for it cannot be ruled out that a concentration of 0·35 M NaCl already inhibits dissolution and 0·18 M $CaCl_2$ does not.

The results given in Fig. 4.2 show how biochemically different components of the same shells can dissolve at different rates. All the GeO_2 bound, by polycondensation, in the shells of *Coscinodiscus asteromphalus* (Werner & Petersen, 1973) is dissolved within 5 minutes at 100°C in distilled water. However, only a small percentage of the SiO_2 from the cells is solubilized.

So far we have no information as to the quantities of other elements such as Al, Fe or Ti naturally bound to the surface of the frustule structures. Ratios of other elements to silicon of less than $1:10^5$ could be significant for the stability of the structure, considering the enormous area of the complex surface, about $3{\cdot}6 \times 10^6$ μm^2 per cell wall of *Coscinodiscus asteromphalus* (110 μm valve diameter). This is about 100 times the surface area of a cylinder with the same linear dimensions.

A mineralization to $Si(OH)_4$ in natural seawater at 30°C of more than 50% was obtained by Kamatani (1969) within 10 days of the silicate of untreated cells of *Skeletonema costatum* and *Chaetoceros gracilis*, but to an extent of less than 10% with *Thalassiosira decipiens*. During these mineralization experiments the dissolution of $Si(OH)_4$ was noted much earlier than the

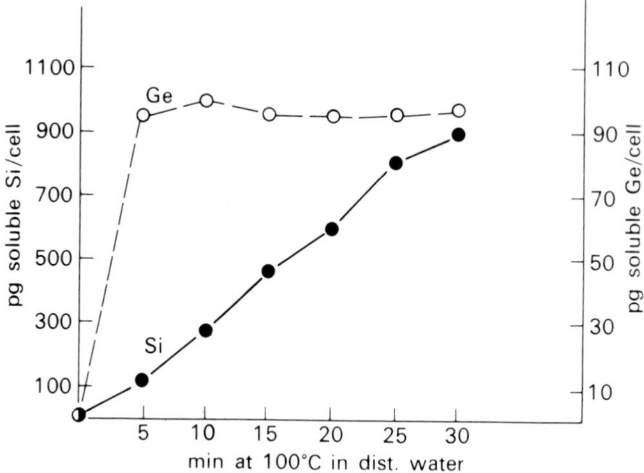

Fig. 4.2. Solubility of silicate (●) and germanate (○) from cells of *Coscinodiscus asteromphalus* (110 μm valve diameter) in water at 100°C in pg/cell. Total Si is about 40 000 pg/cell, total Ge about 100 pg/cell.

occurrence of mineralized phosphate or nitrogenous materials (ammonia, nitrite, nitrate). This means that the Si : P ratio (w/w) in the solution produced in mineralization of *Chaetoceros gracilis* decreased from 102 after 5 days to 17 after 32 days, and with *Thalassiosira decipiens* from 130 after 22 days to 22 after 52 days.

From these experiments we can conclude that at least the smaller diatom species with thin frustules, can have their shells completely dissolved during the sedimentation from the ocean surface to the floor several thousand metres below. This means also that the species comprising the diatomaceous ooze does not reflect the composition of the original diatom plankton samples (Calvert, 1930; Kolbe, 1957).

The further fate of the diatom shells depends on the interaction with other sedimentary material, organic and inorganic, as well as other factors of sedimentology of the deep sea. The interstitial concentrations of siliceous substances from cores under highly productive areas have been determined to be between 600 and 1 000 μM $Si(OH)_4$ (Sievers *et al.* 1965) compared to 210 μM for cores from waters with low productivity (Fanning & Schinck, 1969). The equilibrium concentration for different kinds of clays and sediments with dissolved silicic acid was determined between 100 and 350 μM $Si(OH)_4$ (MacKenzie *et al.* 1967). Thus only under highly productive areas does the concentration of dissolved silicic acid from diatom shells exceed the capacity of clays and sediments for binding the excess. The concentration of

$Si(OH)_4$ in the water of the deep Atlantic is 20–50 μM, still well below the lower limit (100 μM) of binding capacity of clays and sediments.

5 REGULATION OF GENERAL METABOLISM BY SILICATES

The concept of an intimate involvement of silicic acid in the regulation of major pathways of metabolism in diatoms (Werner, 1966) proved to be very productive in the following years. It changed the direct model of $Si(OH)_4$ uptake, transport to vesicles and polycondensation to silicate structures to a more intricate pattern represented in Fig. 4.3. $Si(OH)_4$ biochemistry influences and controls other parts of metabolism of the cells as well as the development of silica structures, just as the pathway of silicates and the morphogenesis of the silica structures are themselves under general metabolic control.

5.1 *$Si(OH)_4$ starvation and recovery experiments*

Results with two of the best studied species in $Si(OH)_4$ deficiency experiments are summarized in Table 4.4. *Navicula pelliculosa* was introduced for studies on silicon metabolism by J. Lewin (1954, 1955a). Concentrations of 10^7 cells/ml were obtained only when the Si concentration was increased to

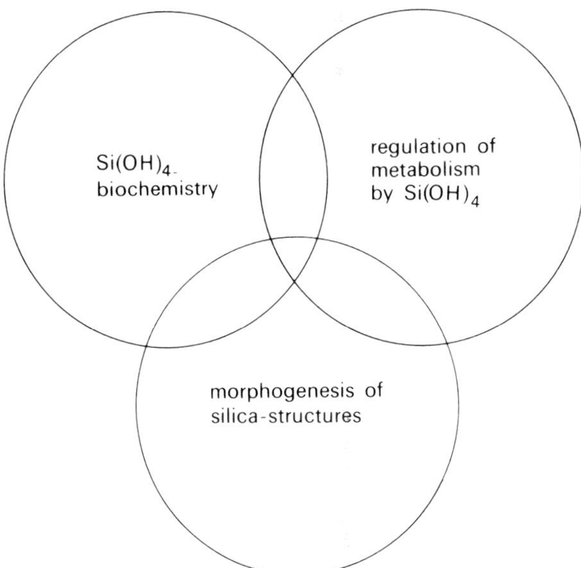

Fig. 4.3. Scheme of interaction of silicate metabolism in diatoms.

Table 4.4. $Si(OH)_4$ starvation and recovery experiments in diatoms.

Time after transfer to a medium without $Si(OH)_4$ in the light	*Metabolic effect in Cyclotella cryptica*	*Reference*
(27·5°C; 10 000 lux) (hours)	86 pg dry weight/cell 12 pg Si/cell 5 hour generation time	Werner (1965, 1966, 1967a, 1968, 1969a, 1970b)
0–1	inhibition of cell division	
1	decrease of pool of 2-oxoglutarate to $\frac{1}{3}$	
2–3	decrease of pool of glutamate to $\frac{1}{3}$	
4	inhibition of net protein synthesis	
4	inhibition of DNA synthesis	
5	inhibition of chlorophyll synthesis	
6–9	increase of fatty acid synthesis by more than 100%	
9	inhibition of carotenoid synthesis	
12	reduction of apparent photosynthesis by 80%	
12	reduction of glycolysis (chrysolaminaran breakdown) by 60–70%	

Time after addition of $Si(OH)_4$ to cells (after 14 hour Si-starvation)	*Metabolic effect in Navicula pelliculosa*	*Reference*
(hours)	24 pg dry weight/cell 1·4 pg Si/cell 6–7 hours generation time	Coombs *et al.* (1967a,b) Coombs & Volcani (1968)
0·25	increase in aspartate label from $^{14}CO_2$	
0·25	increase in glutamate lable from $^{14}CO_2$	
before 0·5	increase in citrate label from $^{14}CO_2$	
before 0·5	increase in 2-oxoglutarate label from $^{14}CO_2$	
before 0·5	increase in nucleotide phosphates (^{32}P-label)	
1	resumed chlorophyll synthesis	
1–4	increase in respiration	
0–6	stop of lipid synthesis	
3	increase in carbohydrate storage	
4	start of net RNA increase	
4	beginning recovery of net protein synthesis	
5–7	start of new cell divisions	

15–30 μg/ml (J. Lewin, 1955b). The SiO_2 content of the species was determined to be between 8·0 (Si-deficient) and 11·6% (regular growth) of the dry weight (Coombs *et al.* 1967a). With *Cyclotella cryptica,* by applying stepwise up to 1 000 μg Si/ml, cell densities of $6{\cdot}5 \times 10^7$/ml were obtained with a maximum SiO_2 content of 30% per dry weight (cells from a 12 hour dark period) and a minimum of 12% (more than 12 hours total Si deficiency, Werner, 1965, 1966).

Besides the variation in the maximum and the minimum Si-content per dry weight, the growth rate (generation time 6–7 hours in *Navicula pelliculosa* and 5 hours in *Cyclotella cryptica*) and the absolute Si content per cell (1·4 pg Si/cell in *Navicula pelliculosa* and 12 pg Si/cell in *Cyclotella cryptica*) are essential criteria to predict how fast the cells will react on depletion and resupply of the nutrient silicic acid.

The summary listed in Table 4.4 shows the gradual decrease and final cessation of major metabolic activity after the transfer of *Cyclotella cryptica* to a medium without silicic acid. The exception is the synthesis of fatty acids, which is in fact increased by more than 100% for several hours. Logarithmically growing cells of *Cyclotella cryptica* synthesized 60–70 μg lipids . mg protein^{-1} . hour^{-1}, $Si(OH)_4$-deficient cells 130–140 μg between 6 and 9 hours after the complete removal of the silicic acid from the medium (Werner, 1970b). During this time, the percentage of fatty acids per dry weight increases from 4·9% to 17% and the percentage of fatty acids in lipids from 38% to 60% (Werner, 1966). The pool of glycerol-1-phosphate also does not decrease during the first 7 hours of $Si(OH)_4$ deficiency (Werner, 1968). The energy requirement for the synthesis of an 18-C fatty acid molecule is 9 ATP (Lehninger, 1965). This means that the energy supply of the cells until the late stages of silicic acid deficiency cannot be affected. This was experimentally confirmed: nucleoside triphosphates, determined as ATP, remained unchanged during the first 7 hours (Werner, 1968). In *Navicula pelliculosa* the pool of nucleoside triphosphates per cell was even slightly higher in Si-starved cells compared with exponentially growing cells (Coombs *et al.* 1967b), and the pool of the triphosphate per cell in *Cylindrotheca fusiformis* was not different in $Si(OH)_4$-deficient cells compared to the controls (Sullivan & Volcani, 1973a).

The complete inhibition of any further increase of the protein content of the cells 4 hours after the removal of $Si(OH)_4$ is another striking response. At this stage, photosynthesis, biosynthesis of chrysolaminaran and glycolysis of sugars were only slightly reduced and were still significantly above zero 8 hours later. A decrease of the pool of 2-oxoglutarate and 1-glutamate to about one third of the control value precedes the cessation of protein synthesis (Werner, 1968). Since the pool of glutamic acid can constitute up to 40% of all the soluble amino acids in diatoms (Werner, 1971b) and glutamic acid serves as N-donor in the synthesis of other amino acids, a shortage of amino acids could account for the cessation of synthesis of protein. However this may only be true for the vast majority of proteins. Perhaps in certain compartments of the cell or for single structural proteins or enzymes, synthesis may continue. The specific activity of aldolase (E.C.4.1.2.13), for instance, increased by a factor of two after the increase of total protein had completely stopped (Werner, 1966).

An early cessation of DNA synthesis in *Cyclotella cryptica* was demonstrated by Werner (1966) and Werner & Pirson (1967), whereas no comparable effect was found in *Navicula pelliculosa* (Coombs *et al.* 1967a,b). Darley & Volcani (1969) demonstrated, in convincing experiments, a very early stop in DNA synthesis after $Si(OH)_4$ depletion in *Cylindrotheca fusiformis*. Protein synthesis and, to a lesser extent, RNA synthesis continued for several hours. Small amounts of $Si(OH)_4$ (0·2 ppm Si) added to the $Si(OH)_4$-deficient cells of *Cylindrotheca fusiformis* ($0{\cdot}45 \times 10^6$ cells/ml) did not satisfy the silicate requirement of DNA synthesis, while 0·4 ppm Si allowed 50% of the cells to complete mitosis and an increase of DNA by 80%. Further experiments by Sullivan & Volcani (1973a,b) with $Si(OH)_4$-starved cells showed that neither the pools of deoxyribonucleotides were different from the controls, nor was any evidence found of an effect of silicic acid on DNA polymerase or thymidine monophosphate kinase *in vitro*. But, since in these experiments the *in vitro* activity was only 12·5% of that calculated for *in vivo* DNA synthesis, the measured DNA polymerase activity cannot account for all the *in vivo* DNA replication. Also, assuming that the low measured activity is representative, we cannot rationalize that a somewhat decreased DNA polymerase activity (from 0·85 pmoles ^{3}H-TTP . mg protein^{-1} . min^{-1} to 0·53 units in $Si(OH)_4$-deficient cells, could account for a complete stop of DNA synthesis. Thus the authors' idea that an inhibition of the initiator proteins in the replicon units of diatom chromosomes might be responsible for the complete stop of DNA synthesis remains a distinct possibility.

The results of some recovery experiments from $Si(OH)_4$-starved cells of *Navicula pelliculosa* given in Table 4.4 are in good agreement with the starvation studies in *Cyclotella cryptica*. An early increase in glutamate, aspartate and 2-oxoglutarate ^{14}C-label precedes the recovery of protein synthesis. The accumulation of lipids is immediately stopped for several hours while carbohydrate storage begins already at 3 hours. A late reduction of glycolysis during starvation corresponds with an early increase in respiration during the recovery. This means that the cells keep their energy supply from respiration available during both phases, the starvation as well as the recovery, from the beginning. The response of the cells in the recovery experiments, of course, depends on the duration of $Si(OH)_4$ starvation. At every distinct stage of the starvation, a recovery experiment would give somewhat different results. Relationships between these findings, cell division and silica structure formation will be treated in part 7 of this chapter.

5.2 *Inhibitor experiments with $Ge(OH)_4$*

The use of the inhibitor germanic acid in studying the regulation of metabolism by silicic acid has some advantages over starvation experiments. By applying high doses the gradual exhaustion of internal SiO_2-$Si(OH)_x$ reserves can be blocked and thus a still faster response of metabolism provoked.

For significant inhibition of cell division, molar ratios of Ge : Si of 0·07 in *Cyclotella cryptica* (Werner, 1965) of 0·08 in *Navicula pelliculosa* and 0·15 in *Phaeodactylum tricornutum* (J. Lewin, 1966c) of 0·05 to 0·5 in *Cylindrotheca fusiformis* (Darley, 1969) and of 0·013 in *Coscinodiscus asteromphalus* (Werner & Petersen, 1973) have been used. Absolute concentration is also important, values between 1·04 and 375 μg GeO_2/ml (1×10^{-5} M to $3{\cdot}60 \times 10^{-3}$M) were used.

In the presence of $1{\cdot}44 \times 10^{-3}$ M $Ge(OH)_4$ chlorophyll synthesis in *Cyclotella cryptica* is rapidly blocked, whereas synthesis of chrysolaminaran and the production of photosynthetic O_2 are not affected (Werner, 1967a). An almost complete inhibition of chlorophyll synthesis is independent of the cell stage; it is inhibited in logarithmic growth phase cells (with all stages in the mitotic life cycle) and in cells just after division. Inhibition of chlorophyll synthesis by $Ge(OH)_4$ was also shown in some higher plants for example *Lemna minor*, whereas no effect either on growth or on chlorophyll formation in several green algae and flagellates was noted (Werner, 1967b). Biosynthesis of a typical chloroplast enzyme (NADP-dependent glyceraldehyde phosphate dehydrogenase (E.C.1.2.1.13) is specifically inhibited by low concentrations of germanic acid ($7{\cdot}2 \times 10^{-5}$ M) whereas the cytoplasmic enzymes NAD-dependent glyceraldehyde phosphate dehydrogenase (E.C.1.2.1.12) and the lactate dehydrogenase (E.C.1.1.1.27) are not affected at all. In contrast to the photosynthetic production of carbohydrates, the breakdown in the dark is reduced to one third by $Ge(OH)_4$ in *Cyclotella cryptica*. But oxygen consumption in *Navicula pelliculosa* (J. Lewin, 1966) was unimpaired.

By application of a Ge : Si ratio of 0·1 Darley & Volcani (1969) were able to inhibit 75% of the cell separation with only small effects on net protein synthesis and DNA synthesis. A ratio of 0·5 inhibited DNA synthesis to the same extent and protein net synthesis still more than $Si(OH)_4$ starvation.

5.3 *Comparison of $Si(OH)_4$ starvation with phosphate and combined nitrogen starvation*

The reaction of diatoms in phosphate and in nitrogen deficiency is characteristically different from $Si(OH)_4$ starvation. Cells without an external phosphate supply continue to divide for more than three generations and then start to accumulate carbohydrates. The lipid content does not increase during this first stage (Werner, 1966). This is also the case in other unicellular algae (Kuhl, 1968). The accumulation of carbohydrates up to 67% of the organic dry weight (Werner, 1966) in the first stages of nitrogen deficiency is still more obvious. Lipid accumulation starts only very gradually. For several days a rate of only 14–16 μg lipids.mg protein^{-1}.hour^{-1} is reached which is only 10% of the rate in $Si(OH)_4$-deficient cells. In general it can be said that many diatom species react faster on $Si(OH)_4$-deficiency than on removal of any other nutrient (except water and carbon source) and that the characteristic pattern of metabolic responses on $Si(OH)_4$ deficiency cannot be mimicked by any other deficiency.

At this point it should be emphasized that there is sufficient experimental evidence to conclude that the characteristic oil droplets found in many diatom cells in their natural environment is a reaction to $Si(OH)_4$ deficiency rather than any other factor. Oil droplets are in any case not storage products of diatoms growing under ideal conditions, as is still mentioned in some textbooks. Further details of the response of diatoms to nutritional deficiencies are discussed in Chapter 7.

5.4 *Physiologically active silicoborate compounds*

The boric acid requirement of diatoms was conclusively established by J. Lewin (1965, 1966a). The inhibition of growth on boric acid deficiency in synthetic seawater medium was much more obvious in the marine species *Cylindrotheca fusiformis, Navicula incerta, Nitzschia angularis, Nitzschia frustulum, Melosira nummuloides, Skeletonema costatum* and *Thalassiosira fluviatilis* compared with the freshwater species *Hantzschia amphioxys, Navicula tomoides* and *Nitzschia palea.* Of the marine species tested, the requirement for boric acid was least in *Cyclotella cryptica* and of the freshwater species in *Navicula pelliculosa.* At low H_3BO_3 concentrations (0·01–0·1 ppm B) there was some reduction in the growth rate of *Cylindrotheca fusiformis* with increasing $Si(OH)_4$ quantities in the medium (J. Lewin, 1966b). Some interaction of these two compounds in the physiology of diatom cell wall (silica shell) formation was suggested.

The direct effect of trace amounts of silicoborates (Si-O-B-compounds) on the growth of *Cyclotella cryptica* strain M2 in a medium made from highly purified water and pure nutrient chemicals was demonstrated by Werner (1969b). In the absence of silicoborates the generation time of the strain M2 of *Cyclotella cryptica* slowed down and came to an almost complete stop within one or two weeks. The supplement of the silicoborates (in addition to regularly applied concentrations of boric acid and silicic acid) resulted in an obvious recovery of the cells after 3–4 days. The control cultures, with only H_3BO_3 and $Si(OH)_4$ prepared from 'Aerosil' silica gel or sodium silicate in the medium showed no or only very slight growth, resulting in a difference of cell number between the cultures after 5 days of a factor of 20 or more. The positive effect of the silicoborates on the growth of *Navicula pelliculosa* in a synthetic medium was less marked but still significant. No effect was seen on the growth of *Coscinodiscus asteromphalus* in an enriched seawater medium. Effective silicoborate solutions could not only be produced by melting quartz and boric acid together with sodium carbonate or sodium hydroxide followed by a solution in water but also by sonication of borosilicate glass (Werner, 1969b).

Acid hydrolysis of the silicoborate preparations destroyed the biological activity and increased the amount of free $Si(OH)_4$ by a factor of almost three. The effective concentration of boron in the borosilicate preparations from glass was as low as 0·008 ppm with *Cyclotella cryptica*. The results indicate that silicoborates are essential compounds in the silicate metabolism of some diatoms, which the cells themselves, under certain conditions, fail to produce from the components H_3BO_3 and $Si(OH)_4$. Thus, they appear to function as a type of inorganic vitamin. It is so far unknown in which part of silicate metabolism or silica structure formation the silicoborates are used.

5.5 *Conclusions*

From the preceding it can be concluded that there are at least four different interactions of $Si(OH)_4$ biochemistry with the general metabolism of the cells (besides cell wall formation)

1 In the citric acid cycle between acetyl-CoA (used for fatty acid synthesis) and 2-oxoglutarate, probably at one of the isoenzymes of isocitrate dehydrogenase, which is an allosteric enzyme, regulated for example also by ADP.

2 In the synthesis of special proteins, localized in cell organelles (chromosomes, chloroplasts).

3 In the regulation of respiration and chrysolaminaran utilization.

4 In the regulation of chlorophyll synthesis.

6 BIOCHEMISTRY OF SILICIC ACID

6.1 *Basic chemical and related studies*

Silicon dioxide and water are the two most intensively studied substances in chemistry (Rostow, 1973). A huge literature is therefore available on the inorganic and ceramic chemistry of silica, silicates and silicic acids (for reviews see Eitel, 1964; Levin *et al.* 1961). But our knowledge about accepted biochemical reactions and pathways of silicic acid is still rather scarce.

Only a few chemical studies will be mentioned here, those which have produced some basic progress towards an understanding of the interaction of silicic acids with physiologically interesting compounds. Clark & Holt (1957); Clark *et al.* (1957) and Holt & Went (1959) studied the complexing of silicic acid with proteins such as albumin, gelatin and pepsin.

Weiss *et al.* (1957, 1961b) demonstrated the formation of water stable esters between silicic acid and catechol, quercetol and tannic acid. These authors showed the configuration

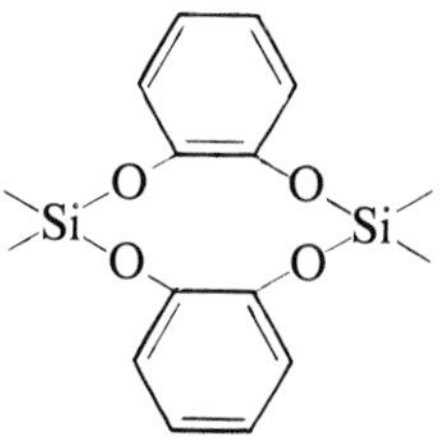

to be more stable in water than —Si—O—Si—

whereas the esters of silicic acid with aliphatic dienols such as ascorbic acid are much less stable. Diatomite reacted in water to the extent of about 30% with catechol within 10 days at pH 8·2 in forming tricatecholic-silicic acid esters. However no reaction was found with glycerol, mannitol or sorbitol. The addition of amino acids and urea markedly changed the dialysis rate of silica from intensively shaken quartz powder (Holzapfel, 1953).

Comprehensive information is available concerning the adsorptive properties of silica gel due to the practical importance of these compounds in purification technology and as catalysts in the hydrocarbon industry. Many of these results might also be useful as models for the interaction of anions, cations and organic molecules with silica gel and silica. It has recently been shown that the catalytic properties of Ni and Co bound to special types of kieselguhr were different from those when the metals were bound to a more acidic support (Guidot & Germain, 1974).

The formation of stable complexes with organic substances is complicated by the delicate polycondensation-hydrolysis equilibrium of orthosilicic acid ($Si(OH)_4$) and 'disilicic acid' $H_2Si_2O_5$ (or (Si_6O_6) $O_{6/2}$ (OH_6)-lepidoid) (Kautsky *et al.* 1962). Different stages of the marginal zones of this molecule are shown in Fig. 4.4. Only the groups marked, B, C and D are 'hydrolysis-active'. In a 1-year-old solution of silicic acid (21 mg SiO_2/ml, pH 4·5) about 4% of all the Si atoms are hydrolyzed more rapidly (within 10 minutes) after reaction of the produced $Si(OH)_4$ with molybdate and were assumed to be located in such positions. Even with quartz the solubility depends very much on particle size. Since no data were given on the size or the specific surface of their quartz suspensions, the claim of active biodegradation of quartz by *Proteus mirabilis* (Lauwers & Heinen, 1974) though very small might be merely an effect of a better dispersion of the quartz particles on addition of bacteria and thereby an increase in solubilization. The same authors reported significant effects of *Equisetum arvense* on solubilization of silica gel.

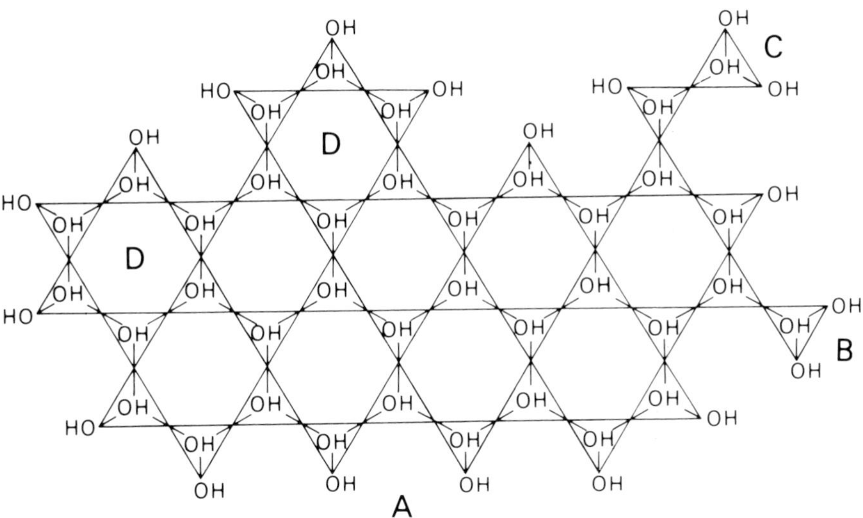

Fig. 4.4. Net of silicic acid $H_2Si_2O_5$ with different phases of growth (B,C,D). From Kautsky *et al.* (1962).

Studies on silicosis are another valuable source of information about the interaction of silica, especially quartz powders with physiological components. Weiss *et al.* (1961a) reported the formation of an iron phosphatosilicate in the lungs of rats, exposed to quartz powders. After repeated intratracheal application of coal dust in rats it was shown that fibrosis increased with SiO_2 retention and does not depend on total dust uptake (Einbrodt, 1965). The elimination of Fe_2O_3 increased, that of Al_2O_3 decreased but Zn remained on the same level with increased total dust deposition (Rosmanith, 1975). Polyvinylpyridine-*N*-oxide could prevent an experimentally produced silicosis (Weller, 1973). Siliceous dust itself can cause a marked increase in hydroxyproline content in rat lungs compared with the exposure to other dusts (Senczuk & Mlynarczyk, 1974).

Epitaxy phenomena (a type of intimate surface interaction) of amino acids and peptides on quartz surfaces were studied by Seifert (1959, 1961a, 1961b). No solubilizing effect, as claimed by Schlentrich (1965) of rat tissues on implanted quartz platelets (30 μm thick and with a size of 0·5 to 1 mm^2) was found with repeated measurements (Mosebach, 1966). Also studies with higher plants (Engel, 1953; Umemura *et al.* 1961; see also J. Lewin & Reimann, 1969) helped to develop some valuable methods for investigating silicate metabolism in diatoms.

6.2 *Silica fractions from diatom cells*

Two major attempts have so far been made to fractionate diatom cells to determine the silica content in cell components separated from the shells (Werner, 1966; Mehard *et al.* 1974). The two schemes applied are presented in Fig. 4.5. A particular problem is the fact that any method of cell disruption produces shell fragments that co-sediment with the fractions. By combining fractional centrifugation and precipitations with ammonium sulphate, one special fraction, 'plasmatic silicic acid' (PSA) was characterized from *Cyclotella cryptica* with the following properties (Werner, 1966)

1 A constant $Si(OH)_x$: protein ratio despite further $(NH_4)_2SO_4$ precipitation (10–13 µg PSA/mg protein) in sediment V (Fig. 4.5, A).
2 A more than ten times increased solubility in water at 100°C, compared to shell fragments sedimenting at the same or even higher g forces.
3 A response to $Si(OH)_4$ deficiency and recovery of the cells, especially under conditions with prolonged starvation, when the protein content of the cells decreased (variation by a factor of 10).

The characterization of a silicate fraction by using two different extraction procedures (water and dilute acid) (Werner, 1966) seems, after more recent experiments (Werner & Petersen, 1973) less useful, since the solubility of the shell silica in water can vary with nutritional conditions.

With their scheme, Mehard *et al.* (1974) found a decrease of ^{31}Si per fraction in the general order: mitochondrial fraction, chloroplast fraction, vesicle fraction and microsome fraction using cells of *Cylindrotheca fusiformis* and *Nitzschia alba*. Electron micrographs of the organelle fractions showed that the mitochondrial fraction of *Nitzschia alba* and the chloroplast fraction of *Cylindrotheca fusiformis* still contained wall fragments. The unit volume of these fragments could be estimated to be less than 1% of the volume of a single organelle in the fraction. Assuming the same ratio also for the vesicle fraction or the microsome fraction, this would certainly mean that invisible tiny wall fragments could also be present in these fractions. Nevertheless this scheme of fractionation of cell organelles from diatoms has proved to be valuable. From 10^6 cells of *Cylindrotheca fusiformis* (exposed for 10 minutes to $^{31}Si(OH)_4$) 0·96 ng Si were found in the chloroplast fraction, 1·44 ng in the mitochondrial fraction, 0·15 ng in the vesicle fraction and 0·12 ng in the microsome fraction. From 10^6 cells of *Nitzschia alba,* 10·0 ng Si were found in the mitochondrial fraction, 0·51 ng Si in the vesicle fraction and 0·07 ng Si in the microsome fraction. All fractions from *Cylindrotheca fusiformis* had 20–90% of their label extractable with water, whereas the label from the mitochondrial fraction of *Nitzschia alba* (with very obvious wall fragments) was extractable only to 3·5%.

Together with the physiological evidence (page 131), these data support the hypothesis of a special silicic acid metabolism in cell organelles of diatoms.

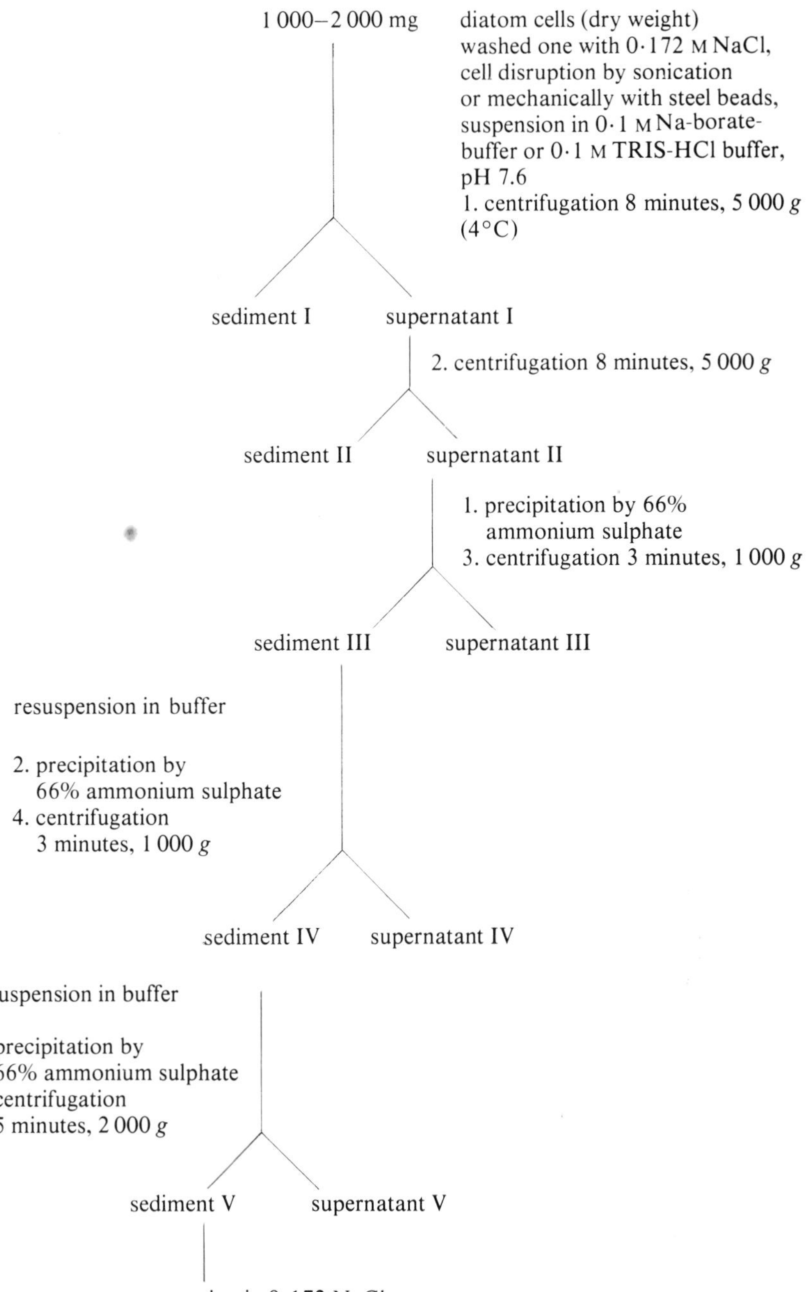

Fig. 4.5, A. Fractionation of diatom cells (Werner, 1966).

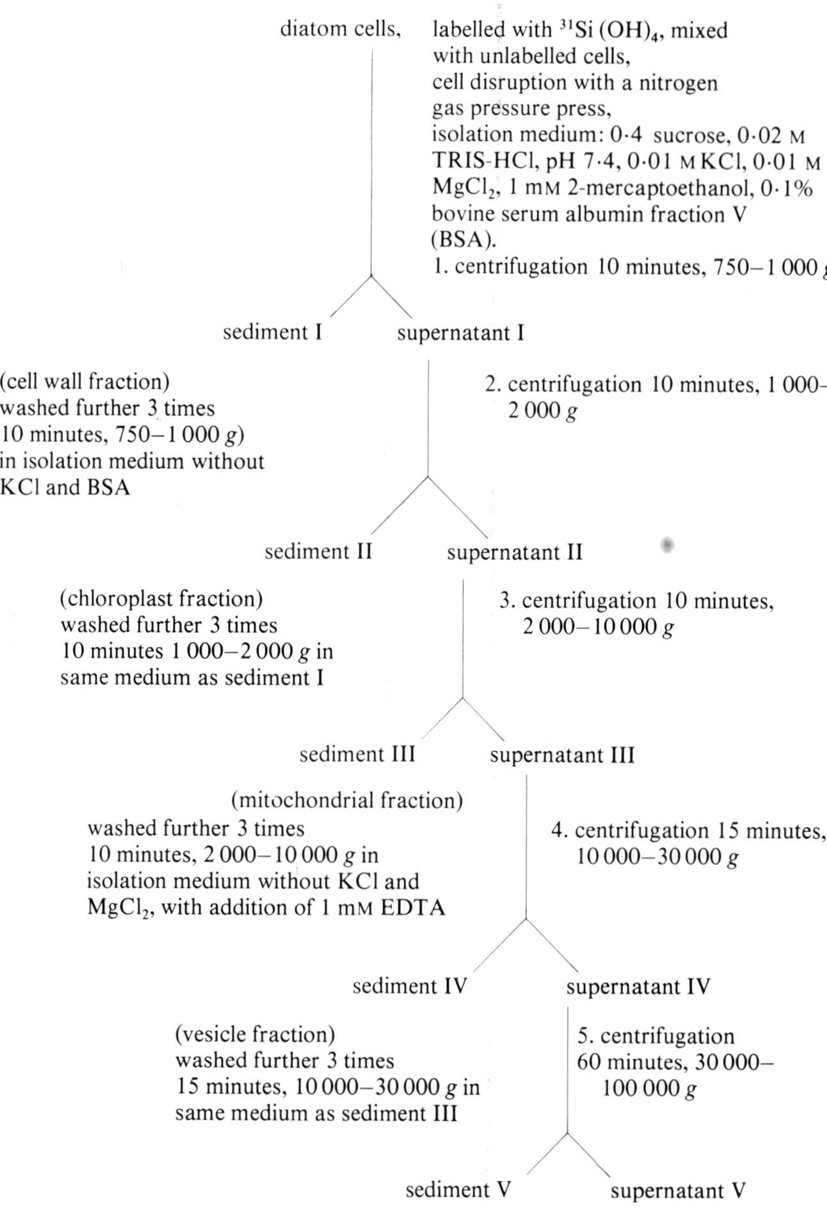

Fig. 4.5, B. Fractionation of diatom cells (Mehard *et al.*, 1974).

6.3 *Cell-free systems*

Further support for this same hypothesis comes from experiments with incorporation of labelled $Si(OH)_4$ and $Ge(OH)_4$ in cell-free systems. Mehard *et al.* (1974) obtained incorporation of 3·1 μg Si . mg protein^{-1} . hour^{-1} in the chloroplast fraction of *Cylindrotheca fusiformis,* 2·2 μg Si in the mitochondrial fraction, 0·9 μg Si in the vesicle fraction and 0·5 μg Si in the microsome fraction. The incorporation in the fractions of *Nitzschia alba* with values between 1·1 μg Si and 1·3 Si μg in all the fractions was much less specific.

Results from some recent experiments with cell-free extracts of *Cyclotella cryptica* and separation of the incubated mixtures by thin-layer chromatography (Roth, 1975; Werner & Roth, 1976) are shown in Fig. 4.6. Using two-dimensional DEAE-cellulose thin-layer chromatograms, the effect of cell-free extracts (supernatant of 38 000 × *g*) in binding ^{68}germanate can be seen in the location of spots both in the centre as well as on the baseline. The baseline spots (R_f in the first direction with 0·15 M NaCl is zero) show further

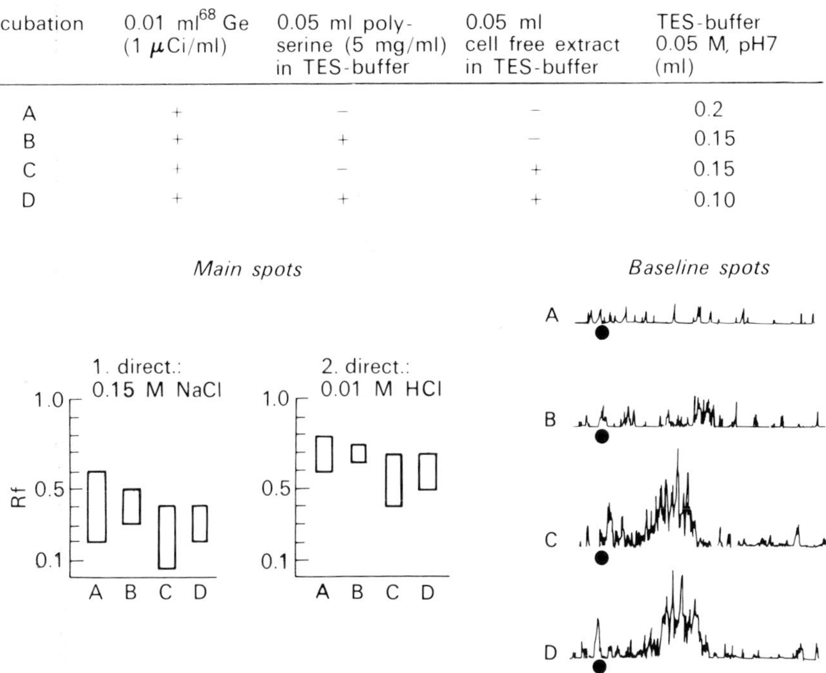

incubation	0.01 ml^{68} Ge (1 μCi/ml)	0.05 ml polyserine (5 mg/ml) in TES-buffer	0.05 ml cell free extract in TES-buffer	TES-buffer 0.05 M, pH7 (ml)
A	+	–	–	0.2
B	+	+	–	0.15
C	+	–	+	0.15
D	+	+	+	0.10

Fig. 4.6. Thin-layer chromatogramms (baseline and main spots) of incubation of cell-free extracts (supernatant 38 000 × *g*) from *Cyclotella cryptica* with polyserine and $^{68}Ge(OH)_4$. From Werner & Roth (1977).

(B) that polyserine alone binds a small fraction of the label. A replacement of polyserine by polymethionine gives no such spots. Polyserine also reduces the spreading of the main spots compared to incubation with cell-free extract, or germanate alone. The thin-layer plates were measured not earlier than 12 hours after the development of the chromatogram to allow virtually all the ^{68}gallium originating from ^{68}germanium before the separation, to decay to the stable ^{68}zinc.

7 METABOLISM AND SILICATE STRUCTURES

Several aspects of cell wall and silica shell formation in diatoms are covered in Chapter 3. In addition the following points will be treated here in more detail: experimental studies concerning the influence of metabolic factors on silica shell formation, incorporation of other cell wall components, models of shell formation and hypotheses about the physiological and ecological significance of silica structures.

7.1 *Experimental approaches*

The sequence of events preceding cell division in normally growing diatoms was described by Geitler (1953a,b, 1962). The following sequence occurs: cytoplasmic cell expansion, separation of the connecting bands, mitosis, cytokinesis and valve formation; within this sequence the dependence of a single event from the preceding is experimentally difficult to evaluate. The generation time of the species is determined principally by the duration of cytoplasmic cell expansion. With a full supply of all nutrients and energy this is limited by the rate of protein synthesis. The growth rate itself can vary over a wide range (see Chapter 2, Table 2.1). The fastest growing species so far reported are *Chaetoceros gracilis* with a 4·4 hours generation time (Thomas, 1966) and *Cyclotella cryptica* with a 5 hours generation time (Werner, 1966). Very large species only divide after 24 hours or more (Werner, 1971a). Compared to these periods, mitosis occurs rapidly in diatoms (v. Denffer, 1950) and also valve formation can occur in some species within 20 minutes (Reimann, 1960).

Our knowledge about the relationship of distinct nutritional and environmental factors to silica structures and shell formation in general is hampered by the fact that many effects have been observed in only one or a few species.

In *Achnantes longipes* manganese deficiency caused the inhibition of the formation of new shells and several other effects (v. Stosch, 1942). In prolonged Mn-deficient cultures the protoplasts become detached from the shells, and eventually separated from them as isolated naked protoplasts. Some of these protoplasts developed into plasmodia with several hundred nuclei (v. Stosch, 1942). After addition of manganese to individual naked protoplasts, new shells were formed; however, irregular cell forms were common.

The capacity of diatoms to form new shells inside the old cell wall was demonstrated by Küster–Winkelmann (1938) using plasmolyzed cells in hypertonic media. A special effect of osmotic pressure on the valve morphology of *Skeletonema subsalum* was shown by Paasche *et al.* (1975). This brackish-water diatom reduced the connecting processes in length and caused flat-shaped valves at salinities of 0·3% or less, while at concentrations of 0·5% or more the valves were dome-shaped and the connecting processes elongated. Several effects of nutritional and envrionmental factors on shell morphology were observed and discussed in detail by Geissler (1970). She emphasizes that some genera such as *Coscinodiscus* and *Pinnularia* show a marked variability of certain characteristics of cell morphology. Also some species, for example *Synedra tabulata* and *Navicula mutica* are especially variable.

Which cytological event gives the decisive signal for the formation of the valve is still not firmly decided. Geitler's theory (Geitler, 1953a, 1962), that formation of each valve is always preceded by a mitosis in diatoms was discussed by v. Stosch & Kowallik (1969) and supported by other observations. Two modes of valve formation in addition to the normal cell division were noted: (1) Unequal cytokinesis, after which only the bigger daughter cell forms a new valve; the smaller one, with or without a nucleus, forms no new valve. (2) Acytokinetic division (nuclear division without cytokinesis), after which the cells can plasmolyze and form new valves.

Cytokinesis is inhibited in *Cyclotella cryptica* by Ca^{2+} or Mg^{2+} deficiency, but these cells form two lateral shells surrounding the median region (Badour, 1968). These lateral shells have the same fine structure as the valves, but are curved. After inhibition of mitosis by application of colchicine or β-phenetylalcohol the much elongated cells with a high silica content (Werner, 1965) produced more than one pair of these lateral shells (Badour, 1968). These experiments indicate that neither complete cytokinesis nor mitosis are compulsory prerequisites for the formation of silica shells. Oey & Schnepf (1970) demonstrated in *Cyclotella cryptica* that the formation of lateral shells depends on DNA replication and is probably induced by factors arising in the S phase or in the G_2 phase. In this connection our knowledge is incomplete, in particular concerning the function of the paracentrosomes (Manton *et al.* 1970).

There are many questions still to be answered as to the relation of these different cytological events with the formation of the various parts of the frustules in addition to the valves such as the valvocopular bands, the connecting bands and the intercalary bands of the girdle (Round, 1972; v.

Stosch, 1974). The large differences in the extension, in the subunits and the fine structure of the girdle, in different species (v. Stosch, 1974) lead to the same fundamental and still open question as do the different valve patterns of the more than one thousand species (Hendey, 1959; Helmcke *et al.* 1963–75) studied so far by electron microscopy: that is, which informational matrix determines the fine structure of the silica shells?

It is obvious that the regulation of any other component of diatom cell walls can also influence the formation of the silica shells (J. Lewin, 1957; Werner & Stangier, 1976). Especially helpful might be the further use of mutants (Werner, 1970c) which have so far been rarely experimentally generated, the production of significant quantities of naked vegetative cells (Hendey, 1946; v. Stosch, 1965) or gametes (Werner, 1971c) for biochemical studies and further analyses of the plasma and smooth membranes of diatoms (Sullivan & Volcani, 1974).

The period of $Si(OH)_4$ uptake preceding shell formation is accompanied by an increased incorporation of $^{14}CO_2$ into protein amino acids (53%) of cell walls of *Navicula pelliculosa* (Coombs & Volcani, 1968), while the other half of the label was found in sugars and sugar derivatives (17% in mannose, the rest in glucuronic acid, xylose, fucose, glucose, galactose and rhamnose). During the cell separation period, only 3% of the ^{14}C label was incorporated into protein amino acids of the cell walls, but 40% into xylose and fucose. The non-protein amino acids, 3·4 dihydroxyproline and *N*-trimethyl hydroxylysine, have been isolated from diatom cell walls (Nakajima & Volcani, 1969, 1970). The amino acid composition of the cell wall fraction (sediment of $1\,800 \times g$) and the cell content (supernatant of $1\,800 \times g$) was compared in *Navicula pelliculosa, Melosira granulata, Cyclotella stelligera, Cyclotella cryptica,* and *Nitzschia brevirostris* by Hecky *et al.* (1973). The serine content in the cell wall fraction revealed the highest value compared to the cell content: a surplus of 156% in *Cyclotella stelligera* and of 147% in *Cyclotella cryptica*, but only 20% in *Navicula pelliculosa*. Glutamic and aspartic acids on the other hand were much less common in the cell wall fraction than in the cell content, aromatic and sulphur containing amino acids were also reduced in relative quantity. The relative proportions of sugars in the same species in both fractions is less consistent. Glucose content in the cell wall fraction was always reduced compared to the cell content, while xylose was always increased. Due to significant contamination by fragments from the other fraction (cell rupture was by sonication) the true differences in cell wall composition and that of cytoplasmic fractions might be still higher, as the authors suggested.

7.2 *Models and hypotheses*

Supported by some experiments in cell-free systems described in part 6.3 (Fig. 4.6) and more general information on cell wall formation in plants

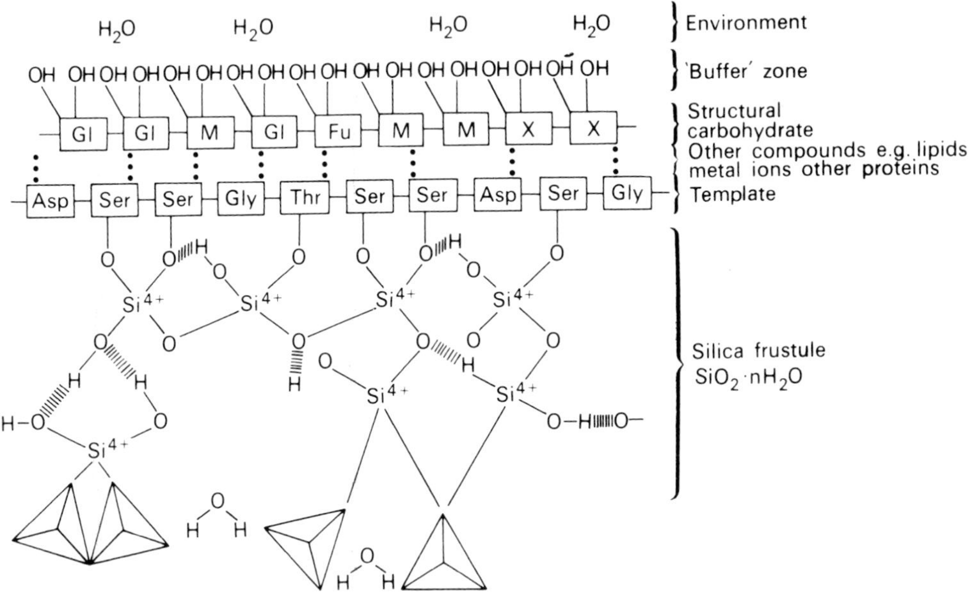

Fig. 4.7. Model of arrangement of layers in the cell wall of diatoms. Layer of polysaccharides with Gl (glucose), M (mannose), Fu (fucose) and X (xylose). Layer of template protein for polycondensation of $Si(OH)_4$ with Ser (serine), Thr (threonine), Asp (aspartic acid) and Gly (glycine). From Hecky *et al.* (1973).

(Lamport & Miller, 1971; Albersheim & Anderson-Prouty, 1975) a promising model is that proposed by Hecky *et al.* (1973) shown in Fig. 4.7.

The template, determining the arrangements of the first layer of $SiO_2.(H_2O)$ would be distinguished by its high serine content and secondly by threonine and glycine. The formation of this layer is perhaps influenced by the sequence of amino acids. Over how many Si–O bonds this template can stretch its influence, considering an interatomic distance of 0·16 nm, and angle of 137–141° for the Si–O bond (Eitel, 1964) and a valve 150 nm thick, remains unknown. Especially important would be in any new findings if the various parts of the frustule (valves, girdle, craticula) and the structural differentiations of the valves (valve face, costae, hyaline fields, ocelli, pseudonoduli, setae, spines and raphes (see Anonymous, 1974) are in fact connected to different sequences of template proteins. The sequencing of these proteins would be a definitive basis for discrimination of species, varieties and forms.

Two other ideas about silica cell walls compared to other cell wall types are outlined below.

The first concerns energetics. Compared to other cell wall materials, such as cellulose, pectin, chitin, mannans, xylans or alginic acid, polycondensation of $Si(OH)_4$ might be energetically more economical by a factor of 5–10 (see page 117) than the use of other material, possibly besides calcite or aragonite (Darley, 1974). Considering that silicates can contribute up to 30% of the dry weight of cells, this could have been of decisive ecological and evolutionary advantage.

The second idea concerns the quality of silica structures (like silica gel) to adsorb various substances from low concentrations (see page 132). Living diatoms are also more effective in uptake from very low concentrations of nutrients (inorganic and organic) than many other phytoplankton classes (Werner, unpublished results). Together these observations lead to the notion, that in addition to the periplasmatic binding proteins (Boos, 1974) and specific transport-ATPases, the impressive inner surface area of the silica frustule of diatoms might also be involved in the competition for nutrients at low concentration.

8 REFERENCES

ALBERSHEIM P. & ANDERSON-PROUTY A.J. (1975) Carbohydrates, proteins, cell surfaces and the biochemistry of pathogenesis. *Ann. Rev. Pl. Physiol.* **26,** 31–52.

ANONYMOUS (1974) Proposals for a standardization of diatom terminology and diagnoses. *Beih. Nova Hedwigia* **52,** 323–54.

ARMSTRONG F.A.J. (151) The determination of silicate in seawater. *J. mar. biol. Ass. U.K.* **30,** 149–60.

AZAM F. (1974) Silicic acid uptake in diatoms studied with [^{68}Ge]-germanic acid as tracer. *Planta (Berl.)* **121,** 205–12.

AZAM F., HEMMINGSEN B.B. & VOLCANI B.E. (1973) Germanium incorporation into the silica of diatom cell walls. *Arch. Microbiol.* **92,** 11–20.

AZAM F., HEMMINGSEN B.B. & VOLCANI B.E. (1974) Role of silicon in diatom metabolism. V. Silicic acid transport and metabolism in the heterotrophic diatom *Nitzschia alba. Arch. Microbiol.* **97,** 103–14.

BADOUR S.S. (1968) Experimental separation of cell division and silica shell formation in *Cyclotella cryptica. Arch. Microbiol.* **62,** 17–33.

BARBER D.A. & SHONE M.G.T. (1966) The absorption of silica from aqueous solutions by plants. *J. exp. Botany* **17,** 569–78.

BOOS W. (1974) Bacterial transport. *Ann. Rev. Biochem.* **43,** 123–46.

BRAARUD T. (1948) On variations in form of *Skeletonema costatum* and their bearing on the supply of silica in cultures of diatoms. *Nytt Mag. Naturvid.* **86,** 31–44.

BUSBY W.F. & LEWIN J. (1967) Silicate uptake and silica shell formation by synchronously dividing cells of the diatom *Navicula pelliculosa* (Bréb) Hilse *J. Phycol.* **3,** 127–31.

CALVERT R. (1930) Diatomaceous earth. *Am. chem. Soc. Monograph Ser.* No. 52, New York, 251 pp.

CALVIN M. (1974) Biopolymere: Entstehung, Chemie und Biologie. *Angew. Chem.* **86,** 111–20.

CARLISLE E. (1972) Silicon: an essential element for the chick. *Science* **178,** 619–21.

CHEN C.H. & LEWIN J. (1969) Silicon as a nutrient element for *Equisetum arvense. Can. J. Bot.* **47,** 125–31.

CLARK S.G. & HOLT P.F. (1957) The interaction of silicic acid with collagen and gelatin monolayers. *Trans. Faraday Soc.* **53,** 1509–15.

CLARK S.G., HOLT P.F. & WENT C.W. (1957) The interaction of silicic acid with insulin, albumin and nylon monolayers. *Trans. Faraday Soc.* **53,** 1500–8.

COOMBS J., DARLEY W.M., HOLM-HANSEN O. & VOLCANI B.E. (1967a) Chemical composition of *Navicula pelliculosa* during silicon-starvation synchrony. *Pl. Physiol.* **42,** 1601–6.

COOMBS J., HALICKI P.J., HOLM-HANSEN O. & VOLCANI B.E. (1967b) Changes in concentration of nucleoside triphosphates in silicon-starvation synchrony of *Navicula pelliculosa* (Bréb.) Hilse. *Expl. Cell Res.* **47,** 315–28.

COOMBS J. & VOLCANI B.E. (1968) Silicon-induced metabolic transients in *Navicula pelliculosa* (Bréb.) Hilse. *Planta (Berl.)* **80,** 264–79.

COOPER L.H.N. (1952) Factors affecting the distribution of silicate in the North Atlantic Ocean and the formation of North Atlantic deep water. *J. mar. biol. Ass. U.K.* **30,** 511–36.

COTTRELL T. (1958) *The strength of chemical bonds.* 2nd. ed., 317 pp. Butterworths, London.

DARLEY W.M. (1969) Silicon requirements for growth and macromolecular synthesis in synchronized cultures of the diatoms *Navicula pelliculosa* (Brébisson) Hilse and *Cylindrotheca fusiformis* Reimann and Lewin. Thesis. 148 pp. University of California, San Diego.

DARLEY W.M. (1974) Silification and calcification. In *Algal Physiology and Biochemistry* (ed. W.D.P. Stewart) pp. 655–75. Blackwell Scientific Publications, Oxford.

DARLEY W.M. & VOLCANI B.E. (1969) Role of silicon in diatom metabolism. A silicon requirement for deoxyribonucleic acid synthesis in the diatom *Cylindrotheca fusiformis* Reimann and Lewin. *Expl. Cell Res.* **58,** 334–42.

DAVIS C.O., HARRISON P.J. & DUGDALE R.C. (1973) Continuous culture of marine diatoms under silicate limitation. I. Synchronized life cycle of *Skeletonema costatum. J. Phycol.* **9,** 175–80.

DENFFER V. D. (1950) Die planktische Massenkultur pennater Grunddiatomeen. *Arch. Mikrobiol.* **14,** 159–202.

DUGDALE R.C. (1967) Nutrient limitation in the sea: dynamics, identification, and significance *Limnol. Oceanogr.* **12,** 685–95.

EINBRODT H.J. (1965) Quantitative and qualitative Untersuchungen über die Staubretention in der menschlichen Lunge. *Beitr. Silikose-Forsch.* **87,** 1–105.

EINSELE W. & GRIM J. (1938) Über den Kieselsäuregehalt planktischer Diatomeen und dessen Bedeutung für einige Fragen ihrer Ökologie. *Z. Bot.* **32,** 545–90.

EITEL W. (1964) *Silicate science.* Vol. I. Silicate structures. Academic Press, New York, 666 pp.

ENGEL W. (1953) Untersuchungen über die Kieselsäureverbindungen im Roggenhalm. *Planta (Berl.)* **41,** 358–90.

ENGEL W. & HOLZAPFEL L. (1960) Kieselsäurenachweis in Gegenwart von Phosphorsäure und Eiweißkomponenten: Analytischer Nachweis und Trennungsmethoden. *Beitr. Silikose-Forsch.* **4,** 67–71.

FAIRBRIDGE R.W. (1966) Marine Sediments. In *The Encyclopedia of Oceanography* (ed. R.W. Fairbridge) pp. 469–74. Reinhold, New York.

FANNING K.A. & SCHINK D.R. (1969) Interaction of marine sediments with dissolved silica. *Limnol. Oceanogr.* **14,** 59–68.

FRIEDEN E. (1972) The chemical elements of life. *Scient. Am.* **227,** 52–64.

FRITZ G. (1963) Zur Chemie der Siliciumverbindungen. Eine Entwicklung auf der Basis pyrochemischer Untersuchungen. *Fortschr. chem. Forsch.* **4,** 459–553.

GEISSLER U. (1970) Die Schalenmerkmale der Diatomeen—Ursachen ihrer Variabilität und Bedeutung für die Taxonomie. *Beih. Nova Hedwigia* **31,** 511–35.

GEITLER L. (1953a) Abhängigkeit der Membranbildung von der Zellteilung bei Diatomeen und differentielle Teilungen im Zusammenhang mit der Bildung der Innenschalen. *Planta (Berl.)* **43,** 75–82.

GEITLER L. (1953b) Das Auftreten zweier obligater metagamer Mitosen ohne Zellteilung während der Bildung der Erstlingsschalen bei den Diatomeen. *Ber. dt. bot. Ges.* **66,** 222–7.

GEITLER L. (1962) Alle Schalenbildungen der Diatomeen treten als Folge von Zell- oder Kernteilungen auf. *Ber. dt. bot. Ges.* **75,** 393–6.

GLOCKLING F. (1969) *The Chemistry of Germanium.* 234 pp. Academic Press, London & New York.

GOERING J.J., NELSON D.M. & CARTER J.A. (1973) Silicic acid uptake by natural population of marine phytoplankton. *Deep Sea Res.* **20,** 777–89.

GUIDOT G. & GERMAIN J.E. (1974) Hydrogenolysis and isomerization of indan on metallic catalysts. *Annls. Chim. (Paris)* **9,** 191–201.

GUILLARD R.R.L., KILHAM P. & JACKSON T.A. (1973) Kinetics of silicon limited growth in the marine diatom *Thalassiosira pseudonema* Hasle and Heimdahl (=*Cyclotella nana.* Hust.). *J. Phycol.* **9,** 233–7.

HARRISON P.J. (1974) Continuous culture of the marine diatom *Skeletonema costatum* (Grev.) Cleve under silicate limitation. Ph.D. dissertation, 140 pp. University of Washington, Seattle.

HECKY R.E., MOPPER K., KILHAM P. & DEGENS E.T. (1973) The amino acids and sugar composition of diatom cell walls. *Mar. Biol.* **19,** 323–31.

HEINEN W. (1963) Siliziumverbindungen. In *Moderne Methoden der Pflanzenanalyse* Vol. 6 (ed. K. Paeck, M.V. Tracey, H.F. Linskens) p. 4–20. Springer-Verlag, Berlin-Göttingen-Heidelberg.

HELMCKE J.G., KRIEGER W. & GERLOFF J. (1963–75) (eds.) *Diatomeenschalen im elektronenmikroskopischen Bild.* Vol. I–X. 1023 Plates. Cramer, Lehre.

HENDEY N.I. (1946) Diatoms without siliceous frustules. *Nature* (Lond.) **158,** 588.

HENDEY N.I. (1959) The structure of the diatom cell wall as revealed by the electron microscope. *J. Quekett micros. club* **5,** 147–75.

HESLOP R.B. & ROBSINSON P.L. (1967) *Inorganic chemistry* 3 ed., 774 pp. Elsevier, Amsterdam, London & New York.

HOLT P.F. & WENT C.W. (1959) The interaction of silicic acid with surface films of pepsin, laminarin and cellulose acetate. *Trans. Farday Soc.* **55,** 1435–50.

HOLZAPFEL L. (1953) Organische Kieselsäureverbindungen XXIII. Untersuchungen zur Dialysierbarkeit organischer Quarz-Kieselsäureverbindungen. *Z. anorg. allg. Chem.* **273,** 186–92.

ILER R.K. (1955) *The Colloid Chemistry of Silica and Silicates.* 324 pp. Cornell University Press, Ithaca N.Y.

ISHIBASHI H. (1936) The effects of silicic acid on the growth of rice plants. *J. Sci. Soil Manure (Tokyo)* **10,** 244–56.

JOLLES A. & NEURATH FR. (1898) Eine colorimetrische Methode zur Bestimmung der Kieselsäure im Wasser. *Z. angew. Chem.* **11,** 315–6.

JØRGENSEN E.G. (1953) Silicate assimilation by diatoms. *Physiologia Pl.* **6,** 301–15.

JØRGENSEN E.G. (1955) Variations in the silica content of diatoms. *Physiologia Pl.* **8,** 840–5.

KAMATANI A. (1969) Regeneration of inorganic nutrients from diatom decomposition. *J. oceanogr. Soc. Japan* **25,** 63–74.

KAMATANI A. (1971) Physical and chemical characteristics of biogenous silica. *Mar. Biol.* **8,** 89–95.

KATCHALSKY A. (1973) Prebiotic synthesis of biopolymers on inorganic templates. *Naturwissenschaften* **60,** 215–20.

KAUTSKY H., PFLEGER H., REISE R. & VOGELL W. (1962) Lösungsgleichgewicht, Lösungsvorgang und Kolloidstruktur der Kieselsäure. Lepidoide VII, *Z. Naturf.* **17b,** 491–9.

KESTER D.R., DUEDALL I.W., CONNORS D.N. & PYTKOWICZ R.M. (1967) Preparation of artificial sea water. *Limnol. Oceanogr.* **12,** 176–8.

KILHAM P. (1971) A hypothesis concerning silica and the fresh water planktonic diatoms. *Limnol. Oceanogr.* **16,** 10–8.

KING E.J. (1939) The biochemistry of silicic acid VIII. The determination of silica. *Biochem. J.* **33,** 944–54.

KOLBE R.W. (1957) Diatoms from equatorial Indian ocean cores. *Rep. Swed. deep Sea Exped.* **9,** Nr. 1, 1–50.

KRAUSKOPF K.B. (1956) Dissolution and precipitation of silica at low temperatures. *Geochim. cosmochim. Acta* **10,** 1–26.

KREY J. (1942) Nährstoff- und Chlorophylluntersuchungen der Kieler Förde. *Kieler Meeresforsch.* **4,** 1–17.

KÜSTER-WINKELMANN G. (1938) Über die Doppelschalen der Diatomeen. *Arch. Protistenk.* **91,** 237–66.

KUHL A. (1968) Phosphate metabolism of green algae. In *Algae, Man and the Environment* (ed. D.F. Jackson) pp. 37–52. Syracuse University Press, Syracuse.

LAMPORT D.T.A. & MILLER D.H. (1971) Hydroxyproline arabinosides in the plant kingdom. *Pl. Physiol.* (Wash.) **48,** 454–6.

LANGER K. & FLÖRKE O.W. (1974) Near infrared absorption spectra (4 000–9 000 cm^{-1}) of opals and the role of 'water' in these $SiO_2 \cdot n\ H_2O$ minerals. *Fortschr. Miner.* **52,** 17–51.

LAUWERS A.M. & HEINEN W. (1974) Bio-degradation and utilization of silica and quartz. *Arch. Microbiol.* **95,** 67–78.

LEHNINGER A.L. (1965) *Bioenergetics, the molecular basis of biological energy transformations,* 255 pp. W.A. Benjamin, New York & Amsterdam.

LEVIN E.M., ROBBINS C.R. & MC MURDIE H.F. (1961) Phase diagrams for ceramists. Am. Ceram. Soc., Columbus/Ohio.

LEWIN J.C. (1954) Silicon metabolism in diatoms. I. Evidence for the role of reduced sulfur compounds in silicon utilization. *J. gen. Physiol.* **37,** 589–99.

LEWIN J.C. (1955a) Silicon metabolism in diatoms. II. Sources of silicon for growth of *Navicula pelliculosa. Pl. Physiol.* **30,** 129–34.

LEWIN J.C. (1955b) Silicon metabolism in diatoms. III. Respiration and silicon uptake in *Navicula pelliculosa. J. gen. Physiol.* **39,** 1–10.

LEWIN J.C. (1957) Silicon metabolism in diatoms. IV. Growth and frustule formation in *Navicula pelliculosa. Can. J. Microbiol.* **3,** 427–33.

LEWIN J.C. (1961) The dissolution of silica from diatom walls. *Geochim. Cosmochim. Acta* **21,** 182–98.

LEWIN J.C. (1965) The boron requirement of a marine diatom. *Naturwissenschaften* **52,** 70.

LEWIN J.C. (1966a) Boron as a growth requirement for diatoms. *J. Phycol.* **2,** 160–3.

LEWIN J.C. (1966b) Physiological studies of the boron requirement of the diatom *Cylindrotheca fusiformis* Reimann and Lewin. *J. expl. Bot.* **17,** 473–9.

LEWIN J.C. (1966c) Silicon metabolism in diatoms. V. Germanium-dioxide, a specific inhibitor of diatom growth. *Phycologia* **6,** 1–12.

LEWIN J. & CHEN C. (1968) Silicon metabolism in diatom VI. Silicic acid uptake by a colorless marine diatom, *Nitzschia alba* Lewin and Lewin. *J. Phycol.* **4,** 161–6.

LEWIN J.C. & REIMANN B.E.F. (1969) Silicon and plant growth. *Ann. Rev. Pl. Physiol.* **20,** 289–304.

LISITZIN A.P. (1971) Distribution of siliceous microfossils in suspension and in bottom sediments. In *The Micropaleontology of Oceans* (ed. B.M. Funnel & W.R. Riedel) pp. 173–96. Cambridge University Press, Cambridge.

LUND J.W.G. (1950) Studies on *Asterionella formosa* Hass. II. Nutrient depletion and the spring maximum. Part III. *J. Ecol.* **38,** 15–35.

MACKENZIE F.T., GARRELS R.M., BRICKER O.P. & BICKLEY F. (1967) Silica in sea water: control by silica minerals. *Science* **155,** 1404–5.

MANTON I., KOWALLIK K. & STOSCH V' H.A. (1970) Observations on the fine structure and development of the spindle at mitosis and meisos in a marine centric diatom (*Lithodesmium undulatum*) IV. The second meiotic division and conclusion. *J. Cell Sci.* **7,** 407–43.

MEHARD C.W., SULLIVAN C.W., AZAM F. & VOLCANI B.E. (1974) Role of silicon in diatom metabolism. IV. Subcellular localization of silicon and germanium in *Nitzschia alba* and *Cylindrotheca fusiformis*. *Physiologia Pl.* **30,** 265–72.

MOSEBACH R. (1966) Anlösung implantierter Quarz-Plättchen durch Körperflüssigkeiten und Gewebe von Ratten? *Z. Naturf.* **21b,** 417–20.

MULLIN J.B. & RILEY M.P. (1955) The colorimetric determination of silicate with special reference to sea and natural waters. *Analytica chim. Acta.* **12,** 162–76.

NAKAJIMA T. & VOLCANI B.E. (1969). 3,4-Dihydroxyproline: A new amino acid in diatom cell walls. *Science* **164,** 1 400–1.

NAKAJIMA T. & VOLCANI B.E. (1970) *N*-Trimethyl-α-hydroxylysinephosphate and its nonphosphorylated compound in diatom cell walls. *Biochem. Biophys. Res. Comm.* **39,** 28–33.

NOLL W. (1956) Charakteristika der Siloxanbildung in polymeren Anionen (Silikaten) und polymeren Molekülen (Siliconen). *Fortschr. Mineral.* **34,** 52–84.

OEY J.L. & SCHNEPF E. (1970) Über die Auslösung der Valvenbildung bei der Diatomee *Cyclotella cryptica*. *Arch. Mikrobiol.* **71,** 199–213.

PAASCHE E. (1973a) Silicon and the ecology of marine plankton diatoms. I. *Thalassiosira pseudonana (Cyclotella nana)* growth in a chemostat with silicate as limiting nutrient. *Mar. Biol.* **19,** 117–26.

PAASCHE E. (1973b) Silicon and the ecology of marine plankton diatoms. II. Silicate uptake kinetics in five diatom species. *Mar. Biol.* **19,** 262–9.

PAASCHE E. (1973c) The influence of cell size on growth rate, silica content and some other properties of four marine diatom species. *Norw. J. Bot.* **20,** 197–204.

PAASCHE E., JOHANNSON S. & EVENSEN D.L. (1975) An effect of osmotic pressure on the valve morphology of the diatom *Skeletonema subsalum* (A. Cleve) Bethg. *Phycologia* **14,** 205–11.

PAECHT-HOROWITZ M. (1973) Die Entstehung des Lebens. *Angew. Chem.* **85,** 422–30.

PAECHT-HOROWITZ M., BERGER J. & KATCHALSKY A. (1970) Prebiotic synthesis of polypeptides by heterogeneous polykondensation of amino-acid adenylates. *Nature* **228,** 636–9.

PYTKOWICZ R.M. & KESTER D.R. (1971) The physical chemistry of sea water. *Oceanogr. Mar. Biol. Ann. Rev.* **9,** 11–60.

REIMANN B. (1960) Bildung, Bau und Zusammenhang der Bacillariophyceenschalen. *Nova Hedwigia* **2,** 349–73.

RICHTER O. (1906) Zur Physiologie der Diatomeen. I. Mitteilung. *Sber. Akad. Wiss. Wien, math. nat. Klasse* **115,** 27–119.

ROSMANITH J. (1975) Verlauf von Elimination aus der Lunge und Leber von Eisen, Zink, Kieselsäure und Aluminium nach wiederholter intratrachealer Kohlenstaubapplikation (Tierversuche) *Beitr. Silikose-Forsch. (Pneumokon.)* **27,** 47–61.

ROSTOW E.G. (1973) Silicon. In *Comprehensive Inorganic Chemistry* (eds J.C. Bailar, H.J. Emeleus, R. Nyholm & A.F. Trotman-Dickenson) Vol. III, pp. 1 323–1 467. Pergamon Press, Oxford.

ROTH R. (1975) Radiodünnschichtchromatographische Untersuchungen an Extrakten aus *Cyclotella cryptica* mit ^{68}Ge und ^{71}Ge. Thesis. University of Marburg/L.

ROUND F.E. (1972) The formation of girdle, intercalary bands and septa in diatoms. *Nova Hedwigia* **23,** 449–63.

SCHLENTRICH A. (1965) Tierexperimentelle Untersuchungen zur Frage der Löslichkeit von Bergkristall-Implantaten in tierischem Gewebe. *Naturwissenschaften* **52,** 562.

SEELMAN-EGGEBERT W., PFENNIG G. & MÜNZEL H. (1974) Chart of Nuclides. Ges. f. Kernforschung, Karlsruhe.

SEIFERT H. (1959) Über die orientierte Aufwachsung von Ologopeptiden auf Quartzoberflächen. *Naturwissenschaften* **46,** 261.

SEIFERT H. (1961a) Epitaxie von biologisch bedeutsamen Stoffen auf Quarzoberflächen. *Naturwissenschaften,* **48,** 713–4.

SEIFERT H. (1961b) Biokristallographie, ein junges Grenzgebiet I/II. *Naturw. Rdsch.* **14,** 7–11, 60–5.

SENCZUK W. & MLYNARCZYK W. (1974) Estimation of biological activity of some industrial dusts. I. Biochemical study of pneumoconiosis inducing properties of industrial dusts. *Bromatol. Chem. Toksykol.* **7,** 181–7.

SIEVERS R., BECK K.C. & BERNER R.A. (1965) Composition of interstitial waters of modern sediments. *J. Geol.* **73,** 39–73.

SORRENTINO F.A. & PAUL J. (1970) Simultaneous determination of arsenic, germanium, phosphorus and silicon. *Microchem. J.* **15,** 446–51.

STEGEMANN H. & FITZEK J. (1954) Analyse von Lungenstäuben. *Beitr. Silikose-Forsch.* **31,** 30–40.

STOSCH V. H.A. (1942) Form und Formwechsel der Diatomee *Achnanthes longipes* in Abhängigkeit von der Ernährung, mit besonderer Berücksichtigung der Spurenstoffe. *Ber. dt. bot. Ges.* **60,** 2–16.

STOSCH V. H.A. (1965) Manipulierung der Zellgröße von Diatomeen im Experiment. *Phycologia* **5,** 21–44.

STOSCH V. H.A. (1974) An amended terminology of the diatom girdle. *Beih. Nova Hedwigia* **53,** 1–28.

STOSCH V. H.A. & KOWALLIK K. (1969) Der von L. Geitler aufgestellte Satz über die Notwendigkeit einer Mitose für die Schalenbildung von Diatomeen. Beobachtungen über die Reichweite und Überlegungen zu seiner zellmechanischen Bedeutung. *Öst. bot. Z.* **116,** 454–74.

STRICKLAND J.D.H. & PARSONS T.R. (1968) A practical handbook of seawater analysis. *Bull. Fish. Res. Bd. Can.* No. 167, 311 pp.

SULLIVAN C.W. & VOLCANI B.E. (1973a) Role of silicon in diatom metabolism. II.

Endogenous nucleoside triphosphate pools during silicic acid starvation of synchronized *Cylindrotheca fusiformis*. *Biochim. Biophys. Acta* **308,** 205–11.

SULLIVAN C.W. & VOLCANI B.E. (1973b) Role of silicon in diatom metabolism. III. The effects of silicic acid on DNA polymerase, TMP kinase and DNA synthesis in *Cylindrotheca fusiformis*. *Biochim. Biophys. Acta* **308,** 212–299.

SULLIVAN C.W. & VOLCANI B.E. (1974) Isloation and characterization of plasma and smooth membranes of the marine diatom *Nitzschia alba*. *Arch. Biochem. Biophys.* **163,** 29–45.

TESSENOW U. (1966) Untersuchungen über den Kieselsäurehaushalt der Binnengewässer. *Arch. Hydrobiol. Suppl.* **32,** 1–136.

TEICHNER S. & PERNOUX E. (1951) The texture and surface area of kieselguhrs after various treatments. *Clay Miner. Bull.,* **1,** 145–50.

THOMAS W.H. (1966) Effects of temperature and illuminance on cell division rate of three species of tropical oceanic phytoplankton. *J. Phycol.* **2,** 17–22.

T̊UMA J. (1962) Optimum conditions for the colorimetric microdetermination of silicon. *Mikrochim. Acta* **3,** 513–23.

UMEMURA Y., NISHIDA J., AKAZAWA T. & URITANI J. (1961) Effect of silicon compounds on plant enzymes involved in phosphorous metabolism. *Arch. Biochem. Biophys.* **92,** 392–8.

WACKER A. (1958) Zur Entstehung des Lebens auf der Erde. *Angew. Chem.* **70,** 519–26.

WAGNER F. (1940) Die Bedeutung der Kieselsäure für das Wachstum einiger Kulturpflanzen, ihren Nährstoffhaushalt und ihre Anfälligkeit gegen echte Mehltaupilze. *Phytopath. Z.* **12,** 427–79.

WEISS A. & HOFMANN U. (1951) Batavit. *Z. Naturf.* **6b,** 405–9.

WEISS A., KLOSTERKÖTTER W. & STRECKER F.J. (1961a) Über die Bildung des quellungsfähigen Eisenphosphatosilikates der Silikoselungen aus Quarz im Tierversuch. *Naturwissenschaften* **48,** 163–4.

WEISS A., REIFF G. & HOFMANN G. (1957) Untersuchungen an Fremdstäuben aus Lungen und an wasserbeständigen Kieselsäureresten. In *Die Staublungenerkrankungen* (eds K.W. Jötten & H. Gärtner) pp. 79–91. Steinkopf Verlag, Darmstadt.

WEISS A., REIFF G. & WEISS A. (1961b) Zur Kenntnis wasserbeständiger Kieselsäurerester. *Z. anorg. allg. Chem.* **311,** 151–79.

WELLER W. (1973) Prüfung der Wirkung von D-Penicillamin auf eine experimentelle Silikose im Intraperitonealtest. *Beitr. Silikose-Forsch. (Pneumokon.)* **25,** 39–63.

WERNER D. (1965) Die Kieselsäure im Stoffwechsel von *Thalassiosira fluviatilis* Hust. Dissertation Universitat Göttingen.

WERNER D. (1966) Die Kieselsäure im Stoffwechsel von *Cyclotella cryptica,* Reimann, Lewin und Guillard *Arch. Mikrobiol.* **55,** 278–308.

WERNER D. (1967a) Hemmung der Chlorophyllsynthese und der $NADP^+$—abhängigen Glycerinaldehyd-3-Phosphat-Dehydrogenase durch Germaniumsäure bei *Cyclotella cryptica*. *Arch. Mikrobiol.* **57,** 51–60.

WERNER D. (1967b) Untersuchungen über die Rolle der Kieselsäure in der Entwicklung höherer Pflanzen. I. Analyse der Hemmung durch Germanium saüre. *Planta (Berl.)* **76,** 25–36.

WERNER D. (1968a) Stoffwechselregulation durch den Zellwandbaustein Kieselsäure: Poolgrößenänderungen von α-Ketoglutarsäure, Aminosäuren und Nukleosidphosphaten. *Z. Naturforsch.* **23,** 268–72.

WERNER D. (1968b) Beiträge zur Physiologie und Biochemie der Kieselsäure. *Ber. dt. bot. Ges.* **81,** 425–9.

WERNER D. (1969) Silicoborate als erste nicht C-haltige Wachstumsfaktoren. *Arch. Mikrobiol.* **65,** 258–74.

WERNER D. (1970a) Bestimmung niederpolymerer Kieselsäuren aus *Cyclotella cryptica* durch Atomabsorptionsspektroskopie für Silizium, *Hoppe-Seyler's Z. physiol. Chem.* **351,** 134–5.

WERNER D. (1970b) Productivity studies on diatom cultures. *Helgoländer wiss. Meeresunters.* **20,** 97–103.

WERNER D. (1970c) Isolierung einer Gallertkolonien bildenden Mutante von *Cyclotella cryptica* (Diatomeae nach Temperaturschockbehandlung. *Int. Revue ges. Hydrobiol.* **55,** 403–7.

WERNER D. (1971a) Der Entwicklungszyklus mit Sexualphase bei der marinen Diatomee *Coscinodiscus asteromphalus.* I. Kultur und Synchronisation von Entwicklungsstadien. *Arch. Mikrobiol.* **80,** 43–9.

WERNER D. (1971b) Der Entwicklungszyklus mit Sexualphase bei der marinen Diatomee *Coscionodiscus asteromphalus.* II. Oberflächenabhängige Differenzierung während der vegetativen Zellverkleinerung. *Arch. Mikrobiol.* **80,** 115–33.

WERNER D. (1971c) Der Entwicklungszyklus mit Sexualphase bei der marinen Diatomee *Coscinodiscus asterophalus.* III. Differenzierung und Spermatogenese. *Arch. Mikrobiol.* **80,** 134–46.

WERNER D., PAWLITZ H.D. & ROTH R. (1975) The separation of ^{32}silicon from contaminating ^{3}H and ^{60}Co by incorporation into diatoms. *Z. Naturf.* **30c,** 423–4.

WERNER D. & PETERSEN M. (1973) Traceruntersuchungen mit ^{71}Germanium im Silikatstoffwechsel von Diatomeen. *Z. Pfl. Physiol.* **70,** 54–65.

WERNER D. & PIRSON A. (1967) Über reversible Speicherung von Kieselsäure in *Cyclotella cryptica. Arch. Mikrobiol.* **57,** 43–50.

WERNER D. & ROTH R. (1977) Bindung von Germanat an Polyserin in zellfreien Systemen aus *Cyclotella cryptica* (in preparation for *Z. Pfl. Physiol.*).

WERNER D. & STANGIER E. (1976) Silica and temperature dependent colony size of *Bellerochea maleus f. biangulata. (Centrales, Diatomeae) Phycologia* **15,** 73–7.

CHAPTER 5

PHOTOSYNTHESIS

ERIK G. JØRGENSEN

Plantephysiologisk Institut,
University of Copenhagen,
Øster Farimagsgade 2 A, Denmark

1 INTRODUCTION

It is not the purpose of this chapter to give a general account of photosynthesis, but to discuss in principle those aspects in which diatoms are known to differ from other algae, or for which diatoms may be studied as convenient models.

2 PHOTOSYNTHETIC APPARATUS

The photosynthesis of diatoms takes place in the chloroplasts, which differ greatly in shape and size in different diatom species. Some species possess numerous chloroplasts in the form of granules or discs. In other species we find one or more plates (see Mereschkowsky, 1903). In spite of the wide variation in morphology, current evidence shows the ultrastructure of the chloroplasts to be very uniform. The chloroplasts are nearly always traversed by thylakoid bands, where three or more thylakoids are stacked together (Gibbs, 1970). Grana are always absent while pyrenoids are present in most diatom species.

3 PIGMENTS

The chloroplasts contain the following pigments: chlorophyll *a* and *c*, β-carotene, fucoxanthin, diadinoxanthin and diatoxanthin (Hager & Stransky, 1970). The chemical structure of chlorophyll *c* has been recently solved (Dougherty *et al.* 1970) and it has been shown to exist in two forms, i.e. c_1 and c_2, in diatoms, Phaeophyceae and some other algae (Strain *et al.* 1971, Jeffrey, 1972). The chemical structure of fucoxanthin has also recently been elucidated (Jensen, A., 1966). According to Stransky & Hager (1970) the light-induced xanthophyll interconversions take place between diadinoxanthin and diatoxanthin in diatoms. The absorption of *in vivo* pigments of *Phaeodactylum* is shown in Fig. 5.1.

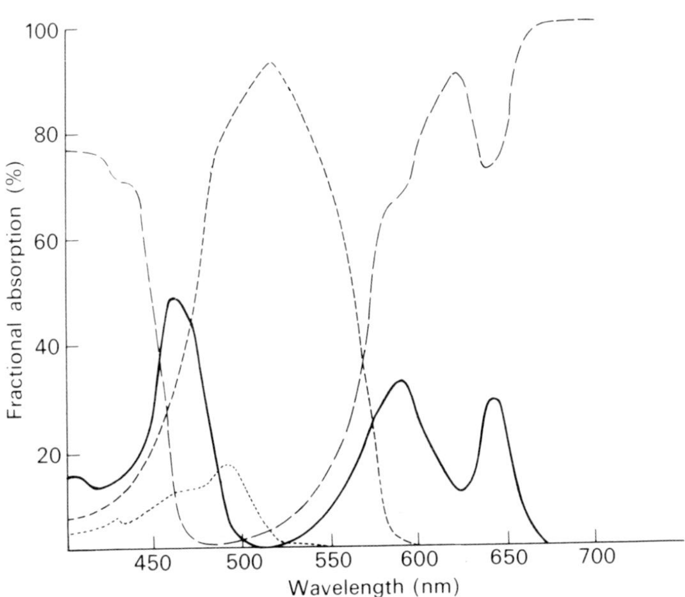

Fig. 5.1. Fractional absorption of *in vivo* pigments of *Phaeodactylum tricornutum*. Chlorophyll *a* (———), chlorophyll *c* (———), fucoxanthin (– – –), and other carotenoids, principally β-carotene (· · · · ·). The ordinate has dimensions of percent of total absorbed quanta absorbed by each pigment. After Mann & Myers, 1968.

4 FUNCTION OF THE PIGMENTS

Cultures of algae have been used in several studies concerned with biophysical and biochemical analyses of photosynthesis; however, green algae are usually preferred and diatoms have been used only occasionally. The favourite diatom species used is *Phaeodactylum tricornutum*. This species was isolated by Allen & Nelson (1910) in Plymouth and referred to *Nitzschia closterium f. minutissima* but was reclassified by Lewin (1958) as *Phaeodactylum tricornutum*.

The principal pigments of diatoms are as in brown algae, chlorophyll *a*, chlorophyll *c* and fucoxanthin. With regard to the pigments participating in the xanthophyll cycle the two groups of plants are quite different. While diatoms

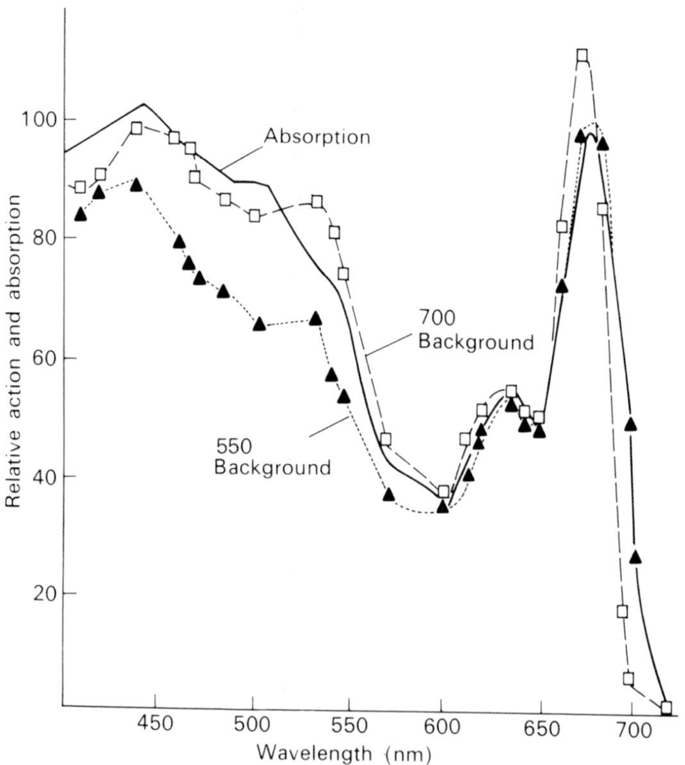

Fig. 5.2. Action spectra of *Phaeodactylum tricornutum* obtained with constant background of 200 rate units of 550 nm (▲) or 700 nm (□). Added light signals at indicated wavelengths were adjusted in intensity to give about 50 rate units below 680 nm and about 30 rate units above 680 nm. Absorption was normalized to 100 at 674 nm. After Mann & Myers, 1968.

contain diatoxanthin and diadinoxanthin the following three xanthophylls are present in brown algae: violaxanthin, antheraxanthin and zeaxanthin (Hager & Stransky, 1970). The precise function of these xanthophyll cycles in the plants is not fully understood.

On the basis of qualitative experiments on the photosynthesis of algae in coloured light, using motile aerostatic bacteria as oxygen indicators, Engelmann (1883, 1884) proposed as early as the 1880s the involvement in the photosynthetic process of pigments other than chlorophyll *a*. He concluded from the experiments that the fucoxanthin of brown algae was as effective as chlorophyll in sensitizing photosynthesis.

Dutton & Manning (1941) concluded from measurements of the quantum yield of photosynthesis in *Phaeodactylum* that fucoxanthin, which absorbs light of longer wavelengths than do most carotenoids, was as effective as chlorophyll *a*.

Tanada (1951) working with *Navicula minima* confirmed the activity of fucoxanthin, and also demonstrated that chlorophyll *c* is an effective sensitizer of photosynthesis was found to be practically constant from 680 nm down to absorption of fucoxanthin differs markedly from that in solution, being displaced to longer wavelengths by an estimated 40 nm. The quantum yield of photosynthesis was found to be practically constant from 680 nm down to 530 nm and was followed by a depression at about 480 nm just in the region of the absorption by the residual carotenoids. It was therefore concluded that whereas fucoxanthin is about as effective as the chlorophylls in sensitizing photosynthesis, other carotenoids appeared not to be involved in this process. This view is further confirmed by comparison of absorption and action spectra of a diatom (Fig. 5.2).

Accessory pigments such as chlorophyll *c* and fucoxanthin function by transferring their absorbed energy to chlorophyll *a* before this energy is available for CO_2 reduction. This is shown most clearly by fluorescence studies of living cells. For diatoms such studies have been carried out by Dutton *et al.* (1943), and Wassink & Kersten (1944, 1946).

5 PIGMENTS OF THE TWO PHOTOSYSTEMS

Analyses of the photosynthetic apparatus during the last two decades suggest that *in vivo* chlorophyll *a* exists in at least four chlorophyll forms with different absorption maxima (Brown, 1972) and that these different forms are split between two pigment systems, one of which (photosystem II) is responsible for water decomposition and oxygen production, and the other (photosystem I) for production of high energy compounds needed for the reduction of CO_2.

Measurements of fluorescence emission spectra and fluorescence action spectra provide some information concerning the organization of the two photosystems. Based on measurements of this kind, Goedheer (1970) suggested that fucoxanthin and chlorophyll *c* in diatoms and brown algae participate in photosystem II and that there is a close connection between the β-carotene and the chlorophyll *a* of the system I of photosynthesis.

Diatoms and brown algae differ from other plants in their fluorescence emission spectra. While chlorophyll *a* has a fluorescence maximum at 685 nm in nearly all plants measured at room temperature, this maximum is found at 681 nm in diatoms and brown algae. At $-196°C$ the difference is much more pronounced. At this temperature we find three fluorescence maxima in green, blue-green and red algae and in chloroplasts of higher plants: 685, 695 and 715–735 nm, while only two maxima are found in diatoms and brown algae: 690–695 and 705–715 nm (Sugahara *et al.* 1971).

Goedheer (1970) found a very sharp absorption maximum at 669 to 670 nm in brown algae, when measured at $-196°C$, which is not found in the chlorophyll absorption bands of red, blue-green and green algae. As a chlorophyll *a* form with an absorption maximum at 670 nm is generally assumed to be connected with system II of photosynthesis, while a chlorophyll *a* with a maximum at about 680 nm is assumed to be connected with system I, the absorption spectrum indicates that in brown algae and diatoms, in contrast to most other photosynthetic organisms, a predominant fraction of chlorophyll *a* is connected with system II.

Both in brown algae collected late in the season and in old cultures of *Phaeodactylum* a higher ratio 680/670 nm absorption at $-196°C$ and correspondingly a higher fluorescence at 715 nm were found (Goedheer, 1970; French, 1955, 1967). This implies an increase in ageing cells in system I chlorophyll relative to that of system II.

Phaeodactylum was further investigated by Brown (1967). She found that the fluorescence emission spectrum measured at room temperature at about 710 nm reported by French (1955) is stronger in cells grown in light of low intensity. A light intensity of 2000 lux produced the highest value.

Similar observations were made with the diatom *Detonula*. Jupin & Giraud (1971) and Jupin (1973a,b) compared cultures of this diatom grown in white light and in red light, respectively. They found that cultures from red light, as for *Phaeodactylum* at low light intensity, were enriched in chlorophyll F 705. This suggests a shift in the ratio between photosystem I and II in favour of the former.

Diatoms, like other plants, show reduced quantum efficiency in the spectral region beyond 680 nm. If light of shorter wavelengths is supplied simultaneously, the photosynthesis by light of wavelengths greater than 680 nm is enhanced. Studies of this 'second Emerson effect' in *Navicula minima* gave evidence for the existence of at least three forms of chlorophyll *a* in this diatom (Emerson & Rabinowitch, 1960; Govindjee & Rabinowitch,

1960). Mann & Myers (1968) have studied the photosynthetic enhancement in *Phaeodactylum*.

There appear to be no studies of photosynthesis by whole diatom chloroplasts. However, some studies using subchloroplast fragments provide certain information concerning the construction of the two photosystems. Ogawa *et al.* (1968) separated photosystem I and II chloroplast particles from *Phaeodactylum* by sonication, differential centrifugation, sodium dodecyl sulphate treatment and electrophoresis. They found that the particle II fraction was more enriched with chlorophyll *c*, fucoxanthin, diatoxanthin and diadinoxanthin while the particle I fraction was most enriched with β-carotene.

In a few algae, including *Ochromonas* and *Phaeodactylum*, some of the chlorophyll forms are destroyed by rupture of the cells. The shoulder on the absorption spectrum at 685 nm, the small band near 705 nm, and the large fluorescence emission peak at 712 nm all disappeared when the cells were ruptured (Brown, 1969).

Apart from the photosynthetic pigments of the photosystems of diatoms, their components have been very little studied. No studies of cytochromes, quinones and other components of the photosynthetic electron-transfer chain are available. Nor do mutant strains of diatoms appear to have been used to solve the problems of the electron-transfer chain as in the case of green algae (Levine, 1969). Nor has the size of the photosynthetic unit in diatoms been investigated.

6 DARK REACTIONS OF PHOTOSYNTHESIS

During the light reactions of photosynthesis the light energy is converted to chemical energy as the 'assimilatory-power'. This chemical energy is used during the so-called 'dark reactions' to reduce CO_2 to carbohydrate. Little attention has been paid to these processes in diatoms.

We know at present with certainty that free CO_2, and also bicarbonate in certain species such as *Coccolithus huxleyi* and *Scenedesmus quadricauda*, can be utilized as a direct carbon source by the photosynthesis of aquatic plants (Steemann Nielsen, 1966; Raven, 1970). There is no information about the carbon sources in diatoms. Steemann Nielsen (1966) gives some evidence for the utilization of free CO_2 in *Skeletonema costatum*. Hood & Park (1962) claimed that their experiments demonstrated the utilization of bicarbonate in a series of marine algal species—one of which was *Nitzschia closterium*, but both Steemann Nielsen (1963) and Watt & Paasche (1963) were able to demonstrate that their experiments do not provide a basis for such an assumption.

The carbon dioxide fixation of diatoms has never been examined in detail. Barker (1935) examined the photosynthetic quotients of *Nitzschia palea* and *(N. closterium=) Phaeodactylum* and Wassink & Kersten (1944) examined *Nitzschia dissipata*. The authors found values for the assimilatory quotient CO_2/O_2 of about 0·9 which was almost identical to the quotient observed in *Chlorella*. This does not suggest any unusual features in the photosynthesis of diatoms. For many years lipids were believed to constitute the only food reserve in diatoms. However, investigations of v. Stosch (1951), and Collyer & Fogg (1955) suggest that diatoms may contain substantial amounts of polysaccharide and there appears to be no compelling reason for supposing that their photosynthetic mechanism differs in any fundamental way from that in other algae and higher plants.

v. Denffer (1950) examined the growth curve for a culture of *Nitzschia palea* and compared it with the lipid formation in the diatom cells. The lipid formation occurred mainly in the stationary growth phase and hence is due to unfavourable conditions. In fixation experiments with ^{14}C and the diatom *Navicula pelliculosa* Fogg (1956) showed that in cells which had ceased to grow due to nitrogen deficiency, a major portion of the carbon fixed was incorporated in fats. Werner (1966, 1970) found in experiments with *Cyclotella cryptica* that nitrogen-deficient cells accumulated lipid over a period of many days while under conditions of silicic acid deficiency, immediately fat storage occurred. With regard to fat accumulation diatoms therefore do not differ markedly from algae like *Chlorella* (see Collyer & Fogg, 1955).

7 RATE OF PHOTOSYNTHESIS IN DIATOMS

It is very difficult to compare the rates of photosynthesis of different diatoms. This is due to the different ways of measuring the rate of photosynthesis and the many different ways of expressing the results. One can determine the rate of photosynthesis as the amount of oxygen produced or as the amount of CO_2 fixed, and the two different determinations cannot be compared with any certainty. Much more difficult, or impossible, is to compare the parameters used by different authors to express the unit of diatom cells. One can express the photosynthesis per volume of culture, per number of cells, per dry weight of cells, per unit of chlorophyll, per packed cell volume (PCV) etc. As different authors usually provide only few details of the parameters of the cells they have worked with, it is often impossible to compare the results given in different papers. When comparing the rate of photosynthesis of diatoms it is also important to consider the effect of environmental factors on the rate of photosynthesis. Not only are the environmental factors during the measurements of the rate of photosynthesis important, but also the conditions under which the diatom cells were grown prior to the experiment.

7.1 *Rate of photosynthesis and the cell cycle*

Vegetative growth of most centric diatoms affects their size in two ways. Firstly, the length of the pervalvar axis increases to almost twice its value from one cell division to another, and secondly the diameter decreases by a constant decrement at each cell division. During both types of changes in cell size the rate of photosynthesis is affected. The changes taking place during the growth of a single diatom cell can be studied in synchronized cultures. In a synchronized culture of *Skeletonema costatum* growing for alternating periods of 12 hours in the light (3 klux) and 12 hours in the dark Jørgensen (1966) found that the light-saturated photosynthesis was lowest in cells which had just divided and highest in fully grown cells, differing from one extreme to the other by a factor of two.

With regard to the second case of changes in cell size, the decrease in cell diameter during several cell generations results in changes in the rate of photosynthesis. In a culture of *Skeletonema* where the cell volume decreased from 200 μ^3 to 50 μ^3 and the chlorophyll *a* content in $\mu g/10^6$ cells decreased from 1·0 to 0·44, the rate of light-saturated photosynthesis decreased at the same time from 6·0–2·0 μg C $hour^{-1}.10^6$ cells (Jørgensen, 1969).

The photosynthetic activity will also change with the growth phase of the cells. Thus Ebata & Fujita (1971) show for batch cultures of *Phaeodactylum tricornutum, Nitzschia closterium* and *Chaetoceros* sp. that the rate of saturated photosynthesis is highest during the log phase of growth and that a rapid decrease occurs in the interphase from the log phase to the stationary phase. The low concentration of the phosphorus source was suggested as the main cause for the change in photosynthetic activity.

8 ENVIRONMENTAL FACTORS INFLUENCING PHOTOSYNTHESIS

8.1 *Light intensity*

The rate of photosynthesis of diatoms has the same relation to light intensity as in other algae and higher plants. This relation can be seen in light intensity–photosynthesis curves (Fig. 5.3).

8.2 *Adaptation to different light intensities*

The light intensity at which the diatoms have been cultured affects both the morphology and the physiology of the cells. The effect on the morphology has not been studied to any great extent. Brown & Richardson (1968) showed that the cell volume of *Nitzschia closterium* depends on the light intensity, the cell volume being greatest at light intensities below 2 klux. In the same species the

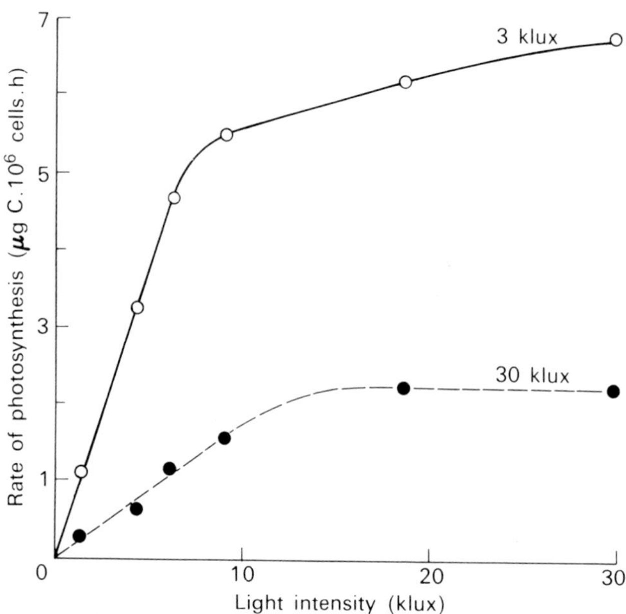

Fig. 5.3. The '*Chlorella*-type' of adaptation. Light intensity–photosynthesis curves for *Chlamydomonas moevusii* grown at 3 klux and 30 klux, respectively. After Jørgensen, 1969.

same authors found the size of the chloroplasts to decrease with increasing light intensities.

The light intensity also seems to affect the number of thylakoids in the chloroplasts in such a way that the highest number of thylakoids is found in cells cultured at the lowest light intensity as it is found in *Chlorella* and some higher plants (Treharne *et al.* 1964 and Goodchild *et al.* 1972). The effect of light intensity on the ultrastructure of chloroplasts in diatom species has been studied in *Detonula* only by Jupin (1973c). Studies by Myers & Graham (1971) and Sheridan (1972) in *Chlorella* species show further that the light intensity in this alga affects the size of the photosynthetic unit with the highest number of chlorophyll molecules per unit at low light intensity and the number of chlorophyll molecules per unit decreasing with increasing light intensity. This problem, however, has not been studied in diatoms.

The changes in the physiology of the cells which take place when the cells adapt from one light intensity to another is best illustrated by the changes in the light intensity–photosynthesis curves. The three photosynthetic variables P_m, I_k and K^1 first proposed by Talling (1957) give a good description of the

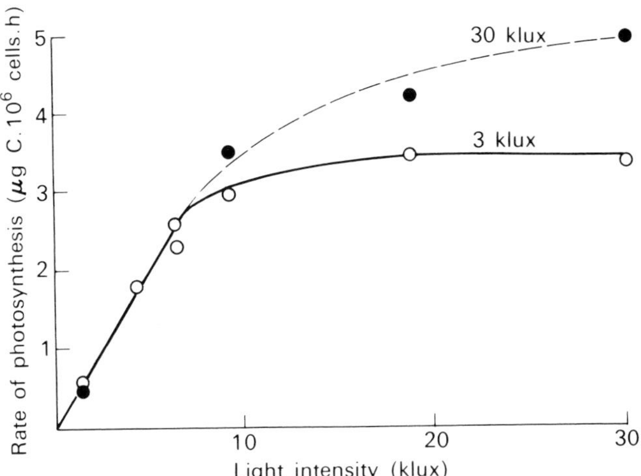

Fig. 5.4. The '*Cyclotella*-type' of adaptation. Light intensity–photosynthesis curves for *Cyclotella meneghiniana* grown at 3 and 30 klux, respectively. After Jørgensen, 1964.

curves. For I_k see Fig. 5.7. The physiological adaptation from one light intensity to another may be caused by changes in the content of photosynthetically active pigments or by changes in the photosynthetically active enzymes, or both. Jørgensen (1964, 1969) studied the adaptation of several algal species to different light intensities. He distinguishes between two main types of adaptation: The *Chlorella*-type and the *Cyclotella*-type. The *Chlorella*-type is characterized by changes in chlorophyll content per cell with changes in light intensity. More chlorophyll per cell is formed at low light levels than at high (Fig. 5.3). In the *Cyclotella*-type the chlorophyll content is about the same in cells grown at high and at low levels of light, while the actual photosynthetic rate is considerably higher in cells developed at high intensities. It is assumed that in the latter case increased contents of photosynthetically active enzymes in the dark reaction steps of photosynthesis has caused the rise in the rate of photosynthesis (See Fig. 5.4).

All of the species of diatoms mentioned in the papers of Jørgensen (1964, 1969) were referred to the *Cyclotella*-type. However, the author has also found species of diatoms belonging to the *Chlorella*-type.

It is further emphasized that Jørgensen's cultures were grown in incandescent light. Note that if *Skeletonema costatum* is cultured in fluorescent light it adapts as the *Chlorella*-type; but cultured in incandescent light it adapts as the *Cyclotella*-type (unpublished observations).

8.3 *Effect of high light intensities*

When algal cells adapted to a low light intensity are transferred to a higher light intensity there is a lag period before the cells adapt to the new light intensity (about 24 hours). During this period a decrease in the rate of light-saturated photosynthesis and of 'chlorophyll inactivation' occurs in many algal species (e.g. *Chlorella vulgaris* cf. Steemann Nielsen, 1962). Diatoms, or at least some diatoms, are able to tolerate a transfer even to the extremely high light intensity of 100 klux without showing 'chlorophyll inactivation' and without any decrease in the rate of light-saturated photosynthesis (Jørgensen, 1964). It must be assumed that cells of this species have a special mechanism protecting them against photo-oxidation.

8.4 *Light quality*

The action spectrum for photosynthesis in diatoms has already been described (see Fig. 5.2) and shows the relation between the rate of photosynthesis and the light quality. For diatoms this is characterized in relation to most other plants by the high rate of photosynthesis at wavelengths between 500 and 560 nm. It was also shown by Gabrielsen & Steemann Nielsen (1938), Mothes & Sagromsky (1941) and Baatz (1941) that light energy of these wavelengths is utilized for growth and photosynthesis.

Adaptation to different light qualities

Wallen & Geen (1971a,b,c) have recently studied the adaptation of *Cyclotella nana* to blue, green and white light. They found the highest rate of photosynthesis in the cells grown in blue light at all light intensities. The higher rate of photosynthesis is due to a higher content of chlorophylls *a* and *c* in cells grown in blue light.

They found further that blue and green light influenced algal metabolism in favour of protein synthesis, while white light caused carbohydrate synthesis; this has been found also by several other authors for other groups of algal and higher plants (Voskresenskaya, 1972).

In the diatom *Detonula* sp., Jupin (1973a) observed considerable differences both in the ultrastructure of the chloroplast (as mentioned above) and in the content of pigments in cells grown in red and in white light. Cells grown in red light are enriched with chlorophyll *Ca* 705 while the contents of chlorophyll *c* and fucoxanthin are lower than in cells grown in white light.

8.5 *Light periodicity*

The effect of light–dark cycles on the growth of plankton algae has been studied by several authors but few have concerned themselves with the effect

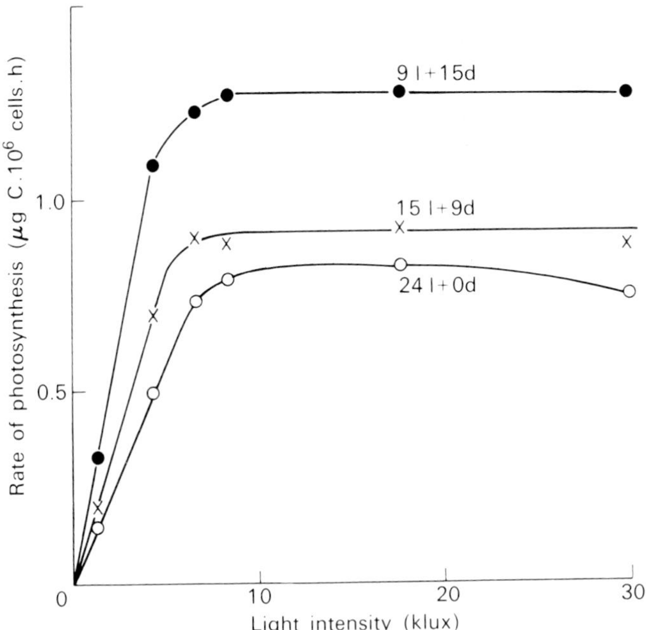

Fig. 5.5. Light intensity–photosynthesis curves for *Skeletonema costatum* grown at 3 klux and 2°C but with different light–dark period. The light was given either continuously or in periods of 15 hours light and 9 hours dark or 9 hours light and 15 hours dark. After Jørgensen & Steemann Nielsen, 1965.

on the pigment content and on the rate of photosynthesis in diatoms. In experiments with *Ditylum brightwellii* and *Nitzschia turgidula,* Paasche (1968) found that the chlorophyll concentration in the cells was higher the shorter the photoperiod. The same was the case in experiments by Jørgensen & Steemann Nielsen (1965), with *Skeletonema costatum* grown at 3 klux and only 2°C. In these experiments the light-saturated rate of photosynthesis was higher in a light–dark cycle with 9 hours of light and 15 hours of dark than with 15 hours of light and 9 hours of dark. The lowest rate was found in continuous light (see Fig. 5.5).

8.6 *Temperature*

When discussing the influence of temperature on the photosynthesis of diatoms the time course is extremely important. Immediately after a change in

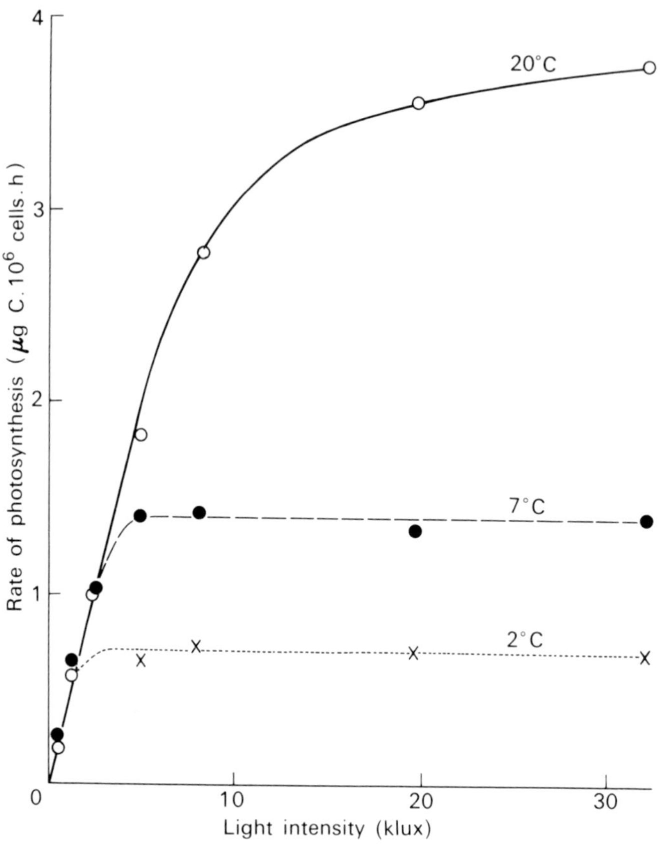

Fig. 5.6. The effect of temperature on the rate of photosynthesis. Light intensity–photosynthesis curves for *Skeletonema costatum* grown at 3 klux, 20°C, but transferred for 30 minutes to 20°C, 7°C and 2°C, respectively. After Steemann Nielsen & Jørgensen, 1968.

temperature we observe a simple effect on all chemical processes, including the enzymatic processes of respiration and photosynthesis. In the latter process it is only the rate of light saturation which is affected. At low light intensities, where temperature-insensitive photochemical part processes limit the rate of photosynthesis, temperature is of no importance. Several workers have convincingly shown this (see Fig. 5.6). However, for longer periods the algae gradually adapt to the new temperature. The adaptation will usually be completed in a few days. In experiments with *Skeletonema costatum,* Steemann Nielsen & Jørgensen (1968) showed for algae adapted to 20°C that whereas the light-saturated rate of photosynthesis dropped immediately after

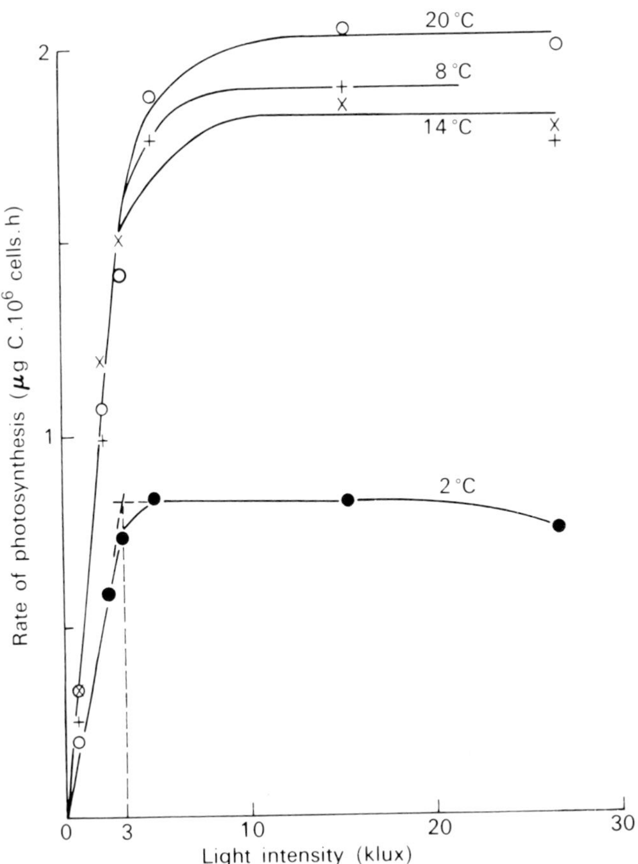

Fig. 5.7. Adaptation to different temperatures in *Skeletonema costatum*. Light intensity–photosynthesis curves for cells grown at 3 klux and at 20°C, 14°C, 8°C and 2°C, respectively. Temperature during experiment is the same as during growth. I_K is shown for the 2°C curve. After Jørgensen & Steemann Nielsen, 1965.

transfer from 20°C to 8°C to about one-third (cf. Fig. 5.6), the light-saturated rate is practically the same as at 20°C after the completion of the adaptation to 8°C (cf. Fig. 5.7).

The authors explain this adaptation to low temperature by suggesting an increase in the content of enzymes per cell at the lower growth temperature. Jørgensen (1968) has further given indirect support for this theory by showing that *Skeletonema* cells grown at 7°C contained about twice as much protein per cell as cells grown at 20°C.

The temperature adaptation was also studied by Morris & Farrell (1971). They also found that the content of protein per cell was higher in cells grown at lower temperatures than in those grown at 20°C both in the flagellate *Dunaliella tertiolecta* and in *Phaeodactylum tricornutum*. In cells of *Dunaliella* grown at 12°C they found further that the activity of ribulose-1,5-diphosphate carboxylase was more than 50% greater than in cells grown at 20°C.

8.7 *Chlorophyll content*

In another section we have discussed the effect of different light qualities on the rate of photosynthesis and at the same time the influence on the formation of pigment. In this section the relation between the rate of photosynthesis and the pigment content of cells is discussed.

Steemann Nielsen (1961) has stressed the importance of studying the influence of chlorophyll content on the rate of photosynthesis at low light intensities where the rate of the overall process of photosynthesis is limited by the rate of the photochemical part of the processes.

The rate of photosynthesis in mg C per mg chlorophyll *a* per hour at a light intensity of 1 klux has been found to vary between 0·40 and 0·60 for different planktonic green algae and diatoms (Steemann Nielsen & Jørgensen, 1968). The average value, for 60 experiments with *Skeletonema costatum* was 0·56 mg C per mg chlorophyll *a* per hour.

The amount of photosynthesis per unit of chlorophyll at saturating light intensity is influenced by temperature, CO_2 level, and many other factors. Ryther & Yentsch (1957) gave a mean value of 3·7 mg C assimilated per mg chlorophyll per hour, based on studies of cultures of several marine diatoms and other algae and also of natural populations.

As mentioned above, the contents of chlorophyll *c* and fucoxanthin in diatom cells are influenced by the quality but also by the intensity of the light. In several experiments with *Skeletonema costatum* where a great variation in the cell content of fucoxanthin was found, it was not possible to show any real correlation between fucoxanthin content and the rate of photosynthesis (Jørgensen, 1970).

9 REFERENCES

Allen E.J. & Nelson E.W. (1910) On the artificial culture of marine plankton organisms. *J. mar. biol. Ass. U.K.* **8,** 421–74.

Baatz I. (1941) Die Bedeutung der Lichtqualität für Wachstum und Stoffproduktion planktonischer Meeresdiatomeen. *Planta* **31,** 726–66.

Barker H.A. (1935) Photosynthesis in diatoms. *Arch. Mikrobiol.* **6,** 141–56.

BROWN J.S. (1967) Fluorescence emission from the forms of chlorophyll *a*. *Carnegie Institution Year Book* **65,** 483–7.

BROWN J.S. (1969) Absorption and fluorescence of chlorophyll *a* in particle fractions from different plants. *Biophys. J.* **9,** 1 542–52.

BROWN J.S. (1972) Forms of chlorophyll *in vivo*. *Ann. Rev. Pl. Physiol.* **23,** 73–86.

BROWN T.E. & RICHARDSON F.L. (1968) The effect of growth environment on the physiology of algae: Light intensity. *J. Phycol.* **4,** 38–54.

COLLYER D.M. & FOGG G.E. (1955) Studies on fat accumulation by algae. *J. expl. Bot.* **6,** 256–75.

DENFFER V.D. (1950) Die Planktische Massenkultur pennater Grunnddiatomeen. *Arch. Mikrobiol.* **14,** 159–202.

DOUGHERTY R.C., STRAIN H.H., SVEC W.A., UPHANS R.A. & KATZ J.J. (1970) The structure, properties, and distribution of chlorophyll *c*. *J. amer. chem. Soc.* **92,** 2 826–33.

DUTTON H.L. & MANNING W.M. (1941) Evidence for carotenoid sensitized photosynthesis in the diatom *Nitzschia closterium*. *Am. J. Bot.* **28,** 516–26.

DUTTON H.L., MANNING W.M. & DUGGAR B.M. (1943) Chlorophyll fluorescence and energy transfer in the diatom *Nitzschia closterium*. *J. phys. Chem.* **47,** 308–13.

EBATA T. & FUJITA Y. (1971) Changes in photosynthetic activity of the diatom *Phaeodactylum tricornutum* in a culture of limited volume. *Pl. Cell Physiol.* **12,** 533–41.

EMERSON R. & RABINOWITCH E. (1960) Red drop and role of auxiliary pigments in photosynthesis. *Pl. Physiol.* **35,** 477–85.

ENGELMANN T.W. (1883) Farbe und Assimilation. *Bot. Ztg.* **41,** 1–13.

ENGELMANN T.W. (1884) Untersuchungen über die quantitativen Beziehungen zwischen Absorption des Lichtes und Assimilation in Pflanzenzellen. *Bot. Ztg.* **42,** 81–93 und 97–105.

FOGG G.E. (1956) Photosynthesis and the formation of fats in a diatom. *Ann. Bot.* **10,** 265–85.

FRENCH C.S. (1955) Fluorescence spectrophotometry of photosynthetic pigments. In Johnson F.H., *The Luminescence of Biological Systems, AAAS, Washington D.C.,* 51–4.

FRENCH C.S. (1967) Changes with age in the absorption spectrum of chlorophyll *a* in a diatom. *Arch. Mikrobiol.* **59,** 93–103.

GABRIELSEN E.K. & STEEMANN NIELSEN E. (1938) Kohlensäure-assimilation und Lichtqualität bei den marinen Planktondiatomeen. *Rapp. P.-v. Réun. Cons. perm. int. Explor. Mer.* **108,** 20–1.

GIBBS S.P. (1970) The comparative ultrastructure of the algal chloroplast. *Ann. N.Y. Acad. Sci.* **175,** 454–73.

GOEDHEER J.C. (1970) On the pigment system of brown algae. *Photosynthetica* **4(2),** 97–106.

GOODCHILD D.J., BJÖRKMAN O. & PYLIOTIS N.A. (1972) Chloroplast ultrastructure, leaf anatomy, and content of chlorophyll and soluble protein in rainforest species. *Carnegie Inst. Yb.* **71,** 102–7.

GOVINDJEE R. & RABINOWITCH E. (1960) Two forms of chlorophyll *a in vivo* with distinct photochemical functions. *Science* **132,** 355–6.

HAGER A. & STRANSKY H. (1970) Das Carotinoidmuster und die Verbreitung des lichtinduzierten Xanthophyllcyclus in verschiedenen Algenklassen. V. Einzelne Vertreter der Cryptophyceae, Euglenophyceae, Bacillariophyceae, Chrysophyceae und Phaeophyceae. *Arch. Mikrobiol.* **73,** 77–89.

HOOD D.W. & PARK K. (1962) Bicarbonate utilization by marine phytoplankton in photosynthesis. *Physiologia Pl.* **15**, 273–82.

JEFFREY S.W. (1972) Preparation and some properties of crystalline chlorophyll c_1 and c_2 from marine algae. *Biochim. Biophys. Acta* **279**, 15–33.

JENSEN A. (1966) Carotenoids of Norwegian brown seaweeds and of seaweed meals. *Norweg. Inst Seaweed Res Report* No. 31.

JØRGENSEN E.G. (1964) Adaptation to different light intensities in the diatom *Cyclotella Meneghiniana* Kütz. *Physiologia Pl.* **17**, 136–45.

JØRGENSEN E.G. (1966) Photosynthetic activity during the life cycle of synchronous *Skeletonema* cells. *Physiologia Pl.* **19**, 789–99.

JØRGENSEN E.G. (1968) The adaptation of plankton algae. II. Aspects of the temperature adaptation of *Skeletonema costatum. Physiologia Pl.* **21**, 423–7.

JØRGENSEN E.G. (1969) The adaptation of plankton algae IV. Light adaptation in different algal species. *Physiologia Pl.* **22**, 1 307–15.

JØRGENSEN E.G. (1970) The adaptation of plankton algae V. Variation in the photosynthetic characteristics of *Skeletonema costatum* cells grown at low light intensity. *Physiologia Pl.* **23**, 11–7.

JØRGENSEN E.G. & STEEMANN NIELSEN E. (1965) Adaptation in plankton algae. *Memorie Ist. ítal. Idrobiol.* 18 Suppl., 37–46.

JUPIN H. (1973a) Modification pigmentaires et ultrastructurales chez la diatomée. *Detonula* sp. cultivé en lumière rouge. *Arch. Mikrobiol.* **91**, 19–27.

JUPIN H. (1973b) Modification de l'importance relative des deux systèmes photochimiques chez la diatomée *Detonula* sp. cultivé en lumière rouge. *Physiol. Veg.* **11**, 507–17.

JUPIN H. (1973c) Modifications pigmentaires et ultrastructurales chez la diatomée *Detonula* sp. cultivée en lumière rouge. *Arch. Mikrobiol.* **91**, 19–27.

JUPIN H. & GIRAUD G. (1971) Modification du spectre d'absorption dans le rouge lointain d'une diatomée cultivée en lumière rouge. *Biochim. Biophys. Acta* **226**, 98–102.

LEVINE R.P. (1969) The analysis of photosynthesis using mutant strains of algae and higher plants. *Ann. Rev. Pl. Physiol.* **20**, 523–40.

LEWIN J.C. (1958) The taxonomic position of *Phaeodactylum tricornutum. J. gen. Microbiol.* **18**, 427–32.

MANN J.E. & MYERS J. (1968) Photosynthetic enhancement in the diatom *Phaeodactylum tricornutum. Pl. Physiol.* **43**, 1 991–5.

MERESCHKOWSKY C. (1903) Les types de l'endochrome chez les diatomées. *'Scripta Bontanica' Horti Universitatis Petropolitanae,* **Fas.XXI**, 107–93.

MORRIS J. & FARRELL K. (1971) Photosynthetic rates, gross patterns of carbon dioxide assimilation and activities of ribulose diphosphate carboxylase in marine algae grown at different temperatures. *Physiologia Pl.* **25**, 372–7.

MOTHES K. & SAGROMSKY H. (1941) Über experimentelle chromatische Adaptation bei grünen und braunen Meeresalgen. *Naturwissenschaften* **29**, 271–2.

MYERS J. & GRAHAM J. (1971) The photosynthetic unit in *Chlorella* measured by repetitive short flashes. *Pl. Physiol.* **48**, 282–6.

OGAWA T., KANAI R. & SHIBATA K. (1968) Distribution of carotenoids in the two photochemical systems of higher plants and algae. In *Comparative Biochemistry and Biophysics of Photosynthesis* (Shibata *et al.* eds.) 22–35. University of Tokyo Press, Tokyo.

PAASCHE E. (1968) Marine plankton algae grown with light–dark cycles. II. *Ditylum brightwellii* and *Nitzschia turgidula. Physiologia Pl.* **21**, 66–77.

RAVEN J.A. (1970) Exogenous inorganic carbon sources in plant photosynthesis. *Biol. Rev.* **45,** 167–221.

RYTHER J.H. & YENTSCH C.S. (1957) The estimation of phytoplankton production in the ocean from chlorophyll and light data. *Limnol. Oceanogr.* **2,** 281–6.

SHERIDAN R.P. (1972) Adaptation to quantum flux by the Emerson photosynthetic unit. *Pl. Physiol.* **50,** 355–9.

STEEMANN NIELSEN E. (1961) Chlorophyll concentration and rate of photosynthesis in *Chlorella vulgaris. Physiologia Pl.* **14,** 868–76.

STEEMANN NIELSEN E. (1962) Inactivation of the photochemical mechanism in photosynthesis as a means to protect the cells against too high light intensities. *Physiologia Pl.* **15,** 161–71.

STEEMANN NIELSEN E. (1963) On bicarbonate utilization by marine phytoplankton in photosynthesis. With a note on carbamino carboxylic acids as a carbon source. *Physiologia Pl.* **16,** 466–9.

STEEMANN NIELSEN E. (1966) The uptake of free CO_2 and HCO_3^- during photosynthesis of plankton algae with special reference to the *Coccolithophorid Coccolithus huxleyi. Physiologia Pl.* **19,** 232–40.

STEEMANN NIELSEN E. & JØRGENSEN E.G. (1968) The adaptation of plankton algae I. General Part. *Physiologia Pl.* **21,** 401–13.

STOSCH V.H.A. (1951) Über das Leukosin, den Reservestoff der *Chrysophyten. Naturwissenschaften* **38,** 192–3.

STRAIN H.H., COPE B.T., MCDONALD G.N., SVEC W.A. & KATZ J.J. (1971) Chlorophyll c_1 and c_2. *Phytochemistry* **10,** 1 109–14.

STRANSKY H. & HAGER A. (1970) Das Carotinoidmuster und die Verbreitung des lichtinduzierten Xanthophyllcyclus in verschiedenen Algenklassen. II. *Xanthophyceae. Arch. Mikrobiol.* **71,** 164–90.

SUGAHARA K., MURATA N. & TAKAMIYA A. (1971) Fluorescence of chlorophyll in brown algae and diatoms. *Pl. Cell Physiol.* **12,** 377–85.

TALLING J.F. (1957) Photosynthetic characteristics of some freshwater plankton diatoms in relation to underwater radiation. *New Phytol.* **56,** 29–50.

TANADA T. (1951) The photosynthetic efficiency of carotenoid pigments in *Navicula minima. Am. J. Bot.* **38,** 276–83.

TREHARNE R.W., MELTON C.W. & ROPPEL R.M. (1964) Electron spin resonance signals and cell structure of *Chlorella pyrenoidosa* grown under different light intensities. *J. mol. Biol.* **10,** 57–62.

VOSKRESENSKAYA N.P. (1972) Blue light and carbon metabolism. *Ann. Rev. Pl. Physiol.* **23,** 219–34.

WALLEN D.G. & GEEN G.H. (1971a) Light quality in relation to growth, photosynthetic rates and carbon metabolism in two species of marine plankton algae. *Mar. Biol.* **10,** 34–43.

WALLEN D.G. & GEEN G.H. (1971b) Light quality and concentration of protein, RNA, DNA and photosynthetic pigments in two species of marine plankton algae. *Mar. Biol.* **10,** 44–51.

WALLEN D.G. & GEEN G.H. (1971c) The nature of photosynthate in natural phytoplankton populations in relation to light quality. *Mar. Biol.* **10,** 157–68.

WASSINK E.C. & KERSTEN J.A.H. (1944) Observations sur la photosynthèse et la fluorescence chlorophyllienne des diatomées. *Enzymologia* **11,** 282–312.

WASSINK E.C. & KERSTEN J.A.H. (1946) Observations le spectre d'absorption et sur le rôle des caroténoids dans la photosynthèse des diatomées. *Enzymologia* **12,** 3–32.

WATT W.D. & PAASCHE E. (1963) An investigation of the conditions for distinguishing between CO_2 and bicarbonate utilization by algae according to the methods of Hood and Park. *Physiologia Pl.* **16,** 674–81.

WERNER D. (1966) Die Kieselsäure im Stoffwechsel von *Cyclotella cryptica. Arch. Mikrobiol.* **55,** 278–308.

WERNER D. (1970) Productivity studies on diatom cultures. *Helgoländer wiss. Meeresunters.* **20,** 97–103.

CHAPTER 6

HETEROTROPHIC NUTRITION

JOHAN A. HELLEBUST

Department of Botany, University of Toronto,
Toronto, Ontario M5S 1A1, Canada

and

JOYCE LEWIN

Department of Oceanography, University of Washington,
Seattle, Wash. 98195, USA

1 INTRODUCTION

The ability of photosynthetic diatoms to grow on organic substances in the dark is intuitively of great advantage over an obligatory photoautotrophic mode of existence in those aquatic environments where the cells may periodically be subjected to complete, or almost complete, darkness. Many facultative and a few obligate heterotrophic diatom species have been discovered. Most of these are species of pennate diatoms frequently found in inshore, shallow-water, benthic environments. No typically open-water, centric species has yet been found capable of existing heterotrophically.

In order to grow in the dark on an organic substrate as a sole carbon and energy source, a cell must effectively take up the substrate from the surrounding medium and then assimilate it into all components essential for growth *via* light-independent metabolic reactions. Although some diatoms have been reported to be unusually permeable to organic substances such as sugars, sugar alcohols (Stadelmann, 1962), and amino acids (Cholnoky-Pfannkuche, 1965), the concentrations of organic substances in natural waters are rarely so high that entry *via* passive diffusion can provide the cells with organic metabolites at sufficient rates and concentrations for effective growth. On the contrary, a high permeability to common metabolites must be of distinct disadvantage to cells living in aquatic environments with extremely low concentrations of the same metabolites, since this would result in rapid losses of organic carbon from the cells. The evolution of specific, and often ingeniously controlled, transport systems for organic substrates—coupled with low permeability to metabolites in general—appears to be the mechanism by which the uptake problem has been resolved.

The nature of such transport systems, the distribution and characteristics of obligate and facultative heterotrophic nutrition among diatom species, and the relatively unknown metabolic features determining heterotrophic capabilities in diatoms will be discussed in sequence in this chapter. Finally, the importance of heterotrophic nutrition and the ability to take up various organic substances to diatoms in their natural environments will be considered.

2 UPTAKE MECHANISMS FOR ORGANIC SUBSTANCES

Specific uptake systems for substrates supporting heterotrophic growth in diatoms have been found in every case investigated so far. The uptake kinetics for such substrates either obey simple Michaelis–Menten kinetics, indicating carrier-mediated transport involving one carrier system, or more complex kinetics, indicating the involvement of more than one carrier system for the same substrate. Complex uptake kinetics may be due to conformational changes in components of a carrier system at certain substrate levels (Nissen, 1974). The ability of diatom cells to take up specific substrates and to regulate their transport systems in response to environmental parameters appears to be of fundamental importance in diatom heterotrophy. Furthermore, uptake systems for substrates that cannot support growth in the dark have been found in certain diatom species. In such cases, the uptake system may still be of ecological significance in supplying cells with additional organic carbon and energy sources, when light conditions only allow very low rates of photosynthesis, or in promoting survival for longer periods of time in complete darkness. Nitrogen-containing organic compounds may in cases of deficiencies of inorganic nitrogen serve as nitrogen sources for cells which possess uptake systems for these compounds.

Uptake systems for amino acids have now been reported for a number of diatom species (Table 6.1). The centric diatom *Melosira nummuloides,* which frequently occurs in benthic marine environments where light intensities are low, possesses constitutive and active uptake systems for amino acids, but for no other class of common substrates (sugars, organic acids, or sugar alcohols) (Hellebust & Guillard, 1967; Hellebust, 1970). Even the amino acid analogue α-amino-isobutyric acid is actively taken up by this clone. Several amino acids were concentrated more than 100-fold, when presented in the medium at a concentration of 0·1 mM (Hellebust, 1970). *M. nummuloides* utilizes several amino acids as nitrogen sources for growth in the absence of inorganic nitrogen (Hellebust & Guillard, 1967), and glutamate can contribute a very large fraction of the cell's carbon at low light intensities (Table 6.1). However, this clone of *M. nummuloides* appears to be an obligate photoautotroph (Lewin, 1963; Guillard, unpublished data).

Amino acids appear to be taken up by at least three different transport systems by diatoms: one for basic, one for acidic, and at least one for neutral amino acids (Hellebust, 1970; North & Stephens, 1972; Liu & Hellebust, 1974a,b). In several cases, the complex uptake kinetics for neutral amino acids indicate the involvement of more than one carrier, and the concentration range over which the kinetics is determined is important to the accuracy of the determination of half-saturation constants (K_s) for the systems (cf. Fig. 6.1 for

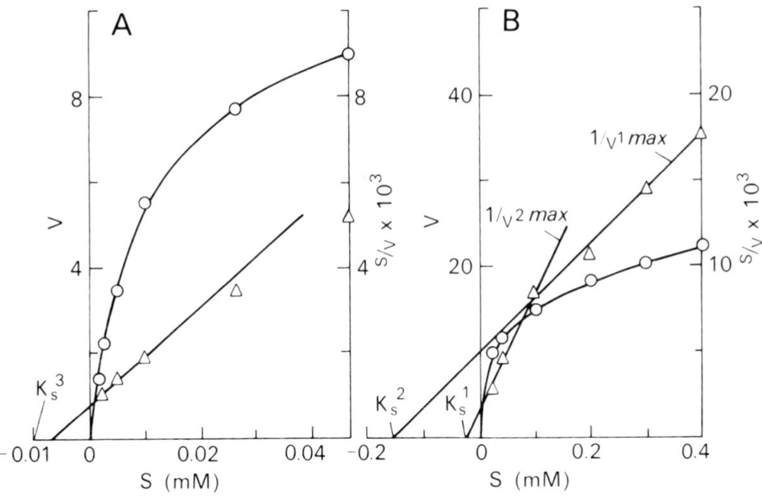

Fig. 6.1. Rate of uptake of proline by *Cyclotella cryptica* (Clone T-13-L) as a function of substrate concentration (circles). S/V plotted as a function of S (triangles). K_s is half-saturation constant for uptake, and V_{max} the maximum uptake rate. From Liu & Hellebust, 1974a.

Table 6.1. Uptake systems for sugars, amino acids and organic acids described for various diatom species. Half-saturation constants (K_S) for uptake are presented, and comments made about the nature of uptake kinetics, and presence of uptake system in light (L) or dark (D) cells. If not present in L-cells, a dark period with or without substrate is necessary for formation of uptake system. (Complex kinetics: kinetics differing from simple Michaelis–Menten kinetics. Figures in brackets refer to K_S values for low affinity uptake systems in cases of complex kinetics.)

Diatom	*Origin*	*Substrate*	K_s (mM)	*Comments*	*Reference*
Centric					
Coscinodiscus sp. (Clone NY-82-15, White)	Dredge spoil dumping area, New York Bight, NY, USA	Glucose	0·24	In D-cells	White (1974b)
Cyclotella cryptica (Clone WT-1-8, White)	Shallow brackish pond, Martha's Vineyard, Mass., USA	Glucose Galactose	0·1 0·7	In D-cells In D-cells	White (1974b) White (1974b)
Cyclotella cryptica (Clone 0-3A, Guillard)	Brackish pond, Martha's Vineyard, Mass., USA	Glucose	0·06	In D-cells	Hellebust (1971)
Cyclotella cryptica (Clone T-13-L, Guillard)	Brackish pond, Martha's Vineyard, Mass., USA	Arginine Glutamate Proline	0·003 (0·03) 0·036 (0·15) 0·006 (0·14)	In L- and D-cells	Liu & Hellebust (1974a,b)
Melosira nummuloides (Clone Mel-3, Guillard)	Brackish pond, Martha's Vineyard, Mass., USA	Arginine Valine	0·008 0·04 (1·4)	In L- and D-cells (complex kinetics)	Hellebust (1970)
Pennate					
Cocconeis diminuta (U. of Miami, Bunt)	Mud-sediments of Biscayne Bay, Fla., USA	Glucose Lactate Acetate	0·006 0·01 5	Cells incapable of dark growth (complex kinetics)	Bunt (1969) Cooksey (1972)

Cylindrotheca fusiformis (Strain # 13, Watson)	Marine littoral, Woods Hole, Mass., USA	Lactate	0·1–0·2	In D-cells (complex kinetics). Substrate required for induction of C-4 acid transport	Hellebust & Lewin (1972)
		Malate	0·1–0·2		
		Succinate	0·1–0·2		
		Fumarate	0·3		
Navicula pavillardi (Clone T3100, Swift)	Sargasso sea, probably inshore contaminant	Glutamate	0·02	In L- and D-cells	Lewin & Hellebust, 1975
Navicula pelliculosa (Clone NP2A, Jolley)	On vegetation in salt marsh, Aberystwyth, Wales	Glucose	0·05	Low uptake capacities in L-cells	Jolley & Hellebust (1974)
		Glutamate	0·1		
		Aspartate	0·1		
		Lactate	0·05		
Nitzschia alba (Clone Link 001, Linkins)	On seaweeds in estuarine areas, Mass., USA	Glucose	0·05 (0·2)	Glucose uptake influenced by acetate	Linkins (1973)
		Acetate	1·0		
Nitzschia angularis (Strain # 35-M, Lewin & Lewin)	On *Condrus crispus*, Herring Cove, Nova Scotia, Canada	Glutamate	0·02	In L- and D-cells	Lewin & Hellebust (1976)
		Glucose	0·03	In D-cells grown with substrate	
Nitzschia laevis (Strain # 72-M, Lewin & Lewin)	Seawater tank, Woods Hole, Mass., USA	Glucose	0·03	In L- and D-cells	Lewin & Hellebust (in prep.)
		Glutamate	0·03	In L- and D-cells	
		Alanine	0·02	In L- and D-cells	
		Lactate	0·4	In L- and D-cells	
Nitzschia ovalis (Strain # 680, Lewin)	On *Ectocarpus* sp. in benthic tidepool Nova Scotia, Canada	Arginine	0·002	In L- and D-cells regulated by N availability	North & Stephens (1972)

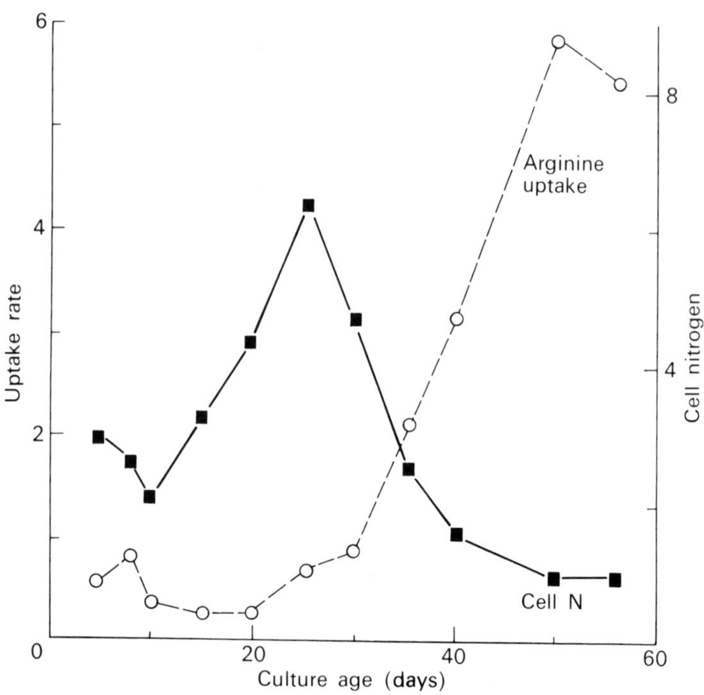

Fig. 6.2. Rate of arginine uptake and nitrogen content of *Nitzschia ovalis* during growth in nitrogen-limiting medium (uptake rates expressed in relative units per cell). Adapted from North & Stephens, 1972.

the uptake of proline by *Cyclotella cryptica*: K_s for the low proline range (high affinity uptake system) is 6 μM (K_s^3) from Fig. 6.1A, and 14 μM (K_s^1) from 6.1,B). K_s values from high affinity uptake systems for amino acids range from 0·002–0·040 mM (Table 6.1).

Although amino acid uptake systems appear to be constitutive for most diatoms studied so far, i.e. present in both light- and dark-incubated cells in the absence of amino acids, in the diatoms possessing such systems the actual capacity for amino acid uptake may be regulated to some extent by the nitrogen economy of the cells (Fig. 6.2; North & Stephens, 1972; Liu & Hellebust, 1974a,b). Wheeler *et al.* (1974) have recently studied the ability of a large number of inshore, benthic, pennate diatoms as well as some planktonic centric diatoms to take up and utilize amino acids as nitrogen sources for growth in the light. Most of the pennate diatoms were found to possess significant capacities for amino acid uptake as well as the capability of utilizing several of these amino acids as nitrogen sources. In contrast, the centric diatoms, except for a tychoplanktonic *Melosira* species, showed little ability to take up amino acids. The very limited capacities of several common planktonic, centric diatoms for amino acid uptake have also been demonstrated by Hellebust (1970) (see also Guillard, 1963; Hayward, 1965). Wheeler *et al.* (1974) demonstrated that the capacity for amino acid uptake in diatoms frequently increases by transfer of cells to nitrogen-limiting conditions. In some cases, transfer of cells from high to low light conditions has a similar effect on amino acid uptake capacity (Lewin & Hellebust, unpublished; Jolley & Hellebust, unpublished data).

Uptake mechanisms for organic acids have been demonstrated for a few pennate diatoms. *Cylindrotheca fusiformis*, a common marine littoral diatom of very wide distribution, when grown at high light intensities does not take up any organic acid (Hellebust & Lewin, 1972). However, transfer of cells from light to dark is sufficient for the development of a lactate uptake system, even in the absence of lactate. Transport systems serving 4-carbon, dicarboxylic acids (Table 6.1) are only induced in the presence of either lactate or any one of the dicarboxylic acids: fumarate, malate, or succinate. Only the completely dissociated forms of these organic acids appear to be taken up, and the process requires metabolic energy. The uptake abilities of this diatom are very specific, in that other organic acids, such as acetate or pyruvate, amino acids, or glucose are not taken up. Since glucokinase is present in *C. fusiformis,* its inability to take up glucose must be due to the lack of a transport system for this sugar.

Bunt (1969) investigated lactate uptake by a marine *Cocconeis* species (*C. minuta,* cf. Cooksey, 1972), isolated from the sediments of Biscayne Bay, Florida. The complex uptake kinetics indicate the presence of more than one carrier system for lactate uptake. The uptake system has a K_s of only 0·01 mM indicating effective uptake at low substrate concentrations. However, this species can only utilize lactate to aid growth at low light intensities; no growth occurs in the dark.

Although acetate can support heterotrophic growth of a few diatom species (Lewin & Lewin, 1960), the uptake mechanism for this organic acid has only been studied for the apochlorotic (non-photosynthetic) pennate diatom *Nitzschia alba* (Linkins, 1973), and for the obligate photoautotroph, *Cocconeis minuta* (Cooksey, 1972). Acetate is taken up by *N. alba via* a constitutive single passive uptake system of relatively low affinity (K_s= 1 mM; Table 6.1). When the cells have been preadapted to glucose, and glucose is simultaneously present with acetate, acetate appears to be taken up *via* the glucose transport system (Linkins, 1973). *Cocconeis minuta* also takes up acetate *via* a low affinity passive uptake system.

Glucose is one of the most common substrates for heterotrophic growth of diatoms. In several cases, transport systems for glucose have been demonstrated to be controlled by light conditions. The small euryhaline centric

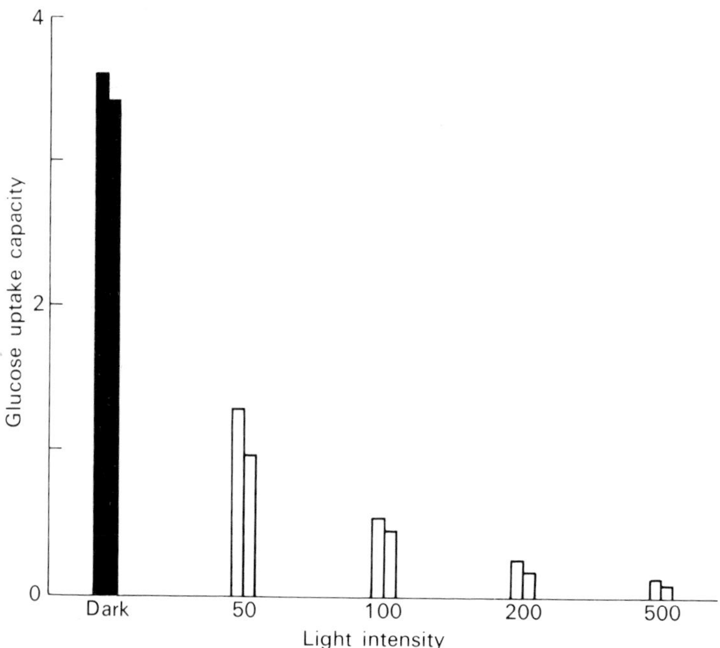

Fig. 6.3. Rate of glucose uptake of *Cyclotella cryptica* (Clone 0-3A) cells following incubation for 3 days in glucose-free media in the dark, or at different light intensities (units: ft.c.). From Hellebust, 1971b.

diatom *Cyclotella cryptica,* when grown at high light intensities has practically no ability to take up glucose (Hellebust, 1971b; White, 1974b). When cells are incubated at lower light intensities, or in complete darkness, the glucose transport capacity increases (Fig. 6.3). The time course for increase in glucose transport capacity for cells transferred to the dark and the decrease of transport capacity for dark-incubated cells transferred to the light are shown in Fig. 6.4. The presence of glucose is not necessary for dark induction of the glucose transport system, and the action spectrum for light inactivation of the transport system is identical to that for photosynthesis, indicating that some product of photosynthesis must be involved in the inactivation process (Hellebust, 1971b). Enzyme studies of light- and dark-grown cells demonstrate clearly that light conditions control the transport system for glucose, rather than glucose assimilation following glucose entry (Hellebust, 1971b). White (1974b) has shown similar light control for galactose uptake in another clone of *C. cryptica.* Glucose and galactose appear to share the same transport system. Other hexoses or pentoses are not taken up. The affinity of the transport system is much greater for glucose than for galactose (Table 6.1).

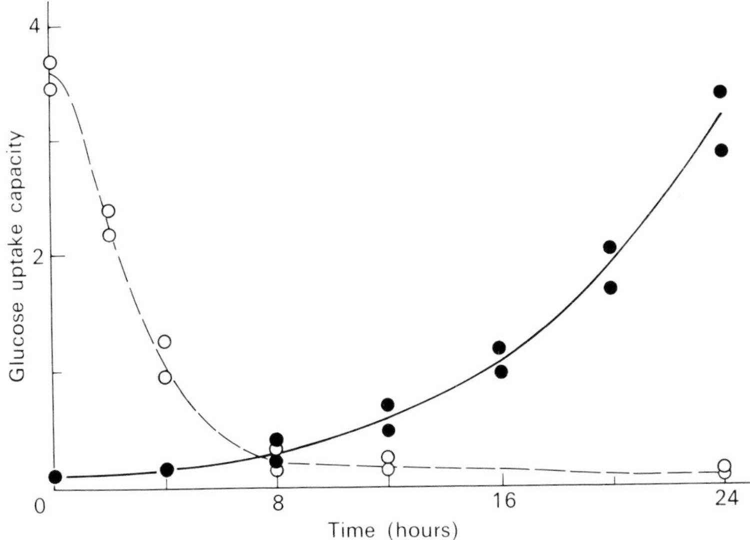

Fig. 6.4. Time course for glucose uptake capacities of *Cyclotella cryptica* (Clone 0-3A) cells previously kept in the dark for 1 day and then transferred to a light intensity of 500 ft.c. (open circles), and by cells previously grown at 800 ft.c. and then transferred to the dark (closed circles). From Hellebust, 1971b.

The glucose transport system must consist, at least in part, of a protein, since its induction in the dark is prevented by cycloheximide, an inhibitor of protein synthesis (Hellebust, 1971b). Glucose taken up by the cells is almost quantitatively phosphorylated within 10 minutes, either through the transport process itself, or by the high affinity glucokinase present constitutively in the cells. Energy metabolism, demonstrated to be required for glucose uptake, is probably related to this phosphorylation process.

White (1974b) has demonstrated essentially the same type of transport mechanism and light control for glucose uptake by a *Coscinodiscus* species, isolated from a relatively polluted area of the New York Bight, as that described above for *C. cryptica*. The affinity for glucose by this species (K_s=0·24 mM) is somewhat less than for *C. cryptica* (K_s=0·1 or 0·06 mM; Table 6.1).

Although glucose-6-phosphate is not taken up by light-grown *Cyclotella cryptica*, the induction of non-specific extracellular phosphatases under phosphate-limiting conditions (Kuenzler & Perras, 1965) will result in the formation of free glucose, which can be taken up by cells incubated in the dark or at low light intensities, i.e. conditions favouring the formation of a glucose transport system.

The non-photosynthetic pennate diatom *Nitzschia alba* possesses an active transport mechanism serving three different hexoses; glucose, galactose, and fucose (Linkins, 1973). Glucose uptake appears to be very complex in this diatom and may be due to two separate transport systems separated in time; (a) an initial temporal rapid uptake system (K_s=0·2 mM) serving to fill metabolic pools; and (b) a subsequent low capacity, high affinity system (K_m=0·05 mM), which is persistent and probably responsible for supply of sugar for general metabolism.

Transport systems for glucose have recently been reported for several other pennate diatoms (Table 6.1). Glucose is taken up *via* a simple uptake mechanism of very high affinity (K_s=0·6 μM) by *Cocconeis minuta* (Bunt, 1969). No information is available about possible light control of this transport system. *Navicula pelliculosa* (Jolley & Hellebust, unpublished data) and *Nitzschia angularis* (Lewin & Hellebust, 1976) have light-regulated glucose transport systems. Glucose transport by *Nitzschia laevis*, on the other hand, appears to occur *via* a constitutive system (Lewin & Hellebust, in preparation). Another marine benthic diatom species, *Amphora* sp., which is capable of growing on glucose or fructose in the dark, rapidly develops a glucose uptake system in response to the presence of glucose in the medium (Cooksey & Chansang, 1974; personal communication). The regulation of the glucose uptake system in this diatom is thus different from that described for other diatoms.

No uptake system has so far been found for alcohols in diatoms, although a marine pennate species, which can grow on glycerol in the dark (J. Lewin, unpublished data), may possess an uptake system for this sugar alcohol.

In cases where diatoms utilize macromolecules for heterotrophic growth, they presumably secrete enzymes which hydrolyze the macromolecules prior to uptake of their constituent amino acids or sugars. This at least must be the case for the non-photosynthetic diatom *Nitzschia alba* which can grow on microcrystalline cellulose, agar, or chitin (Linkins, 1973). Two other *Nitzschia* species, *N. frustulum* and *N. filiformis*, are able to digest agar, as evidenced by pit-formation on the agar surface at sites of individual colonies (Lewin & Lewin, 1960), and must, therefore, secrete enzymes which hydrolyze agar.

3 GROWTH ON ORGANIC SUBSTANCES IN THE DARK

Even though it has been clearly shown that many diatom species are capable of taking up a wide spectrum of organic molecules both in the light and in the dark, there are only *relatively few* organic compounds that can act as substrates for support of continuous growth in the dark, and, similarly, only *certain species* that possess the capability of utilizing one or more organic compounds as substrates for dark growth, in addition to their normal autotrophic mode of growth in the light. Although facultative heterotrophy is fairly widespread among the pennate diatoms, it is a rare occurrence among species of centric diatoms, which are predominantly photoautotrophs (Lewin, 1963; Lylis & Trainor, 1973). Even species that are closely related to one another (belonging to the same genus) show quite different behaviour from one another as far as their nutritional capabilities are concerned. Thus, it is impossible to make very many generalizations; rather, each species must be considered as a separate case and its nutritional characteristics studied individually. Diatoms are not unique in this respect; the same remarks apply as well to algal species belonging to other classes.

Among those diatom species capable of growing in the dark by utilizing organic substrates, there is considerable variation among the species with regard to the rate at which they can grow on a given substrate. Some species may utilize a particular compound very efficiently and multiply in the dark as rapidly or almost as rapidly as they are able to multiply when growing autotrophically in the light, while other species are capable of growing only at much reduced rates in the dark as compared with their maximum growth rate in the light. Here again, it is impossible to make generalizations.

In the following paragraphs, experimental results of studies of dark growth of diatoms with various types of organic compounds as substrates will be discussed (see Table 6.2 for list of known heterotrophic diatoms).

3.1 *Amino acids*

Studies of the utilization of amino acids as possible substrates for dark growth are really only just beginning. It has been known for many years that apochlorotic diatom species could be found in marine environmental habitats associated with decaying seaweeds. Subsequently, studies made with pure cultures of these colourless diatom species revealed that glutamate could serve as an excellent carbon and/or nitrogen source for both *Nitzschia alba* and *Nitzschia leucosigma* (Lewin & Lewin, 1967).

Also, it has been known for some time that commercially available preparations of digested proteins (such as tryptone, which is an enzymatic digest of casein consisting of amino acids and peptides) were useful additions to culture media for maintenance of vigorous diatom cultures in the light and also were even capable of supporting growth of certain selected pennate diatom species in the dark (see Lewin, 1963). Complex mixtures, such as yeast extract and casamino acids acted similarly (Lewin & Hellebust, 1970). Thus, it seemed logical to expect that individual amino acids, peptides, or mixtures of these might be compounds of importance to heterotrophic nutrition.

Table 6.2. Diatoms capable of growing on organic substrates in the dark.

Substrate	*Species and clones*
Glucose	*Amphora coffaeiformis* (34, 47, 51, 56, D2, D106, Rice)* (G)†. *A. delicatissima* (OP-2) (L). *Coscinodiscus* sp. (NY-82-15) (M). *Cyclotella cryptica* (0-3A, WT-1-8, WT-2) (B, M, L). *C. meneghiniana* (White; 9 clones, Lylis & Trainor; 4 clones) (L, J). *Melosira nummuloides* (OP-1-18) (L). *Navicula incerta* (66, 69) (G). *N. minima* (Lewin) (A). *N. pelliculosa* (7 clones) (A). *N.* spp. (4 unknowns) (A). *Nitzschia alba* (177-M, Link 001) (H, I). *N. angularis* var. *affinis* (35-M) (G). *N. filiformis* (36) (G). *N. fonticola* (Lewin, OP-1) (A, L). *N. frustulum* (13, 53, 68) (G). *N. kützingiana* (2 clones) (L). *N. laevis* (72-M) (G). *N. leucosigma* (180-M) (H). *N. marginata* (10) (G). *N. obtusata* var. *scapelliformis* (OP-1) (L). *N. punctata* (52) (G). *N. tenuissima* (Rice) (G).
Galactose	*Cyclotella cryptica* (WT-1-8) (M). *Nitzschia alba* (Link 001) (I).
Fructose, gluconate, mannose or rhamnose	*Nitzschia alba* (Link 001) (I).
Glucose + glutamate	*Nitzschia angularis* var. *affinis* (35-M) (E).
Glucose + tryptone	*N. thermalis* (Lewin) (B). *Amphora coffaeiformis* (8 isolates) (B). *Nitzschia frustulum* (3 isolates) (B).
Glycerol	Pennate species (J. Lewin, unpublished).
Acetate	*Amphora coffaeiformis* (D106) (G). *Nitzschia alba* (Link 001). (I). *N. tenuissima* (Rice) (G).
Lactate	*Amphora coffaeiformis* (47, Provasoli) (G). *Cylindrotheca fusiformis* ('*Nitzschia closterium*'; Watson-13, D105, Starr, Rice) (C, G). *Nitzschia alba* (177-M, Link 001) (H, I). *N. curvilineata* (5) (G). *N. frustulum* (13) (G). *N. laevis* (72, 5-Watson) (G). *N. leucosigma* (180-M) (H). *N. putrida* (176-M) (H). *N. tenuissima* (Rice) (G).
Succinate	*Cylindrotheca fusiformis* (Watson-13) (C). *Nitzschia alba* (177-M) (H). *N. leucosigma* (180-M) (H). *N. putrida*) (176-M) (H).
Fumarate or malate	*Cylindrotheca fusiformis* (Watson-13) (C).
Alanine	*Nitzschia angularis* var. *affinis* (35-M) (E).
Glutamate	*Navicula minima* (WT-1) (L). *N. pavillardi* (T3100) (K, D). *Nitzschia alba* (177-M) (H). *N. angularis* var. *affinis* (35-M). *N. frustulum* var. *subsalina* (OP-1) (L). *N. kützsingiana* (WT-1) (L). *N. laevis* (72-M) (E). *N. leucosigma* (180-M) (H). unidentified pennate diatoms (11 clones) (L).
Cellulose, agar, alginate or chitin	*Nitzschia alba* (Link 001) (I).

The heterotrophic nutrition of three species of photosynthetic pennate diatoms capable of growing on glutamate in the dark has recently been investigated (Lewin & Hellebust, 1975, 1976a,b,c). The three species behave quite differently from one another in their patterns of response to substrates. *Navicula pavillardi* is capable of growing heterotrophically on glutamate, but cannot utilize glucose as a substrate. The ability of this species to utilize glutamate was originally discovered by Swift (1967). *N. pavillardi* has a constitutive uptake system for glutamate, but none for glucose. *Nitzschia angularis* grows in the dark on glutamate, but only at about 15% of its optimal photoautotrophic rate. Glucose by itself does not support heterotrophic growth of this species, but rapid growth occurs in the dark when both glutamate and glucose are present (Fig. 6.5). Glutamate appears necessary for effective glucose uptake, and possibly also for metabolic reasons. A few other amino acids also allow some growth in the dark in the presence of glucose. Alanine by itself allows very slow, but definite heterotrophic growth of this diatom. The third pennate diatom investigated, *Nitzschia laevis*, is capable of good heterotrophic growth using either glutamate or glucose as substrates.

Exponentially growing cells of all three species are capable of immediate growth when transferred from photoautotrophic conditions to the dark with glutamate. In no case was a lag phase observed, indicating that the cells were immediately able to take up and assimilate glutamate. Also, it was noted that all three species showed more rapid initial growth at the lower, rather than at the higher glutamate concentrations used. This may have important ecological implications and should be further investigated in chemostat experiments.

Thus far, no other amino acids appear to function as efficiently as glutamate in supporting growth of pennate diatoms in the dark, but further studies may reveal that other amino acids are effective for other species.

Although amino acids are taken up very efficiently by the marine centric diatom *Melosira nummuloides* (Clone Mel-3) both in the light and in the dark, this species is not able to grow in the dark in the presence of amino acids (Lewin, 1963; Guillard, unpublished observations). Protein synthesis from alanine and other amino acids takes place in the dark as well as in the light (see Table 6.3) while synthesis of glucose occurs only in the light (Hellebust, 1970). Thus, a possible explanation for the inability of *M. nummuloides* to grow on amino acids in the dark is the incapability of gluconeogenesis of the cells in the

---

* Clones: If no clone or strain designation is supplied in reference, clone designated by name of isolator or supplier.

† References: (A) Lewin, 1953. (B) Lewin, 1963. (C) Lewin & Hellebust, 1970. (D) Lewin & Hellebust, 1975. (E) Lewin & Hellebust, 1976. (F) Lewin & Hellebust, in preparation. (G) Lewin & Lewin, 1960. (H) Lewin & Lewin, 1967. (I) Linkins, 1973. (J) Lylis & Trainor, 1973. (K) Swift, 1967. (L) White, 1972. (M) White, 1974a.

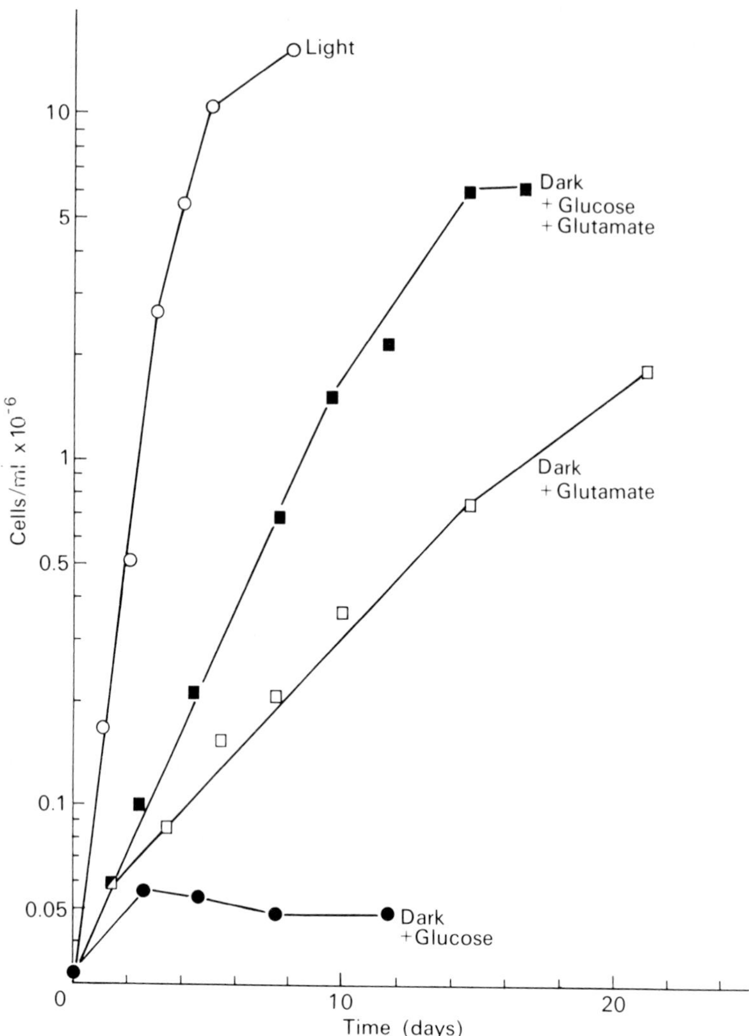

Fig. 6.5. Autotrophic and heterotrophic growth of *Nitzschia angularis* var. *affinis* (Clone 35-M). (Light intensity 2 000 lux; temperature 20°C; Glucose concentration 0·6 mM; Glutamate concentration 2 mM). Inoculum for dark cultures taken from cultures growing in the light. Adapted from Lewin & Helleburst, 1976).

dark. Similarly, although glutamate, alanine, and arginine are effectively taken up and respired in the dark by *Cyclotella cryptica* and a *Coscinodiscus* species, no growth results from the metabolism of the amino acids by these diatoms (White, 1974b).

Table 6.3. Comparison of rates of photoassimilation of CO_2 and of uptake and assimilation of glutamate by *Melosira nummuloides* at different light intensities (4 hour experiment). From Hellebust, 1970

Experimental system	*Rates of uptake and assimilation (µg atoms $^{14}C.cell^{-1}.hour^{-1} \times 10^9$)*					
	Dark		*0·5 klux*		*10 klux*	
	Total	*Protein*	*Total*	*Protein*	*Total*	*Protein*
$^{14}CO_2$ + Glutamate	—*	—	15	12	160	75
CO_2 + ^{14}C-Glutamate	85	12	110	25	100	45

* Not determined (dark uptake of $^{14}CO_2$ probably negligible).

3.2 *Organic acids*

In earlier investigations where a large number of organic substrates were tested as possible substrates for dark growth of diatoms, it became apparent that among organic acids lactate is by far the most commonly used substrate (Lewin & Lewin, 1960). Only a few diatoms species are known to utilize other organic acids, such as acetate, fumarate, malate, or succinate for dark growth (see Table 6.2).

The utilization of lactate, succinate, fumarate and malate as heterotrophic substrates by the marine pennate diatom *Cylindrotheca fusiformis* has been investigated in some detail by Lewin & Hellebust (1970). The minimum doubling time in the dark on any of these substrates is about 30 hours as compared with about 13 hours for photoautotrophic growth. Cells taken from cultures growing in the light and used as inoculum for growth with organic acids in the dark only begin to multiply after a lag period of about 2 days. This represents the time necessary for induction of carrier systems for transport of the organic acids into the cells (Hellebust & Lewin, 1972). Growth rate as a function of lactate concentration shows a complex relationship, indicating the possible involvement of two different saturable concentration-dependent processes with different affinities and capacities for lactate. Other organic acids tested such as acetate, pyruvate, and citrate, do not support dark growth of *C. fusiformis* (Lewin & Hellebust, 1970).

Glycolate was found not to support dark growth of any of 39 strains of algae, including two diatoms, studied by Droop & McGill (1966).

Three species of apochlorotic pennate diatoms, *Nitzschia putrida, N. alba,* and *N. leucosigma,* were found capable of utilizing either lactate or succinate as sole sources of organic carbon (Lewin & Lewin, 1960), while a different isolate of one of the species, *N. alba*, was shown to be capable of utilizing acetate as well as lactate (Linkins, 1973). Thus, organic acids are important substrates for colourless diatoms, in addition to amino acids as discussed above.

3.3 *Sugars*

Several sugars, pentoses as well as hexoses, have been tested as possible substrates for support of dark growth of various diatom species. Glucose is clearly the most suitable sugar, since it supports growth of a considerable number of species indefinitely in the dark (Lewin, 1953; Lewin & Lewin, 1960; Lewin, 1963; Lewin & Hellebust, 1970; Lylis & Trainor, 1973; White, 1974a). Galactose was found not to support heterotrophic growth when tested with 15 species of marine diatoms (Lewin & Lewin, 1960); however, its utilization for dark growth by *Cyclotella* cryptica has recently been reported by White (1974a). Fructose supports growth of the freshwater pennate diatom *Navicula pelliculosa* in the light in the absence of CO_2, but not heterotrophic growth (Lewin, 1963). Sucrose is reported to support heterotrophic growth of *Amphora* sp. (Cooksey & Chansang, 1974). No other sugars have been reported to serve as substrates for dark growth of photosynthetic diatoms. The colourless diatom *Nitzschia alba* (Clone Link 001) utilizes glucose, fructose, mannose, galactose, fucose, or rhamnose as carbon sources for growth, but not sucrose, maltose, or xylose (Linkins, 1973).

Cyclotella cryptica, a centric brackish-water diatom, utilizes glucose for dark growth (Lewin & Lewin, 1960; Lewin, 1963). More recently, this species has been extensively studied with respect to the mechanisms involved in glucose uptake and assimilation (Hellebust, 1970; 1971a,b; White, 1974b). It grows in the dark at a maximum rate (doubling time: 27 hours) considerably lower than its photoautotrophic rate (doubling time: 17 hours) (Lewin, 1963) (Fig. 6.6). The dark rate is not limited by the capacity of the cells for glucose uptake, but presumably by some metabolic steps of glucose assimilation (Hellebust, 1971a). A strain of *C. cryptica* isolated more recently by White (1974a) grows on glucose in the dark at a rate similar to its optimal photoautotrophic rate. This strain appears to be more efficient both in its processes of uptake and dark assimilation of glucose than the earlier strains studied. White suggests that facultative heterotrophs kept under photoautotrophic laboratory conditions for long periods of time may lose some of their heterotrophic capacity.

When exponentially growing cells are transferred from light into the dark in the presence of glucose, there is always a lag period of 1–2 days before growth commences (Fig. 6.6; Lewin, 1963). This lag period represents the time needed for development of the transport system necessary for glucose uptake by the cells (Hellebust, 1970, 1971b). Essentially the same conclusion was reached by White (1974a) for a more recent isolate of *C. cryptica*, and by Lylis & Trainor (1973) for a closely related freshwater diatom, *Cyclotella meneghiniana*.

The lag phase mentioned above which occurs upon light-to-dark transfer of *Cyclotella cryptica* and *C. meneghiniana* is not characteristic of all species.

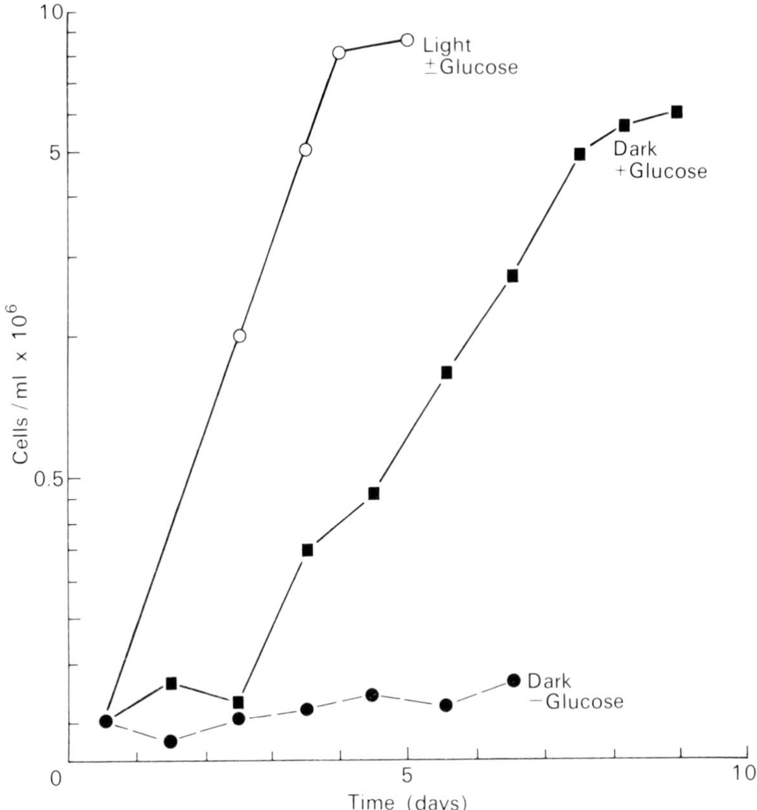

Fig. 6.6. Autotrophic and heterotrophic growth of *Cyclotella cryptica* (Clone 0-3A) in aerated cultures. (Light intensity 3 000 lux; temperature 20°C.) Inoculum for dark cultures taken from cultures growing in the light. From Lewin, 1963.

The pennate diatom *Nitzschia laevis*, which can grow in the dark in the presence of glucose, does not exhibit a lag phase when transferred from light into the dark in a glucose medium (Lewin & Hellebust, in preparation).

Melosira nummuloides (Clone OP-1-18), recently isolated from a brackish-water pond, is capable of limited heterotrophy on glucose (White, 1972). With a relatively high concentration (1 mM) of glucose the cells multiplied about 4 times in one month in the dark. The cells remained darkly pigmented and formed long chains. In the absence of glucose, no cell division took place; the chains fragmented and the cells lost pigments. This clone was isolated from glucose-enriched agar after a 3-week dark incubation.

Since a recent isolate of *Cyclotella cryptica* is able to grow in the dark both on glucose and galactose, it is possible to compare the two sugars in their efficiency as growth substrates (White, 1974a). Growth rate and yield with 1 mM glucose is greater than with 1 or even 5 mM galactose. The low growth rates observed with galactose are adequately explained by the low galactose uptake capacity of this diatom.

3.4 *Alcohols and other substrates*

There has been very little published on alcohols as potential substrates for heterotrophic growth. Glycerol supports growth of the freshwater diatom *Navicula pelliculosa* in the light in the absence of CO_2, but not in the dark (Lewin, 1963). Another study shows that high concentrations of this polyol enhance the growth of *Phaeodactylum tricornutum* at low light intensities, but do not support dark growth (Cheng & Antia, 1970). Glycerol was found to serve as a substrate for dark growth for only one out of several marine pennate diatom species recently investigated (Lewin, unpublished). Various alcohols and sugar alcohols, such as ethyl alcohol, mannitol, inositol, sorbitol, and dulcitol, have been tested by various workers without success (Lewin, 1953; Lewin & Hellebust, 1970; Linkins, 1973; Lylis & Trainor, 1973; White, 1974a).

A characteristic of diatoms which are facultative heterotrophs is their ability to retain both their pigmentation and photosynthetic capacity when grown in the dark on organic substrates for considerable periods of time (for at least 1 year in some cases; Lewin & Hellebust, 1970; White, 1974a).

Since diatoms are rather prolific producers of extracellular polymeric materials of various sorts, the question has been raised concerning the possible use of such materials as substrates, presumably by extracellular digestion and resorption. So far, there is no evidence for the occurrence of such activity. Cells of the freshwater diatom *Navicula pelliculosa* are enclosed in a capsular material which is a polymer of glucuronic acid; neither the polymer nor glucuronic acid itself supports heterotrophic growth (Lewin, 1955). Similarly, extracellular fibres consisting of chitin are produced by the marine centric diatoms *Cyclotella cryptica* and *Thalassiosira fluviatilis*, but neither the polymer nor glucosamine itself is used by these species for growth in the dark (McLachlan & Craigie, 1966).

Linkins (1973) has made the very interesting observation that the colourless diatom *Nitzschia alba* is able to utilize alginate, cellulose, and agar for growth. Chitin was found to be a less suitable substrate, and soluble starch is not used at all. Growth is slower with cellulose than glucose; presumably the breakdown of cellulose is rate-limiting. When presented with microcrystalline cellulose, the cells produce an extracellular acid sulphate-containing polysaccharide which surrounds the substrate causing attachment of the cells to the cellulose granules. Presumably β-1,4-glucanases are secreted to hydrolyze the macromolecules. These observations suggest the possibility of similar behaviour among some of the pigmented diatoms.

4 ASSIMILATION AND RESPIRATION OF ORGANIC SUBSTANCES

Diatoms capable of growing heterotrophically on specific substrates obviously are able to utilize the substrates both in respiratory pathways to generate sufficient energy for growth, and in biosynthetic pathways to produce carbon skeletons for all necessary cell components. In a few cases, the efficiency of substrate assimilation for heterotrophic growth of diatoms has been determined. *Navicula pavillardi* was found to assimilate glucose into cell carbon with 41% efficiency in the dark at 20°C (Lewin & Hellebust, 1975). *Nitzschia angularis* assimilates glutamate with an efficiency of 43% under the same conditions. As mentioned earlier, this species is only able to utilize glucose for heterotrophic growth when glutamate also is present, and to a lesser extent when alanine, aspartate, or asparagine are present. From growth experiments with excess glucose and different levels of glutamate present in the medium, it was estimated that glucose only supplies about 50% of the cell's carbon; the rest must come from glutamate (or presumably from the other amino acids that allow growth on glucose) (Lewin & Hellebust, 1976). Another facultative heterotroph, *Nitzschia laevis*, utilizes glucose with an efficiency of 48%, and glutamate with an efficiency of 43% when incubated in the dark at 20°C (Lewin & Hellebust, in preparation). Lactate was found to be assimilated with a 54% efficiency by *Cylindrotheca fusiformis* (Hellebust & Lewin, 1972). It should be noted that temperature is an important factor in determining assimilation efficiency. Thus, according to Ng (1969), *Escherichia coli* assimilates glucose carbon with a 53% efficiency at 30°C; at 10°C the efficiency is only 37%.

Very little information is available about specific pathways of assimilation of organic substances by diatoms. When glucose is taken up by *Cyclotella cryptica*, it is almost completely phosphorylated or assimilated into other cell constituents within 10 minutes (Hellebust, 1971b). Linkins (1973) similarly demonstrated rapid phosphorylation and metabolism of glucose by the apochlorotic pennate diatom *Nitzschia alba*. This diatom, which produces a high proportion of its cell constituents as lipids, respires glucose at approximately the same rate *via* the hexose monophosphate shunt as *via* the Embden–Meyerhof pathway. In agreement with this finding, gluconate is also readily respired and assimilated by this alga. Reducing power in the form of NADPH, generated by operation of the hexose monophosphate shunt, is presumably used for lipid synthesis. In short-term experiments, about 40% of the total glucose taken up in the dark by the centric diatoms *Cyclotella cryptica* and *Coscinodiscus* sp. is completely respired and released as CO_2 (White, 1974b). Phosphofructokinase, a key regulatory enzyme of the Emden–Meyerhof pathway is controlled by the ATP concentration in *Cyclotella cryptica* (Hellebust, 1971b).

Organic acids are very rapidly assimilated and respired by diatoms capable of utilizing them for heterotrophic growth. After only 10 minutes uptake of lactate or succinate by *Cylindrotheca fusiformis*, more than 95% of these organic acids are converted to other cell constituents, particularly metabolites immediately derived from the tricarboxylic acid cycle (Hellebust & Lewin, 1972). Enzymes necessary for the immediate assimilation of lactate, malate, fumarate, and succinate, i.e. the respective dehydrogenases of these organic acids are present in *C. fusiformis* cells at levels sufficient to account for the observed heterotrophic growth rates on these substrates, regardless of whether the cells have been preincubated in the light or in the dark. Thus, while growth conditions play a major role in regulating transport capacities for organic acids in *C. fusiformis*, they only induce minor changes in levels of assimilatory enzymes. Lactate dehydrogenase has been found present at very high levels also in *Cyclotella cryptica* and *Navicula pelliculosa* grown under photoautotrophic (aerobic) conditions (Werner, 1967), and photoautotrophically grown *Nitzschia closterium* was found to contain about 1·3 mM lactate (Rho, 1959). This enzyme must, therefore, play an important role in diatom metabolism also under these conditions.

Navicula pavillardi, which can grow in the dark on glutamate, is able to assimilate this amino acid rapidly into other amino acids as well as glucose (Lewin & Hellebust, 1975). Other amino acids, which do not support heterotrophic growth, such as alanine and isoleucine, appear to be metabolized only to a limited extent, and not to be assimilated into glucose.

As already discussed, diatoms exhibit a great deal of variation in their trophic abilities. In some cases, the inability of a diatom species to utilize a specific organic substrate may simply be due to its lack of a transport system for the substrate. For instance, *Cylindrotheca fusiformis,* while possessing an active glucokinase, does not possess a glucose transport system (Hellebust & Lewin, 1972). In other cases, diatoms may possess transport systems for organic substances, which they cannot utilize for heterotrophic growth. They may utilize substances for respiration, photosynthesis in the absence of CO_2, or to stimulate growth under limiting light conditions, and be incapable of growth in the dark on the same compounds. Very little biochemical information is available about such contradictory behaviour.

One likely explanation for the failure to grow in the dark on substances which can be taken up and utilized by the cells for respiration, is that certain essential metabolic steps may only occur in the light. As already mentioned, alanine, which is readily taken up and converted by cells of *Melosira nummuloides* to other amino acids in the dark as well as light, is only converted to glucose under light conditions (Hellebust, 1970). This indicates that photosynthetic generation of reducing power and/or ATP is essential for gluconeogenesis to occur, and the cells are, therefore, unable to produce carbohydrates from alanine in the dark. Similar conclusions were reached for the inability of the green alga *Pandorina morum* to utilize acetate for heterotrophic growth (Palmer & Togasaki, 1971). Acetate is rapidly taken up and respired in the dark. However, only in the light are significant amounts of acetate assimilated into polysaccharides.

The metabolism of acetate and lactate has been studied in the marine pennate diatom *Cocconeis diminuta* (Cooksey, 1972). Although this species metabolizes these organic acids both in the light and the dark, it is not capable of dark growth in the presence of either compound. The acids are oxidized to CO_2 in the dark, but respiration is apparently not effectively coupled to ATP production. Cooksey therefore concluded that inadequate ATP levels may be the reason for the inability of this species to grow heterotrophically. *Phaeodactylum tricornutum,* another obligate photoautotroph, behaves similarly to *C. diminuta* in that it oxidizes acetate to CO_2 in the dark but is unable to grow on it (Cooksey, 1974).

In general, light stimulates assimilation of organic substrates into macromolecules. For instance, light strongly stimulates the amount of glutamate carbon incorporated into proteins by *Melosira nummuloides* (Table 6.3; Hellebust, 1970), the amount of lactate carbon into macromolecules by *Cylindrotheca fusiformis* (Darley & Ohlman, 1974; personal communications), and the amount of acetate carbon assimilated into lipids and other macromolecules by *Cocconeis diminuta* (Cooksey, 1972) and *Phaeodactylum tricornutum* (Table 6.4; Cooksey, 1974). The slow rate of conversion of certain metabolites derived from substances taken up by diatoms, to macromolecules in the dark, may be the main reason why facultatively heterotrophic diatoms usually grow more slowly in the dark than in the light. If there is an absolute dark block of one or more of such metabolic steps necessary for the synthesis of essential cell constituents, the species is unable to grow heterotrophically.

Table 6.4. Relative rates of incorporation of ^{14}C-labelled acetate or lactate into various cell fractions by the marine pennate diatom *Cocconeis diminuta*. From Cooksey, 1972

	Acetate-2-^{14}C		*Lactate-U-^{14}C*	
Cell fraction	*Light*	*Dark*	*Light*	*Dark*
Soluble fraction	19	62	7	9
Lipids	151	40	29	6
Nucleic acids	4	4	1	1
Protein and residue	67	57	40	20

Light can also strongly decrease that fraction of a substrate taken up by diatom cells which is completely respired (Cooksey, 1972, 1974; Darley, personal communication). This phenomenon is probably due to light inhibition of key steps in respiratory pathways by products of photosynthetic reactions (cf. Hirt *et al.* 1971; Heber, 1974).

The fact that many diatoms are able to utilize amino acids as nitrogen sources for growth or as respiratory substrates (Guillard, 1963; Hayward, 1965; Thomas, 1966; Hellebust & Guillard, 1967; North & Stephens, 1972; Liu & Hellebust, 1974a, 1974b; Wheeler *et al.* 1974; White, 1974b), and that sugars and organic acids can be used for respiration and synthesis of many cell constituents (Cooksey, 1972, 1974), even if these substances do not support dark growth, may nevertheless be important for growth or survival of cells under conditions of limiting light conditions or limiting supplies of inorganic nitrogen. Stimulation of growth of diatoms at low light intensities has been reported for *Cocconeis diminuta* (Bunt, 1969), and for *Cyclotella cryptica* and a *Coscinodiscus* species (White, 1974a). Lewin (1953) has shown that glucose, glycerol, or fructose can be used as carbon sources for photoautotrophic growth of *Navicula pelliculosa* in the absence of CO_2.

When amino acids are utilized by diatoms as sources of nitrogen for growth in the absence of inorganic nitrogen, it appears that intact amino acid molecules usually are taken up by the cells (Hellebust & Guillard, 1967; Liu & Hellebust, 1974a). When small amounts of non-nitrogenous radioactive compounds appeared in the medium of *Nitzschia ovalis* supplied with ^{14}C-labelled amino acids, there was no simultaneous appearance of free ammonia, indicating intracellular deamination and subsequent excretion of the deaminated products (Stephens & North, 1971). Only in the case of *Nitzschia thermalis* is there definite evidence for extracellular deamination of amino acids (Saubert, 1957).

Amino acids, supplied singly at low concentrations in the medium, are in some cases inhibitory to growth of diatoms (e.g. threonine and isoleucine for *Melosira nummuloides*, Hellebust, 1970; or glutamic acid for *Phaeodactylum tricornutum*, Hayward, 1965). It seems likely that certain amino acids, when highly concentrated in the cells due to active transport systems, may interfere with the biosynthesis of other amino acids essential for growth (Hellebust, 1970; Antia *et al.*, 1975).

5 ECOLOGICAL CONSIDERATIONS

There has been considerable speculation and discussion concerning the possible role of facultative heterotrophy among algal species in their natural habitats (e.g. Lylis & Trainor, 1973; Lee *et al.*, 1975). From results of relevant studies presently available we arrive at the following conclusions: (1)

heterotrophic capabilities are fairly widespread among pennate diatoms, but rare among centric diatoms. (2) Pennate diatoms are predominantly benthic or epiphytic, and most often inhabit environments rich in dissolved organic matter (tidal estuaries, tide pools, mud flats, salt marshes, brackish-waters, sand flats). (3) Centric diatoms are predominantly planktonic, and are most often found in open waters with relatively low concentrations of dissolved organic matter. (4) The few centric diatom species presently known to have heterotrophic tendencies have been isolated from habitats rich in dissolved organic matter.

5.1 *Comparison of pennate and centric diatoms*

Among the pennate diatoms, members of the genus *Nitzschia* and *Amphora* show the most pronounced development of heterotrophic abilities. All diatoms presently known to be obligate heterotrophs are *Nitzschia* species. They occur in habitats strongly favouring heterotrophic growth, such as decaying piles of seaweeds where light is absent or limiting and organic substrates in rich supply. *N. alba*, which has been isolated from such habitats (Linkins, 1973), has developed the most extreme heterotrophic capabilities so far observed in a diatom. It appears to secrete hydrolytic enzymes, attaches itself to suitable macromolecular substrates by secretion of sticky polysaccharides, and utilizes a large number of sugars as well as organic acids for growth (see Table 6.2). Another obligate heterotroph, *Nitzschia putrida*, has been shown to occur abundantly in strongly polluted sites of the Kieler Fiord, Germany (Benecke, 1901). Cholnoky (1968), in his extensive monograph on the ecology of freshwater diatoms, makes frequent reference to his personal studies demonstrating that many *Nitzschia* species develop optimally in highly eutrophic waters, especially where the organic nitrogen content is high (40–50 mg organic N/l). He refers to these species as nitrogen heterotrophs. *N. thermalis* and *N. fonticola* prefer highly eutrophic, often low oxygen conditions (Cholnoky, 1968). *N. fonticola* is capable of heterotrophic growth on glucose, and *N. thermalis* on glucose in the presence of tryptone (Lewin, 1963). In contrast, *N. palea*, which characteristically avoids eutrophic, low oxygen conditions (Cholnoky, 1968), is unable to grow on glucose in the dark (27 separate isolates tested; Lewin, 1963).

Several *Nitzschia* species observed in slime casings of frog eggs (*Rana temporaria*) by Cholnoky (1929), e.g. *Nitzschia kützingiana*, now are believed to benefit from this exceptional microhabitat by available organic nitrogen (Cholnoky, 1968). Cholnoky (1968) has demonstrated that *N. kützingiana* needs amino acids as sources of nitrogen for growth, and that this species is frequently found in highly eutrophic waters. White (1972) has isolated two clones of *N. kützingiana* which can utilize glucose for heterotrophic growth, and one which utilizes glutamate. Pennate diatoms are frequently observed in slime casings produced by various invertebrates, e.g. those of many aquatic insect larvae, and surrounding faecal pellets of sea urchins (Geitler, 1968; Hellebust, personal observations). The large slime balls formed by the green symbiotic protozoan *Ophrydium versatile* contains large numbers of pennate diatoms, such as *Cymbella cesati* and *Nitzschia flexoides* (Geitler, 1968). Diatoms living in such associations may, at least at times, exist as facultative heterotrophs.

Cyclotella meneghiniana, which is known to have strong heterotrophic capabilities (Cholnoky, 1968; Lylis & Trainor, 1973; White, 1972) was found to be one of the dominant diatom species (15% of total) in a sewage maturation pond in South Africa (Schoeman, 1972). Other dominant species in this highly eutrophic pond were *Amphora tenerrima, A. subacutiuscula* and *Nitzschia fonticola. Nitzschia thermalis* was also found, but in lower concentrations, presumably due to the relatively high salt content of the pond. Thus, in a number of cases there is good agreement between heterotrophic capabilities of diatoms and their occurrence in environments favouring heterotrophy. A recent study by Lee *et al.* (1975) also points in this direction.

5.2 *Survival and growth*

It is well known from laboratory studies that diatom cells are capable of survival for weeks and even months when kept either in dim light or in complete darkness (Schreiber, 1927; Matsue, 1954; Lewin, 1961; Lewin & Guillard, 1963; Antia & Cheng, 1970; Ignatiades & Smayda, 1970; Bunt & Lee, 1972; Umebayashi, 1972; Smayda & Mitchell-Innes, 1974), and this survival does not depend on the presence of any external organic substances in the medium. Antia & Cheng (1970) have used the expression 'dormancy induced by darkness', while Schreiber (1927) spoke of 'physiological resting cells' to describe this interesting phenomenon of dark survival. Thus, survival through prolonged periods of darkness would not appear to be of any particular problem for diatom species in their natural habitats. The presence of organic substances in the surrounding environment might, however, serve to prolong dark survival for even longer periods of time compared with what is possible in the absence of organic compounds (Hellebust & Guillard, 1967; White, 1974a).

The occurrence of cell multiplication in darkness is quite a different matter from mere dark survival. We now know that many species have the ability to take up molecules of organic substances in the dark, and the ability to respire and metabolize such substances, but not grow in the absence of light. Other species are capable of heterotrophic growth on the same substances. There appears to be considerable variability among species with regard to the possible metabolic pathways that are operative. As more species are studied, more new patterns of response are uncovered; in fact, among those species studied so far, no two have responded to organic substances in quite the same way. This high degree of specificity is most remarkable, and is probably good indirect evidence for the evolutionary adaptation of species to various specific environmental habitats and situations.

The fact that some species can take up organic molecules upon transfer into the dark and begin to multiply immediately, whereas others undergo a lag of 1 or 2 days while transport systems are induced, presents another interesting difference in species behaviour, for which the ecological significance is not immediately obvious. However, the ability of the latter group of species to develop transport systems completely independent of the presence of substrates would be a mechanism offering considerable ecological advantage. Cells depending on external substrates for induction of transport systems would require relatively high concentrations, since the substrates can enter such cells initially only by passive diffusion. However, since concentrations of organic substances in natural environments are usually very low (see e.g. Bohlin, 1970; Andrews & Williams, 1971), dependence on substrate induction for development of transport systems is unlikely to be an effective mechanism. The very fact that several diatom species develop transport systems for a limited number of organic substrates—and precisely those most likely to be commonly abundant, e.g. glucose, lactate and glutamate—in response to low light or dark conditions (Hellebust, 1970; 1971b; Lewin & Hellebust, 1970, 1975, 1976) strongly supports the assumption that these species have evolved highly sophisticated and ecologically significant mechanisms for facultative heterotrophy.

It has been demonstrated that active transport systems for organic substances are operative in many diatoms, but with half-saturation constants for uptake that are usually high in relation to the concentrations of such substances in natural waters. Since most of the uptake systems studied, as well as heterotrophic growth rates appear to be effective only at relatively high substrate concentrations, it is likely that diatom species possessing these particular metabolic and transport systems make use of them in micro-environments where concentrations of potential substrates are considerably higher than the levels generally quoted for samples of natural waters (Andrews & Williams, 1971; Clark *et al.* 1972). It appears unlikely that diatoms are able to compete successfully with obligate heterotrophs at low or non-existent light intensities in open waters where dissolved organic compounds are present at very low concentrations (Wright & Hobbie, 1965; Williams, 1970). In marine or freshwater sediments which often have high contents of organic material, the release of suitable organic metabolites due to the action of extracellular microbial enzymes and the low light intensities frequently prevailing would favour facultative heterotrophy of benthic and tychoplanktonic diatoms. However, even in these environments the most important heterotrophic organisms are probably bacteria (Munro & Brock, 1968). Diatoms probably utilize their facultative heterotrophic abilities mainly for survival or slow growth during periods of time when light is limiting or unavailable. In highly eutrophic bodies of water, such as sewage treatment ponds or river systems carrying organic pollutants, and where light conditions are often poor, facultative heterotrophy is probably also important for the survival of photosynthetic diatoms. Amino acids may be important for diatom growth in some environmental situations, when inorganic nitrogen is extremely low (Wheeler *et al.* 1974).

Our inability to reach more definite conclusions about the role of heterotrophy in the ecology of diatoms is due to: (a) the great complexity of the natural environment, our lack of knowledge of the identity and concentrations of organic substances in the many possible microhabitats, and the changes in availabilities of such substrates with time; (b) our incomplete knowledge of the range of heterotrophic capabilities of diatoms and their regulation in response to environmental factors; some surprising species and clonal differences have already been discovered; and (c) insufficient knowledge of the possible role of antibacterial and antifungal secretions by diatoms in habitats where there might be competition for organic substrates (cf. Gauther, 1969; Aubert, 1971; Linkins, 1973).

Contribution No. 810 from the Department of Oceanography, University of Washington, Seattle, Wash. 98195, USA.

Support for the preparation of this chapter was given by a grant A 3062 from the National Research Council of Canada and grant GA 27498 from the National Science Foundation (USA).

6 REFERENCES

ANDREWS P. & WILLIAMS P.J. LEB. (1971) Heterotrophic utilization of dissolved organic compounds in the sea. III. Measurement of the oxidation rates and concentrations of glucose and amino acids in sea water. *J. mar. biol. Ass. U.K.* **50,** 859–70.

ANTIA N.J. & CHENG J.Y. (1970) The survival of axenic cultures of marine planktonic algae from prolonged exposure to darkness at 20°C. *Phycologia* **9,** 179–83.

ANTIA N.J., KRIPPS R.S. & DESAI I.D. (1975) L-threonine deaminase in marine planktonic algae. V. Studies on the specificity and pH-dependence of feedback inhibition by L-isoleucine. Evidence for feedback regulation by L-valine. *J. Phycol.* **10,** 60–9.

AUBERT M. (1971) Télémédiateurs chimiques et équilibre biologique océanique. Deuxième Partie. Nature chimique de l'inhibiteur de la synthèse d'un antibiotique produit par un diatomée. *Rev. Int. Océanogr. Méd.* **21,** 17–22.

BENECKE W. (1900) Über farblose Diatomeen der Kieler Förde. *Pringsheim's Jahrb. d. wiss. Bot.* **35,** 535–47.

BOHLIN H. (1970) Untersuchungen über freie gelöste Aminosäuren in Meerwasser. *Mar. Biol.* **6,** 213–25.

BUNT J.S. (1969) Observations on photoheterotrophy in a marine diatom. *J. Phycol.* **5,** 37–42.

BUNT J.S. & LEE C.C. (1972) Data on the composition and dark survival of four sea-ice microalgae. *Limnol. Oceanogr.* **17,** 458–61.

CHENG J.Y. & ANTIA N.J. (1970) Enhancement by glycerol of photoheterotrophic growth of marine planktonic algae and its significance to the ecology of glycerol pollution. *J. Fish. Res. Bd. Canada* **27,** 335–46.

CHOLNOKY B.J. (1929) Symbiose zwischen Diatomeen. *Arch. f. Protistenkunde* **68,** 523–30.

CHOLNOKY B.J. (1968) *Die Ökologie der Diatomeen in Binnengewässern.* J. Cramer, Lehre, Germany.

CHOLNOKY-PFANNKUCHE K. (1965) Protoplasmatische Untersuchungen an Salz- und Süsswasserdiatomeen. *Protoplasma* **40,** 345–54.

CLARK M.E., JACKSON G.A. & NORTH W.J. (1972) Dissolved free amino acids in southern California coastal waters. *Limnol. Oceanogr.* **17,** 749–58.

COOKSEY K.E. (1972) Metabolism of organic acids by a marine pennate diatom. *Pl. Physiol.* **50,** 1–6.

COOKSEY K.E. (1974) Acetate metabolism by whole cells of *Phaeodactylum tricornutum* Bohlin. *J. Phycol.* **10,** 253–7.

COOKSEY K.E. & CHANSANG H. (1974) Glucose metabolism by a marine diatom of the genus *Amphora. Pl. Physiol.* **51** (Supplement, June, 1974), 12 (Abstract).

DARLEY W.M. & OHLAMN C.T. (1974) Utilization of lactate by the marine pennate diatom *Cylindrotheca fusiformis* Reimann & Lewin. *J. Phycol.* 10 (Supplement, June, 1974), 5 (Abstr.).

DROOP M.R. & MCGILL S. (1966) The carbon nutrition of some algae. The inability to utilize glycollic acid for growth. *J. mar. biol. Ass. U.K.* **46,** 679–84.

GAUTHER M. (1969) Activité antibactérienne d'une diatomé marine: *Asterionella notata* (Grun.). *Rev. Int. Océanogr. Méd.* **35–36,** 103–71.

GEITLER L. (1968) Auxosporenbildungen bei einigen pennaten Diatomeen und *Nitzschia flexoides* n. sp. in der Gallerte (colloidal mass) von *Ophrydium versatile. Oesterr. Bot. Z.* **115,** 482–90.

GUILLARD R.R.L. (1963) Organic sources of nitrogen for marine centric diatoms. In *Marine Microbiology* (ed. C.H. Oppenheimer) pp. 93–104. C.C. Thomas, Springfield, Ill.

HAYWARD J. (1965) Studies on the growth of *Phaeodactylum tricornutum* (Bohlin) I. The effect of certain organic nitrogenous substances on growth. *Physiol. Pl.* **18,** 201–7.

HEBER U. (1974) Metabolite exchange between chloroplasts and cytoplasm. *Ann. Rev. Pl. Physiol.* **25,** 393–42.

HELLEBUST J.A. (1970) The uptake and utilization of organic substances by marine phytoplankters. In *Symposium on Organic Matter in Natural Waters* (ed. D.W. Hood) pp. 223–56. *Inst. Mar. Sci. Occas. Publ.* No. 1, Univ. of Alaska, Fairbanks.

HELLEBUST J.A. (1971a) Kinetics of glucose transport and growth of *Cyclotella cryptica* Reimann, Lewin & Guillard. *J. Phycol.* **7,** 1–4.

HELLEBUST J.A. (1971b) Glucose uptake by *Cyclotella cryptica*: dark induction and light inactivation of transport system. *J. Phycol.* **7,** 345–9.

HELLEBUST J.A. & GUILLARD R.R.L. (1967) Uptake specificity for organic substrates by the marine diatom *Melosira nummuloides. J. Phycol.* **3,** 132–6.

HELLEBUST J.A. & LEWIN J. (1972) Transport systems for organic acids induced in the marine pennate diatom *Cylindrotheca fusiformis*. *Can. J. Microbiol.* **18,** 225–33.

HIRT G., TANNER W. & KANDLER O. (1971) Effect of light on the rate of glycolysis in *Scenedesmus obliquus*. *Pl. Physiol.* **47,** 841–3.

IGNATIADES L. & SMAYDA T.J. (1970) Autecological studies on the marine diatom *Rhizosolenia fragilissima* Bergon. II. Enrichment and dark viability experiments. *J. Phycol.* **6,** 357–64.

JOLLEY E.T. & HELLEBUST J.A. (1974) Preliminary studies on the nutrition of *Navicula pelliculosa* (Breb.) Hilse., and an associated bacterium, *Flavobacterium* sp. *J. Phycol.* **10** (Supplement, June, 1974), **7** (abstr.).

KUENZLER E.J. & PERRAS J. (1965) Phosphatases of marine algae. *Biol. Bull.* **128,** 271–84.

LEE J.J., MCENERY M.E., KENNEDY E.M. & RUBIN H. (1975) A nutritional analysis of a sublittoral diatom assemblage epiphytic on *Enteromorpha* from a Long Island salt marsh. *J. Phycol.* **11,** 14–49.

LEWIN J.C. (1953) Heterotrophy in diatoms. *J. gen. Microbiol.* **9,** 305–13.

LEWIN J.C. (1955) The capsule of the diatom *Navicula pelliculosa*. *J. gen. Microbiol.* **13,** 162–9.

LEWIN J.C. (1961) The dissolution of silica from diatom walls. *Geochim. Cosmochim. Acta* **21,** 182–98.

LEWIN J.C. (1963) Heterotrophy in marine diatoms. In *Symposium on Marine Microbiology* (ed. C.H. Oppenheimer) pp. 229–35. C.C. Thomas, Springfield, Ill.

LEWIN J.C. & GUILLARD R.R.L. (1963) Diatoms. *Ann. Rev. Microbiol.* **17,** 373–414.

LEWIN J. & HELLEBUST J.A. (1970) Heterotrophic nutrition of the marine pennate diatom *Cylindrotheca fusiformis*. *Can. J. Microbiol.* **16,** 1123–9.

LEWIN J. & HELLEBUST J.A. (1975) Heterotrophic nutrition of the marine pennate diatom *Navicula pavillardi* Hustedt. *Can. J. Microbiol.* **21,** 1335–42.

LEWIN J. & HELLEBUST J.A. (1976) Heterotrophic nutrition of the marine pennate diatom *Nitzschia angularis* var. *affinis* (Grun.) Perag. *Marine Biology* **36,** 313–20.

LEWIN J. & HELLEBUST J.A. Heterotrophic nutrition of the marine pennate diatom *Nitzschia laevis* Hust. (in preparation).

LEWIN J.C. & LEWIN R.A. (1960) Auxotrophy and heterotrophy in marine littoral diatoms. *Can. J. Microbiol.* **6,** 127–34.

LEWIN J. & LEWIN R.A. (1967) Culture and nutrition of some apochlorotic diatoms of the genus *Nitzschia*. *J. gen. Microbiol.* **46,** 361–7.

LINKINS A.E. (1973) Uptake and utilization of glucose and acetate by a marine chemoorganotrophic diatom, clone Link 001. Ph.D. Thesis, University of Massachusetts, Amherst.

LIU M.S. & HELLEBUST J.A. (1974a) Uptake of amino acids by the marine centric diatom *Cyclotella cryptica*. *Can. J. Microbiol.* **20,** 1109–18.

LIU M.S. & HELLEBUST J.A. (1974b) Utilization of amino acids as nitrogen sources, and their effects on nitrate reductase in the marine diatom *Cyclotella cryptica*. *Can. J. Microbiol.* **20,** 1119–25.

LYLIS J.C. & TRAINOR F.R. (1973) The heterotrophic capabilities of *Cyclotella meneghiniana*. *J. Phycol.* **9,** 365–9.

MATSUE Y. (1954) On the culture of the marine diatom *Skeletonema costatum* (Grev.) Cleve. *Rev. Fish. Science, Jap. Soc. Adv. Sci.,* Tokyo; 41 pp. (Japanese).

MCLACHLAN J. & CRAIGIE J.S. (1966) Chitan fibres in *Cyclotella cryptica* and growth of *C. cryptica* and *Thalassiosira fluviatilis*. In *Contemporary Studies in Marine Science* (ed. H. Barnes) pp. 511–7. George Allen & Unwin, London.

MUNRO A.L.S. & BROCK T.D. (1968) Distribution between bacterial and algal utilization of soluble substances in the sea. *J. gen. Microbiol.* **51,** 35–42.

NG H. (1969) Effect of decreasing growth temperature on cell yield of *Escherichia coli. J. Bact.* **98,** 232–7.

NISSEN P. (1974) Uptake mechanisms: inorganic and organic. *Ann. Rev. Pl. Physiol.* **25,** 53–79.

NORTH B.B. & STEPHENS G.C. (1972) Amino acid transport in *Nitzschia ovalis* Arnott. *J. Phycol.* **8,** 64–8.

PALMER E.G. & TOGASAKI R.K. (1971) Acetate metabolism by an obligate photo-autotrophic strain of *Pandorina morum. J. Protozool.* **18,** 640–4.

RHO J.H. (1959) Some aspects of the metabolism of the marine diatom *Nitzschia closterium* (Ehrenberg) Wm. Smith. *Internat. Oceanogr. Congr. Preprints,* AAAS, Washington, D.C., pp. 199–200.

SAUBERT S. (1957) Amino acid deamination by *Nitzschia thermalis* and *Scenedesmus bijugatus. South African J. Sci.* **53,** 335–8.

SCHOEMAN F.R. (1972) A further contribution to the diatom flora of sewage enriched waters in southern Africa. *Phycologia* **11,** 239–45.

SCHREIBER E. (1927) Die Reinkultur von marinem Phytoplankton und deren Bedeutung für die Erforschung der Produktionsfähigkeit des Meerwassers. *Wiss. Meeresunters. Helgoland N.F.* **16,** 1–34.

SMAYDA T.J. & MITCHELL-INNES B. (1974) Dark survival of autotrophic, planktonic marine diatoms. *Mar. Biol.* **25,** 195–202.

STADELMANN E.J. (1962) Permeability. In *Physiology and Biochemistry of Algae* (ed. R.A. Lewin) pp. 493–528. Academic Press, New York.

STEPHENS G.C. & NORTH B.B. (1971) Extrusion of carbon accompanying uptake of amino acids by marine phytoplankters. *Limnol. Oceanogr.* **16,** 752–7.

SWIFT E. (1967) Isolation, description, purification, nutrition, and physiology of unicellular algae from the tropical Atlantic Ocean. Ph.D. Thesis. The John Hopkins University. 144 pp. Univ. Microfilms, Ann Arbor, Mich.

THOMAS W.H. (1966) Surface nitrogenous nutrients and phytoplankton in the northeastern tropical Pacific Ocean. *Limnol. Oceanogr.* **11,** 393–400.

UMEBAYASHI O. (1972) Preservation of some cultured diatoms. *Bull. Tokai Reg. Fish. Res. Lab.* **No. 69,** 55–62.

WERNER D. (1967) Hohe Aktivitäten von Laktatdehydrogenase bei aerob wachsenden Diatomeen. *Naturwissenschaften* **54,** 474–5.

WHEELER P.A., NORTH B.B. & STEPHENS G.C. (1974) Amino acid uptake by marine phytoplankters. *Limnol. Oceanogr.* **19,** 249–59.

WHITE A.W. (1972) Facultative heterotrophy in marine diatoms. Ph.D. Thesis. Harvard University. 179 pp. Cambridge, Mass.

WHITE A.W. (1974a) Growth of two facultatively heterotrophic marine centric diatoms. *J. Phycol.* **10,** 292–300.

WHITE A.W. (1974b) Uptake of organic compounds by two facultatively heterotrophic marine centric diatoms. *J. Phycol.* **10,** 433–8.

WILLIAMS P.J. LEB. (1970) Heterotrophic utilization of dissolved organic compounds in the sea. I. Size distribution of population and relationship between respiration and incorporation of growth substances. *J. mar. biol. Ass. U.K.* **50,** 859–70.

WRIGHT R.T. & HOBBIE J.E. (1965) The uptake of organic solutes by planktonic bacteria and algae. *Ocean Sci. Ocean. Eng.* **1,** 116–27.

CHAPTER 7

BIOCHEMICAL COMPOSITION

W. MARSHALL DARLEY

Department of Botany, University of Georgia,
Athens, Georgia 30602, USA

1 INTRODUCTION

Information on the biochemical composition of diatoms is useful for several reasons, most of which relate to their role as primary producers in aquatic food chains. Knowledge of the cellular content of major classes of metabolites and how the content may vary with environmental conditions is relevant to the suitability of diatoms as a food source for primary consumers. In the measurement of diatom standing crops and their primary productivity, it is useful to be able to relate one aspect of cellular composition to other, less easily measured, parameters. Biochemical composition may also aid in the characterization of the physiological condition of natural populations of diatoms since certain deficiencies are known to affect composition in predictable patterns. Finally, biochemical composition may have phylogenetic implications.

Data on cellular composition are reported in the literature on the basis of a variety of parameters and it is not obvious which is to be preferred. Consider the growth of an individual cell reproducing by binary fission as is the case with diatoms. On a per cell basis, every cellular component will vary by at least a factor of two in the course of one cell cycle. In an analysis of a population of cells in log phase growth under constant and optimal conditions, such as may be found in laboratory chemostat cultures, one obtains an average value for the population which represents cell composition at the mid-point of the cell cycle. In laboratory batch culture or in natural populations, however, conditions are seldom constant and one may expect that the population will not be randomly distributed with respect to progress through the cell cycle. Diurnal periodicity in illumination, to take only one example, is known to have a synchronizing effect on cell division and cause periodicity in various physiological parameters (Eppley *et al.* 1971). Analysis of such a population, even if it is healthy and growing vigorously, on a per cell basis at different times of the day will yield different results depending on the phase of the cell cycle occupied by the majority of the cells. The problem becomes more complex if one considers the effect of any number of factors that may limit growth or, in the extreme case, result in a 'stationary phase' population. Cells under such conditions tend to accumulate and eventually will be arrested in different phases of the cell cycle depending on the particular deficiency; further metabolic rearrangement may then occur, again a function of the particular deficiency. On the one hand, it is obvious that data expressed on a per cell basis are difficult to interpret unless the growth conditions are precisely determined, a situation which is seldom encountered. On the other hand, cell number is one of the easiest parameters to measure. In many cases, especially for comparative purposes in laboratory work, an analysis on a per cell basis will be satisfactory or even desirable.

In many cases composition data are expressed relative to some measure of biomass. The use of cell volume is to be avoided with diatoms because of the large vacuole in centric species and because of the diminution in cell size that usually accompanies asexual reproduction. It has been demonstrated, for example, that protein and chlorophyll do not decrease in strict proportion to the decreases in cell volume in *Coscinodiscus asteromphalus* (Werner, 1971b). Strathman (1967) found that plasma volume was more appropriate than total cell volume in estimating cell carbon in planktonic diatoms. Dry weight can also be troublesome in diatoms because of the presence of the siliceous cell wall which is a highly variable percentage of the dry weight in different species (Lewin, 1962; Nalewajko, 1966). Ash-free dry weight is a better value on which to base composition data and approaches in utility what is perhaps the best measure, that of cell carbon (Strickland, 1965). Both of these parameters are a reasonable reflection of the living component of the cell with the possible exception of situations in which large amounts of storage products are accumulated or in organisms with thick organic cell walls, as in armoured dinoflagellates. Of the two, cell carbon was shown to give a better estimate of the caloric content of phytoplankton (Platt & Irwin, 1973). In measurements of natural waters, however, non-living organic particulate matter may easily interfere with biomass determinations. Indirect approximations of living biomass have been obtained through the measurement of chlorophyll *a* (Strickland, 1965; Jeffrey, 1974) or ATP (Holm-Hansen & Booth, 1966; Holm-Hansen, 1970) both of which are rapidly degraded upon the death of the cell. These parameters are subject to variation depending on the physiological condition of the cells, but can serve as a guide to estimate cell biomass in situations where large amounts of detritus are encountered.

2 MAJOR CLASSES OF METABOLITES

Tables 7.1, 7.2 and 7.3 present data on the gross biochemical composition of representative diatoms under both nutrient sufficient and deficient conditions. It is evident from the numerous blank spaces in these tables that few comprehensive studies have been conducted. Even where these attempts have been made (e.g. Parsons *et al.* 1961), the large number of variables that are known to affect cell composition necessitate that only a few of the possible growth conditions are chosen for investigation if the study is to be a detailed one. In addition, culture conditions are not always well defined and the biochemical methods employed are not always the same. It is therefore difficult to come to any meaningful conclusions from these data.

In spite of these misgivings, an inspection of Tables 7.1, 7.2 and 7.3 would seem to indicate that there is a great deal of variation among species, even when analysed in the same laboratory under similar conditions (Parsons *et al.* 1961). In nutrient-sufficient cells, protein content almost always exceeds carbohydrate content which, in turn, usually exceeds the lipid content. Protein would appear to be the least variable of the principal cell components. The biochemical composition of nutrient-sufficient diatoms is very similar to that of other marine phytoplankton (Parsons *et al.* 1961).

2.1 *Protein*

An analysis of the distribution of component amino acids from total cellular protein yields only an average composition of the hundreds of different proteins which reflect a given species' unique genetic information. Therefore, one would expect that consistent differences in amino acid composition between species and between groups of algae would be difficult to distinguish without resorting to the isolation and analysis of individual proteins. Nevertheless, a comparison of 6 diatoms with 25 species of phytoplankton (Chuecus & Riley, 1969b; Chau *et al.* 1967) suggests that diatoms are characterized by greater relative abundance of serine, higher concentrations of glycine than glutamic and aspartic acids and by the presence of arginine and isoleucine at higher concentrations than aspartic acid and lysine respectively.

Table 7.1. Biochemical composition of diatoms as percent dry weight in nutrient sufficient and deficient cells

Organism	Limiting nutrient (if any)	Protein		Carbohydrate		Lipid		Chl a		Ash		Reference
		+	–	+	–	+	–	+	–	+	–	
Chaetoceros sp.		35		6·6		6·9				28		Parsons *et al.* (1961)
Skeletonema costatum		37		20·8		4·7				39		Parsons *et al.* (1961)
Coscinodiscus sp.		17		4·1		1·8				57		Parsons *et al.* (1961)
Phaeodactylum tricornutum		33		24·0		6·6				7·6		Parsons *et al.* (1961)
Phaeodactylum tricornutum		47		23		8				22		Ansell *et al.* (1964)
Navicula pelliculosa	Si	42	39	21	15	25	34	1·8	1·6	8·2	8·0	Coombs *et al.* (1967a)
Cyclotella cryptica	N	33	10	28	50	13	15					Werner (1966)
	P	33	21	26	35	13	13					Werner (1966)
	Si	31	25	26	15	12	30					Werner (1966)
Coastal phyto-plankton*	N	50	25	17	48	27	24					Eppley *et al.* (1971)

* Mostly diatoms, ash-free dry weight.
\+ Indicates nutrient sufficient cells.
– Indicates cells deficient in the indicated nutrient.

Table 7.2. Biochemical composition of diatoms expressed as a ratio involving total cellular carbon

Organism	Limiting nutrient (*if any*)	*Prot:C*		*Carbo:C*		*Lipid:C*		*Chl:C*		*C:P*		*C:N*		
		+	−	+	−	+	−	+	−	+	−	+	−	
Chaetoceros sp.		1·12		0·22		0·21				21		5·5		Parsons *et al.* (1961)
Skeletonema costatum		1·38		0·79		0·17				15		4·4		Parsons *et al.* (1961)
Coscinodiscus sp.		1·08		0·27		0·11				40		5·9		Parsons *et al.* (1961)
Phaeodactylum tricornutum		0·88		0·64		0·17				19		7·0		Parsons *et al.* (1961)
Ditylum brightwellii		1·40		0·13		0·24		0·045		21		4·5		Strickland *et al.* (1969)
Thalassiosira fluviatilis	N							0·043	0·010			6·3	43·5	Hobson & Pariser (1971)
Thalassiosira pseudonana	N			0·12	1·2			0·019	0·002			4·2	11·1	Hobson & Pariser (1971)
Thalassiosira pseudonana	N							0·035	0·011			5·1	12·1	Eppley & Renger (1974)
Navicula pelliculosa	Si	1·0	0·96	0·48	0·37	0·60	0·82	0·043	0·035					Coombs *et al.* (1967a)
Coastal phyto-plankton*	N	1·17	0·55	0·40	1·06	0·57	0·53					5·2	5·9	Eppley *et al.* (1971)
Coastal phyto-plankton*	N	2·0	1·3	0·2	1·1	0·15	0·35	0·04	0·017	20	33	3·0	4·7	Antia *et al.* (1963)

* Mostly diatoms.
+ Indicates nutrient sufficient cells.
− Indicates cells deficient in the indicated nutrient.

Table 7.3. Biochemical composition of diatoms expressed as a ratio involving chlorophyll *a*

Organism	*Limiting nutrient (if any)*	*Carbon : Chl* +	*Carbon : Chl* −	*Protein : Chl* +	*Protein : Chl* −	*Carbo. : Chl* +	*Carbo. : Chl* −	*Lipid : Chl* +	*Lipid : Chl* −	*Carotenoid : Chl* +	*Carotenoid : Chl* −	*Nitrogen : Chl* +	*Nitrogen : Chl* −	*Reference*
Chaetoceros sp.		44		50		10		10		0·26		8		Parsons *et al.* (1961)
Skeletonema costatum		26		36		21		5		0·28		6		Parsons *et al.* (1961)
Coscinodiscus sp.		69		75		19		8		0·31		12		Parsons *et al.* (1961)
Phaeodactylum tricornutum		26		23		17		5		0·44		4		Parsons *et al.* (1961)
Ditylum brightwellii		22		31		3		5·2				4·9		Strickland *et al.* (1969)
Navicula pelliculosa	Si	23	28	23	26	11	10	14	23					Coombs *et al.* (1967a)
Coastal phyto-plankton*	N	25	60	60	100	6	60	5	17	0·95	1·15	10	16	Antia *et al.* (1963)

* Mostly diatoms.
+ Indicates nutrient sufficient cells.
− Indicates cells deficient in the indicated nutrient.

Table 7.4. Amino acid composition of diatoms (percent of total amino acids). Due to acid hydrolysis tryptophan was omitted in the calculations.

	Average (6 species)[1]			*Average (6 species)*[2]	
Amino acids	*Av.*	*Min.*	*Max.*	*Cell contents*	*Cell wall*
Alanine	9·5	7·7	11·5	8·2	8·3
Arginine	8·5	7·1	9·6	1·6	1·8
Aspartic acid	7·2	6·0	8·1	12·0	11·2
Cysteine	1·4	1·2	1·8	0·5	0·2
Glutamic acid	7·2	5·9	9·0	12·4	8·7
Glycine	9·6	8·0	11·0	10·5	14·8
Histidine	1·5	1·1	1·8	0·2	0·5
Isoleucine	4·4	3·6	5·1	4·7	4·1
Leucine	8·6	7·7	10·2	9·1	6·9
Lysine	2·9	2·1	4·1	4·0	2·9
Methionine	1·3	0·5	1·9	1·8	0·6
Phenylalanine	4·1	3·1	4·7	4·4	3·0
Proline	4·0	3·3	4·5	6·5	5·2
Serine	10·4	7·6	12·2	9·0	17·1
Threonine	6·8	6·1	8·1	6·3	7·5
Tyrosine	0·9	0·4	1·4	2·7	1·5
Valine	2·3	2·1	2·8	6·2	5·8

[1] Chuecas & Riley (1969b).
[2] Hecky *et al.* (1973).

These data are reproduced in Table 7.4 along with averaged data from a study comparing the amino acid distribution in the cell walls of representative diatoms to that of cell contents (Hecky *et al.* 1973). Glutamic acid accounts for approximately one-half of the free amino acids in *Coscinodiscus asteromphalus* (Werner, 1971b). There are also several less extensive reports on amino acid composition in diatoms (Fowden, 1954; Parsons *et al.* 1961; Cowey & Corner, 1963; Ogino, 1963; Wallen & Green, 1971a,c). Analyses by different workers, even if on the same organism, often show little similarity, possibly a result of differences in growth conditions and/or analytical methods.

Two other studies of nitrogenous compounds in diatoms are worth noting. Proline accumulation has been reported in *Cyclotella meneghiniana* when placed under high osmotic stress (Schobert, 1974); a similar phenomenon has been reported in higher plants. Lui & Roels (1972) found that the distribution of nitrogen in *Biddulphia aurita* varied as a function of the exogenous nitrogen source. When growing on ammonium, the percentage of soluble nitrogen in the cells almost doubled as compared to cells growing on nitrate. The most marked change in the soluble nitrogen fraction was a large increase in the relative abundance of amide nitrogen.

Table 7.5. Sugar composition of diatoms (percent of total carbohydrate)

		Skeletonema costatum[2]			*Average (6 species)*[3]	
Sugar	*Average (4 species)*[1]	*Whole cells*	*Hot H_2O extract*	*Hot H_2O residue*	*Cell contents*	*Cell wall*
Glucose	53	48	91	6	63	24
Galactose	12	2	—	3	10	11
Mannose	10	31	1	55	13	30
Xylose	2	7	tr	11	4	16
Arabinose	—	3	5	2	tr	tr
Ribose	4	1	1	1	2	1
Fucose	4	5	tr	10	4	15
Rhamnose	14	6	1	12	3	3

[1] Parsons *et al.* (1961).
[2] Handa (1969).
[3] Hecky *et al.* (1973).

2.2 *Carbohydrate*

The principal polysaccharide in diatoms is chrysolaminaran, a water-soluble β-($1 \rightarrow 3$) glucan with a small degree of branching at C-6, containing an average of 21 glucose units (Beattie *et al.* 1961; Ford & Percival, 1965a). It is similar to laminaran found in the brown algae but lacks the terminal mannitol residues. Its function as a storage product is supported by Handa (1969) who observed a rapid decline in the water-soluble carbohydrate fraction when cells of *Skeletonema costatum* were placed in the dark.

The abundance of chrysolaminaran is reflected in the distribution of total monosaccharides in diatoms as well as in the predominance of glucose in the hot water extract of *Skeletonema costatum* (Table 7.5). Allan *et al.* (1972) also fractionated the carbohydrate component of several diatoms and reported similar results, although they reported several unidentified sugars in the hot aqueous insoluble fraction. A similar trend in relative abundance is evident in the sugar distribution of the hot aqueous insoluble fraction (Handa, 1969) and in the cell walls (Hecky *et al.* 1973) suggesting that cell wall polysaccharides represent a major component of the former. The average values listed in Table 7.5 obscure the very large quantitative differences observed among species (Allan *et al.* 1972; Hecky *et al.* 1973; Parsons *et al.* 1961).

There is very little information on free sugars and related low molecular weight compounds in diatoms. *Phaeodactylum tricornutum* contains glucose, laminaribiose, laminaritriose, laminitol, myoinositol and scyllitol (Ford & Percival, 1965a); free glucose has also been reported in other diatoms (Handa, 1969; Wallen & Geen, 1971a,c). More species should be investigated to determine if diatoms have a characteristic soluble product comparable to sucrose in the green algae or mannitol in the brown algae.

Many diatoms also produce extracellular polysaccharides in the form of threads, gelatinous capsules, mucus pads, stalks or tubes. The composition of the macromolecular material varies widely in the organisms examined. The capsule produced by stationary phase cells of *Navicula pelliculosa* is a polyuronide composed of glucuronic acid residues (Lewin, 1955) while the capsules of oval cells of *Phaeodactylum tricornutum* contain xylose, mannose, fucose and galactose (Lewin *et al.* 1958). The mucilage tubes of *Amphipleura rutilans* are a polymer of xylose and mannose with traces of rhamnose and protein (Lewin, 1958). *Thalassiosira fluviatilis* and *Cyclotella cryptica* produce long filaments which have been characterized as a very pure crystalline form of chitin (Falk *et al.* 1966; Blackwell *et al.* 1967). A detailed analysis of the stalk material of *Gomphonema olivaceum* revealed a β-linked polysaccharide composed of D-galactose, D-xylose and an alkali-stable sulphate ester in the ratio of 1:1:1·2 (Huntsman & Sloneker, 1971). Two possible structures were proposed.

2.3 *Lipids*

For many years lipids were believed to be the only storage product in diatoms. It is perhaps for this reason and also because of the importance of diatoms in aquatic food chains that a relatively large number of studies have been directed at their lipid composition.

The overall lipid composition of diatoms closely resembles that of green algae and higher plants (Kates & Volcani, 1966; Opute, 1974a,b,c). The major components are triglyceride, sulphoquinovosyl diglyceride, monogalactosyl diglyceride, digalactosyl diglyceride, lecithin, phosphatidyl glycerol and phosphatidyl inositol. Phosphatidyl ethanolamine is present in low concentrations in autotrophically grown cells, but is more abundant when the cells are grown at low light intensities or in the dark (Opute, 1974b). Phosphatidyl serine has not been detected in diatoms. There are no apparent qualitative differences between marine and freshwater or pennate and centric species, although only one centric species has been examined for total lipids. At least two unknown sulpholipids are present in diatoms; evidence suggests that their synthesis is enhanced in darkness (Kates & Volcani, 1966; Opute, 1974a).

Most studies of diatom lipids concentrate on the fatty acid composition. Diatoms are characterized by an unusual distribution of fatty acids compared to green algae and higher plants (Table 7.6). The C_{14}, C_{16} and C_{20} acids comprise the bulk of the diatom fatty acids while the unsaturated C_{18} acids, particularly linolenic acid, 18:3, are either absent or present at very low levels. Among other groups of algae only the Chrysophyta, Rhodophyta and Pyrrophyta are known to have low 18:3 content (Ackman *et al.* 1968;

Table 7.6. Fatty acid composition in diatoms (percent of total fatty acids)

Fatty acids[1]	*Navicula pelliculosa*[2]	*Nitzschia palea*[3]	*Cylindrotheca gracilis*[4]	*Phaeodactylum tricornutum*[5]	*Nitzschia alba*[6]	*Skeletonema costatum*[5]	*Thalassiosira fluviatilis*[7]
14:0	2·8	6·2	7·0	8·6	30·0	6·2	7·9
16:0	9·1	22·8	16·4	10·7	20·9	11·1	23·2
16:1	30·8	44·7	21·3	27·3	8·5	21·7	44·8
16:2	7·4	3·6	4·2	13·4	13·4	6·1	2·8
16:3	18·3	1·6	—	9·9	ND	11·4	6·5
18:0	tr	tr	1·0	0·1	tr	—	0·3
18:1	6·2	2·5	5·3	4·7	23·6	1·8	0·4
18:2	3·9	tr	2·9	0·5	4·7	2·1	0·5
18:3	2·6	tr	—	0·2	tr	—	0·2
20:4	4·5	6·3	6·2	0·7	tr	3·9	0·6
20:5	14·5	12·0	24·4	18·2	10·8	30·2	8·0

1. The first number indicates chain length, the second number indicates the number of double bonds.
2. Kates & Volcani (1966), 3. Opute (1974c), 4. DeMort *et al.* (1972), 5. Chuecas & Riley (1969a), 6. Ackman *et al.* (1968), 7. Tornabene *et al.* (1974).
ND = not determined; tr = trace.

Chuecas & Riley, 1969a). Linolenic acid is the major unsaturated fatty acid in green plants and has been implicated in chloroplast function, but its role is apparently performed by a combination of unsaturated C_{16} and C_{20} acids in the diatom galactolipids (Kates & Volcani, 1966). This suggestion has been given support by Opute (1974c) who analysed the fatty acid component of individual lipids and found high proportions of unsaturated C_{16} and C_{20} fatty acids in the galactolipids. While a detailed analysis of the lipids of isolated chloroplasts would be interesting in this regard, we do have a study on a non-photosynthetic diatom, *Nitzschia alba* (Tornabene *et al.* 1974). As shown in Table 7.6, the qualitative fatty acid composition is similar to photosynthetic diatoms, but a higher proportion of 14:0 and 18:1 and a lower proportion of 16:1 was noted. The consistent high levels of polyunsaturated C_{20} acids suggests involvement of these acids in non-chloroplast membranes as well.

There are no apparent consistent differences in fatty acid composition among freshwater, marine, pennate or centric diatoms to be seen in the representative species listed in Table 7.6. Differences as great as those between species are found among studies on the same organism from different laboratories (data not shown). Even different strains of the same species, examined in the same laboratory show significant quantitative differences (Tornabene *et al.* 1974). It is useful, however, to compare the fatty acid composition of the same organism grown under different conditions in the same laboratory. Fatty acid composition as a function of culture age has been investigated (Ackman *et al.* 1964; Pugh, 1971), although the factor or factors inducing stationary phase are usually not known. Different mineral deficiencies are known to have different effects on cell composition (see page 213) and therefore it is important to know the limiting factor under the conditions employed. Pugh (1971) found that in *Coscinodiscus eccentricus* the C_{16} group of acids increased their relative importance as the culture aged while the C_{18} acids decreased. C_{20} acids as a group increased their relative abundance until the beginning of the stationary phase after which they declined. There were no major effects of salinity.

Recent work has demonstrated the presence of a unique 21:6 hydrocarbon (heneicosahexane) in diatoms and other, but not all, groups of algae (Blumer *et al.* 1971; Lee & Loeblich, 1971). Among diatoms and dinoflagellates it appears to be limited to photosynthetic species suggesting a function in chloroplast membranes (Lee & Loeblich, 1971). This hydrocarbon has been traced up the food chain to certain species of copepods (Blumer *et al.* 1970; Lee *et al.* 1974). Sterols have also been investigated in diatoms (Low, 1955; Tornabene *et al.* 1974; Orcutt & Patterson, 1974).

2.4 *Pigments*

Diatoms belong to the 'brown line' of algae containing chlorophyll *a* and chlorophyll *c*. Since chlorophyll *a* is often used as a measure of phytoplankton biomass in primary productivity studies, it is well to recognize that pigment content is subject to a great deal of variation, especially as a function of light intensity and nutient deficiencies (see page 212). The chlorophyll *a* content of algae generally lies in the range of 0·3–2% of the dry weight (Meeks, 1974) and diatoms are within this range (Parsons, 1961). The reported values for the chlorophyll *a*: carbon ratio usually fall between 0·01 and 0·06 (Parsons *et al.* 1961; Antia *et al.* 1963; Sloan & Strickland, 1966; Bunt, 1968; Strickland *et al.* 1969).

Chlorophyll *c* has recently been resolved into two spectrally distinct components: Chl c_1 and Chl c_2 (Jeffrey, 1969) both of which are present in diatoms. Their structural formulas have been elucidated (Strain *et al.* 1971). The ratio of Chl c_1: Chl c_2 is usually 1 in diatoms but may be higher or lower in some species. Total chlorophyll *c* accounts for between 11 and 37% of total chlorophyll (Jeffrey, 1972; Bunt, 1968) although higher values have been reported (Parsons, 1961).

The major carotenoids of the diatoms are β-carotene, fucoxanthin, diadinoxanthin, diatoxanthin, and neofucoxanthin (Strain *et al.* 1944). Their structures are given by Goodwin (1974). Fucoxanthin is the principal carotenoid (Parsons, 1961; Healey *et al.* 1967). The ratio of total carotenoids : chlorophyll *a* varies from 0·26–1·2 (Parsons *et al.* 1961; Antia *et al.* 1963; Sloan & Strickland, 1966). Bunt (1968; Bunt & Lee, 1972) reported somewhat higher ratios, in the range of 1·12–2·11, for several psychrophilic diatoms.

2.5 *Nucleic acids*

The DNA content of diatom cells is nearly proportional to cell size as determined by total organic carbon. In spite of a 60-fold difference in cell carbon among the four species examined, DNA remains between 1 and 3% of cellular carbon (Holm-Hansen, 1969). The DNA content per exponential cell of *Navicula pelliculosa* doubled after the induction of polyploid cells with colchicine (Coombs *et al.* 1968). Exponential cultures of the three different cell types of *Phaeodactylum tricornutum* were found to have the same DNA content, suggesting they do not represent an alternation of haploid and diploid generations (Darley, 1968). DNA content per nucleus was used as a measure of progress of the cell through the cell cycle in a study of the events initiating wall formation in *Cyclotella cryptica* (Oey & Schnepf, 1970). It was concluded that some event which occurs during or slightly after the DNA synthetic phase of the cell cycle triggers cell wall formation.

The percentage of guanine + cytosine (G + C) in diatom DNA is generally less than 50, but varies between 37 and 58 in the 11 species examined (Sober, 1970). In this respect diatoms resemble higher organisms more than the other groups of algae which tend to have G + C values above 50. Pennate species tend to have higher values than centric species, but more studies are needed before definite conclusions can be drawn.

Very little information is available on diatom RNA. The ratio of RNA/DNA in *Navicula pelliculosa* varied from 5·1 to 3·8 (RNA/C = 0·046–0·036) as an exponential culture was subjected to silicon starvation (Coombs *et al.* 1967a). The RNA/DNA ratio was 2 in a dark starved culture of *Cyclotella cryptica* and increased to 2·5 after 12 hours of silicon starvation in the light (Werner, 1966). RNA/C in two marine centric diatoms decreased from 0·015 and 0·008 to 0·012 and 0·002 respectively with nitrogen starvation (Hobson & Pariser, 1971). In a study of phylogenetic relationships among the algae, utilizing the molecular weights of ribosomal RNA, Soh (1974) examined three diatoms and reported values of $1{\cdot}27–1{\cdot}28 \cdot 10^6$ daltons and $0{\cdot}70–0{\cdot}71 \cdot 10^6$ daltons for the heavy and light cytoplasmic rRNAs respectively and $1{\cdot}05–1{\cdot}08 \cdot 10^6$ daltons and $0{\cdot}56 \cdot 10^6$ daltons for the heavy and light chloroplast rRNAs respectively. The pennate diatoms could be separated from the centric species examined on the basis of the degradation pattern of the heavy cytoplasmic component.

Reported values for the ATP/C ratio in diatoms fall between 0·0043 and 0·0034 (Strickland *et al.* 1969; Holm-Hansen, 1970; Eppley *et al.* 1971) and are similar to other algae examined (Holm-Hansen, 1970). The levels of total nucleoside triphosphates in two diatoms have been investigated with respect to silicon metabolism (Coombs *et al.* 1967b,c).

2.6 *Cell walls*

The siliceous frustule is formed inside a membrane-bound vesicle which persists, perhaps in a modified form, in the mature cell wall (see page 99). Total organic matter represents 16–90% of the weight of the cell wall in six species examined by Kates & Volcani (1968). Protein and polysaccharides account for the majority of the organic matter; lipid and hexosamines are minor components representing 0·8–1·3% and 0·4–14% respectively in the species examined to date (Kates & Volcani, 1968; Hecky *et al.* 1973). Uronic acids and sulphated sugars are also apparently present, but no data are available (Coombs & Volcani, 1968). The relative proportion of protein and polysaccharide is highly variable, perhaps on account of the sequence of events in wall synthesis. Coombs & Volcani (1968) found that protein is the principal organic component added to cell walls during silica deposition, while carbohydrate is continually being added to mature walls.

Analysis of the relative proportions of amino acids, sugars and fatty acids, shows that the composition of the cell wall is distinct from the cell contents (Hecky *et al.* 1973; Kates & Volcani, 1968). Cell wall protein is enriched in glycine, serine and threonine and relatively diminished in acidic and aromatic amino acids (Table 7.4). Serine is especially abundant, reaching 25 Mol% in *Cyclotella stelligera*. Hecky *et al.* (1973) propose a template function in silica deposition for the high levels of hydroxyl-containing amino acids (serine and threonine). Hydroxyproline, a characteristic component of higher plant cell walls, is present at low levels in diatom cell walls (Gotelli & Cleland, 1968). Nakajima & Volcani (1969, 1970) reported the presence of new derivatives of hydroxyproline and lysine in diatom cell walls, although data on their relative abundance were not included.

Cell wall polysaccharides are enriched in xylose and mannose; glucose is also a major component of the cell wall but is consistently less abundant than it is in the cell contents (Table 7.5). With the exception of fucose, which is more abundant in the cell wall of estuarine species, there is no apparent amino acid or monosaccharide pattern separating pennate and centric or freshwater and marine species (Hecky *et al.* 1973).

Ford & Percival (1965b) have described a hot alkali soluble

glucuronomannan from *Phaeodactylum tricornutum* containing 27% glucuronic acid and 7·5% half ester sulphate in addition to mannose. They suggest the backbone of this molecule consists of 1,3-linked mannopyranose units and that it possibly functions as a component of the cell wall. If so, it would account for the enriched mannose content of the wall noticed by Hecky *et al.* (1973).

3 ENVIRONMENTAL FACTORS AFFECTING COMPOSITION

Just as the physiological state of the cell is a function of its environment, the biochemical composition of the cell will also vary as the environment changes. These variations in cellular characteristics are adaptations which enable the cell to function better or survive longer in a modified environment. In the algae the most significant environmental parameters are nutrient availability and light.

3.1 *Nitrogen source*

It has been firmly established that the form of dissolved inorganic nitrogen determines the level of nitrate reductase in many algae, including diatoms. The form of nitrogen also influences cell composition, although there is no clear explanation for this effect. As shown in Table 7.7 there is a significant increase

Table 7.7. Biochemical composition of *Ditylum brightwellii* as affected by nitrogen source

	Ammonia grown[1]		*Nitrate grown*[1]		*Laboratory culture*[2]	
Constituent	μg	*pg/cell*	μg	*pg/cell*	μg	*pg/cell*
Carbon	100	1 350	100	1 550	100	680
Nitrogen	22·5	300	20·5	320	18·0	120
Phosphorus	4·8	65	5·1		3·4	23
Protein*	140	1 900	128	2 000	112	760
Lipid	23·5	320	38	590	—	—
Carbohydrate	13	180	11·5	180	18·5	125
Chlorophyll *a*	4·5	61	5·0	78	5·6	38
DNA	2·3	31	1·95	30	—	—
ATP	0·49	6·6	0·54	8·4	—	—
Silicon	58	780	47	730	28	190

* Protein = N × 6·25.

[1] Grown at low nutrient levels in a deep tank at 14·5°C; Strickland *et al.* (1969).

[2] Grown at high concentrations of nitrate and other nutrients at 20°C; Eppley *et al.* (1967).

in chlorophyll, phosphorus, ATP and especially lipid in *Ditylum brightwellii* cells grown on nitrate as compared to cells grown on ammonium. The differences are more evident on a per cell basis than on a carbon basis due to accumulation of lipid. These cells were grown at natural nutrient concentrations in a deep tank (Strickland *et al.* 1969) and may be compared to the composition of cells grown in the laboratory at high concentrations of nitrate and other nutrients (Table 7.7; Eppley *et al.* 1967). Although the laboratory grown cells were smaller, a comparison on the basis of carbon shows some differences, especially with respect to carbohydrate content. The nitrogen source also influenced the chemical composition of nutrient enriched samples of coastal phytoplankton (Eppley *et al.* 1971). Not as many parameters were measured, but the results were consistent with the deep tank experiment.

3.2 *Nitrogen and phosphorus deficiency*

The effects of nutrient deficiency in algae have recently been critically reviewed by Healey (1973). The most general effect of nutrient deficiency on the composition of algae is a decrease in protein and photosynthetic pigments and an increase in the storage products, carbohydrate and lipid (Healey, 1973; Strickland, 1965). Most investigations on diatoms have concerned marine forms and as a consequence the effect of nitrogen deficiency has received the most attention because this element rather than phosphorus is usually the limiting nutrient in the sea (Strickland, 1965). Although there are many studies on the effects of nutrient deficiency in diatoms, relatively few of them have dealt with aspects of gross composition, choosing instead to study growth rates or metabolic parameters. Eppley & Renger (1974) give a recent summary of this type of research, much of it with chemostat cultures.

As shown in Tables 7.1, 7.2 and 7.3, nitrogen deficiency increases the carbohydrate content and decreases the chlorophyll and protein content. Lipid synthesis is not initially stimulated by nitrogen deficiency, but does increase upon prolonged nitrogen starvation (Werner, 1966). Opute (1974d) has carried out a more detailed study of fat accumulation in *Nitzschia palea* under nitrogen starvation. The decrease in chlorophyll *a* (Tables 7.1, 7.2 and 7.3) is consistent with the observation that nitrogen deficiency caused a decrease in the number of chromatophores in four marine centric diatoms (Holmes, 1966). The effect of phosphorus deficiency on diatoms is less well known. Holmes (1966) found the effect on chlorophyll *a* and chromatophore number was similar to, but less intense than, that of nitrogen deficiency. The effect on carbohydrate synthesis was not as great as with nitrogen deficiency (Table 7.1; Werner, 1966).

3.3 *Silicon deficiency*

Diatoms are unique among the algae in requiring silicon for growth. Since silicate concentrations in nature may be limiting at certain times, it is of interest to understand how silicon deficiency affects cell composition. In general it may be stated that most of the effects on composition are indirect, resulting from a cessation of growth at later stages in the cell cycle (just prior to mitosis or during cell wall formation). The overall composition of the cell then reflects the composition at that stage of development (Darley & Volcani, 1969). Other effects would seem to be a direct result of silicon deficiency.

When light grown exponential cells of *Cyclotella cryptica* are exposed to silicon deficiency (Werner & Pirson, 1967; Werner, 1969), the cellular content of all major metabolites continues to increase for a time with the synthesis of certain constituents (e.g. protein and chlorophyll) being curtailed before others (e.g. RNA and carotenoids). Similar results were obtained with *Navicula pelliculosa* (Coombs *et al.* 1967a; Healey *et al.* 1967), although the sequence was somewhat different. A sharp decrease in the size of the α-ketoglutaric acid and glutamic acid pools represents a rather rapid and possibly direct effect of silicon deficiency (Werner, 1968). Silicon is also directly involved in DNA synthesis in *Cylindrotheca fusiformis* (Darley & Volcani, 1969), apparently on account of a silicon requirement for the synthesis of DNA polymerase (Sullivan & Volcani, 1973). An interesting feature of silicon deficiency is the continued accumulation of lipid, whereas nitrogen or phosphorus deficiency results in an accumulation of carbohydrate (Tables 7.1, 7.2 and 7.3; Werner, 1966; Coombs *et al.* 1967a). Surplus silica stored during dark incubation in *Cyclotella cryptica* has been implicated as a factor in prolonged synthesis of major metabolites once cells are returned to the light in a silicon-free medium (Werner & Pirson, 1967). Alternatively, their data may be interpreted in terms of an accumulation of cells at different stages of the cell cycle (Darley & Volcani, 1969). Busby & Benson (1973) have investigated the metabolism of the sulphonic acids, cysteinolic acid, sulphonpropanediol and sulpholipid during silicic acid deficiency in *Navicula pelliculosa*.

3.4 *Light*

Light intensity as well as light quality (wavelength) are known to affect plant metabolism (see Soeder & Stengel, 1974; Voskresenskaya, 1972). Both of these parameters can be of importance to diatoms due to the differential attenuation of light of various wavelengths with depth in aquatic environments.

Wallen & Geen (1971a,b) have considered the effect of light quality on *Thalassiosira pseudonana (Cyclotella nana)*. Blue or green light was found to influence cell metabolism in favour of protein synthesis at the expense of carbohydrate synthesis, compared to white light of the same intensity, suggesting that the composition of phytoplankton varies with depth. On a per cell basis the amount of chlorophyll, protein, RNA and DNA was higher in blue and lower in green light than in white light of the same intensity (Wallen & Geen, 1971b). Carbon values are not included but the variable level of DNA, a very conservative feature of cell composition, suggests that the results may be explained in part by differences in the average age of the cells under different treatments. Carotenoids were highest in green light indicating that chromatic adaptation had occurred. A similar study on natural phytoplankton populations was consistent with the results obtained in the above culture studies. In addition, it was observed that the percentage of release of dissolved organic carbon by the cells decreased with depth relative to total fixation (Wallen & Geen, 1971c).

With respect to light intensity, photosynthetic pigments are the most frequently examined aspect of cell composition, but the report is usually incidental to the principal part of the study. In general, cells growing at higher light intensities have a lower chlorophyll concentration per cell and a higher maximum photosynthetic rate (Strickland, 1965). This phenomenon has been reported in several diatoms (Bunt, 1968; Durbin, 1974) and has been correlated with a decrease in chloroplast size in a pennate diatom (Brown & Richardson, 1968). An increased carotenoid : chlorophyll *a* ratio at higher light intensities is consistent with these data (Bunt, 1968). Jørgensen (1969) reports no difference in chlorophyll content in four diatoms grown at 3 and 30 klux, but perhaps 3 klux is not low enough to initiate the response.

Diatoms continue to synthesize their photosynthetic pigments when growing heterotrophically in the dark and are capable of photosynthesis immediately upon return to the light (Lewin & Hellebust, 1970; White, 1974). A loss of diadinoxanthin in heterotrophic cells (Lewin & Hellebust, 1970) is consistent with a proposed photoprotective function for this pigment (Healey *et al.* 1967). White (1974) noted that the chlorophyll *a* to carbon ratio more than doubled in heterotrophic cells of *Coscinodiscus* sp., while Lewin & Hellebust (1970) reported no major change in this ratio for heterotrophic cells of *Cylindrotheca fusiformis*.

Illumination cycles are known to synchronize cell division in laboratory cultures (Pirson & Lorenzen, 1966) and in natural phytoplankton communities (Eppley *et al.* 1971). Cell composition may therefore vary over a 24 hour cycle as a consequence of differential synthesis of cell constituents both during the cell cycle and during light or dark periods. In light–dark synchronized laboratory cultures of diatoms, chlorophyll synthesis is mostly limited to the light period (Jørgensen, 1966; Eppley *et al.* 1967; Paasche, 1968). In a natural phytoplankton sample, however, chlorophyll synthesis was not periodic and as a result the carbon : chlorophyll ratio varied from 50 to 110 during the course of one day (Eppley *et al.* 1971).

Light intensity has an influence on fatty acid composition in diatoms.

Opute (1974c) found quantitative differences in *Navicula muralis* grown autotrophically or heterotrophically. Heterotrophic cultures had higher levels of 14:0, 16:0, 18:0 and 18:1, but much lower levels of 16:1 and 20:5. When an obligate autotroph, *Nitzschia palea*, was grown at light intensities approaching the compensation point, its fatty acid composition approached that typical of heterotrophic cells of *Navicula muralis*. The overall effect is to lower the degree of unsaturation as cells approach the heterotrophic condition. These studies are consistent with the comparison of the non-photosynthetic *Nitzschia alba* and photosynthetic diatoms (Table 7.6). At higher light intensities, however, other shifts in composition may occur. A comparison of the fatty acid composition of *Cylindrotheca fusiformis* at 200 and 2000 foot candles (Orcutt & Patterson, 1974) showed that the percentages of both 16:0 and 16:1 increased at the higher light intensity, while the percentage of 20:5 decreased.

4 LIFE CYCLE EFFECTS ON COMPOSITION

Diatoms not only possess a unique requirement for silicon, but many diatoms also undergo a unique and gradual diminution in cell size with continued asexual cell division. Rejuvenation to the original sized cell is usually accomplished through a sexual cycle. Visual observation of this phenomenon immediately suggests that smaller cells contain less cell components than large cells, but until recently there has been no quantitative information on the comparative biochemistry of the variously sized cells. Werner (1971a,b,c) has reported on a detailed study of the life cycle of *Coscinodiscus asteromphalus* and Paasche (1973) has provided a less intensive study on species of four other genera.

Table 7.8 summarizes the biochemical composition data from the large and small cells of *C. asteromphalus* (Werner, 1971b). On a per cell basis all

Table 7.8. Biochemical composition (ng/cell) of large and small cells of *Coscinodiscus asteromphalus* (Werner, 1971b)

	Valve diameter		*Large cell: small cell*
Constituent	*170–200 μm*	*70–80 μm*	
Protein	10·0	3·5	2·9
Chlorophyll *a*	0·7	0·28	2·5
DNA	0·07	0·07	1·0
Total free amino acids	0·45	0·29	1·6
Potassium	23·0	2·0	11·5
Magnesium	0·105	0·025	4·2
Relative cell volume	6	1	6

cell constituents decrease as cell size diminishes. On the basis of cell volume, however, the smaller cells must have higher levels of most cell constituents than the large cells since these components do not show as large a decrease as does cell volume. The cell vacuole must therefore have undergone more than a 6-fold decrease; the potassium content of the cells provides evidence for this hypothesis (Werner, 1971b). Magnesium, also a component of the vacuole, but important in cytoplasmic function as well, decreases 4-fold. Although the total free amino acid pool size decreases, the relative abundance of amino acids is unchanged. It is interesting that the free amino acids have increased relative to protein by 30% in the smaller cells. It is gratifying to note that the DNA content does not change. Paasche (1973) reported similar changes in chlorophyll and protein per cell, although he disagrees with Werner's (1971b) suggestion that cell surface area is the critical factor controlling cell synthesis at various stages of the life cycle.

5 ORGANIC MATTER EXCRETION

The release of soluble organic matter by algae is now a well known phenomenon. Since Hellebust (1974) has recently reviewed the literature on this subject, including a number of studies on diatoms, only general principles from that review and more recent work will be discussed here.

Diatoms apparently do not differ significantly from other algae in the type of extracellular product released. Polysaccharides and glycolic acid are the most frequently observed excretory products. Release tends to be highest under conditions in which cell division is inhibited and photoassimilation continues, although healthy, actively growing cells also may release significant quantities. Extracellular release is related to the physiological condition of the cells and undoubtedly also to cell composition, especially soluble components. There is little known, however, about the effect of excretion on cell composition beyond Wallen & Geen's (1971c) observation that the relative amount of released carbon was directly related to the size of the ethanol-soluble fraction which decreased with depth in natural phytoplankton samples.

In addition to the extracellular polysaccharides of structural significance (page 206), diatoms release soluble polysaccharides. Hydrolysates of such material from several pennate and centric species yielded rhamnose and mannose as the most abundant sugars (Allan *et al.* 1972). Galactose and fucose also represented a substantial component of this material in some, but not all of these species. In *Nitzschia frustulum* production of this extracellular material was found to coincide with the onset of the stationary phase of growth; its composition argues against its origin as a product of cell lysis. Its composition varied with salinity with rhamnose predominating at high salinity and mannose being the predominant sugar at low salinities (Allan *et al.* 1972).

A recent study (Ignatiades & Fogg, 1973) with cultures of *Skeletonema costatum* found that excretion by exponential phase cells amounted to 2·1–4·0% of total carbon assimilated. A detailed survey of factors affecting this release revealed that exposure to dark conditions and nutrient deficiency increased this percentage by 65 to 87%. Other factors such as increased bicarbonate ion in the medium, cell senescence and inhibitory light intensities has less of an effect. Increased population density decreased the relative amount released. These observations were consistent with results obtained with natural phytoplankton samples in which *Skeletonema costatum* was the dominant species (Ignatiades, 1973). Glycolic acid was suggested to be the major component of the extracellular material in this study. In another study (Bell *et al.* 1974) two marine bacteria were tested for their response to the extracellular products of *S. costatum.* One isolate achieved a steady state population two orders of magnitude higher in the presence of the alga than in its absence, while another isolate appeared to be inhibited by the presence of the alga. Thus algal excretory products may have a variety of roles in the ecology of natural waters.

Vitamins are another ecologically important class of compounds and are known to be released by various phytoplankters, including planktonic diatoms (Carlucci & Bowes, 1970a,b). Mixed culture experiments demonstrated the release of a vitamin by one phytoplankter and its utilization by another. A marine diatom–marine bacterium interaction has also been demonstrated (Haines & Guillard, 1974) in which the bacterium utilizes diatom excretory products for growth and in turn releases vitamin B_{12} which can satisfy the requirement for this vitamin in several marine diatoms. Carlucci & Bowes (1972) have measured the vitamin content of two diatoms. *Phaeodactylum tricornutum,* a non-vitamin requirer, contained 0·29–0·96 ng B_{12}, 5–15 ng thiamine and 0·45–1·70 ng biotin/mg C. A B_{12} requirer, *Skeletonema costatum,* contained 0·06 ng B_{12}, 5–36 ng thiamine and 0·16–2·10 ng biotin/mg C.

6 REFERENCES

ACKMAN R.G., JANGAARD P.M., HOYLE R.J. & BROCHERHOFF H. (1964) Origin of marine fatty acids. I. Analysis of the fatty acids produced by the diatom *Skeletonema costatum. J. Fish. Res. Bd. Can.* **21,** 747–56.

ACKMAN R.G., TOCHER C.S. & MCLACHLAN J. (1968) Marine phytoplankton fatty acids. *J. Fish. Res. Bd. Can.* **25,** 1 603–20.

ALLAN G.G., LEWIN J. & JOHNSON P.G. (1972) Marine polymers. IV. Diatom polysaccharides. *Bot. Mar.* **15,** 102–8.

ANSELL A.D., COUGHLAN J., LANDER K.F. & COOSMORE F.A. (1964) Studies on the mass culture of *Phaeodactylum.* IV. Production and nutrient utilization in outdoor mass culture. *Limnol. Oceanogr.* **9,** 334–42.

ANTIA N.J., MCALLISTER C.D., PARSONS T.R., STEPHENS K. & STRICKLAND J.D.H. (1963) Further measurements of primary production using a large-volume plastic sphere. *Limnol. Oceanogr.* **8,** 166–83.

BEATTIE A., HIRST E.L. & PERCIVAL E. (1961) Studies on the metabolism of the Chrysophyceae. Comparative structural investigations on leucosin (chrysolaminarin) separated from diatoms and laminarin from the brown algae. *Biochem. J.* **79,** 531–7.

BELL W.H., LANG J.M. & MITCHELL R. (1974) Selective stimulation of marine bacteria by algal extracellular products. *Limnol. Oceanogr.* **19,** 833–9.

BLACKWELL J., PARKER K.D. & RUDALL K.M. (1967) Chitin fibers of the diatoms *Thalassiosira fluviatilis* and *Cyclotella cryptica*. *J. molec. Biol.* **28,** 383–5.

BLUMER M., GUILLARD R.R.L. & CHASE T. (1971) Hydrocarbons of marine phytoplankton. *Mar. Biol.* **8,** 183–9.

BLUMER M., MULLIN M.M. & GUILLARD R.R.L. (1970) A polyunsaturated hydrocarbon (3,6,9,12,15,18-heneicosahexane) in the marine food web. *Mar. Biol.* **6,** 226–35.

BROWN T.E. & RICHARDSON F.L. (1968) The effect of growth environment on the physiology of algae: light intensity. *J. Phycol.* **4,** 38–54.

BUNT J.S. (1968) Some characteristics of microalgae isolated from Antarctic sea ice. *Ant. Res. Ser.* **11,** 1–14.

BUNT J.S. & LEE C.C. (1972) Data on the composition and dark survival of four sea ice microalgae. *Limnol. Oceanogr.* **17,** 458–61.

BUSBY W.F. & BENSON A.A. (1973) Sulfonic acid metabolism in the diatom *Navicula pelliculosa*. *Pl. Cell Physiol.* **14,** 1 123–32.

CARLUCCI A.F. & BOWES P.M. (1970a) Production of vitamin B_{12}, thiamine and biotin by phytoplankton. *J. Phycol.* **6,** 351–7.

CARLUCCI A.F. & BOWES P.M. (1970b) Vitamin production and utilization by phytoplankton in mixed cultures. *J. Phycol.* **6,** 393–400.

CARLUCCI A.F. & BOWES P.M. (1972) Vitamin B_{12}, thiamine and biotin contents of marine phytoplankton. *J. Phycol.* **8,** 133–7.

CHAU Y.K., CHUECAS L. & RILEY J.P. (1967) The component combined amino acids of some marine phytoplankton species. *J. mar. biol. Ass. U.K.* **47,** 543–54.

CHUECAS L. & RILEY J.P. (1969a) Component fatty acids of the total lipids of some marine phytoplankton. *J. mar. biol. Ass. U.K.* **49,** 97–116.

CHUECAS L. & RILEY J.P. (1969b) The component combined amino acids of some marine diatoms. *J. mar. biol. Ass. U.K.* **49,** 117–20.

COOMBS J., DARLEY W.M., HOLM-HANSEN O. & VOLCANI B.E. (1967a) Studies on the biochemistry and fine structure of silica shell formation in diatoms. Chemical composition of *Navicula pelliculosa* during silicon-starvation synchrony. *Pl. Physiol.* **42,** 1 601–6.

COOMBS J., HALICKI P.J., HOLM-HANSEN O. & VOLCANI B.E. (1967b) Studies on the biochemistry and fine structure of silica shell formation in diatoms. Changes in concentration of nucleoside triphosphates during synchronized division of *Cylindrotheca fusiformis* Reimann & Lewin. *Expl. Cell Res.* **47,** 302–14.

COOMBS J., HALICKI P.J., HOLM-HANSEN O. & VOLCANI B.E. (1967c) Studies on the biochemistry and fine structure of silica shell formation in diatoms. II. Changes in concentrations of nucleoside triphosphates in silicon-starvation synchrony of *Navicula pelliculosa* (Bréb) Hilse. *Expl. Cell Res.* **47,** 315–28.

COOMBS J., LAURITIS J.A., DARLEY W.M. & VOLCANI B.E. (1968) Studies on the biochemistry and fine structure of silica shell formation in diatoms. V. Effects of colchicine on wall formation in *Navicula pelliculosa* (Bréb) Hilse. *Z. Pfl. Physiol.* **59,** 124–52.

COOMBS J. & VOLCANI B.E. (1968) Studies on the biochemistry and fine structure of silica shell formation in diatoms. Chemical changes in the wall of *Navicula pelliculosa* during its formation. *Planta* **82,** 280–92.

COWEY C.B. & CORNER E.D.S. (1963) Amino acid composition of *Skeletonema costatum*. *J. mar. biol. Ass. U.K.* **43,** 495–511.

DARLEY W.M. (1968) Deoxyribonucleic acid content of the three cell types of *Phaeodactylum tricornutum* Bohlin. *J. Phycol.* **4,** 219–20.

DARLEY W.M. & VOLCANI B.E. (1969) A silicon requirement for deoxyribonucleic acid synthesis in the diatom *Cylindrotheca fusiformis* Reimann & Lewin. *Expl. Cell Res.* **58,** 334–42.

DEMORT C.L., LOWRY R., TINSLEY I. & PHINNEY H.K. (1972) The biochemical analysis of some estuarine phytoplankton species. I. Fatty acid composition. *J. Phycol.* **8,** 211–6.

DURBIN E.G. (1974) Studies on the autecology of the marine diatom *Thalassiosira nordenskioldii* Cleve. I. The influence of daylength, light intensity and temperature on growth. *J. Phycol.* **10,** 220–5.

EPPLEY R.W., HOLMES R.W. & PAASCHE E. (1967) Periodicity in cell division and physiological behavior of *Ditylum brightwellii,* a marine planktonic diatom, during growth in light–dark cycles. *Arch. Mikrobiol.* **56,** 305–23.

EPPLEY R.W., CARLUCCI A.F., HOLM-HANSEN O., KIEFER D., MCCARTHY J.J., VENRICK E. & WILLIAMS P.M. (1971) Phytoplankton growth and composition in shipboard cultures supplied with nitrate, ammonia or urea as the nitrogen source. *Limnol. Oceanogr.* **16,** 714–51.

EPPLEY R.W. & RENGER E.H. (1974) Nitrogen assimilation of an oceanic diatom in nitrogen-limited continuous culture. *J. Phycol.* **10,** 15–23.

FALK M., SMITH D.G., MCLACHLAN J. & MCINNES A.G. (1966) Studies on chitan (β-($1 \rightarrow 4$)-linked 2-acetamido-2-deoxy-D-glucan) fibers of the diatom *Thalassiosira fluviatilis* Hustedt. II. Proton magnetic resonance, infrared and X-ray studies. *Can. J. Chem.* **44,** 2 269–81.

FORD C.W. & PERCIVAL E. (1965a) The carbohydrates of *Phaeodactylum tricornutum*. I. Preliminary examination of the organism and characterization of low molecular weight material and of a glucan. *J. chem. Soc.* 7 035–41.

FORD C.W. & PERCIVAL E. (1965b) The carbohydrates of *Phaeodactylum tricornutum*. II. A sulphated glucuronomannan. *J. chem. Soc.* 7 042–6.

FOWDEN L. (1954) A comparison of the compositions of some algal proteins. *Ann. Bot.* **18,** 257–66.

GOODWIN T.W. (1974) Carotenoids and biliproteins. In *Algal Physiology and Biochemistry* (ed. W.D.P. Stewart) pp. 176–205. Blackwell Scientific Publications, Oxford.

GOTELLI I.B. & CLELAND R. (1968) Differences in the occurrence and distribution of hydroxyproline-proteins among the algae. *Am. J. Bot.* **55,** 907–14.

HAINES K.C. & GUILLARD R.R.L. (1974) Growth of vitamin B_{12}-requiring marine diatoms in mixed laboratory cultures with vitamin B_{12}-producing marine bacteria. *J. Phycol.* **10,** 245–52.

HANDA N. (1969) Carbohydrate metabolism in the marine diatom *Skeletonema costatum*. *Mar. Biol.* **4,** 208–14.

HEALEY F.P. (1973) The inorganic nutrition of algae from an ecological viewpoint. *CRC Critical Rev. Microbiol.* **3,** 69–113.

HEALEY F.P., COOMBS J. & VOLCANI B.E. (1967) Changes in pigment content of the diatom *Navicula pelliculosa* (Bréb) Hilse in silicon-starvation synchrony. *Arch. Mikrobiol.* **59,** 131–42.

HECKY R.E., MOPPER K., KILHAM P. & DEGENS E.T. (1973) The amino acid and sugar composition of diatom cell walls. *Mar. Biol.* **19,** 323–31.

HELLEBUST J.A. (1974) Extracellular products. In *Algal Physiology and Biochemistry* (ed. W.D.P. Stewart) pp. 838–63. Blackwell Scientific Publications, Oxford.

HOBSON L.A. & PARISER R.J. (1971) The effect of inorganic nitrogen on macromolecular synthesis by *Thalassiosira fluviatilis* Hustedt and *Cyclotella nana* Hustedt grown in batch culture. *J. exp. mar. Biol. Ecol.* **6,** 71–8.

HOLMES R.W. (1966) Light microscope observations on cytological manifestations of nitrate, phosphate and silicate deficiency in four marine centric diatoms. *J. Phycol.* **2,** 136–40.

HOLM-HANSEN O. (1969) Algae: amounts of DNA and organic carbon in single cells. *Science* **163,** 87–8.

HOLM-HANSEN O. (1970) ATP levels in algal cells as influenced by environmental conditions. *Pl. Cell Physiol.* **11,** 689–700.

HOLM-HANSEN O. & BOOTH C.R. (1966) The measurement of adenosine triphosphate in the ocean and its ecological significance. *Limnol. Oceanogr.* **11,** 510–19.

HUNTSMAN S.A. & SLONEKER J.H. (1971) An exocellular polysaccharide from the diatom *Gomphonema olivaceum. J. Phycol.* **7,** 261–4.

IGNATIADES L. (1973) Studies on the factors affecting the release of organic matter by *Skeletonema costatum* (Greville) Cleve in field conditions. *J. mar. biol. Ass. U.K.* **53,** 923–35.

IGNATIADES L. & FOGG G.E. (1973) Studies on the factors affecting the release of organic matter by *Skeletonema costatum* (Greville) Cleve in culture. *J. mar. biol. Ass. U.K.* **53,** 937–56.

JEFFREY S.W. (1969) Properties of two spectrally different components in chlorophyll *c* preparations. *Biochim. Biophys. Acta* **177,** 456–67.

JEFFREY S.W. (1972) Preparation and some properties of crystalline chlorophyll c_1 and c_2 from marine algae. *Biochim. Biophys. Acta* **279,** 15–33.

JEFFREY S.W. (1974) Profiles of photosynthetic pigments in the ocean using thin-layer chromatography. *Mar. Biol.* **26,** 101–10.

JØRGENSEN E.G. (1966) Photosynthetic activity during the life cycle of synchronous *Skeletonema* cells. *Physiologia Pl.* **19,** 789–99.

JØRGENSEN E.G. (1969) The adaptation of plankton algae. IV. Light adaptation in different algal species. *Physiologia Pl.* **22,** 1 307–15.

KATES M. & VOLCANI B.E. (1966) Lipid components of diatoms. *Biochim. Biophys. Acta* **116,** 264–78.

KATES M. & VOLCANI B.E. (1968) Studies on the biochemistry and fine structure of silica shell formation in diatoms. Lipid components of the cell walls. *Z. Pfl. Physiol.* **60,** 19–29.

LEE R.F. & LOEBLICH A.R. III (1971) Distribution of 21:6 hydrocarbon and its relationship to 22:6 fatty acid in algae. *Phytochemistry* **10,** 593–602.

LEE R.F., NEVENZEL J.C. & LEWIS A.G. (1974) Lipid changes during life cycle of marine copepod, *Euchaeta japonica* Marukawa. *Lipids* **9,** 891–8.

LEWIN J.C. (1955) The capsule of the diatom *Navicula pelliculosa. J. gen. Microbiol.* **13,** 162–9.

LEWIN J.C. (1962) Silicification. In *The Physiology and Biochemistry of Algae* (ed. R.A. Lewin) pp. 445–55. Academic Press, New York.

LEWIN J.C. & HELLEBUST J.A. (1970) Heterotrophic nutrition of the marine pennate diatom, *Cylindrotheca fusiformis. Can. J. Microbiol.* **16,** 1 123–9.

LEWIN J.C., LEWIN R.A. & PHILPOTT D.E. (1958) Observations on *Phaeodactylum tricornutum. J. gen. Microbiol.* **18,** 418–26.

LEWIN R.A. (1958) The mucilage tubes of *Amphipleura rutilans. Limnol. Oceanogr.* **3,** 111–3.

LOW E.M. (1955) Studies on some chemical constituents of diatoms. *J. mar. Res.* **14,** 199–204.

LUI N.S.T. & ROELS O.A. (1972) Nitrogen metabolism of aquatic organisms. II. The assimilation of nitrate, nitrite and ammonia by *Biddulphia aurita. J. Phycol.* **8,** 259–64.

MEEKS J.C. (1974) Chlorophylls. In *Algal Physiology and Biochemistry* (ed. W.D.P. Stewart) pp. 161–75. Blackwell Scientific Publications, Oxford.

NAKAJIMA T. & VOLCANI B.E. (1969) 3,4 dihydroxyproline: a new amino acid in diatom cell walls. *Science* **164,** 1400–1.

NAKAJIMA T. & VOLCANI B.E. (1970) ε-*N*-trimethyl-L-δ hydroxylysine phosphate and its nonphosphorylated compound in diatom cell walls. *Biochim. Biophys. Res. Comm.* **39,** 28–33.

NALEWAJKO C. (1966) Dry weight, ash and volume data for some freshwater planktonic algae. *J. Fish. Res. Bd. Can.* **23,** 1285–88.

OEY J.L. & SCHNEPF E. (1970) Über die Auslösung der Valvenbildung bei der Diatomee *Cyclotella cryptica.* Versuche mit Colchicin, Actinomycin D und Fluordesoxyuridin. *Arch. Mikrobiol.* **71,** 199–213.

OGINO C. (1963) Studies on the chemical composition of some natural foods of aquatic animals. *Bull. Jap. Soc. scient. Fish.* **29,** 459–62.

OPUTE F.I. (1974a) Physiological studies of the sulfolipids of diatoms. *J. Expl. Bot.* **25,** 798–809.

OPUTE F.I. (1974b) Physiological studies of the phospholipids of diatoms. *J. Expl. Bot.* **25,** 810–22.

OPUTE F.I. (1974c) Lipid and fatty-acid composition of diatoms. *J. Expl. Bot.* **25,** 823–35.

OPUTE F.I. (1974d) Studies on fat accumulation in *Nitzschia palea* Kutz. *Ann. Bot.* **38,** 889–902.

ORCUTT D.M. & PATTERSON G.W. (1974) Effect of light intensity upon lipid composition of *Nitzschia closterium (Cylindrotheca fusiformis). Lipids* **9,** 1000–3.

PAASCHE E. (1968) Marine plankton algae grown with light-dark cycles. II. *Ditylum brightwellii* & *Nitzschia turgidula. Physiologia Pl.* **21,** 66–77.

PAASCHE E. (1973) The influence of cell size on growth rate, silica content and some other properties of four marine diatom species. *Norweg. J. Bot.* **20,** 197–204.

PARSONS T.R. (1961) On the pigment composition of eleven species of marine phytoplankton. *J. Fish. Res. Bd. Can.* **18,** 1017–25.

PARSONS T.R., STEPHENS K. & STRICKLAND J.D.H. (1961) On the chemical composition of eleven species of marine phytoplankters. *J. Fish. Res. Bd. Can.* **18,** 1001–16.

PIRSON A. & LORENZEN H. (1966) Synchronized dividing algae. *Ann. Rev. Pl. Physiol.* **17,** 439–58.

PLATT T. & IRWIN B. (1973) Caloric content of phytoplankton. *Limnol. Oceanogr.* **18,** 306–10.

PUGH P.R. (1971) Changes in the fatty acid composition of *Coscinodiscus eccentricus* with culture-age and salinity. *Mar. Biol.* **11,** 118–24.

SCHOBERT B. (1974) Influence of water stress on metabolism of diatoms. I. Osmotic resistance and proline accumulation in *Cyclotella meneghiniana. Z. Pfl. Physiol.* **74,** 106–20.

SLOAN P.R. & STRICKLAND J.D.H. (1966) Heterotrophy of four marine phytoplankters at low substrate concentrations. *J. Phycol.* **2,** 29–32.

SOBER H.A. (ed.) (1970) *Handbook of Biochemistry*. Selected Data for Molecular Biology, 2nd ed. Chemical Rubber Company, Cleveland, Ohio.

SOEDER C.J. & STENGEL E. (1974) Physico-chemical factors affecting metabolism and growth rate. In *Algal Physiology and Biochemistry* (ed. W.D.P. Stewart) pp. 714–40. Blackwell Scientific Publications, Oxford.

SOH G.L. (1974) Nucleic acid homologies among the algae: a study emphasizing ribosomal ribonucleic acids. Ph.D. Dissertation, University of Georgia, Athens, Georgia.

STRAIN H.H., COPE B.T. JR., MCDONALD G.N., SVEC W.A. & KATZ J.J. (1971) Chlorophylls c_1 and c_2. *Phytochemistry* **10,** 1 109–14.

STRAIN H.H., MANNING W.M. & HARDIN G.J. (1944) Xanthophylls and carotenes of diatoms, brown algae, dinoflagellates and sea anemones. *Biol. Bull.* **86,** 169–91.

STRATHMANN R.R. (1967) Estimating the organic carbon content of phytoplankton from cell volume or plasma volume. *Limnol. Oceanogr.* **12,** 411–8.

STRICKLAND J.D.H. (1965) Production of organic matter in the primary stages of the marine food chain. In *Chemical Oceanography,* Vol. I (eds. J.P. Riley & G. Skirrow) pp. 478–610. Academic Press, New York.

STRICKLAND J.D.H., HOLM-HANSEN O., EPPLEY R.W. & LINN R.J. (1969) The use of a deep tank in plankton ecology. I. Studies of the growth and composition of phytoplankton crops at low nutrient levels. *Limnol. Oceanogr.* **14,** 23–34.

SULLIVAN C.W. & VOLCANI B.E. (1973) The effects of silicic acid on DNA polymerase, thymidylate kinase and DNA synthesis in the marine diatom, *Cylindrotheca fusiformis*. *Biochim. Biophys. Acta* **308,** 212–29.

TORNABENE T.G., KATES M. & VOLCANI B.E. (1974) Sterols, aliphatic hydrocarbons and fatty acids of a nonphotosynthetic diatom, *Nitzschia alba*. *Lipids* **9,** 279–84.

VOSKRESENSKAYA N.P. (1972) Blue light and carbon metabolism. *Ann. Rev. Pl. Physiol.* **23,** 219–34.

WALLEN D.G. & GEEN G.H. (1971a) Light quality in relation to growth, photosynthetic rates and carbon metabolism in two species of marine plankton algae. *Mar. Biol.* **10,** 34–43.

WALLEN D.G. & GEEN G.H. (1971b) Light quality and concentration of proteins, RNA, DNA and photosynthetic pigments in two species of marine plankton algae. *Mar. Biol.* **10,** 44–51.

WALLEN D.G. & GEEN G.H. (1971c) The nature of the photosynthate in natural phytoplankton population in relation to light quality. *Mar. Biol.* **10,** 157–68.

WERNER D. (1966) Die Kieselsäure im Stoffwechsel von *Cyclotella cryptica* Reimann, Lewin & Guillard. *Arch. Mikrobiol.* **55,** 278–308.

WERNER D. (1968) Stoffwechselregulation durch den Zellwandbaustein Kieselsäure: Poolgrössenänderungen von α-Ketoglutarsäure, Aminosäuren und Nucleosidphosphaten. *Z. Naturf.* **23b,** 268–72.

WERNER D. (1969) Beiträge zur Physiologie und Biochemie der Kieselsäure. *Ber. dt. bot. Ges.* **81,** 425–9.

WERNER D. (1971a) Der Entwicklungscyclus mit Sexualphase bei der marinen Diatomee *Coscinodiscus asteromphalus*. I. Kultur und Synchronisation von Entwicklungsstadien. *Arch. Mikrobiol.* **80,** 43–9.

WERNER D. (1971b) Der Entwicklungscyclus mit Sexualphase bei der marinen Diatomee *Coscinodiscus asteromphalus*. II. Oberflächenabhängige Differenzierung während der vegetativen Zellverkleinerung. *Arch. Mikrobiol.* **80,** 115–33.

WERNER D. (1971c) Der Entwicklungscyclus mit Sexualphase bei der marinen Diatomee *Coscinodiscus asteromphalus*. III. Differenzierung und Spermatogenese. *Arch. Mikrobiol.* **80,** 134–46.

WERNER D. & PIRSON A. (1967) Über reversible Speicherung von Kieselsäure in *Cyclotella cryptica*. *Arch. Mikrobiol.* **57**, 43–50.

WHITE A.W. (1974) Growth of two facultatively heterotrophic marine centric diatoms. *J. Phycol.* **10**, 292–300.

CHAPTER 8

MOVEMENTS

MARGARET A. HARPER

169 Pembroke Road,
Northland, Wellington 5, New Zealand

1 INTRODUCTION

Zoologists have tackled the movement of practically all motile algae except diatoms. No other protozoans have the unusual restriction of a rigid silica frustule whose only uncovered openings are two long narrow slits, the raphes. Diatoms without raphes are not mobile. Some observers thought they saw cilia sticking out through the raphe (see Fritsch, 1935 p. 593, Hopkins, 1967), but now cross-sections photographed in the electron microscope show that the raphe is an empty slit only a fraction of a micrometre wide (Drum *et al.* 1966). An argument still continues as to whether diatoms always leave slime trails or not. Certainly these have only been seen by a few people as they are transparent and their width is virtually submicroscopic. Even if one is convinced of their existence, their rôle in diatom locomotion is not self-evident. Recently more than one mechanism has been suggested in which the diatom's unique raphe slit becomes a remarkable device making slime secretion an economic and efficient means of propulsion. Thus diatoms move much faster than most 'gliding' organisms, although the most active blue-green alga *Oscillatoria princeps* can travel at 11 μm/sec (Halfen & Castenholz, 1971).

Why should these unicellular plants move? Not all diatoms do. It is principally those diatoms which live on the surface of sand and mud which have raphes. Further sediment is continually being deposited in these habitats whose surfaces are disturbed by tides or wind and currents. Diatoms migrate to the top of sediment directed by a set of responses to light, air and gravity. Mass movement of diatom populations downward before incoming tides causes the rapid disappearance of the golden brown 'blushes' of blooms on estuarine mud. These movements have been observed for nearly half a century longer (Fauvel & Bohn, 1907) than the daily migrations of motile diatoms on freshwater sediment (Round & Happey, 1965).

2 OBSERVATIONS

2.1 *Microscope observations*

Motile diatoms can be observed moving on coverslips or tissues removed from the surface of collected sediment on which they were left overnight (Eaton & Moss, 1966). Diatoms push through sediment, sometimes jerking; they glide on the glass slide, often stopping and always reversing their direction every few minutes. Close examination of a raphe may reveal particles being carried along by it (or particles may be added). At first particle transport appears to take place equally often either forwards or backwards with respect to the direction in which the diatom is travelling. This seems anomalous (Martens, 1940) but Nultsch (1962) observed diatoms in girdle view moving against lumps of sediment. He found that any one raphe pushed in the same way against particles and the lump at the same time. Martens (1940) cannot have focused on the lower raphe system which is in contact with the slide. In 300 inverted microscope observations of particle transport *below* diatoms, only 33 brief movements took place in the direction the diatom was travelling. None was in the focal plane of the slide surface on which the diatom was travelling (Harper, 1967, also Drum & Hopkins, 1966). Mean speeds of particle transport and diatom movement are usually similar, but the maximum speed of particle movement is greater because there is less reaction from the small light particles than the slide. Particles would move more slowly than diatoms if driven by a viscous raphe fluid (Müller, 1889; Nultsch, 1962) as the surrounding water would tend to roll them backwards.

Some slow movements have been noticed in species without raphes (Hendey, 1964). These are due to flotation in semi-liquid mud or to mucilage exudation from pores as suggested for desmids (see Jarosch, 1962). Hopkins (1969) found *Synedra tabulata* moved a few micrometres at a time at a rate of 0·1–1 μm/sec leaving a 5–9 μm wide trail. Proof that active raphes are essential for true locomotion comes from the observation that *Phaeodactylum*

Table 8.1. Speeds, pushing forces and adhesion of various moving diatoms. From Harper & Harper, 1967 and Harper, 1967.

Species	*Speed* (μm/sec)		*Traction* (nN)		*Adhesion* (nN)	
	mean	*max*	*mean*	*max*	*mean*	*max*
Amphora ovalis	2·2	4·5	480*	480**	2250	4800
Nitzschia linearis	11	24	11	210**	19	66
N. sigmoidea	9·4	17	0·8	19	13	93
Cymatopleura solea	5*	14	11*	22	2·5	48
Navicula oblonga	11	20	1·5	5·7	8	42
Gyrosigma acuminatum	5	15	65*	100**	11*	13
Surirella biseriata	6*	10	—	—	1·1	7·3
S. ovalis	—	—	47*	73**	1·6	2·9

* Indicates mean of less than 10 observations.

** Exceeds 25 nN (nanoNewtons) tractive force: see discussion of Gordon & Drum (1970) in text.

tricornutum is only motile when it is in the phase in which the cells have frustules with raphes (Lewin *et al.* 1958). Diatoms may be immobilized by their raphes being enclosed by a mucilage pellicle as sometimes in *Navicula pelliculosa* (Lewin, 1955), this often occurs in liquid cultures and during auxospore formation (Fritsch, 1935 p. 525).

Recorded average speeds of different species, moving normally, range from 0·2 to 25 μm/sec, with the maximum rate for any species about twice its average (Drews & Nultsch, 1962; Harper, 1967; Table 8.1). They only move about two-thirds of the time even during periods of maximum motility. Two tube-dwelling species, *Frustularia asymmetrica* and *Navicula obtusa,* have been observed by Williams (1965) withdrawing at 500 and 240 μm/sec respectively but such high speeds might be caused by abnormal elastic or osmotic phenomena (Harper, 1967). Hendey (1964) considers most colonial species are non-motile or sluggish, as are the small solitary diatoms living on sand grains (Round, 1965a). Round's 'slightly motile' diatoms averaged less than 2 μm/sec and his 'semi-motile' ones 2 to 4 μm/sec (Harper, 1969). Faster moving diatoms all lived on mud. Their speeds depended on temperature (Hopkins, 1963; Jarosch, 1962), light (Nultsch, 1956, 1971), oxygen (Müller, 1894) and time of day (Harper, 1967). Adding neutral red (Wagner, 1934) or 0·01% caffeine citrate (Drum & Hopkins, 1966; Drum, 1969) may increase motility, but *l*-histidine (Nultsch, 1962) does not. Dimethyl sulphoxide (used to dissolve cytochalasin B) inhibits the movement of diatoms, but not of other cells (Spangle & Armstrong, 1973). Various other drugs (Jarosch, 1962; Drum & Hopkins, 1966) also decrease diatom movement. Diatoms usually take several seconds to reach their maximum speed from rest; perhaps it takes time

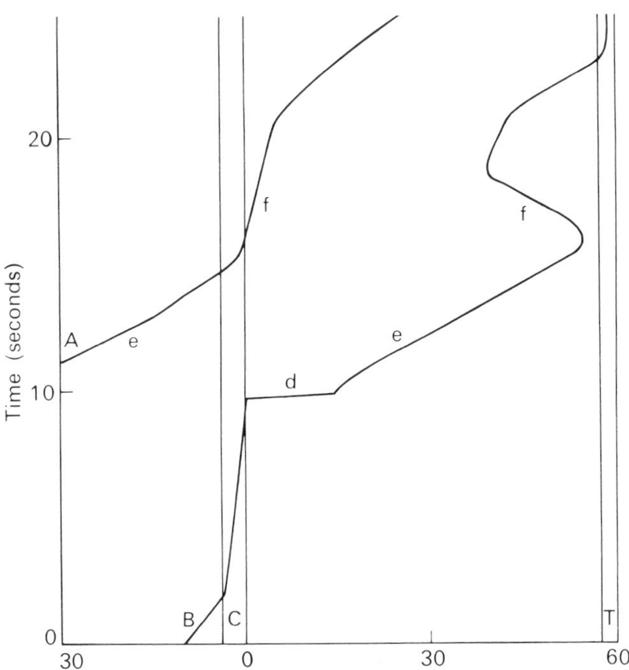

Fig. 8.1. Particle transport along the raphes of a *Navicula oblonga* cell in girdle view. A, a single polystyrene latex sphere 1·3 μm diameter; B, a group of spheres about 5 μm diameter; C, a central gap in raphe system (here particles slow down); d, jump from central gap; ee, 2 particles in parallel motion; ff, 2 particles in contrary motion; T, termination of raphe.

for a cell to fully coordinate its locomotory effort. Part of the 10-second delay in reaction to light by *Navicula peregrina* (Nultsch & Wenderoth, 1973) may be caused by this too. Acceleration of particles transported along a *N. oblonga* raphe did not exceed 7 μm/sec (Harper, 1967) and their reversal took 1 second (Fig. 8.1). The author has never observed any diatom reversing itself faster than this. But elastic recoil of one filmed group of particles exceeded 100 μm/sec for a fraction of a second. The speed and direction of particle transport can vary along a single raphe but even on a large diatom like *Nitzschia sigmoidea* the particles only move towards each other and never pass.

Diatoms are incapable of moving in a continuous straight line, they appear to be forced to alternate the direction any one raphe drives in. So their routes are a series of curves or zig-zags depending on the shape of their raphe system (Nultsch, 1956). *Eunotia lunaris* can move in three different types of curve

(Zauer, 1950). These patterns are partly an artefact of movement on a flat surface, as is the spiral path followed by *Pleurosigma angulatum* moving spirally in a fine glass cylinder (Drum & Hopkins, 1966). The patterns can be seen when the trails left behind by diatoms are marked by a suspension of animal charcoal particles (Drum & Hopkins, 1966) or by polystyrene spheres in the presence of salts of polyvalent metals (Walsby, 1968). The particles are trapped by the slime trail and it seems that every part of an active raphe which touches the slide surface leaves a trail (Harper, 1969).

Hopkins (1967) has stained the slime trails with several dyes (Brilliant Cresyl, Methylene, Toluidine, Astra and Alcian blues, Alcian green, Bismarck brown and Magneson I). The Cresyl fast violet used by Jarosch (1958) has since proved unsatisfactory (Fischer & Turck, 1969). The droplets stained by Jarosch were not trail. Hopkins (1969) obtained droplets by briefly warming cells (in 50°C water for 20 seconds): he suggests they are synonymous with Bütschli globules and volutin bodies (Hopkins, 1967; Fritsch, 1935 p. 599) and contain partly synthesized trail substance; he also found extruded lipoid bodies (Lauritis *et al.* 1968). Jarosch droplets may be the raphe liquid of Harper & Harper (1967); see Fig. 8.6,B. Mucilage lumps and occasional trails can be stained easily by gentle irrigation with a fresh saturated solution of Alcian blue. Mucilage aggregations show up very clearly in diatomite in which large diatoms have been moving for a while. Trails seem to consist of a polysaccharide made up of mostly glucose units (Hopkins, 1967), although Allan *et al.* (1972) found only a trace of glucose in the exudates of some motile species.

Diatoms move on their raphes so that naviculoid ones usually appear in valve view on a flat surface. But sometimes *Pinnularia* species and others move while lying on their girdle bands. How can their raphes thrust against the substrate then? Staining shows lumps of trail substance accumulated at the central nodule, and these act as an 'artificial substratum' like a peg against which the diatom moves once the lump has stuck to the slide (Harper & Harper, 1967). Raphe contact is essential for locomotion: when a *Pinnularia nobilis* individual was sucked into a micropipette only just wider than itself, its raphes did not touch the curved walls and it could not move (Drum & Hopkins, 1966). Similarly *Bacillaria paxillifer* (formerly *B. paradoxa*) individuals are immobilized by flat surfaces, but can move among particles, because the ridges alongside their raphes prevent contact with a flat surface (Drum & Pankratz, 1965). These ridges normally interlock neighbouring cells and act as 'rails' for the raphes to slide along. How does the colony avoid being gummed up by its own slime secretion? The movement of the colony appears to be synchronized but this is due to the rhythm of reversal of each cell. Drum *et al.* (1971) tested this, by killing the cells on each side of a given one with a laser and finding that the frequency of reversals of the living cell remained normal. If only one of the two raphe systems of touching cells is usually active at any one time then the working one might stretch the trail thread along the other when the colony was prevented from expanding by a mechanical constraint. This would explain the elastic jump from a curved into an extended straight line when the constraint is removed (Jarosch, 1958, 1962). Some *Nitzschia* species sometimes resemble a *Bacillaria* colony as several cells slide on top of each other in a semi-orderly array. Some cells can move faster *en masse* over agar than singly (Wagner, 1934; Schorr, 1939; Peteler, 1940) but they also alter their substratum from agar to living diatoms or accumulated trail.

2.2 *Ultrastructure*

Some of the first electron micrographs of pennate diatoms showed that the raphe was a simple slit (Toman & Rosival, 1948) and not two channels as suggested by Müller (1889). A membrane layer covers the whole frustule, including the punctae (or pores), leaving only the raphes open (Reimann *et al.* 1965; Reimann & Lewin, 1966). In section the raphes usually appear empty (Drum *et al.* 1966) though rarely a few 'fibrils' have been found extending outwards from just inside it (Drum & Hopkins, 1966; Lauritis *et al.* 1967). Sometimes membrane profiles are left near or across the raphe after cell division (Drum, 1963; Drum & Pankratz, 1964). Diatoms may exude mucilage pellicles (Lewin, 1955), tubes (Drum, 1969) or stalks from the raphe or mucilage pores (Hufford & Collins, 1972) as well as trail substance. Drum & Hopkins (1966) found a 'fibrillar bundle' running parallel to the raphe just inside the protoplast. They suggested it was a contractile organelle like similar structures in some slime moulds, but Frey-Wyssling (1972) has pointed out that microtubules in different organisms differ widely in function and origin so no conclusions can be drawn from their similar appearance. Perhaps the bundle might be a skeletal rod, which helps keep the protoplast in contact with the raphe system. This could hold the protoplasm against the raphe when a diatom was plasmolyzed, so it remained motile. Its function might be easiest to study in *Phaeodactylum tricornutum* as these cells are half naked. In nitzschioid diatoms the protoplast is interlaced by costae forming the 'raphe canal' which would hold the protoplast in place against the raphe. These diatoms do not appear to have fibrillar bundles, but special mitochondria near their raphes (Drum *et al.* 1966). Certainly fibrillar bundles are not found in all motile diatoms as claimed by Drum & Hopkins (1966). They also suggested that diatom trails are secreted by 'crystalloid bodies'. In electron micrographs the 'crystalloid bodies' are *not* crystalline, their contents being a little slightly stained whorled, or occasionally striated matter (Drum *et al.* 1966). Inadequately fixed lipoid bodies also remain unstained (Lauritis *et al.* 1968). Dawson (1973) distinguished lipoid droplets from polyphosphate bodies and described six types of cytoplasm vesicle. More study of the secretion and chemistry of trail is required to settle its origin.

2.3 *Adhesion*

The slime trails which trap particles also stick the diatom cells onto surfaces. When a piece of cellophane was placed on top of diatoms by Hofmeister (1940) they only resisted its movement when one raphe system was in contact with the slide and the other with the cellophane. Drum & Hopkins (1966) found that moving diatoms were attached. The adhesion of these diatoms can be measured from the bending of a fine glass fibre pressed against the frustule (see Table 8.1). Moving diatoms always resisted being pushed off their substrata (Drum & Hopkins, 1966; Harper & Harper, 1967). It is also hardly surprising to find that diatoms are stuck to vertical surfaces whether moving or not, but it is worth noting that they can still move at their normal speed. Diatoms do not move as fast on agar as on glass (Richter, 1903; Wagner, 1934; Peteler, 1940). Hopkins (1969) finds that diatoms can move at nearly normal speeds on agar submerged in water but only slowly in dry agar. *Navicula oblonga* sometimes skidded on water-covered 2% agar and only adhered to it half as firmly as to glass; adhesion to clear polystyrene was also less than to glass (Harper & Harper, 1967). Detergents can stop diatom locomotion without killing the cells (Drum & Hopkins, 1966). Even 10 ppm of a cationic one stopped *N. oblonga,* but 100 ppm of a non-ionic and 1 000 ppm of two anionic ones could be added before the diatoms halted (Harper & Harper, 1967). Hopkins (1969) found that six other species behaved in a similar way. Positively charged sols have an effect analogous to cationic detergent as they aggregate on the frustule (Geissler, 1958) and trail.

Most diatoms stop moving in hypertonic solutions of sucrose or urea at lower tensions than those which visibly plasmolyze their cells (Höfler, 1940); some cells become loosened and others more firmly stuck on (Harper, 1967). In saline streams Round (1964) found immobile forms of freshwater species which normally are motile. Detergent pollution can result in such forms, e.g. filaments of *Pinnularia viridis* occurred in a foaming urban stream (Hopkins, 1969). Some diatoms, particularly brackish-water species, can move with an 'empty' front end after plasmolysis or centrifugation or puncture if the protoplast is undisturbed along the active raphe, otherwise they may just 'pirouette' on their apices (Legler & Schindler, 1939; Hofmeister, 1940; Höfler *et al.* 1950; Drews & Nultsch, 1962; Nultsch, 1962; Drum & Hopkins, 1966). Adhesion is affected by heat: at 40°C *Pleurosigma angulatum* cells were unable to readhere after being knocked off a surface, until they were cooled (Drum & Hopkins, 1966). *Navicula radiosa* was killed at this temperature (Nultsch, 1956). *Surirella gemma* stopped migrating at temperatures below

0°C (Hopkins, 1966a) but recovered when warmed. *Nitzschia putrida* cells subjected to −26°C for 15–35 minutes moved at normal speeds on recovery (Wagner, 1958).

2.4 *Resistance to flow*

Adhesion prevents the many motile diatoms which live on rocks and stones in streams being washed away (Douglas, 1958). Resistance to flow can be accurately measured by subjecting diatoms in a horizontal capillary tube to different rates of laminar flow (Harper & Harper, 1967). Small diatom species which often live attached to sand grains, stones and rocks could withstand a shear stress of 100 dyne/cm^2. Species from silt did not resist 10 dyne/cm^2 apart from some *Nitzschia* species which have long raphes in proportion to the size of their frustules. The strongest species seemed to be *Amphora ovalis* which could also push harder on the fibre than any other diatom (Table 8.1). This diatom could push sand grains apart amongst which larger diatoms were trapped, and its frustule was never broken during micromanipulation (Harper, 1969). One habitat of this diatom was a sandy lake beach which was churned up by waves and arthropods. Diatoms also avoid being crushed by aggregating in small hollows in the sand grains (Round, 1965a; Meadows & Anderson, 1968). Single cells of complicatedly shaped species such as *Campylodiscus noricus* found on silt in streams do not resist flow, but together they make the sediment cohere with their sticky trails and 'interwoven' frustules forming a skin which withstands flow (Harper, 1967). When this skin is broken by small water snails ploughing through it or ruptured by trapped bubbles (of oxygen evolved during photosynthesis), chunks of slime-bound sediment and diatoms are often washed away (Hopkins, 1969). Holland *et al.* (1974) have compared the binding of sediment by six marine species; *Navicula directa* was best at stabilizing sand, *Hantzschia amphioxys* clay, *Cylindrotheca closterium* was less good at either, and three colonial species did not stabilize sediments.

Laboratory measurements of resistance to shear stress cannot yet be correlated with catches of diatoms and flow rates in streams. It would help if ecologists studying streams would make more than one observation at a station within a short period of time so that seasonal changes did not mask changes owing to flow rate, and also if they would report the type of stream bed and either shear stress if possible, or failing that any two of the following: maximum speed, volume flow rate, and cross-sectional area. The number of cells (of both motile and non-motile species) floating downstream fluctuates daily, being greatest mid-morning and least at midnight. The distance most cells travel before settling is less than 40 m (Mueller-Haeckel, 1973).

3 BEHAVIOUR

3.1 *Phototaxis*

The light responses of diatoms living on sediment change during the course of the day. The alternation from movement towards light in the morning to movement away later was shown by Fauré-Fremiet (1951) and Palmer (1960) in *Hantzschia virgata* from the sea shore (note that their identification of this diatom as *H. amphioxys* has been corrected by Palmer & Round, 1967, details given in Round, 1970). A complete cycle has been found by Hopkins (1966a) in natural populations of *Surirella gemma* (see Fig. 8.2) and *Pleurosigma angulatum* (Hopkins, 1969). He clamped a Petri dish of glutinous estuarine mud in a vertical position and covered all of it except one vertical face with foil; thus ingeniously separating the directions of gravitation and light. The negative phototactic reactions of *Navicula radiosa* cells collected from a

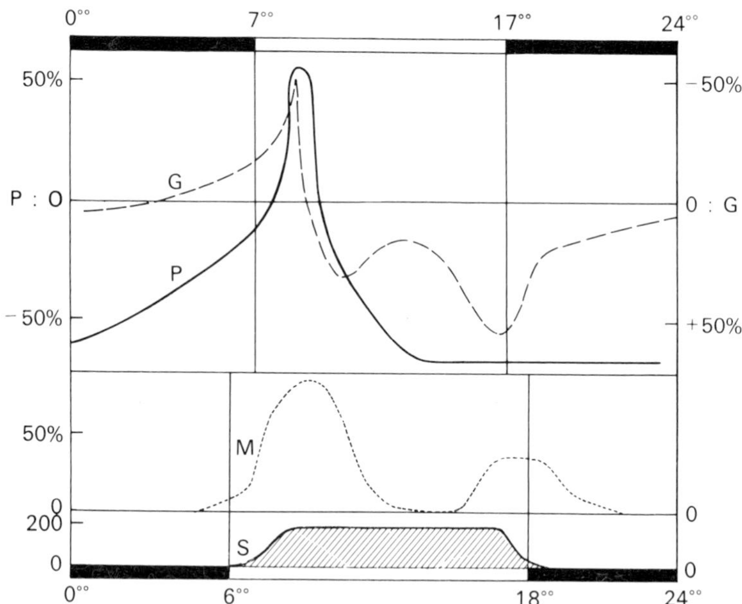

Fig. 8.2. The daily cycles for *Surirella gemma*. G, net percentage of cells showing geotactic response (scale at right, negative response plotted upwards; P, net percentage of cells showing phototactic response (scale at left, positive response plotted upwards); M, percentage of cells found moving; S, number of diatoms found on 1 mm^2 of mud. The horizontal scale is time, from midnight to midnight, with thick black lines indicating periods before sunrise or after sunset. Redrawn from Hopkins, 1969.

freshwater habitat contrasted with the positive reactions of five cultured species (Nultsch, 1956), probably because the natural material had entered the downward phase of its migration cycle. Certainly *N. radiosa* migrated to and from the surface of river silt in an alternate light (1 300 lux, 12 hours) dark (12 hours) regime (Round & Happey, 1965), some cells leaving the surface before noon. Nultsch (1971) has studied the photobiology of a strain of *Nitzschia communis* in axenic culture and found movement of groups of cells towards a light from one side (*topotaxis*) at intensities from 2 to 40 000 lux, with a maximum response at 200 lux. *N. communis* of a different strain had its maximum in the range 1 000 to 2 500 lux (Nultsch, 1956), as did *Navicula buderi* and *Amphora montana,* while for *Nitzschia palea* the range was wider (from 500 to 5 000 lux) and for *N. stagnorum* the maximum was lower (500 to 1 000 lux).

Aggregation in small illuminated areas is a different response known as *phobotaxis,* virtually a 'shock' reaction of cells crossing a light/dark boundary. The number of *N. communis* (Nultsch, 1971) cells in a light patch increases with increasing light intensity to 1 000 lux. This reaction was not tested at higher intensities as the light gradient then extended into the surrounding dark area. The strain of *N. communis* used by Nultsch in 1956 behaved in a similar way and responses were observed to higher light intensities of 2 500, 5 000 and 10 000 lux. *N. palea, Navicula buderi,* and *A. montana* had the same reactions but *Nitzschia stagnorum* avoided brightly lit areas (⩾2 500 lux) which curiously *Navicula radiosa* did not (Nultsch, 1956). A negative response to higher light intensities may account for the observation that pond populations remain on silt surface at noon only on dull days (c. 2 000 lux) but not on bright days (>10 000 lux, Round & Eaton, 1966). Hopkins (1969) found that the population on the surface is determined by the intensity of daylight during the period of upward migration. In winter *P. angulatum, S. gemma* and *N. crucigera* did not surface when the light intensity exceeded 800 lux. In summer they were more sensitive and failed to appear when light values exceeded 300, 500 and 650 lux respectively. Pomeroy (1959) found that high light intensity inhibited photosynthesis, and that salt-marsh diatoms became 'trapped' on the surface. Possibly they were trapped by desiccation (Hopkins, 1963). The sensitivity of diatoms to light depends partly on their metabolic needs, as the phototactic reaction of *Nitzschia communis* (Nultsch, 1971) was increased by dark treatment of 6 to 8 hours before making observations. Diatoms tend to become temporarily immobile after 4 to 10 hours dark treatment, their continued movement depending on previous culture conditions (Nultsch, 1956). Movement during migration periods must use stored carbohydrates as some migration continued for 3 days in darkness, whether interrupted for sampling (Round & Happey, 1965; Round, 1965b; Round & Eaton, 1966), or continuous (Harper, 1969). Hopkins (1966) found that surviving diatoms were motile during migration periods even after 40 days of dark treatment.

Nultsch & Wenderoth (1973) elegantly investigated the photophobic reactions of a clone of *Navicula peregrina* to 1 000 lux. When leaving a lit-up area 95% of the cells reversed when the leading end was between 0·2 and 1·1 cell lengths outside it, the mean being 0·65 lengths. These diatoms were 120 μm long, and travelled at an average speed of 10·5 μm/sec. The average reaction time of 10 seconds before the cells reversed in this experiment was also found when the rear ends of these diatoms moving in shadow were exposed to light, or when the front ends of cells moving in light were darkened. In either case it made little difference how much of the diatom's length was exposed to the stimulus, if between 10% and 75% including an end. The reaction time lengthens as an experimentally lit region covering 25% of the cell length is moved towards the nucleus, becoming equal to the autonomous reversal period of 60 to 70 seconds at the cell centre (Fig. 8.3). If only the front

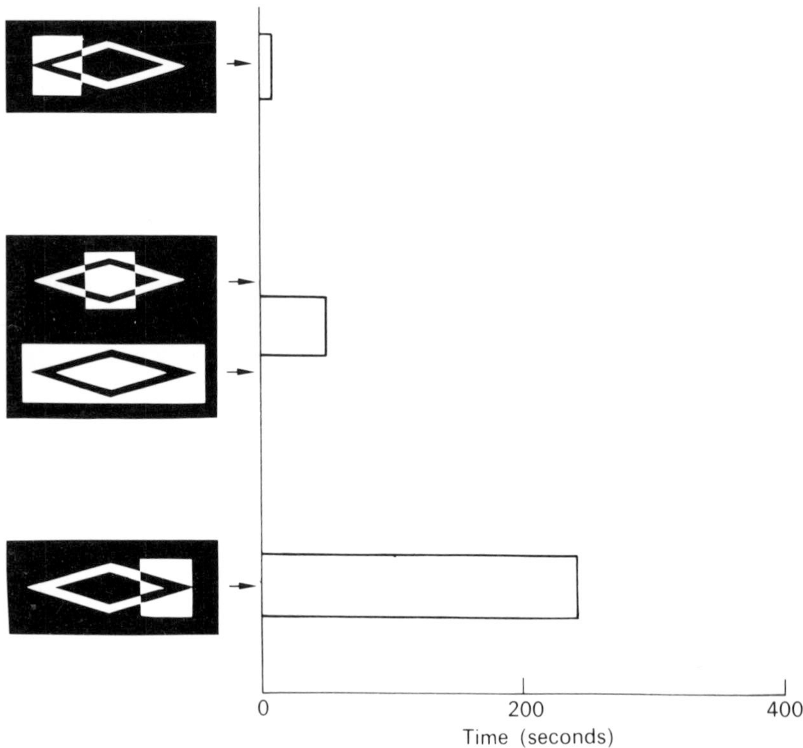

Fig. 8.3. Lengths of time taken before *Navicula peregrina* cells reverse, when partially illuminated. The arrows indicate their direction of movement, and each diagram beside a histogram bar indicates the part of the diatom kept illuminated. Redrawn from Nultsch & Wenderoth, 1973.

end is illuminated the time before the diatom reverses can be extended to 240 to 260 seconds. Possibly the reserves of trail substance at the driving end are used up in this time.

Nultsch & Wenderoth (1973) consider the reaction to be an 'all or nothing' one, in which the number of photoreceptor molecules exposed to light is apparently not significant, but their distance from the cell centre is critical. The poles were found to be most sensitive to light, but the authors suggest, that this is because of the polar position of the raphe ends, where the driving force for locomotion appears to lie, and not because the poles may contain more photoreceptor molecules. Hopkins (1969) also found the poles of *Pleurosigma angulatum* more sensitive to light than the rest of the cell. Nultsch (1971) and Nultsch & Wenderoth (1973) suggest that the photoreceptor may be a carotenoid in the chromatophores, but negative reactions to red light (Nultsch, 1956) suggest, that chlorophyll *a* may be also active. Confirmation that the photoreceptors are carotenoids rather than flavins comes from the observation that the carotenoid free diatom *Nitzschia alba* does not respond to light (Nultsch & Wenderoth, 1973).

Nitzschia communis (Nultsch, 1971) has an optimum of light intensity of 500 lux for both speed and topotaxis. Earlier Nultsch (1956) had observed two other photokinetic reactions: the time lag until movement restarted in light after dark treatment and the length of time for which diatoms were motile in darkness after light treatment. Optimum light intensity for these was the same as for culturing (2 000 to 2 500 lux), and 5 000 lux immobilized the diatoms (5 species) in either experiment (Nultsch, 1956). *Bacillaria paxillifer* also moves more slowly at high light intensities although the period of its reversing rhythm remains constant (Jarosch, 1962). Its colonies on *Caulerpa* sp. tips are extended by day and retracted at night (Funk, 1919).

The response of organisms to light depends on its wavelength. *Nitzschia communis* (see Fig. 8.4 and Nultsch, 1971) is no exception. Its photokinetic action spectrum measured at 28×10^{-11} Einstein/cm^2 using intereference filters is similar to that for whole cell absorption but blue light is less active than red light. Nultsch (1971) considers that photokinesis is directly related to photosynthesis and ATP metabolism, but phobotaxis and topotaxis are not, as their action spectra do not extend into red light. Although he notes some avoidance of red light areas (Nultsch, 1956, 1971), he did not observe any individuals reversing on entering them and so he was not convinced that this was phobotaxis. Heidingsfeld and Nultsch (in Nultsch, 1956) observed that a topotactic reaction only took place when the diatom cell became nearly parallel to the light beam; this suggests that topotaxis could be a form of phobotaxis in which the cell contents shade the rear pole. If so, it seems odd that the action spectrum for topotaxis (Fig. 8.4) is lower than that for phobotaxis near the 440 nm absorption maximum, unless as Nultsch (1971) suggests, the carotenoids act as the photoreceptors. However some results

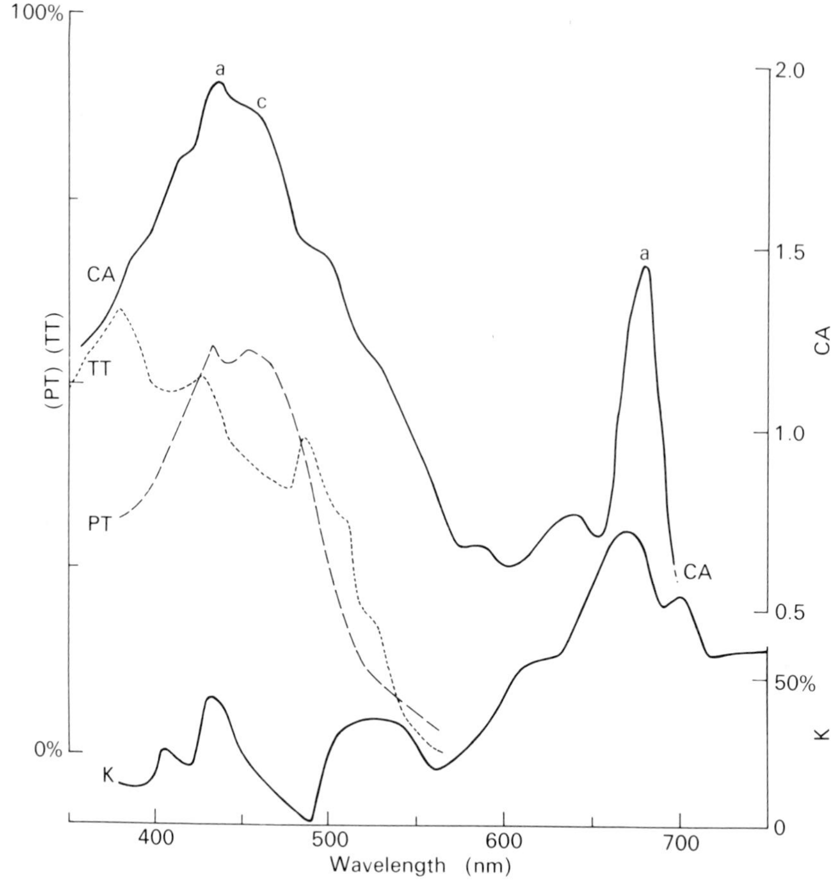

Fig. 8.4. Action spectra for topotaxis (TT), phobotaxis (PT) and kinesis (K) in *Nitzschia communis,* compared with living cell absorption (CA). Absorption maxima of chlorophyll *a* and carotenoids are shown by (a) and (c). Redrawn from Nultsch, 1971.

(Hopkins, 1969) suggest that *Pleurosigma angulatum* cells may turn towards their lit-up side. Hopkins (1966a) found fewer *Surirella gemma* cells appeared on the mud surface during the day under red filters than blue ones. But he found the total population had increased most under a red filter which had a transmission maximum near the 640 nm maximum for chlorophyll *a*. This is surprising, as Tanada (1951) found the quantum yield of photosynthesis in *Navicula minima* varied little with wavelength. Red light penetrates turbid water (Sverdrup *et al.* 1942) and mud further than blue light, and diatoms under 3 mm of mud in low latitudes may be photosynthesizing at a near-maximum rate whilst the photosynthesis of diatoms on the surface is partially inhibited (Taylor & Palmer, 1963).

Single non-motile diatoms on stalks and colonies growing as filaments on sand grains bent towards incident light (phototropism) (Harper, 1969). This photoreaction might be governed by the same mechanism as topo- and phobotaxis because it appears that the same organelles may be active in locomotion and stalk formation in *Cymbella cistula* (Hufford & Collins, 1972).

3.2 *Geotaxis*

Figure 8.2 shows that *Surirella gemma* has a circadian rhythm in geotaxis (Hopkins, 1966a). Negative geotaxis or upward movement before dawn explains anticipation of dawn by river and pond silt populations (Round & Happey, 1965; Round & Eaton, 1966). Diatoms do not sink through the flocculent pond silt when they stop moving as suggested by Round & Eaton (1966). But they actively move down before sunset in the laboratory (Harper, 1969) and in their natural habitat (Harper, 1976). Negative geotaxis helps buried diatoms surface, e.g. *Surirella gemma* through 20 mm of mud in 20 hours (Hopkins, 1966a) or several species through 150 mm pond silt in a week (personal communication, B. Moss). These long-term average upward speeds, of 0·28 and 0·25 μm/sec, are much less than the values in Table 8.1. The measurements there were made over periods of several seconds to a minute along the actual path of a moving diatom.

3.3 *Chemotaxis*

When an oxygen-filled capillary tube was opened in a preparation of diatoms, they collected at the open end in either darkness or red light (Nultsch, 1956). Nultsch (1956) suggested that aggregation under high intensities of red light might be caused by oxygen from photosynthesis. There is some evidence that an increased sensitivity to oxygen around sunset may prevent *Surirella gemma* entering the lower anaerobic layers of mud (Hopkins, 1966a). *S. gemma* and *P. angulatum* both move towards air during their migration periods (Hopkins, 1969). Can diatoms relocate their sediment by chemotaxis? A bit of silt was placed at one side of a small flat dish and when the diatoms from it had moved out to a central patch of daylight, the silt was moved to the other side. It was intriguing to find all the diatoms back in the silt at sunset (Harper, 1967). Hopkins (1969) also noticed that *P. angulatum* avoided cedar wood oil.

3.4 *Thermotaxis*

Some species have high optimal temperatures for speed, e.g. *Navicula radiosa* (28°C: Reimers, 1928), *N. ammophila* (30°C) and *Nitzschia linearis* (36°C:

Reimers, 1928). Other species, viz. *Pleurosigma aestuarii, P. balticum, P. angulatum, Nitzschia sigma* and *N. closterium* (Hopkins, 1963) move fastest at about 15 °C. Their speeds decrease at higher temperatures (Hopkins, 1963) and the period of the reversal rhythm of *B. paxillifer* does too (Jarosch, 1962). *Navicula radiosa* and *Nitzschia linearis* move towards warm places with the same optimum temperature as for speed, but *N. sigmoidea, Pinnularia major, P. stauroptera* and *P. brebissonii* were indifferent to the range 2°–28°C (Reimers, 1928).

3.5 *Thigmotaxis*

Nitzschia putrida stops moving when touched (Wagner, 1934), but not all diatoms do, as six species (see Table 8.1) pushed against a glass fibre. The fast retraction of two tube-dwelling species was provoked by touch (Williams, 1965). Vibration usually stops diatoms moving. Stirring may cause a thigmotactic response in *Surirella gemma* (Hopkins, 1966b).

3.6 *Migration*

Most diatoms depend on photosynthesis and so their carbon assimilation is increased in daylight hours. A rhythm in photosynthetic capacity continued for 3 days in *Phaeodactylum tricornutum* at 430 and at 850 lux (Palmer *et al.* 1964). But periods of increased motility are not those of maximum photosynthesis, as migration upwards takes place around sunrise and downwards around sunset. Variation in motility can be observed in both the laboratory (Hopkins, 1966a; Harper, 1969 and Fig. 8.2) and the field (Harper, 1976) and may be correlated with rhythms of cell division and chloroplast contraction (Round & Eaton, 1966; Hopkins, 1966a, 1969). Most species in the tidal zone can surface twice in a day but *Hantzschia virgata* only appears on top of littoral sand once a day (Palmer & Round, 1967; Round, 1970). This is a unique species, for unlike twenty other species of diatoms in the tidal zone whose behaviour has been studied, it retains its tidal rhythm away from tidal influence for several days (Fauré-Fremiet, 1951, Ganapati *et al.* 1959). This happens both in alternate light–dark conditions and under constant light (11 days at 1 200 lux, Palmer & Round, 1967). A special property of this species' tidal rhythm is that it is self-adjusting. As the time of afternoon low tide approaches sunset, cells rephase in a few days to emerge during morning low tides only. This rephasing took place in the laboratory both in constant light and the light–dark régime. Palmer & Round (1967) suggest that a 24·8 hour (lunar) clock interacts with a 24 hour (solar) clock, the latter inhibiting migration of cells at night and the former determining two 4·5 hour periods per lunar day, spaced 12·4 hours apart and governing positive phototactic movement onto the surface. But the data hardly suffice yet to determine the precision of these 'clocks' (see Enright, 1965). When *H. virgata* cells become negatively phobotactic before high tide they produce a lot of mucilage and stick together themselves and the sand grains (Fauré-Fremiet, 1951; Ganapati *et al.* 1959). This agglutination may immobilize the cells until the mucilage is biodegraded. If the sand is shaded the cells burrow down, but re-emerge when the shade is removed unless the tide would cover them within an hour (Palmer & Round, 1967). Presumably cells which re-emerged had not stuck themselves down. Damming the beach so as to retain water cover also caused them to stay down (Palmer, 1960). Sometimes species behave very differently on firm sand or mud when the sea is clear, not turbid, and stay on the surface when the tide comes in (Herdman, 1921; Callame & Debyser, 1954; Pomeroy, 1959; Perkins, 1960; Leach, 1970). *Pleurosigma aestuarii* has also been seen on the surface of sand before it was exposed by tide (Fauvel & Bohn, 1907). More than 27 lux was needed to produce surfacing in Perkins' (1960) laboratory, and less light induced movement into the sand.

On estuarine muds *Surirella gemma* and other species appear to have a tidal rhythm, which is not retained away from tidal influence. Hopkins (1966a,b) has found that when estuarine mud with *S. gemma* on its surface was stirred or wetted at a particular time one day the diatoms left its surface just before that time the next day. Thus their tidal rhythm in the field may well depend on the mechanical shock of breaking waves or on increases in the water content or salinity of the surface film (Taylor, 1967). However Callame & Debyser (1954) considered that the response was too accurate to be accounted for in this way. The rhythm in *S. gemma* was otherwise a circadian one of phototaxis and geotaxis (Fig. 8.2) which was temporarily suppressed by dark treatment (Hopkins, 1966a).

Besides *S. gemma,* 19 other estuarine species (Aleem, 1950; Callame & Debyser, 1954; Round & Palmer, 1966; Hopkins, 1969; Leach, 1970) have been observed and sometimes found to have a tidal response in the field, but only a circadian rhythm after a day or two away from tides. Fauvel & Bohn (1907) apparently found a more persistent rhythm in *P. aestuarii.* Round & Palmer (1966) found that daily migration continued in constant light (1 200 and 380 lux). As Round & Eaton (1966) pointed out the light intensity is not constant for the diatoms. Estuarine mud (unlike freshwater silt) is so compact that they go down into darkness. In darkness interrupted only for sampling, or when the mud was darkened during the period of upward migration, Round (1965b) found that fewer diatoms surfaced than in the light.

Cold treatment (−5°C for 7 hours) stopped migration of *S. gemma,* and on return to 15°C the peaks for geotaxis and phototaxis appeared delayed (Hopkins, 1966a). Similar rephasing of migration was observed in *Cylindrotheca signata* after being at 2°C for 4 days, on being placed at 12°C (Round & Palmer, 1966). But in *Nitzschia tryblionella* more cells surfaced after the change to 12°C, in an erratic fashion (Round & Palmer, 1966).

Similarly greater numbers of small diatoms on lake sand appeared on the surface at 25°C than at 9°C (Harper, 1967). This may be due to increased motility at the higher temperatures (Hopkins, 1963).

There is of course no tide to disturb freshwater diatoms, but those on silt and sand nevertheless burrow down to spend the night buried in it. Unlike marine species these diatoms (21 species observed) lose their diurnal rhythm in constant bright light, remaining on the surface (1 210 lux, Round & Happey, 1965; 800 and 400 lux Round & Eaton, 1966; and 750 lux, Harper, 1969). Some reduced rhythms were observed in dim light (75, 7·5, 0·5 lux Harper, 1969) and darkness (Round & Happey, 1965; Round & Eaton, 1966; Harper, 1969). The very small diatoms on sand grains probably do not retreat as far as the large estuarine species: less than 2 mm for the former (Harper, 1969); 1 to 4 mm for the latter (Hopkins, 1963, 1966a; Round & Palmer, 1966). All the quantitative work on rhythms of freshwater species was confined to the laboratory, but Harper (1976) has checked their existence in the field by direct observation. Using a microscope tube supported by an improvised wire frame, she watched *Pinnularia viridis* cells moving on mud in a small shallow pool in a stream. It was possible to see movement to and from the mud surface. Very limited periods of motility are probably a feature of tidal diatoms (e.g. *S. gemma,* Hopkins, 1966a, 1969), but diatoms from pond silt remained active throughout the day in the laboratory (Harper, 1969), as did the *P. viridis* in the pool. However, these diatoms, and those on sand grains, are distinctly more active for an hour or two around sunrise and before sunset (Harper, 1969). At these times they are both moving for more of the time and moving faster, also they are reversing more frequently (Harper, 1967). Therefore migration is governed by the interaction of rhythms of motility, phototaxis, geotaxis and disturbance. *Hantzschia virgata* is, it seems, the only truly tidal diatom.

4 LOCOMOTION THEORIES

The early suggestions that diatoms were driven by pseudopodia or cilia, osmotic or cytoplasmic currents, gas jets or by the motion of an external layer are considered elsewhere. (Taylor, 1929; Fritsch, 1935; Drews & Nultsch, 1962 and Hendey, 1964). Liebisch (1929) suggested that the protoplast undulated, communicating thrust to the outside via a pectin membrane which he thought went across the raphe slit. No such membrane lies across an active raphe, and so the protoplast's thrust would have to go direct to the outside. But the raphe is so deep and so narrow that viscous friction would prevent any significant amount of thrust being exerted externally along it.

A general cytoplasmic theory was proposed by Jarosch (1962) to explain all cell motions. Waving actin fibrils are fixed to the cell wall at one end, if outside the cell they drive it, if inside they cause streaming in the cytoplasm.

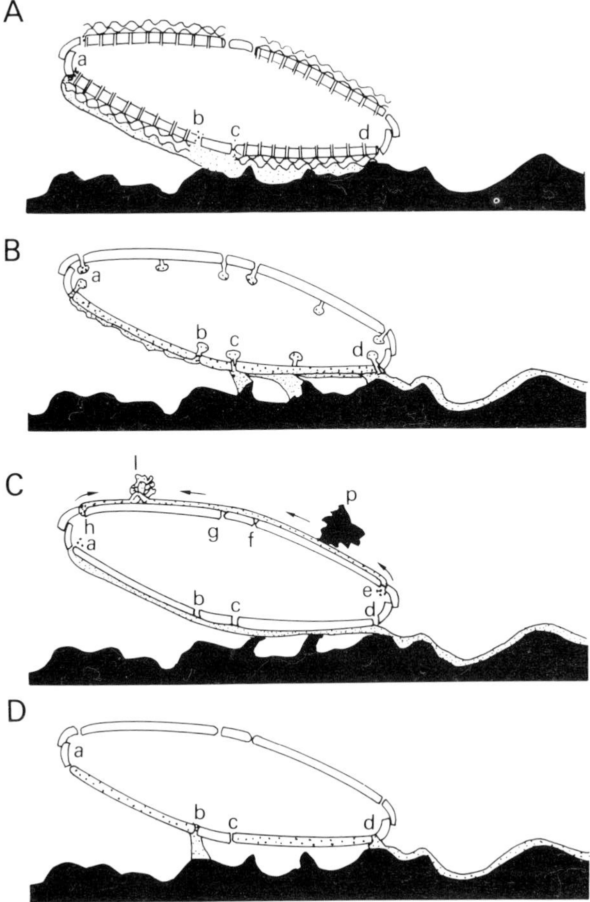

Fig. 8.5. Four locomotion theories, represented on schematic sections of pennate diatoms in their apical planes. Trail precursor is shown by coarse stipple, trail by fine stipple.

(A) Jarosch (1962). Mucilage secreted at pores a and c being driven by undulating actin filaments, connected to the protoplast.
(B) Hopkins & Drum (1966). Trail secreted through entire lengths of raphes ab and cd and expanding on leaving the raphe system.
(C) Harper & Harper (1967). Motion due to trail secretion from pore A. Secretion from upper pores e and h forming lump at l and particle p being carried forward.
(D) Gordon & Drum (1970). Capillary flow of a liquid along raphes ab and cd, conversion to trail at pores b and d. A, B, and C redrawn from their respective references.

Active actin attached to cell membranes and myosin have been found in non-muscle cells (Adelstein, 1974). The actin filaments were supposed to be connected through the raphe to the protoplast (Fig. 8.5,A) but protein fibrils are not found in the raphe. Rotating filaments appear in Hopkins & Drum's (1966) first theory. They suggest that polysaccharide fibrils are produced by the 'crystalloid bodies' near the raphe slits. They describe these fibrils as expanding and spiralling as they pass through the slit, pushing the diatom away when they stick against a surface (Fig. 8.5,B). This is a form of mucilage propulsion somewhat different from that first suggested by Lauterborn (1894): his threads appear to act like extensible ski sticks. Alternatively Hopkins & Drum (1966) proposed that the 'fibrillar bundle' is contractile and pumps mucilage out of the raphe. But surely exudation from a slit in an object would push away at right angles to the long axis of the slit, unless it were constrained by surface tension or in some other way. Diatoms usually move along a line parallel to their long axis. Hopkins (1969) implies that trail fibrils are held in place by penetrating right through the raphe fissure and being driven along it on the inside. Raphe friction and trail flexibility would interfere with this mechanism.

Now diatoms can be thought of as minuscule caterpillar tractors which leave their track behind them, instead of rolling it around. Harper & Harper (1967) suggested two forces which might hold a mucilage thread, driven by secretion from a terminal pore, along a raphe (Fig. 8.5,C). Osmosis into the protoplast above the raphe could lower the pressure in the slit so as to hold the thread in position along a raphe (Fig. 8.6,A). About 0·4 atmospheres would suffice for the measured forces exerted by diatoms. But the thread readily sticks to the frustule. What stops the system getting gummed up? The raphe may contain a special liquid which does not mix with water (Fig. 8.6,B). Its interfacial tension (surface of one liquid to another) would then keep the

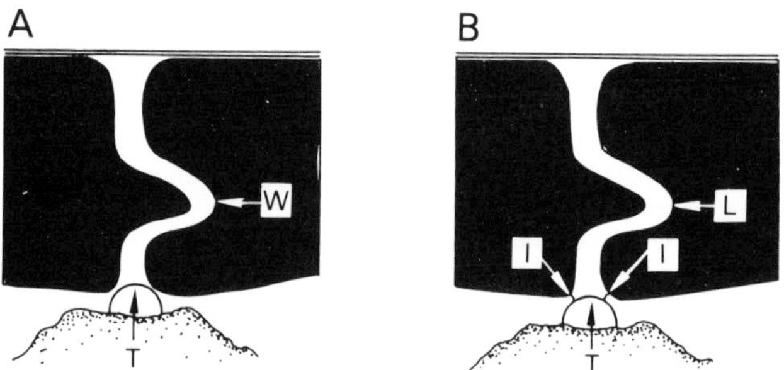

Fig. 8.6. A, The osmotic hypothesis; B, The interfacial tension hypothesis. Schematic raphe cross-sections. Substratum stippled. I, water–liquid interface; L, liquid; T, trail; W, water at low pressure being sucked into the diatom by osmosis. From Harper & Harper, 1967.

mucilage track in position (surface tension of only $\frac{1}{2}$ dyne/cm being needed). This liquid would also lubricate the raphe system, allowing the mucilage to slip along easily. Possibly Jarosch (1958) observed droplets of this liquid which rarely comes out of the raphe. Mucilage propulsion presents a problem if the thrust is provided by its secretion from the cell. The trail is probably secreted through minute pores at the ends of the raphes and these would present a large amount of frictional resistance to the viscous slime. But the trail polysaccharide must come from the protoplast, so a mucilage precursor may either polymerize outside this constriction or pass it in an aligned crystalline form (Gordon & Drum, 1970). In either case the trail may transmit its secretion force by swelling against the sides of the pores. Hopkins (1969) has found that polysaccharide exudation by cultured diatoms is more than sufficient to account for observed trail production during motility periods.

The latest theory is that of Gordon & Drum (1970). They propose that the precursor is a liquid which wets the raphe walls, and that it is moved along the raphe by a capillary pressure difference between the ends. They propose that the pressure difference be maintained by the liquid reacting with water to form trail substance as it leaves the diatom from the far end of a raphe (Fig. 8.5,D). But this theory fails to account for the measured strength of some diatoms (Harper & Harper, 1967). This arises from the following reasoning (personal communication J.F. Harper): If a diatom is driven according to this theory, then the capillary pressure P will be of the order $2\gamma/w$, where γ is the raphe fluid's interfacial tension against water and w is the raphe width. If P changes by an amount of that order from one end of the raphe to the other, Gordon & Drum (1970) show that the maximum speed of flow, V_{max}, is of order $w\gamma/6\eta L$, and the driving force F is of order $2\gamma h$, where L and h are respectively the length and depth of the raphe and η is the viscosity of the raphe fluid. Most of the force F would be used in driving fluid along inside the deep, narrow raphe slit, only an amount F_{ext} of order $2\gamma w$ would be available externally to drive the diatom along. This seems insufficient: γ is unlikely to exceed 50 dyne/cm, w ranges from 20 to 250 nm in different diatom species, and so F_{ext} would be less than 25 nN. Harper & Harper (1967) and Harper (1967) found four different species in which tractive forces (see Table 8.1) were sometimes much higher than this. *Amphora ovalis* pushed the fibre with a force of 480 nN on two occasions. Gordon & Drum's theory also fails to explain how one diatom can successfully climb over the raphe of another without being stopped by its raphe fluid leaking away, unless there are two raphe substances, one a liquid which wets the raphe and the other forms the sealing mucilage trail. But we are then almost back to the theory of Harper & Harper (1967) (see Fig. 8.6,C). The only advance is that if the curvature of the interface I is sharper at the trailing end of the raphe than the leading end, then the associated pressure difference would *help* drive the motion. Such a configuration would arise if the liquid L were sucked back into the diatom at the trailing end. This would also conserve raphe fluid, allowing diatoms to move continuously for several hours to reach the sediment surface after being buried several centimetres. There is some evidence that a sticky thread of trail runs along the whole length of an active raphe system and whichever part touches the slide leaves a bit of trail on it. Also when groups of polystyrene spheres near raphes had been stained with ruthenium red they appeared to be linked by loops of thread in the electron microscope (Harper, 1967). Hopkins (1969) notes loops of trail at the end of non-driving raphes. Analysis of filmed particle transport along the raphes of a stationary *Navicula oblonga* cell lying in girdle view suggested they were stuck to a thread moving along the raphe, and lumps were formed when the trail was either being secreted along both raphes towards the central nodule or from both ends of one raphe (see Fig. 8.5,C, upper raphes). The behaviour of these mucilage threads seems to explain the wide variety of phenomena observed: particle transport, particle flicking, movement on girdle bands, dangling like spiders, and jerky movement on a rough surface waving about on an end is accounted for if the continuation grooves beyond the polar nodules (Dawson, 1972) are used as 'quasi raphes'. There appears to be no reason why raphe liquid could not flow in under the mucilage thread and hold it in place for the short distance required, even though the continuation groove is not open to the interior of the diatom and raphes are (Fig. 8.6,B). Finally as Mann (1905) said a little earlier in this controversy, 'Whatever be the final explanation, it must explain not swimming but creeping; for these organisms are perfectly inert and helpless unless in contact with some fixed surface. . . . The power too must be considerable; for diatoms will push aside in their course inert matter many times their bulk'.

The ecological advantage of motility to plant cells living on unstable sediments is evident but locomotion also confers a genetic advantage to most pennate species as Fritsch (1935, p. 622) points out. It enables distinct individuals rather than contiguous daughter cells to unite to form auxospores.

ACKNOWLEDGEMENTS

I wish to thank Dr J.F. Harper for support and discussions, Dr U.V. Cassie, Dr F.J. Taylor and Mr B. Young for helpful criticism of the manuscript, Professor J.A. Rattenbury for the use of a laboratory, and Dr J.T. Hopkins for the loan of a copy of his thesis.

5 REFERENCES

ADELSTEIN R. (1974) Actin and myosin in non-muscle cells. *New Scient.* **61,** (884) 346–8.

ALEEM A.A. (1950) The diatom community inhabiting the mud flats at Whitstable. *New Phytol.* **49,** 174–88.

ALLAN G.G., LEWIN J. & JOHNSON P.G. (1972) Marine polymers IV. Diatom polysaccharides. *Bot. Mar.* **15,** 102–8.

CALLAME B. & DEBYSER J. (1954) Observations sur les mouvements des diatomées a la surface des sediments marins de la zone intercotidale. *Vie Milieu* **5,** 242–9.

DAWSON P.A. (1972) Observations on the structure of some forms of *Gomphonema parvulum* Kütz. I. Morphology based on light microscopy, transmission and scanning electron microscopy. *Br. phycol. J.* **7,** 255–71.

DAWSON P.A. (1973) Observations on the structure of some forms of *Gomphonema parvulum* Kütz. II. The internal organization. *J. Phycol.* **9,** 165–75.

DOUGLAS B. (1958) The ecology of the attached diatoms and other algae in a small stony stream. *J. Ecol.* **46,** 295–322.

DREWS G. & NULTSCH W. (1962) Spezielle Bewegungsmechanismen von Einzellern (Bakterien, Algen). In *Handbuch der Pflanzenphysiologie* (ed. W. Ruhland) 17(2) pp. 876–919. Springer–Verlag, Berlin.

DRUM R.W. (1963) Cytoplasmic fine structure of the diatom *Nitzschia palea. J. Cell. Biol.* **18,** 429–40.

DRUM R.W. (1969) Light and electron microscope observations on the tube dwelling diatom *Amphipleura rutilans* (Trentepohl) Cleve. *J. Phycol.* **5,** 21–6.

DRUM R.W. & HOPKINS J.T. (1966) Diatom locomotion, an explanation. *Protoplasma* **62,** 1–33.

DRUM R.W. & PANKRATZ H.S. (1964) Pyrenoids, raphes and other fine structures in diatoms. *Am. J. Bot.* **51,** 405–18.

DRUM R.W. & PANKRATZ H.S. (1965) Locomotion and raphe structure of the diatom *Bacillaria. Nova Hedwigia* **10,** 315–7.

DRUM R.W., PANKRATZ H.S. & STOERMER E.F. (1966) Electron microscopy of diatom cells. In *Diatomeenschalen im Electronenmikroskopischen Bild.* (eds. J.G. Helmcke & W. Kreiger) VII pp. 1–25 and 613 plates. Cramer. Weinheim.

DRUM R.W., GORDON R., BERDEN R. & GOEL N.S. (1971) Weakly coupled diatomic oscillators—Bacillaria's paradox resolved. *J. Phycol. Suppl.* **7,** 13.

EATON J.W. & MOSS B.M. (1966) The estimation of numbers and pigment content in epipelic algal populations. *Limnol. Oceanogr.* **11,** 584–95.

ENRIGHT J.T. (1965) The search for rhythmicity in biological time series. *J. theor. Biol.* **8,** 426–68.

FAURÉ-FREMIET E. (1951) The tidal rhythm of the diatom *Hantzschia amphioxys. Biol. Bull. mar. biol. Lab. Woods Hole.* **100,** 173–7.

FAUVEL P. & BOHN G. (1907) Le rythme des marées chez les diatomées littorales. *C. r. Séanc. Soc. Biol.* **62,** 121–3, 242.

FISCHER H. & TURCK K. (1969) Zum Nachweis einer Bewegungssubstanz bei Diatomeen mit Cresylechtviolett. *Ber. dt. bot. Ges.* **82,** 167–73.

FREY-WYSSLING A. (1972) Ueber Mikrotubuli. *Revue Suisse Zool.* **79,** suppl. 29–34.

FRITSCH F.E. (1935) *The Structure and Reproduction of the Algae.* Vol. I pp. 564–651. Cambridge University Press, Cambridge.

FUNK G. (1919) Notizen über Meeresdiatomeen. 1–3. *Ber. dt. bot. Ges.* **37,** 187–92.

GANAPATI P.N., LAKASHMANA RAO M.V. & SUBBA RAO D.V. (1959) Tidal rhythms of some diatoms and dinoflagellates inhabiting the intertidal sands of the Visakhapatnam beach. *Curr. Sci.* **28,** 450–1.

GEISSLER U. (1958) Das Membranpotential einiger Diatomeen und seine Bedeutung für die lebende Kieselalgenzelle. *Mikroskopie* **13,** 145–71.

GORDON R. & DRUM R.W. (1970) A capillarity mechanism for diatom gliding locomotion. *Proc. Nat. Acad. Sci.* **67,** 338–44.

HALFEN L.N. & CASTENHOLZ R.W. (1971) Gliding motility in the blue–green alga *Oscillatoria princeps. J. Phycol.* **7,** 135–45.

HARPER M.A. (1967) Locomotion of diatoms and 'clumping' of blue–green algae. Ph.D. Thesis, University of Bristol. pp. iii–xiii, 1–171.

HARPER M.A. (1969) Movement and migration of diatoms on sand grains. *Br. phycol. J.* **4,** 97–103.

HARPER M.A. (1976) The migration rhythm of the benthic diatom *Pinnularia viridis* on pond silt. *New Zeal. J. mar. freshw. Res.* **10,** 381–4.

HARPER M.A. & HARPER J.F. (1967) Measurements of diatom adhesion and their relationship with movement. *Br. phycol. Bull.* **3,** 195–207.

HENDEY N.I. (1964) An introductory account of the smaller algae of British Coastal Waters. *Fishery Invest. Lond. Series IV.* pp. 1–317 and plates; Min. Ag. Fish. Food.

HERDMAN E.C. (1921) Notes on dinoflagellates and other organisms causing discolouration of the sand at Port Erin. I. *Proc. Trans. Lpool. biol. Soc.* **35,** 59–63.

HÖFLER K. (1940) Aus der Protoplasmatik der Diatomeen. *Ber. dt. bot. Ges.* **58,** 97–120.

HÖFLER K., URL W. & DISKUS A. (1950) Zellphysiologische Versuche und Beobachtungen an Algen der Lagune von Venedig. *Boll. Mus. civ. Stor. nat. Venezia* **9,** 63–94.

HOFMEISTER L. (1940) Mikrochirugische Studien an Diatomeen. *Z. wiss. Mikrosk.* **57,** 259–73.

HOLLAND A.F., ZINGMARK R.G. & DEAN J.M. (1974) Quantitative evidence concerning stabilization of sediments by marine benthic diatoms. *Mar. Biol.* **27,** 191–6.

HOPKINS J.T. (1963) A study of diatoms of the Ouse estuary, Sussex. I. The movement of the mud-flat diatoms in response to some chemical and physical changes. *J. mar. biol. Ass. U.K.* **43,** 653–63.

HOPKINS J.T. (1966a) Some light induced changes in behaviour and cytology of an estuarine mud-flat diatom. In *Light as an Ecological Factor*. Symp. Br. Ecol. Soc. **5,** 335–58. Blackwell Scientific Publications, Oxford.

HOPKINS J.T. (1966b) The role of water in the behaviour of an estuarine mud-flat diatom. *J. mar. biol. Ass. U.K.* **46,** 617–26.

HOPKINS J.T. (1967) The diatom trail. *J. Quekett microsc. Club* **30,** 209–17.

HOPKINS J.T. (1969) Diatom motility: its mechanism, and diatom behaviour patterns in estuarine mud. Ph.D. Thesis University of London, pp. 1–251.

HOPKINS J.T. & DRUM R.W. (1966) Diatom motility: an explanation and a problem. *Br. phycol. Bull.* **3,** 63–7.

HUFFORD T.L. & COLLINS G.B. (1972) The stalk of the diatom *Cymbella cistula. J. Phycol.* **8,** 208–10.

JAROSCH R. (1958) Zur Gleitbewegung der niederen Organismen. *Protoplasma* **50,** 277–89.

JAROSCH R. (1962) Gliding. In *Physiology and Biochemistry of Algae* (ed. R.A. Lewin) pp. 573–81. Academic Press, New York.

KÜSTER E. (1936) Die Gallertbildungen der *Amphipleura rutilans*. *Arch. Protistenk.* **88,** 211–35.

LAURITIS J.A., HEMMINGSEN B.B. & VOLCANI B.E. (1967) Propagation of *Hantzschia* sp. Grunow daughter cells by *Nitzschia alba* Lewin & Lewin. *J. Phycol.* **3,** 236–7.

LAURITIS J.A., COOMBS J. & VOLCANI B.E. (1968) Biochemistry and fine structure of silica shell formation in diatoms IV. *Nitzschia alba*. *Arch. Mikrobiol.* **62,** 1–16.

LAUTERBORN R. (1894) Zur Frage der Ortsbewegung der Diatomeen. Bemerkungen zu der Abhandlung des O. Müller: Die Ortsbewegung der Bacillariaceen betreffend. *Ber. dt. bot. Ges.* **12,** 73–8.

LEACH J.H. (1970) Intertidal mud flat production. *Limnol. Oceanogr.* **15,** 514–21.

LEGLER R. & SCHINDLER H. (1939) Zentrifugierungsversuche an Diatomeenzellen. *Protoplasma* **35,** 469–73.

LEWIN J.C. (1955) The capsule of the diatom, *Navicula pelliculosa*. *J. gen. Microbiol.* **13,** 162–9.

LEWIN J.C., LEWIN R.A. & PHILPOTT D.E. (1958) Observations on *Phaeodactylum tricornutum*. *J. gen. Microbiol.* **18,** 418–26.

LIEBISCH W. (1929) Experimentelle und kritische Untersuchungen über die Pektinmembran der Diatomeen. *Z. Bot.* **22,** 1–65.

MANN A. (1905) Diatoms, the jewels of the plant world. *Smithson. misc. Collns.* p. 56.

MARTENS P. (1940) La Locomotion des Diatomées. *Cellule* **48,** 277–306.

MEADOWS P.S. & ANDERSON J.G. (1968) Micro-organisms attached to marine sand grains. *J. mar. biol. Ass. U.K.* **48,** 161–78.

MUELLER-HAECKEL A. (1973) Experimente zum Bewegungsverhalten von einzelligen Fliesswasser-Algen. *Hydrobiologia* **41,** 221–46.

MÜLLER O. (1889) Durchbrechungen der Zellwand in ihren Beziehungen zur Ortsbewegung der Bacillariaceen. *Ber. dt. bot. Ges.* **7,** 169–80.

MÜLLER O. (1894) Die Ortsbewegung der Bacillariaceen II. *Ber. dt. bot. Ges.* **12,** 136–43.

NULTSCH W. (1956) Studien über die Phototaxis der Diatomeen. *Arch. Protistenk.* **101,** 1–68.

NULTSCH W. (1962) Über das Bewegungsverhalten der Diatomeen. *Planta* **58,** 22–3.

NULTSCH W. (1971) Phototactic and photokinetic action spectra of the diatom *Nitzschia communis*. *Photochem. Photobiol.* **14,** 705–12.

NULTSCH W. & WENDEROTH K. (1973) Phototaktische Untersuchungen an einzelnen Zellen von *Navicula peregrina* (Ehrenberg) Kützing. *Arch. Mikrobiol.* **90,** 47–58.

PALMER J.D. (1960) The role of moisture and illumination on the expression of the rhythmic behaviour of the diatom *Hantzschia amphioxys*. *Biol. Bull. mar. biol. Lab. Woods Hole* **119,** 330.

PALMER J.D., LIVINGSTON L. & ZUSY D. (1964) A persistent diurnal rhythm in photosynthetic capacity. *Nature, Lond.* **203,** 1 087–8.

PALMER J.D. & ROUND F.E. (1967) Persistent vertical-migration rhythms in benthic microflora. VI. The tidal and diurnal nature of the rhythm in the diatom *Hantzschia virgata*. *Biol. Bull. mar. biol. Lab. Woods Hole.* **132,** 44–55.

PERKINS E.J. (1960) The diurnal rhythm of the littoral diatoms of the River Eden Estuary, Fife. *J. Ecol.* **48,** 725–8.

PETELER K. (1940) Bewegungserscheinungen und Gruppenbildungen bei *Nitzschia closterium*. *Ber. oberhess. Ges. Nat -u. Heilk.* **19,** 122–61.

POMEROY L.R. (1959) Algal productivity of the salt marshes of Georgia. *Limnol. Oceanogr.* **4,** 386–97.

REIMERS H. (1928) Über die Thermotaxis niederer Organismen. *Jb. wiss. Bot.* **67,** 242–90.

REIMANN B.E.F., LEWIN J.C. & VOLCANI B.E. (1965) Studies on the biochemistry and fine structure of silica shell formation in diatoms. I. The structure of the cell wall of *Cylindrotheca fusiformis*. *J. cell. Biol.* **24,** 39–55.

REIMANN B.E.F. & LEWIN J.C. (1966) Studies on the biochemistry and fine structure of silica shell formation in diatoms. II. Fine structure of the cell wall of *Navicula pelliculosa* (Bréb) Hilse. *J. Phycol.* **2,** 74–84.

RICHTER O. (1903) Reinkulturen von Diatomeen. *Ber. dt. bot. Ges.* **21,** 493–506 and plate.

ROUND F.E. (1964) The ecology of benthic algae. In *Algae and Man* (ed. D.F. Jackson) pp. 138–84. Plenum Press, New York.

ROUND F.E. (1965a) The epipsammon; a relatively unknown freshwater algal association. *Br. phycol. Bull.* **2,** 456–62.

ROUND F.E. (1965b) Persistent, vertical-migration rhythms in benthic microflora. V. The effect of artificially imposed light and dark cycles. *Proc. V Int. Seaweed Symp.* Halifax pp. 197–203.

ROUND F.E. (1970) The genus *Hantzschia* with particular reference to *H. virgata* var *intermedia* (Grun.) comb. nov. *Ann. Bot.* **34,** 75–91.

ROUND R.E. & EATON J.W. (1966) Persistent, vertical-migration rhythms in benthic microflora. III. The rhythm of epipelic algae in a freshwater pond. *J. Ecol.* **54,** 609–15.

ROUND R.E. & HAPPEY C.M. (1965) Persistent vertical-migration rhythms in benthic microflora. IV. A diurnal rhythm of the epipelic diatom association in non-tidal flowing water. *Br. phycol. Bull.* **2,** 463–71.

ROUND F.E. & PALMER J.D. (1966) Persistent vertical-migration rhythms in benthic microflora. II. Field and laboratory studies on diatoms from the banks of the River Avon. *J. mar. biol. Ass U.K.* **46,** 191–214.

SCHORR L. (1939) Beiträge zur Bewegungsphysiologie der Diatomeen. *Arch. Protistenk.* **92,** 273–82.

SPANGLE L. & ARMSTRONG T.B. (1973) Gliding motility of algae unaffected by cytochalasin B. *Expl. Cell Res.* **80,** 490–3.

SVERDRUP H.F., JOHNSON M.W. & FLEMING R.H. (1942) *The Oceans, their Physics, Chemistry and General Biology*. pp. xii, 1–1 087. Prentice Hall, New York.

TANADA T. (1951) The photosynthetic efficiency of carotenoid pigments in *Navicula minima*. *Am. J. Bot.* **38,** 276–83.

TAYLOR R.B. (1929) *Notes on Diatoms*. pp. 1–269 and plates I–V. Guardian Press, Bournemouth.

TAYLOR F.J. (1967) The occurrence of *Euglena deses* on the sands of the Sierra Leone peninsula. *J. Ecol.* **55,** 345–59.

TAYLOR W.R. & PALMER J.D. (1963) The relationship between light and photosynthesis in intertidal benthic diatoms. *Biol. Bull. mar. biol. Lab. Woods Hole.* **125,** 395.

TOMAN M. & ROSIVAL M. (1948) The structure of the raphe of *Nitzschiae*. *Stud. Bot. Cechosl.* **9,** 26–9.

WAGNER J. (1934) Beiträge zur Kenntnis der *Nitzschia putrida* Benecke insbesondere ihrer Bewegung. *Arch. Protistenk.* **82,** 86–113.

WAGNER J. (1958) Über das Verhalten der *Nitzschia putrida* Benecke nach Abkühlung auf −26°C. *Protoplasma* **49,** 98–115.

WALSBY A.E. (1968) Mucilage secretion and the movements of blue–green algae. *Protoplasma* **65,** 223–38.

WILLIAMS R.B. (1965) Unusual motility of tube-dwelling pennate diatoms. *J. Phycol.* **1,** 145–6.

ZAUER C.M. (1950) Movement of *Eunotia lunaris* (Ehr) Grun. in connection with the problem of locomotion of diatoms in general. *Dokl. Akad. Nauk. SSSR* **72,** 1 131–3 (in Russian).

CHAPTER 9

SEXUALITY

G. DREBES

Biologische Anstalt Helgoland (Litoralstation),
D-2282 List/Sylt, Germany

1 INTRODUCTION

Sexuality is a common feature in the life history of centric and pennate diatoms. Normally, sexual reproduction involves genetic recombination; in diatoms however, it involves additionally restoration of cell size. The first information regarding the nuclear cycle of diatoms is owed to the cytological studies of Klebahn (1896) and especially Karsten (1899, 1912). They discovered the reduction division during the formation of gametes in some pennate diatoms. In the early 1930s the cardinal points in the sexual cycle of the pennate diatoms were already well known (Geitler, 1932). Essentially, our knowledge of the sexual reproduction (and life history) of the Pennales is largely based on the descriptive work of Geitler and his co-workers. The numerous results obtained in this field are covered in the reviews of Geitler (1957, 1969, 1973).

As to the sexual process of the centric diatoms there have been misinterpretations for a long time. Schmidt (1927) and Hofker (1928) described meiotic nuclear divisions during formation of the so-called microspores in *Biddulphia* and *Coscinodiscus*. Although copulation of microspores was never observed, they were considered to be isogametes, and in contrast the formation of auxospores was a merely vegetative process. This was a great step forward from the work of Persitsky (1929), who observed meiotic stages in auxospore mother cells of *Chaetoceros*. However, his conclusion that the auxospores result from an autogamy was not held to be convincing by Geitler (1932). Based on some almost unnoticed observations of Went (1925), Geitler drew attention to the possibility of oogamy. The first definite evidence of oogamous reproduction was furnished by v. Stosch (1950, 1951a,b). Further investigations revealed this type of reproduction as the common process in the Centrales (Geitler, 1952a; v. Stosch, 1954, 1956, 1958a). With improved methods of culturing and *in vivo* observations detailed accounts of the complete ontogenetic cycle including the sexual cycle of the marine planktonic diatoms, *Stephanopyxis turris, S. palmeriana* and *Chaetoceros didymum* were presented (v. Stosch & Drebes, 1964; Drebes, 1966; v. Stosch *et al.* 1973). The sexual reproduction of *Stephanopyxis turris* has been cinematographically documented (Drebes, 1969b). Comprehensive studies were conducted on the ultrastructure of the male gametogenesis in *Lithodesmium undulatum* (Manton & v. Stosch, 1966; Manton *et al.* 1969a,b; 1970a,b).

The number of investigations motivated by physiological interests is rather limited. In three intensive studies, which have been made with *Melosira nummuloides* and *Rhabdonema adriaticum,* detailed reviews of earlier observations of environmental effects on sexual reproduction are included (Bruckmayer-Berkenbusch, 1954; Erben, 1959; Rozumek, 1968). As Dring (1974) points out, most studies of algal reproduction under different environmental conditions have been conducted by workers interested primarily in obtaining the complete life cycle of selected species in culture. Although the results gained in such studies appear incomplete to the physiologist, they provide at least indications of interesting responses. On the other hand, certain physiological studies are less valuable because of misinterpretations of morphological properties (e.g. Gross, 1937, 1940; see v. Stosch, 1965). For the induction of sexualization the influence of parameters, cell size, light, temperature, salinity and substrate were tested. Werner (1971a,b) applied quantitative biochemical methods to study the development from large, sexually non-inducible cells to small sexually inducible cells. There are considerable differences in the physiology of sexual induction and development even within genera, thus, at this stage a detailed comparison between inducing conditions is difficult. One must agree with Werner (1971b) when he states that the developmental physiology in diatoms is still far from being understood at the molecular level. Due to idiosyncrasies in the life cycle of diatoms, they are difficult objects for the physiologist. However, methods have been worked out for certain groups of diatoms permitting a manipulation of the cell size vegetatively without altering the genotype (v. Stosch, 1965). The ability to maintain the strains for any length of time in culture, and thus to investigate the various size-dependent processes makes these diatom species more acceptable for experimental purposes.

At the end of this chapter the problem of sex determination is treated.

2 MORPHOLOGICAL ELEMENTS OF SEXUAL REPRODUCTION

2.1 *Centric diatoms*

This description begins with the Centrales because they are regarded phylogenetically as the more primitive group of diatoms from which the Pennales must have developed in the early Tertiary or late Cretaceous times (Simonsen, 1972). As mentioned above, sexual reproduction in centric diatoms happens as a rule by oogamy in which large non-motile female gametes (eggs) become fertilized by small motile flagellated male gametes (sperms). The Centrales are diplonts with meiosis in the final stage of gametogenesis in both sexes. Some representative examples of the great variation in the details of gametogenesis are illustrated in Figs 9.1 to 9.5.

(a) *Male gametes*

Male determined cells usually undergo a series of successive differentiating mitoses to give small diploid spermatogonia which, in most cases, remain collectively enclosed by the parental frustule. Later, opening of the frustule occurs, and after two meiotic divisions the spermatogonia are each converted into four uniflagellate sperms. Within the spectrum of species, spermatogonia and sperms are produced in different ways. Referring essentially to earlier reviews (v. Stosch, 1954; v. Stosch & Drebes, 1964) the various types, so far known, are listed below without conceding any phylogenetical valuation in the order of appearance. In *Cyclotella* spp. differentiating mitoses are apparently lacking (or without differentiating character). Cells looking like vegetative cells function as spermatogonia (Geitler, 1952a; Schultz & Trainor, 1968). In *Melosira* spp. (Fig. 9.1A) differentiating mitoses give rise to variable number (2, 4, 8, 16, 32) of spermatogonia. For this type of mitoses the term 'depauperizing mitoses' was introduced (v. Stosch & Drebes, 1964). Compared to vegetative cells the spermatogonia of *Melosira* are shorter, thin-walled, and poor in plastids and pigment content. In many other species the spermatogonia are still more simplified and remain enclosed within their

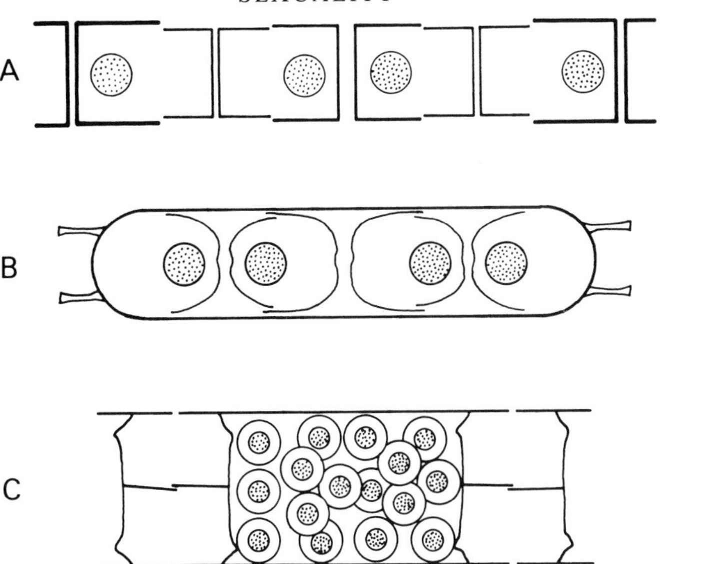

Fig. 9.1. Spermatogonial types in centric diatoms. A, *Melosira* sp., chain with four spermatogonia; B, *Stephanopyxis turris,* four spermatogonia enclosed in their mother cell (spermatogonangium); C, *Lithodesmium undulatum,* spermatogonangium containing 16 naked spermatogonia.

mother cell (spermatogonangium) until meiosis occurs. In *Stephanopyxis turris* (Fig. 9.1B), *S. palmeriana* (Drebes, 1966), *Biddulphia rhombus* and *B. granulata* the spermatogonia are furnished with thin rudimentary silica shells. In the planktonic *Biddulphia* spp. (*B. sinensis, B. regia, B. mobiliensis*) a pair of thin shells is produced only after the first depauperizing mitosis, dividing the mother cell in two half-spermatogonangia. The numerous relatively small spermatogonia are naked like those in *Lithodesmium* (Fig. 9.1C), *Streptotheca, Bellerochea, Rhizosolenia setigera, Chaetoceros* spp. and of the large *Coscinodiscus* spp. in which any formation of spermatogonial shells is lacking. Finally, in *Guinardia flaccida* and *Aulacodiscus argus* the depauperizing mitoses are reduced to nuclear divisions without following cytokineses. The spermatogonangia become multinucleate, in which the spermatogonia are merely represented by a number of nuclei united in a plasmodium.

At the beginning of meiosis the spermatogonia are called spermatocytes. During the meiotic prophase the spermatocytes swell, a process which usually opens the spermatogonangium and/or the spermatogonial shells. Accordingly, the spermatocytes may complete their further development partly free in the medium. Two main types of spermatogenesis are known, the merogenous and the hologenous types (Fig. 9.2). The first type is especially realized in species

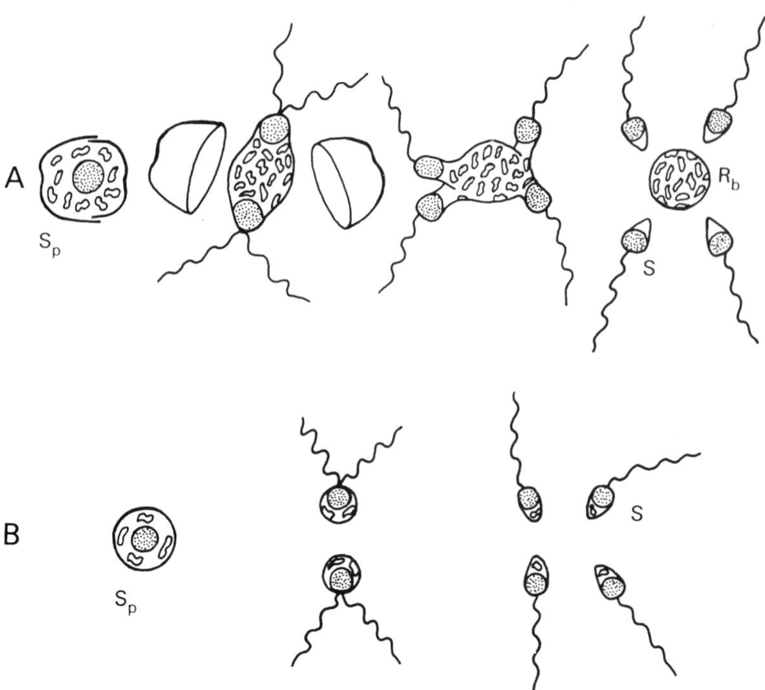

Fig. 9.2. Sperm formation in centric diatoms. A, Merogenous type; a spermatocyte (S_p) forms after two meiotic nuclear divisions four sperms (S), leaving behind a residual body (R_b); B, Hologenous type; by two meiotic nuclear divisions accompanied with cytokinesis in each, the entire spermatocyte is cleaved into four sperms.

with relatively large spermatogonia possessing silica shells. Undergoing the first meiotic nuclear division the spermatocyte gives rise to a binucleate plasmodium because a cytokinesis is lacking. As in all Centrales so far studied, during interkinesis the flagella grow out in pairs near the nuclei. The flagellar bases separate and migrate towards the poles of the second meiotic division figure. When the reduction divisions are completed four uniflagellated colourless sperms detach themselves, leaving behind the mass of cytoplasm with all the plastids in a residual body. The merogenous type is met in *Melosira, Stephanopyxis, Actinocyclus, Actinoptychus, Biddulphia rhombus* and *B. granulata*. In *B. granulata* spermatocytes each produce two residual bodies, as the first reduction division is followed by a cytokinesis (v. Stosch, 1956). In the more common hologenous type both meiotic nuclear divisions are accompanied by a cytokinesis. No residual body is formed, as the whole material of the spermatocyte is equally distributed between the four sperms.

The few plastids in these sperms are changed in shape, reduced in size and are very pale. As in the merogenous type, plastids which are left behind seem to play no further role. A peculiar process of spermatogenesis is found in *Coscinodiscus granii* and *C. concinnus*. The plastids and vacuoles do not remain in a residual body, but are pushed off separately after the swelling phase of the first meiotic prophase (v. Stosch & Drebes, 1964; Drebes, 1974).

The flagellum inserted in front of the sperm cell, is always directed forwards in swimming and the undulatory movement is indicated by the sinuous outline. Internal structure apart, the flagellum resembles the flimmer flagellum familiar in all the heterokont algal groups (v. Stosch, 1953). Observations on the fine structure of the sperms of *Lithodesmium undulatum* revealed an exceptional internal flagellar structure (Manton & v. Stosch, 1966). Contrary to expectation, the flagellar axis does not conform to the usual 9 + 2 pattern, but shows a 9 + 0 pattern lacking central strands (microtubules). This observation was confirmed by Heath & Darley (1972) with the male gametes of the related *Biddulphia levis*.

(b) *Female gametes*

In contrast to the spermatogonia formed by depauperizing mitoses, the oogonia develop directly from vegetative cells and function as mother cells of the auxospores. Only in *Melosira nummuloides* which apparently reproduces by autogamy (Erben, 1959), although sperm production has been observed (see Ettl *et al.* 1967), the formation of an auxospore mother cell is preceded by an unequal vegetative cell division (Rieth, 1953). Young oogonia can be distinguished from vegetative cells by considerable elongation of the cell, a swollen nucleus, and increased number of plastids. Three types of egg formation are recognized (Fig. 9.3):

1 Oogonia containing two eggs (Fig. 9.3A). In the oogonium the first meiotic nuclear division is accompanied by equal cytokinesis resulting in two prospective egg cells. The second reduction division follows without cytokinesis, leaving both eggs with a surviving haploid and a pycnotic nucleus. Representatives of this reproductive type are *Lithodesmium undulatum* (v. Stosch, 1954) and some species of the *Biddulphiales: Biddulphia mobiliensis* (v. Stosch, 1954), *B. regia, B. sinensis* (Drebes, 1974), *B. granulata* (v. Stosch, 1956) and *Attheya decora* (Drebes, unpublished).

2 Oogonia containing a single egg and a polar body (Fig. 9.3B). The polar body, which later disintegrates, results from an unequal cell division following the first reduction division. After the second meiotic karyokinesis in the egg cell one nucleus becomes pycnotic. Sometimes even in the polar body the same process can be observed. Only two species are known to

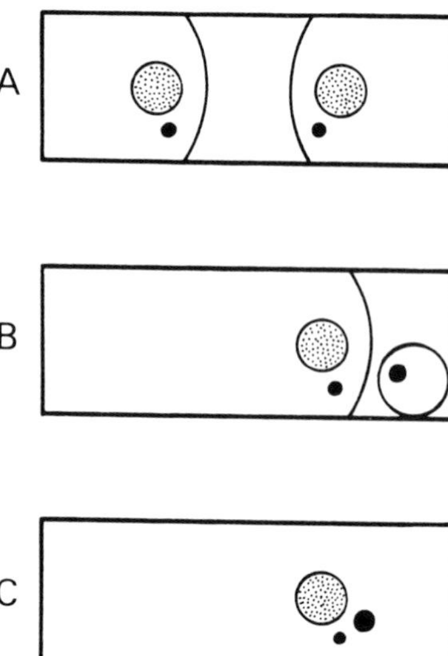

Fig. 9.3. Egg types in centric diatoms. A, Oogonium with two mature eggs, each containing one haploid surviving (punctate) and one pycnotic (black) nucleus; B, Oogonium with one functional egg and one polar body; C, Oogonium with a single egg, containing one haploid surviving and two pycnotic nuclei.

have this oogonial type: *Biddulphia rhombus* and *Cerataulus smithii* (v. Stosch, 1956).

3 Oogonia containing a single egg (Fig. 9.3C). In most centric diatom species (see v. Stosch & Drebes, 1964) cytokineses are lacking during meiosis. After each of the two reduction divisions one nucleus becomes pycnotic. Finally, the mature egg contains a large haploid nucleus surrounded by two pycnotic nuclei of different sizes.

If one considers oocytes and spermatocytes as originally homologous, the first oogonial type would be the primitive one, from which the third type could be derived. Possibly the second type represents a transition, until the cytokinesis was gradually fully suppressed. The most primitive type, an oogonium containing four eggs, has not yet been found and remains therefore, hypothetical.

The comprehensively investigated genus *Stephanopyxis* (v. Stosch & Drebes, 1964; Drebes, 1966; Drebes, 1969a,b) permits live observation down to the level of nuclear processes. Thus, the peculiar process of nuclear abortion (pycnosis) during meiosis could be studied by direct inspection. In the oocyte of *S. turris* the elongating nuclear spindle of the first meiotic division separates the two daughter nuclei—lying in a layer of parietal cytoplasm—with a distance of 25 till 40 μm. After resting for some minutes, the nuclei again approach each other with about the same velocity as the anaphasic stretch, until they become closely associated. After half an hour or more, degeneration of one nucleus and its eventual resorption by the cytoplasm is visible. The second reduction division, which follows an interkinesis lasting 2–3 hours, happens in a similar manner. Questions arise concerning the reasons for nuclear migration and nuclear abortion: Which one of the two nuclei migrates and which becomes pycnotic? In most cases only one nucleus moves, the other remaining more or less stationary. But migration of both is also possible. At this time it is impossible to predict from the position and the moving behaviour of the nuclei which of them will later disintegrate. The migratory as well as the stationary nucleus may degenerate. Abortion of supernumerary nuclei seems to be independent of nuclear migration and nuclear contact. In one case migration failed, nevertheless one nucleus became pycnotic. Even the absence of migration and abortion of both nuclei, ending with death of the cell, has been observed. Conversely, another single observation showed nuclear migration but failure of nuclear pycnosis, followed by an autogamic-like fusion of both nuclei. Unfortunately, the later fate of this irregular development has not been pursued. In *Stephanopyxis* nuclear pycnosis does not happen before reconstruction of both nuclei. However, in *Melosira varians* (v. Stosch, 1951b) and in pennate diatoms (Geitler, 1956) pycnosis starts already in the early telophase. The observations made are still too scanty to tell us the reasons for nuclear pycnosis. Various hypotheses concerning this problem were proposed by Geitler (1956), but so far they are not wholly accepted. As to the role of a 'paracentrosome'-like structure during nuclear abortion, it is referred to in recent investigations by v. Stosch *et al.* (1973).

(c) *Fertilization*

In order to obtain fertilization of eggs surrounded by the oogonial thecae, a site for sperm penetration is needed. As shown in Fig. 9.4, various ways are realized in different species. In *Melosira varians,* between the hypovalva (a hypopleura is lacking) and the free margin of the epipleura the naked plasmatic egg surface is exposed along an annular narrow slit through which the sperm enters (Fig. 9.4A; v. Stosch, 1951b). In *Stephanopyxis turris* and *S. palmeriana* a swelling of the protoplast leads to a slight bending of the oogonium causing a temporary opening for entrance of the sperm (Fig. 9.4B; v. Stosch & Drebes, 1964; Drebes, 1966). In *Chaetoceros* the sperm penetrates into the egg through one of either bristle openings in the girdle zone of the oogonium (Fig. 9.4C; v. Stosch *et al.* 1973). The two-egged oogonia of certain *Biddulphia* spp. offer a greater part of the egg surface to the sperms,

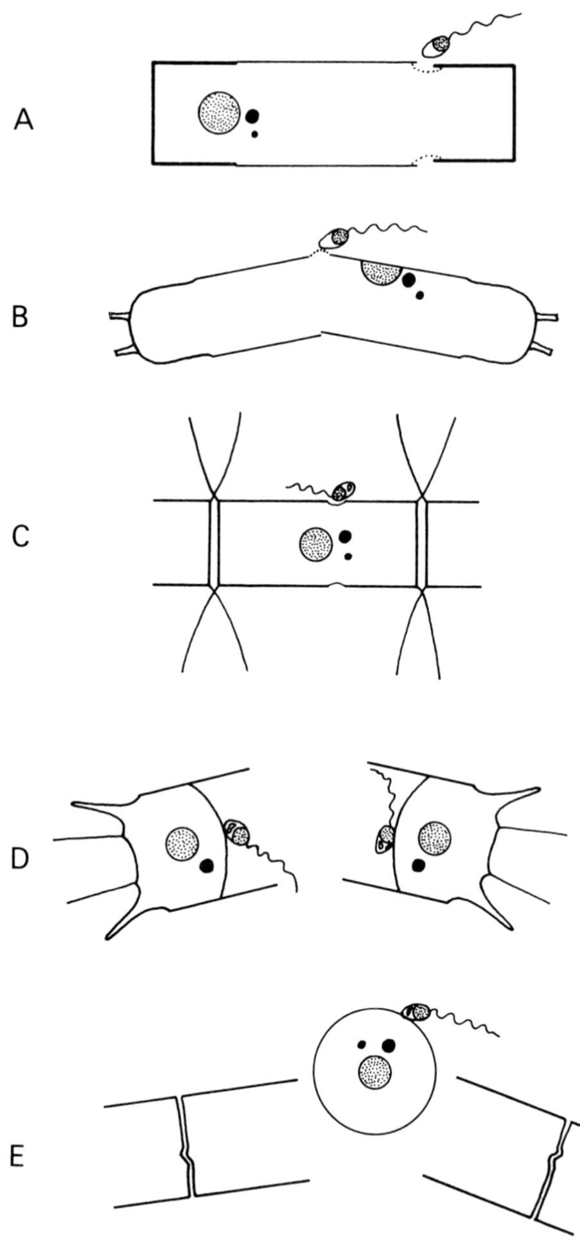

Fig. 9.4. Fertilization in centric diatoms. Various possibilities of sperm penetration into the egg cell are shown in five genera. A, *Melosira*; B, *Stephanopyxis*; C, *Chaetoceros*; D, *Biddulphia*; E, *Streptotheca*.

because the oogonial thecae are fully separated (Fig. 9.4D; v. Stosch, 1954, 1956). By complete shedding of the oogonial shells the eggs of *Lithodesmium, Streptotheca,* and *Bellerochea* become totally naked, allowing fertilization from all sides (Fig. 9.4E; v. Stosch, 1954). In *Biddulphia rhombus* the oogonium remains closed by a weakly silicified membrane joining both thecae (v. Stosch, 1956). For a perforation of this tender membrane by a sperm an enzymatic action is assumed.

Independent from the meiotic nuclear stage of the oocyte, fertilization may happen as soon as the egg surface is mechanically (partly or totally) exposed. However, rendering the eggs fertile for sperms seems also to require a physiological maturation of the female protoplast. As Geitler (1952a) points out for *Cyclotella,* after opening of the oogonia the early egg stages remain unfertilized for a time. Penetration of sperms does not start before late diakinesis. In *Melosira varians,* after egg exposure during interkinesis fertilization is immediately possible (v. Stosch, 1951b). In *Chaetoceros didymum* the sperm may already enter the oocyte during zygotene (v. Stosch *et al.* 1973). In *Biddulphia mobiliensis* and *B. granulata* the eggs are cytologically mature when the oogonia become open and apparently simultaneously ready for conception. (v. Stosch, 1956). Also completely mature, when fertilized, are the eggs of *Stephanopyxis, Lithodesmium,* and *Streptotheca,* because their oogonia become open postmeiotically (v. Stosch & Drebes, 1964). The readiness for conception of eggs may be manifested by a specific chemotactical attraction of the male gametes. (But this idea requires experimental proof.) As a rule during penetration the motile organelle of the sperm, the flagellum, is discarded.

Naturally, in all cases nuclear fusion (karyogamy) does not follow before the female nucleus has reached the mature haploid stage. The transparent cells of *Stephanopyxis* allow an undisturbed live observation of the karyogamic process (v. Stosch & Drebes, 1964; Drebes, 1966). According to the few observations on *S. palmeriana,* the sperm, after passing through the plasmalemma of the egg cell, slowly approaches the large female nucleus and within 2 to 6 minutes attaches itself to it. Meanwhile a contraction of the protoplast closes the oogonium again. In the course of the next half hour the small homogeneously coloured sperm nucleus swells, becomes reticulate, and finally a thread-like structure of chromosomes appears. Then the male chromosomes wander along the periphery of the female nucleus and seem to enter it from all sides. Based on cytological observations this fundamental process is seen to take place in a similar manner in *S. turris.*

(d) *Auxospores, initial cells*

After nuclear fusion the zygote starts to swell to considerable size, reaching generally a multiple of the diameter or apical length of the mother cell. The

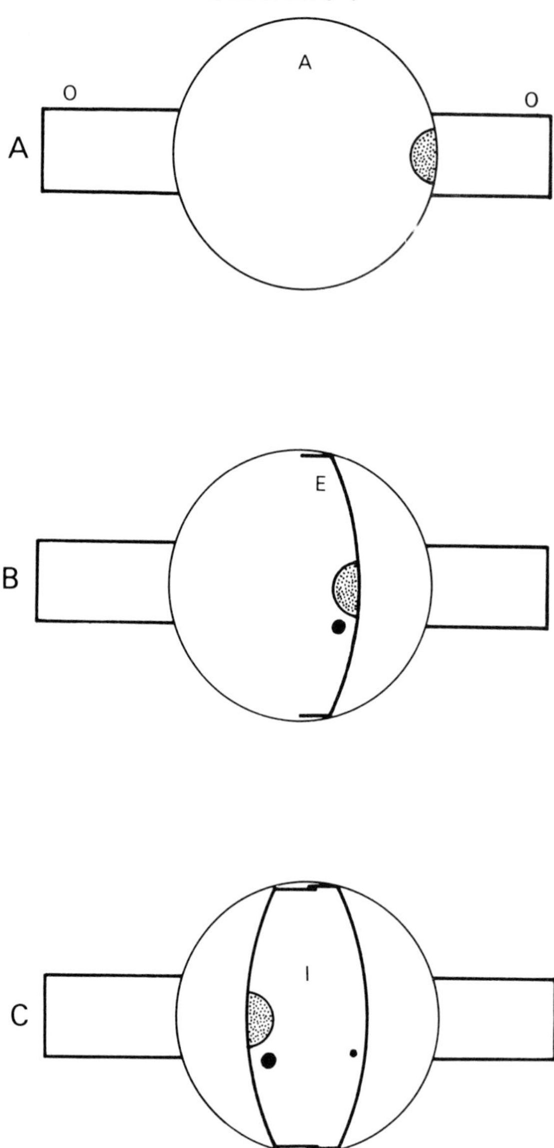

Fig. 9.5. Formation of the initial cell in centric diatoms. A, Intercalary formed auxospore (A) containing the diploid zygotic nucleus, O, oogonial thecae; B, Inside the auxospore membrane the epitheca (E) of the initial cell is formed by a metagamic mitosis, one of both nuclei degenerates; C, Initial cell (I), still surrounded by the auxospore membrane. Beside a functional nucleus there are two pycnotic nuclei resulting from two metagamic mitoses.

fully grown stage is called the auxospore (Fig. 9.5). Since the auxospore appears as a rule to result from a sexual process the term 'auxozygote' is also in use. According to the way in which the auxospores develop within or outside of the oogonial thecae various auxospore types can be distinguished: intercalary, semi-intercalary, lateral, terminal, and free auxospores. (For illustrations of some types see e.g. Karsten, 1928; Drebes, 1974). Intercalary auxospores are widely distributed in the genera, especially in the sub-order Coscinodiscineae (Fig. 9.5). Semi-intercalary auxospores are known from the two-egged oogonia of some *Biddulphia* species. Lateral auxospores, arising laterally on the parent cell by budding, are apparently confined to members of the Rhizosoleniineae, such as *Rhizosolenia, Bacteriastrum,* and *Chaetoceros.* In *Rhizosolenia alata* the auxospore develops terminally from one oogonial end, growing in the same direction as the pervalvar axis of the parent cell. Free auxospores are met in only a few genera, e.g. *Lithodesmium, Streptotheca,* and *Bellerochea,* in which the eggs are liberated completely from their shells before fertilization (v. Stosch, 1954).

As found at first by Reimann (1960) in *Melosira varians,* the very dilatable auxospore membrane (fertilization membrane) consists mainly of organic material with small roundish silica scales incorporated. In the related genus *Stephanopyxis* the auxospore membrane likewise is composed of scale-bearing carbohydrate layers, secreted by apposition during auxozygotic growth (v. Stosch & Drebes, 1964). The isometrically grown, globular-shaped auxospore then may produce the initial cell, indeed occurring in numerous genera and species (Fig. 9.5). But in *Chaetoceros didymum* (and with some Biddulphiaceae), after an isometric growth phase, the zygote passes into an anisometric growth stage (v. Stosch *et al.* 1973). The latter is supported by a system of silica rings, the preperizonium (v. Stosch & Kowallik, 1969), which, in contrast to the homologous perizonium of pennate diatoms, forms a joint layer with the auxospore membrane. Also in *Bacteriastrum hyalinum* a perizonium-like structure could be observed (Drebes, 1972). Compared with the preperizonium, it develops separately without fusing with the auxospore membrane. Therefore, in that case the term 'paraperizonium' has been suggested (v. Stosch *et al.* 1973).

Inside the auxospore envelope the initial cell is formed. The consecutive formation of the first two thecae is preceded and induced by a metagamic (acytokinetic) mitosis in each, corresponding to the law established by Geitler (1963)—and later specified by v. Stosch & Kowallik (1969)—that every formation of a diatom valva is caused by a mitosis. As cytokineses are lacking, supernumerary nuclei become pycnotic (Fig. 9.5). Often, the initial cell varies from the vegetative cells somewhat in shape and silica structure, this can lead to errors in diatom taxonomy. Thus, for example, *Bacteriastrum solitarium* (Mangin) proved to be identical with the initial cell of *B. hyalinum* (Osorio-Tafall, 1935; Drebes, 1967), and the forms of the genus *Debya* Pantocsek correspond well to the initial cells of *Actinoptychus undulatus* (Van Heurck, 1896; Meunier, 1915). In certain species three thecae are formed and give rise to the initial cell, whereas the primary theca is not used and later shed. This is facultatively the case in *Bacteriastrum hyalinum* (Drebes, 1972), or takes place regularly in *Chaetoceros eibenii* (v. Stosch *et al.* 1973). In both species the shed primary theca has a reduced fine structure, e.g. lacking bristles. In addition the initial cell of *C. eibenii* obligately develops into a resting spore (Pavillard, 1921), curiously this does not otherwise occur during the vegetative life cycle. Also in *C. pseudocrinitum* (Ostenfeld) and *Leptocylindrus danicus* the auxospores appear to be transformed directly into resting spores (Gran, 1915). In *C. vanheurcki* (Gran) and *Bacteriastrum hyalinum* a resting spore may be formed at first from the ready initial cell (Ikari, 1926; Drebes, 1967, 1972). During the final events of forming the initial cell the auxospore envelope ruptures, liberating the new enlarged cell.

(e) *Non-oogamous formation of auxospores*

As mentioned above, in the Centrales auxospores are enlarged zygotes resulting, as a rule, from an oogamous sexual process. Until now auxospores arising from automixis (autogamy) or apomixis have been observed in only a few cases, and these are incompletely studied (Iyengar & Subrahmanyan, 1944; Ettl *et al.* 1967). Sexual reproduction by autogamy was found in the fresh- and brackish-water species *Cyclotella meneghiniana* (Iyengar & Subrahmanyan, 1944) and in the marine *Melosira nummuloides* (Erben, 1959). After meiosis in the auxospore mother cell, two of the nuclei unite autogamously. But both species are capable of producing simultaneously flagellated male cells; however fertilization has never been discovered (Schultz & Trainor, 1968; v. Stosch & Drebes, unpublished). Sperm production seems to be a sheer waste in this case. According to Schultz & Trainor (1968) autogamous development may be either a reduced mechanism replacing oogamous reproduction or a facultative adjustment occurring in the event of the oogonium not being fertilized. With an inland clone (salina near Artern/East Germany) of *Melosira nummuloides* auxospores but never sperms were observed (Rieth, 1953). Also, the mediterranean *Melosira moniliformis* var. *octogona* (most probably an autonomous species) forms auxospores without development of male stages (Drebes, unpublished). No meiotic nuclear processes could be detected in the auxospore-producing cells, the auxospores being apparently of pure vegetative origin (Bröer & v. Stosch, unpublished). Experience with several clones of *Actinoptychus undulatus* likewise induces caution before assuming that oogamous reproduction happens solely from the occurrence of auxospores and sperms. In this species no oogamous nuclear fusion was seen. Up to 75% eggs underwent incomplete meiosis followed by parthenogenetic formation of auxospores. About 25% eggs however, completed meiosis. Since they remained unfertilized, however, further development into auxospores failed (Behre & v. Stosch, unpublished).

The cases described here suggest that we are dealing with deficiency phenomena and compensational events of an originally-existing oogamous reproduction.

2.2 *Pennate diatoms*

Centrales and Pennales coincide in certain cardinal points of the sexual cycle. They are diplonts with meiosis during the final stage of gametogenesis, and the zygotes develop into auxospores. In contrast to the Centrales the sexual reproduction of the Pennales generally occurs by morphological isogamy and physiological anisogamy. The gametes are non-flagellated (aplanogametes) and are brought together by a prior pairing of their mother cells (gametangia). For this reason the term 'gametangiogamy' is also in use (Wiese, 1969).

If the pennate diatoms are considered to be derived from the Centrales found in earlier geological strata, this type of isogamy probably developed from oogamy. This appears to be unique in the plant kingdom and may be regarded as a kind of atavism. The derivative nature of the isogametes is, for example, indicated by the loss of flagella, being perhaps analogous to the involution of the wings of certain insular insects (v. Stosch, 1958b).

The various types met in sexual reproduction (allogamy, automixis, apomixis) and auxospore formation are usually represented in a 4-type-system. The system mainly used was established by Karsten (1899), adopted by Hustedt (1930) and substantiated and modified by Geitler (1932, 1957, 1973). The latter contribution of Geitler contains a synopsis of 95 species, varieties and some unnamed intraspecific units, all of which were investigated after 1932. Referring largely to Geitler's descriptive work, the fundamentals of the different reproductive modes are roughly outlined in the following. In addition to the well-known four types a fifth one (transitional type) is introduced here and placed at the beginning in order to demonstrate transitional forms of sexual behaviour between the Centrales and Pennales.

(a) *Type I: Transitional type*

As to transitional reproductive types linking the Centrales with the Pennales, only araphidate diatoms can be considered. Indeed the family of Tabellariaceae occupies such an intermediate position between the two major groups of diatoms. Magne-Simon (1962) points to the dual relationship within this family, leading on the one hand toward the Centrales in *Rhabdonema,* on the other hand toward the Pennales in *Grammatophora*.

As far back as 1885 oogamous reproduction was described in *Rhabdonema arcuatum* (Buffham, 1885). However, the descriptions as well as

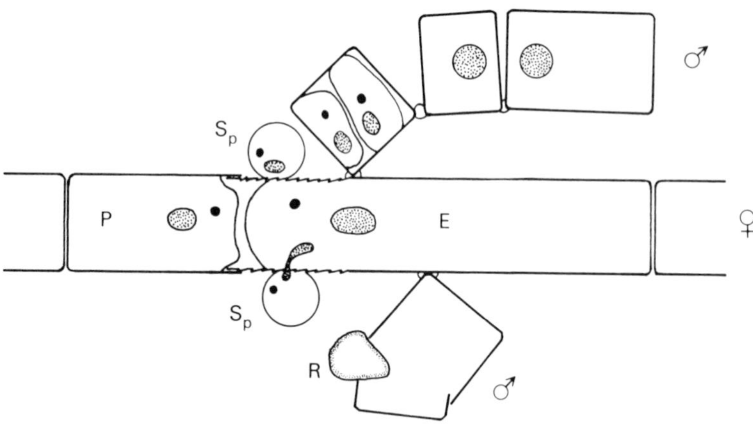

Fig. 9.6. Oogamous reproduction in *Rhabdonema adriaticum*. In the middle a female thread (♀) with an oogonium containing one functional egg (E) and one polar body (P). Attached to the oogonium spermatogonial cells (♂) in different meiotic stages; the lower one has released two spermatia (Sp) and leaves behind a residual body (R). The spermatia are sticking to the girdle region of the oogonium, and the one below is just injecting its nucleus into the egg cell. After v. Stosch, 1958b.

illustrations of this English scientist apparently seemed rather fantastic to his contemporaries and were not seriously regarded. Of course, the fixed material did not allow Buffham to observe details, e.g. sperms and the process of fertilization. Karsten (1899) reported on auxospore formation in two *Rhabdonema* species. Recent studies were made on cultures of *R. adriaticum, R. arcuatum,* and *R. minutum* (v. Stosch, 1958b, 1962). The dioecious *R. adriaticum* was especially thoroughly investigated. As shown in Fig. 9.6, vegetative band-like colonies differentiate male and female gametangia, spermatogonia and oogonia. Like the Centrales, the male gametangia arise from two or more depauperizing mitoses. Accordingly, compared to vegetative cells, the spermatogonia are reduced in cell length, plastids and pigment content. In meiosis the first of both nuclear divisions is followed by cleavage of the protoplast resulting finally in two microgametes per spermatogonium. The naked and globular microgametes, measuring about 25 μm in diameter, contain a surviving haploid and a degenerating pycnotic nucleus each. They are non-flagellated and should better be termed spermatia (cf. Rhodophyta) than sperms. Due to the size of spermatogonia in relation to the two spermatia a residual body may be formed (Schäfer & v. Stosch, unpublished).

Contact between spermatogonia and oogonia is passive and brought about by water movement; attachment however is by mucous pads. The released spermatia creep in amoebic fashion to the fertilizable region of the oogonium.

Curiously, only the male nucleus is injected into the egg. Oogonia are directly formed by considerable elongation of vegetative cells. After polarization in the female protoplast the first meiotic nuclear division, accompanied by unequal cytokinesis, provides a prospective egg cell rich in cytoplasm and plastids and with a relatively large polar body containing only few plastids. The second nuclear division is followed by abortion of one nucleus. After fertilization the zygote secretes a thick-layered gelatinous membrane bearing small silica scales and inflates to an auxospore. When the primarily globular auxospore turns to bipolar directed growth, the so-called perizonium is formed inside the dilating zygotic membrane. The perizonium of *Rhabdonema* has a very complex structure consisting of silicified transverse and longitudinal bands (v. Stosch, 1962). Since the first theca is shed in the development of the initial cell three thecae are formed, each by metagamic mitosis. The last theca results from an extremely unequal cell division, which leaves behind, beside the initial cell, a small aborting, one-valved residual cell. In *R. arcuatum* the residual cell even contains two thecae. This species and also *R. minutum* are monoecious, developing two eggs and two auxospores per oogonium respectively.

In *Grammatophora marina,* closely related structurally to *Rhabdonema,* the oogamy of the Centrales ultimately terminates in a morphological anisogamy. Small male cells stick to the large female cells with the aid of mucous pads and start meiosis in identical fashion. Similar to the events in the oogonium of *Rhabdonema adriaticum,* in the male as well as in the female frustule (gametangium) one haploid gamete and one degenerating polar body develop. Copulation occurs in the female frustule. The zygote rounds-up to an auxospore which, concurrent with considerable elongation of the cell secretes a perizonium. The initial cell arises by two or, as in *Rhabdonema,* three metagamic mitoses. For further details of the sexual process in *Grammatophora* see the work of Magne-Simon (1960, 1962).

Genera belonging to other araphidate families, *Meridion, Diatoma* (Diatomaceae), *Synedra* (Fragilariaceae), sexually reproduce by isogamy (see Geitler, 1973), as will be shown in the following section.

Compared to the large number of studies treating sexual reproduction of pennate diatoms, up till now relatively little attention has been paid to the araphidate species. These species on the border between Centrales and Pennales promise interesting results.

(b) *Type II: Normal type*

In the majority of pennate diatoms the sexual partners cannot be distinguished morphologically. Sexually induced diploid mother cells (gametangia) come together in pairs and surround themselves with a copulation jelly. In certain species a copulation jelly is not conspicuous, or special copulation tubes are developed. Most frequent in the Pennales is the so-called 'normal type' (Figs

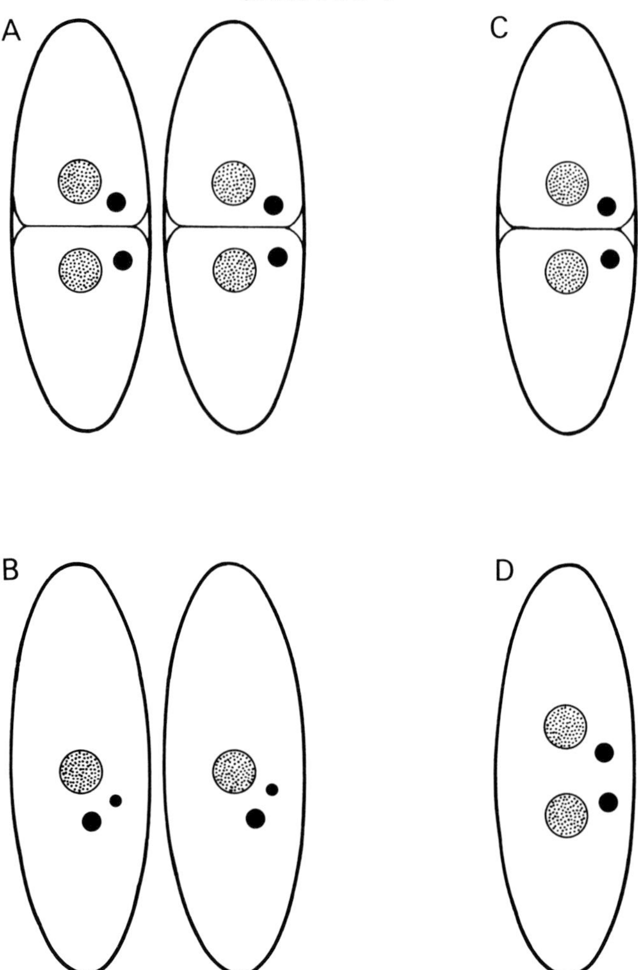

Fig. 9.7. Gametic and copulation types in pennate diatoms. A, Normal type: a pair of gametangia possessing two gametes each. As a result of meiosis every gamete contains one functional haploid and one pycnotic nucleus. Fusion of the gametes by allogamy; B, Reduced type: a pair of gametangia possessing a single gamete each, since no cytokinesis occurs during meiosis. The mature gametes each contain one functional and two pycnotic nuclei. Fusion of the gametes by allogamy. C, Automixis (pedogamy): a solitary gametangium containing two gametes formed like those in (A). Pedogamous fusion of both gametes. D, Automixis (autogamy): a solitary gametangium containing a single gamete. Since nuclear pycnosis is lacking after the first meiotic step, the mature gamete contains two surviving and two pycnotic nuclei. Autogamous fusion of both nuclei occurs.

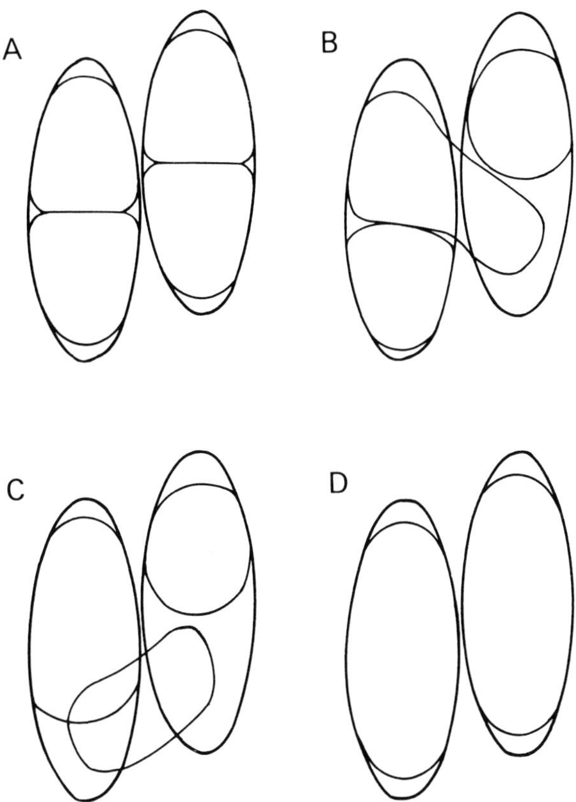

Fig. 9.8. Scheme of a typical allogamous copulation in a pennate diatom (*Gomphonema parvulum*). A, Paired gametangia with mature gametes; B, C, Cross-fertilization between the migratory and stationary gametes; D, Two young zygotes. After Geitler, 1932.

9.7A 9.8, 9.9) in which, combined with meiosis in each of both conjugate cells, two isomorphous aplanogametes are formed. Only the first meiotic nuclear division is accompanied by a cytokinesis whereas after the second a supernumerary nucleus becomes pycnotic. The mature gametes each possess a surviving haploid and a degenerating pycnotic nucleus (Fig. 9.7A). In principle here and with the following 'reduced type' (Type III) the gametes are formed in a similar way to the female gametes (eggs) of the centric diatoms and of the araphidate *Rhabdonema* species (cf. Fig. 9.3, 9.6).

The gametes remain in the parental thecae until they fuse. With respect to their migratory behaviour isogamous and anisogamous fusion of the gametes can be distinguished. In isogamous species the gametes behave similarly and move toward one another. Mostly, however, each mother cell contains one active and one passive gamete. Then, as shown in Fig. 9.8, a crosswise fusion of the migratory with the stationary gametes is effected in two consecutive steps. In other cases one mother cell produces two stationary gametes, the other two active ones. On fusion of the gametes two zygotes start to develop into auxospores.

During the bipolar enlargement of the auxospore its membrane gelatinizes or ruptures and is substituted by the silicified perizonium. The perizonium may be smooth or consists mostly of visible transverse silica rings. Inside the perizonium of the mature rod-shaped auxospore, on the poles of which the remainder of the auxospore membrane appears as caps, the initial cell is formed by two metagamic mitoses. Figure 9.9 represents the final stage of auxospore formation developed into an initial cell. Often the external shape of the initial cell deviates from that of the definite vegetative cell. According to Geitler (1969) a production of four gametes per mother cell, as perhaps observed by Pascher (1932) in a *Nitzschia* species, is doubtful. But in *Navicula*

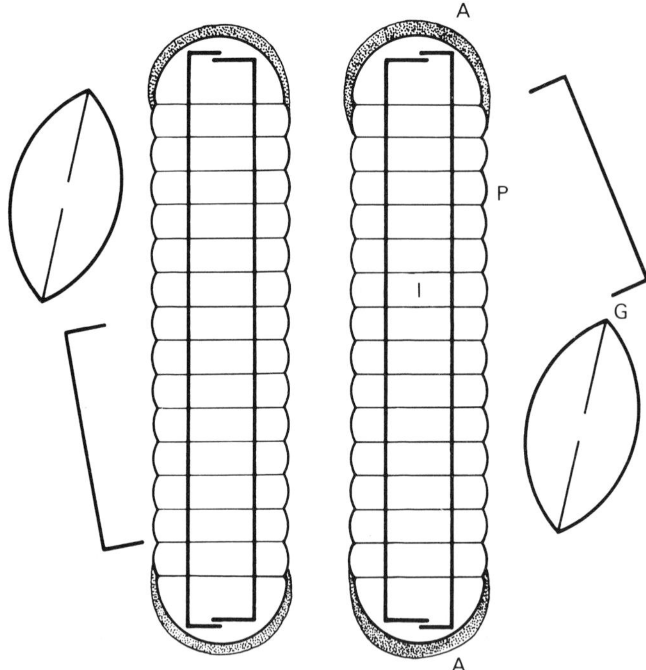

Fig. 9.9. Two initial cells as the final result of allogamous (normal type) auxospore formation in pennate diatoms. A, residuals of the auxospore envelope; P, perizonium; I, initial cell; G, empty gametangial thecae.

radiosa, although a second cytokinesis is lacking, at least all the meiotic tetrad nuclei remain and the zygotes arising contain two pairs of copulating nuclei of which one pair only survives (Geitler, 1952b).

A great diversity in the copulation process has been elucidated by much scientific work done in this field. The limited space available does not permit a full account of the various possibilities in detail.

The 'normal type' can be split up in three sub-types (Geitler, 1973):

1 Gametes anisogamous; migratory and stationary gametes (*Amphipleura pellucida, Cymbella* spp., *Frustulia, Gomphonema* spp., *Navicula halophila, Nitschia* spp., *Synedra ulna*).
2 Gametes orderly isogamous; position of the auxospore axes fixed relating to the mother cells (*Amphora ovalis, Denticula* spp., *Epithemia* spp., *Navicula radiosa, N. cryptocephala* var. 'II', *Surirella ovata*).
3 Gametes ± arbitrarily isogamous; position of the auxospores without firm relation to the mother cells (*Anomoeoneis exilis*).

(c) *Type III: Reduced type*

This type is less frequent and may be derived from the 'normal type'. After the first meiotic nuclear division the cytokinesis is merely rudimentary or even lacking. Thus, the normal number of two gametes formed per mother cell is reduced to one functional gamete. In the case where a rudimentary, extremely unequal, cytokinesis is intercalated, in addition to the single gamete a small degenerating residual cell is formed (*Cocconeis, Eunotia*; Geitler, 1957). The two meiotic nuclear divisions without cytokinesis, each followed by abortion of one nucleus, lead to the formation of a single gamete (Fig. 9.7B). Consequently, from paired mother cells only one auxospore arises. Corresponding to the migratory behaviour of the gametes, isogamous copulation was found in *Cocconeis placentula* races, *Eunotia, Navicula cryptocephala* var. *veneta, Nitzschia amphibia, Surirella* spp. Other races of *C. placentula* and *Navicula seminulum* behave anisogamously (Geitler, 1973).

(d) *Type IV: Automixis*

In some pennate diatoms allogamous reproduction appears obligatorily or facultatively substituted by automixis (Geitler, 1957, 1973; Thaler, 1972). Accordingly, auxospores resulting from automixis arise from isolated (unpaired) or seldomly from pseudo-paired mother cells. Two types of automixis are generally recognized (Geitler, 1973):

1 Pedogamy: In one mother cell two gametes are formed (in the manner of Type II) by cytokinesis following the first meiotic nuclear division (Fig. 9.7C). Thereafter both gametes fuse to form a zygote (*Amphora*

normaniveneta, *Cymbella aspera*, *Gomphonema constrictum* var. *capitatum*).

2 Autogamy: Ab absence of any cytokinesis in one mother cell a single gamete is formed (Fig. 9.7D). Curiously, after the first meiotic nuclear division pycnosis of one nucleus fails and both nuclei survive. Abortion of two nuclei occurs only after the second reduction division, while the two remaining functional nuclei fuse autogamously (*Cymbella ventricosa* var. 'I', *Denticula tenuis*).

According to Geitler (1973) automixis occurs in various species groups and represents perhaps phylogenetically a recent acquisition.

(e) *Type V: Apomixis*

Apomixis includes parthenogenetic as well as purely vegetative formation of auxospores. Auxospores formed parthenogenetically develop from a single diploid cell, either from an isolated mother cell or more often from pseudo-paired mother cells. Pseudo-meiotic events accompanying this process show parthenogenesis obviously as a rudimentary sexual process. Diploid parthenogenesis is known with certainty only in some varieties and minor races of *Cocconeis placentula* and was most thoroughly investigated by Geitler (1973). Normally, the allogamous forms of *C. placentula* produce a single gamete per mother cell and a degenerating polar body originating from an extremely unequal cell division after the first meiotic nuclear division. In *C. placentula* var. *lineata* the reason for parthogenetic development was found in the loss of syndesis (synapsis) in prophase I. The two steps of meiosis are substituted by a single mitosis. In the allogamous *Cocconeis placentula* var. *pseudolineata* and var. *euglyptoides* parthenogenesis may also occur in very exceptional cases. In contrast to a population of *Cocconeis placentula* var. *klinoraphis* from Lunz (Austria), a population from another locality proved to be obligatorily parthenogenetic. Both populations probably constitute races, which can be distinguished only by their reproductive mode. Thus, parthenogenesis seems to be either genotypically fixed, or may be an alternative development.

Earlier reports on apomixis found in other species are incomplete or partly erroneous and require re-investigation. Until now there is no sure evidence for a purely vegetative formation of auxospores in pennate diatoms (Geitler, 1957, 1973).

3 FACTORS INDUCING SEXUALIZATION

3.1 *Cell size*

The striking feature in the diatom life cycle, namely a progressive decrease of cell size concomitant with cell division, strongly influences developmental and physiological processes (v. Stosch, 1965). Accordingly, in contrast to other protists, induction of sexualization in diatoms depends not only on the genotype and specific environmental conditions but also on a suitable cell size as an internal non-genetic factor. Beginning with Pfitzer (1871), in the earlier literature (Ref. see Karsten, 1899) the influence of cell size on auxospore formation is already indicated. Nevertheless, based on comprehensive observations on field material as well as on clonal cultures of pennate diatoms, Geitler (1932) was really the first to show clearly a relation between sexuality and cell size. This applies likewise to the Centrales, as for instance could be already demonstrated by the data of Bergon (1907) made with *Biddulphia mobiliensis*. The cell size is usually expressed as diameter or apical length of the valva. Only relatively small cells, measuring generally about 30 to 40% of maximal valve diameter within a species-specific size range prove to be capable of sexualization. Among the oogamous Centrales the size ranges for the production of males and females are not fully coincident. That means the determination of sex is also a function of cell size. According to available data obtained from field material, within the sexual size range the largest cells develop into oogonia, at a certain intermediate size arbitrarily either male or female cells are produced, and the smallest cells in the range may only form spermatogonia (Fig. 9.10; v. Stosch, 1951b, 1956). In this respect, the pennate but oogamous *Rhabdonema arcuatum* seems to behave similarly, whereas in *R. adriaticum* a lower size limit for the females is lacking (v. Stosch, 1958b; v. Stosch & Drebes, 1964).

Stephanopyxis turris fits the rule established for the Centrales only in a modified manner, since there are no size ranges of unisexual potence. However, although the size ranges of both sexes are totally overlapping, the tendency is still conspicuous for larger cells to preferentially produce females and conversely the smaller ones more abundantly males. In addition the smallest cells, as in the Pennales, remain sterile (v. Stosch & Drebes, 1964). The anomalous behaviour of *Stephanopyxis* may be partly explained by the fact that the data come from cultured material. It seems possible that species having reached their critical minimal size, can be kept in culture for a longer period than they remain alive in the natural environment. Thereby, in culture the size range may be extended but allows merely vegetative growth. On the other hand, differences are obviously species-specific. Thus, for instance in *Actinocyclus ehrenbergii* and *Actinoptychus undulatus,* even in culture the smallest cells of the lower size range are permanently capable of sperm production (v. Stosch & Drebes, 1964). Among the morphologically isogamous Pennales different size ranges of both sexes cannot be distinguished, since both gamete types are isomorphous. According to Geitler (1932), the sexual size range of the pennate diatoms beyond which only vegetative growth occurs has an upper as well as a lower size limit (Fig. 9.10). Comparing the sexual size ranges of the Centrales and Pennales, the latter has a closed range,

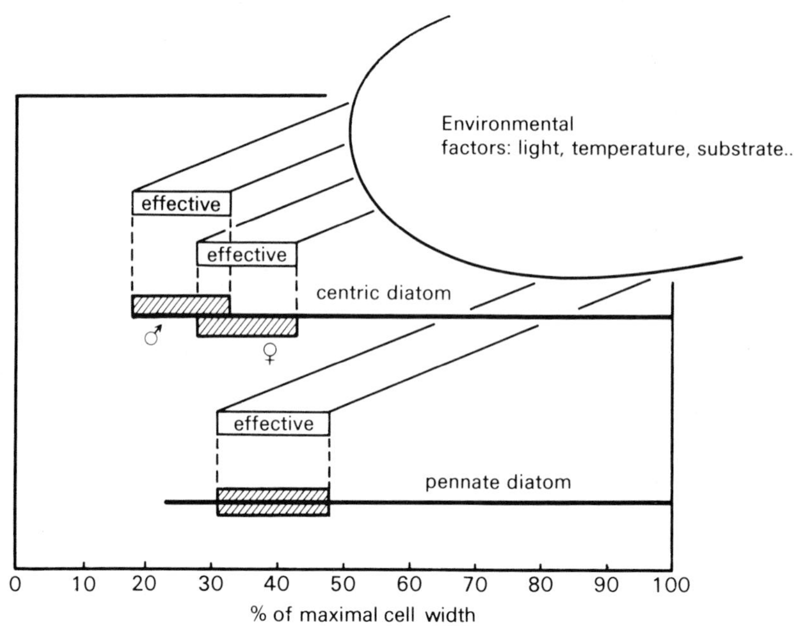

Fig. 9.10. Sexual induction and determination as a function of cell size and environmental factors. The two thick horizontal lines represent the entire size range of a centric diatom (*Melosira varians*) and of a pennate diatom (*Navicula seminulum*). In the centric diatom the hatched rectangle above the size line indicates the size range for male (♂), that below for female (♀) differentiation. In the pennate diatom in which sex differences are not recognizable, the line halves the hatched rectangle representing the size range of the gametangia-forming cells. The 'effective'-indexed rectangles are to demonstrate the cell size-limited effectiveness of environmental factors. After v. Stosch, 1956.

while in that of the former there is no lower limit. Yet a similarity exists, if only the ranges of the centric females and of the pennate gametangia are considered. This may prove to be of importance some day when the phylogenetical relationships between the sexual cells of both diatom groups are further studied.

In diatoms the inner physiological disposition of the cells for sexual reproduction is morphologically expressed by the cell size. Only in cells of the 'right' size are special environmental factors effective in the induction of sexualization (Fig. 9.10). Thus, if the adequate cell size is not realized there are no other conditions which are able to elicit gametogenesis (Geitler, 1932). Apparently only changes of the cell dimensions, normally warranted by the MacDonald-Pfitzer rule, enable the cell metabolism to undergo sexualization in natural environment. The question is raised concerning what parameters alter during decrease of cell size. At first a considerable decrease of the cell volume becomes conspicuous. Usually, the volume ratio between auxospore mother cells and initial cells varies from 1 : 3 to 1 : 5 (Geitler, 1932). In *Melosira varians* maximal cell volume is 20 times the minimum and in *Coscinodiscus curvatulus* 12 times (v. Stosch, 1965). Also changes in the external cell shape are obvious to the eye. Regarding the protoplast, the plasmatic material likewise diminishes proportionally, showing a reduction of size or number of plastids (Geitler, 1932; v. Stosch & Drebes, 1964). In *Cocconeis* and *Gomphonema* even the nucleus becomes slightly smaller (Geitler, 1932), but seems to be nearly size constant in *Stephanopyxis* (v. Stosch & Drebes, 1964). In *Melosira nummuloides* the nuclear size unexpectedly increases during the vegetative cell diminution (Bruckmayer-Berkenbusch, 1954). According to the latter author and Geitler (1932) a shift in the nucleus/plasma ratio in favour of the nucleus may be responsible for sexualization. Bruckmayer-Berkenbusch concluded that anomalously large production of plasmic protein causes meiosis in *Melosira.* To test this suggestion in cultures of *Melosira* the supplies of nitrogen and phosphorus were reduced or omitted from the nutritional medium (Erben, 1959). It transpired that induction of meiosis required no higher nitrogen supply than normal mitoses. Qualitatively, a paper chromatographic analysis of the cell content of *M. nummuloides* gave no evidence of biochemical differences in cells of the fertile and of the non-fertile size range (Erben, 1959). But, if measured quantitatively, the biochemical composition changes in a characteristic way during cell diminution. Referring to recent investigations conducted by Werner (1971a,b,c) with *Coscinodiscus asteromphalus,* in the diatom cell the shift of the nucleus : plasma : vacuole ratio is quantitatively determined by the DNA : protein : potassium ratio. The DNA content remains constant, while other cell parameters measured gradually become reduced in the course of size decrease. Also the sensitivity to metabolic inhibitors changes specifically. Since the generation time of *C. asteromphalus* does not alter with the diminution of the cell surface a reduction of metabolic activity results. It is true that the introduction of new or improved methods allows a characterization of the different developmental stages more exactly, but the fundamental question remains quite open as to why a prior biochemical differentiation is necessary to make cells sexually inducible by external factors.

As an experimental material to resolve this problem diatoms are recommended which permit altering of the cell size by artificial means (v. Stosch, 1965). This applies only to a certain number of diatom species. Avoiding possible clone-specific idiosyncrasies, a vegetative manipulation of the cell size allows the establishment of any number of subclones of different cell size, by maintenance of the genotype. Vegetative cell enlargement can be induced either by nutritional or other cultural manipulations. Narrowing of cells has been performed by nutritional methods in *Stephanopyxis, Rhabdonema arcuatum* and *Achnanthes,* and by operative methods in *Bellerochea* and *Grammatophora.* By abrupt artificial cell diminution fresh descendants of auxospores may be forced to sexual reproduction after only a few cell divisions. Thus inducibility of sexualization is merely a function of cell size and not of cell age measured in number of cell divisions. Cells of the sexual size range immediately lose their sexual inducibility after artificial cell enlargement.

Finally it should be mentioned that not all diatoms so far studied undergo a size decrease according to the MacDonald-Pfitzer rule. In some species unlimited vegetative multiplication without size decrease has been observed (Geitler, 1932; Wiedling, 1948; Locker, 1950). Accordingly, auxospores are never encountered in these species. In this connection the question arises as to whether sexual reproduction may exceptionally occur without subsequent inflation of the zygotes into auxospores.

3.2. *Environmental factors*

In the past century the effect of environmental factors (light, temperature, substrate on the formation of auxospores was already mentioned (Karsten, 1899). As stated above (Introduction), the external factors involved in the induction of sexualization (in general this term is synonymous with auxospore formation) may be considerably different in the vast species spectrum. This can be easily demonstrated by regular observations made on field populations of perhaps 20 species with a cell size in the sexual size range. Usually only two or three species are in auxospore formation at a certain time (Geitler, 1957). At present a generalization about the interplay of the external factors controlling sexualization is no more than speculative. But as a common feature in many species so far studied no significant antagonism between factors promoting vegetative growth and those eliciting gametogenesis seems to exist (Bethge, 1925; Krieger, 1927; Karsten, 1928; Geitler, 1932; Erben, 1959). Also the auxospores of diatoms have, as a rule, no growth arresting stages but continuing vegetative growth. Erben (1959) concluded from culture experiments made with *Melosira nummuloides,* under different nutritional conditions, that auxospores develop as long as substrate conditions allow vegetative growth.

Some culture experiments made with marine planktonic diatoms demonstrate that a sudden increase of light intensity evokes sexualization (v. Stosch & Drebes, 1964; Drebes, 1966). *Stephanopyxis turris,* growing in culture remained merely vegetative under dim light (200–300 lux) at 15°C but became sexual after an exposure to 2 000 lux. An intensified effect was obtained at 4 000 lux combined with a simultaneous rise of the temperature from 15° to 21°C. A similar behaviour was found in *S. palmeriana,* but in the case of this warm water alga the temperature rise was obligatory for the organism to become sexual (Drebes, 1966). In *S. turris,* this change in light intensity and temperature resulted in a cell division rate increasing from 0·33 to 1·5–1·0 divisions per day (Drebes, 1965). This agrees well with the results obtained by Geitler (1932) in pennate diatoms, namely, that the more auxospores develop the stronger vegetative growth occurs.

In cultures of *Coscinodiscus asteromphalus,* kept under optimal growth conditions, spermatogenesis was induced in a small percentage of the cells by changing the illumination from 1 250 to 2 500 lux and the temperature from 24 to 30°C (Werner, 1971c). The rise of temperature caused no significant change in the photosynthetic CO_2 fixation of the cells and the vegetative multiplication rate was nearly the same at 24 and 30°C. However, special attention must be drawn to the result that the conditions employed for sexual induction could be specifically substituted by other factors. Thus, spermatogenesis in *C. asteromphalus* could be induced also, without a rise of temperature, by transferring the cells from cultures, not shaken and additionally aerated to conditions, under which air (+1·5 vol % CO_2) was bubbled through the cultures, and 40 mg methanol/litre was added at the same time. Under these conditions vegetative growth failed.

In *Melosira nummuloides* rather low light intensities (200 lux) were found to be optimal for sexualization (Bruckmayer-Berkenbusch, 1954). A similar effect was noticed by Rozumek (1968) in *Rhabdonema adriaticum,* which showed a strong sexual reaction between only 250 and 500 lux at 15°C, or between 500 and 1 000 lux at 21°C (in both cases 14 hours light daily).

In some investigations, in addition to variations of temperatures and light intensities, the effect of day length has been studied. According to Steele (1965), *Stephanopyxis palmeriana* grew almost completely vegetatively in a short day length (12 : 12 hours). However, in a day length of 16 : 8 hours sexual stages were formed. Holmes (1966) was able to expose potential auxospore cultures of *Coscinodiscus concinnus* (species identification dubious) to 64 different light and temperature combinations simultaneously. The incident light and temperature ranges were the same in each experiment, 0·003–0·5 Langley per min and 9–27°C, respectively, but different photoperiods were used. In the case of the shortest photoperiod (8 hours of illumination daily), gametes appeared over almost the entire range of experimental conditions. Also under the shortest photoperiod between 16 and 24°C the formation of auxospores was accelerated. Conversely, high light intensities for long periods had an inhibitory effect. According to Holmes, the data suggest that in this species sexual reproduction should occur near the sea surface where light intensity is high during that portion of the year when the number of daylight hours is small.

The results of Bruckmayer-Berkenbusch (1954) and Rozumek (1968) obtained with *Melosira nummuloides* and *Rhabdonema adriaticum* respectively, suggest that auxospore formation is proportional to intensity and period of daylight. High light intensities and long photoperiods or continuous illumination inhibited sexual reactions. In *M. nummuloides* formation of auxospores could be totally suppressed with continuous high illumination (3 400 lux) and with high temperatures (24 °C). In *R. adriaticum,* Rozumek found a correlation between light intensity and light period as responsible for sexualization. With low light intensity sexual stages were still formed under continuous illumination, and with sufficiently high light intensities with less than 6 hours light daily. However, under short light periods higher light intensities could be endured as one would expect from the combination of the two factors.

The effects of light quality on sexual reproduction in diatoms is still a wide field open for investigation. Growth experiments with diatom cultures kept under monochromatic light of various wavelengths, indicated a stimulating effect of shorter wavelength (green and blue light) on the formation of auxospores (Baatz, 1941). Blue light has a specific effect on reproduction in a number of algal species (Dring, 1964). For example: the female gametophytes of the brown alga *Laminaria saccharina* grew vegetatively indefinitely in red light, but 12 hours of irradiation with blue light was sufficient to induce egg production (Lüning & Dring, 1972). Possibly also in diatoms blue light may stimulate formation of gametes, with the difference that in diatoms gametogenesis is combined with meiosis. It should be mentioned that it is perhaps less useful or even misleading to enter into a discussion about factors inducing meiosis, as done by Bruckmayer-Berkenbusch (1954), without considering all the morphogenetic processes involved in sexual reproduction.

The separate role of temperature during sexual induction is difficult to assess, unless some other aspects of the diatom metabolism are measured simultaneously. In *Chaetoceros decipiens* and *C. constrictus* formation of 'microspores' (male cells) has been induced by lowering the temperature several degrees from about 10–15 °C down to 5 °C. In *Lithodesmium undulatum, Streptotheca thamensis* and *Bellerochea malleus* a rise of temperature from 15 to 24 °C (combined with increase of light intensity, and inoculation into a fresh medium) caused sexualization (Manton & v. Stosch, 1966; v. Stosch, unpublished). As was already mentioned, similar effects were observed in *Coscinodiscus asteromphalus* and *Stephanopyxis palmeriana* (Werner, 1971c; Drebes, 1966). Well defined temperature optima have been found with *Coscinodiscus concinnus* (?) by Holmes (1966) and for *Rhabdonema adriaticum* by Rozumek (1968). Rozumek made measurements at eight temperatures, ranging over 20 °C, and found that the optimal temperature for sexual reproduction was 15 to 21 °C. As to interaction between temperature and light the studies of both authors indicate that at temperatures above or below the optimum auxospore production is restricted to a narrower range of light energies.

Referring to several reports in the older literature (reviewed by Bruckmayer-Berkenbusch, 1954), osmotic changes were suggested to be responsible for sexualization. Thus, a lowering of salinity or a slight dilution of the culture medium apparently caused formation of auxospores (Allen & Nelson, 1910; Cholnoky, 1928; Schreiber, 1931; Geitler, 1932, 1957; Braarud, 1945). Bruckmayer-Berkenbusch (1954) on the contrary noticed an inhibitory effect when diluting the culture medium of *Melosira nummuloides*. Also with *Cyclotella* formation of auxospores appeared to be induced by changes in salinity (Schultz & Trainor, 1970). In *C. meneghiniana,* a common species in fresh- and brackish-water, sexual stages were observed after transfer from a synthetic freshwater medium to a modified artificial seawater medium (Schultz & Trainor, 1968). A correlation with an increase in the Na^+ concentration in the medium was found. Spermatogenesis could be induced also in *C. cryptica* in artificial seawater with an adjusted sodium level.

In *Ditylum brightwellii* sexual induction is nutritionally controlled (Steele, 1965). Gametes occurred when manganese ions are absent from the medium. In *Melosira nummuloides* auxospore formation was promoted by amino acids, yeast and soil extract (Bruckmayer-Berkenbusch, 1954).

As to a hormonal control of sexuality the experiments made by Rozumek (1968) on the dioecious *Rhabdonema adriaticum,* attract attention. Gametes were only formed when two clones of different sex were cultured in the same vessel. It was supposed that at least two hormones are sharing in sexual induction. However, attempts at inducing gamete formation in one sex by adding filtrates of cultures from the other sex failed.

The data presented here claim neither completeness nor in any way substitute for a study of the original papers. The purpose of this section could only be a compilation of a number of the factors controlling sexual induction. In most cases the data available are too fragmentary to obtain a clear picture, thus having, for the time being, merely the practical value of supplying recipes to obtain sexual reproduction. Increased and detailed physiological investigations, based, as a necessary precondition, on a comprehensive knowledge of the life history and of the vegetative growth conditions of the species concerned, are now particularly desirable.

4 SEX DETERMINATION

The data, especially those obtained from observations on clonal cultures, indicate that the great majority of species are monoecious. In principle, sex determination is assumed to be diplophenotypic. This is the case in the Centrales and, among the Pennales, at least in some araphidate species (v. Stosch, 1955; Geitler, 1957; Ettl *et al.* 1967). Dioecy with possibly diplogenotypic sex determination seems to be rare and is known from only two pennate diatoms, *Rhabdonema adriaticum* (v. Stosch, 1958b) and *Grammatophora marina* (v. Stosch & Drebes, 1964, footnote p. 211). Exceptionally, in *Coscinodiscus granii* several clones behaved unisexually, producing only either male or female gametes (Drebes, 1968). Occasionally, in male clones a few oogonia were formed. This peculiar behaviour was provisionally described as 'subdioecy'. However, some years later newly isolated clones from the same area (Heligoland, North Sea) proved to be truly monoecious (Drebes, 1974). In clonal cultures of *Chaetoceros didymum* predominantly either pure male or female colonies were observed within the sexual size range permitting formation of both sexes (v. Stosch *et al.* 1973). More rarely, colonies occurred containing spermatogonia or oogonia mixed with vegetative cells. Only sometimes hermaphrodite colonies were encountered. At present no explanation for that can be given.

Regarding factors influencing sex determination, the function of cell size has been dealt with above (Fig. 9.10). Its role becomes conspicuous in cases in which both sexes can be morphologically distinguished, e.g. in the oogamous Centrales. Concerning external factors, v. Stosch (1954) found that sex determination in *Lithodesmium undulatum* was influenced by light. A clone of suitable cell size formed in continuous light were almost exclusively females, under a weak light–dark regime only males and, in a strong light–dark rhythm, both males and females. Steele (1965) noted that when a culture of *Stephanopyxis palmeriana* is grown for 16 hours in fairly dim light (30–40 footcandles; 1 fc = 10·76 lux) daily, only females are formed, whereas only males occur in higher light intensities (200–300 footcandles). Similarly, in *S. turris* the ratio of both sexes could be strongly influenced by a combined variation of several factors (Drebes, 1965). A preferred formation of spermatogonia was obtained when very narrow cells (16–20 μm ϕ) of the sexual size range were exposed to long photoperiods with a simultaneous rise in temperature from 15 to 21°C. In suitable cases, exclusively spermatogonia were produced in this way. Conversely, formation of oogonia was favoured in broader cells kept under shorter photoperiods and lower temperatures. Under the latter conditions however, besides oogonia production the percentage of vegetative cells remained relatively high. Although the few data communicated are still rather tentative, they at least indicate that the conditions causing formation of male and female cells are significantly different.

Somewhat problematical are discussions about sex determination in pennate diatoms. Here the lack of recognizable sex differences prevents designation of the gametic stages as (+) and (–). The gametes are in most types morphologically identical. Differences in gamete size or migratory behaviour as well as other diversities are not regarded as true sex characters (Geitler, 1957). Observations on clonal cultures and natural clones brought no evidence for a diplogenotypic sex determination since pairing and copulation occur within the single clones. A diplophenotypic determination may reasonably be assumed for this monoecious diatom group too. But before adopting this generalization, it is useful to refer to some critical discussions derived from reviews of Geitler (1932, 1957); v. Stosch (1955) and Ettl *et al.* (1967). Especially, if one tries to get a comprehensive explanation of allogamous, pedogamous and autogamous reproductive modes as well difficulties arise about the timing of sex determination. In the case of a general assumption of diplophenotypic sex determination, in allogamous species the mother cells (gametangia) of a copulating pair would be already sexually determined during the diplophase. Accordingly, one mother cell would produce only (+) gametes, and the other one only (–) gametes. However, if failure of gametangial pairing causes facultatively pedogamy, then the fixed sexual tendency within the mother cell must be reversible in one of the two gametes. Some evidence supporting this hypothesis, is discussed by v. Stosch (Ettl *et al.* 1967, p. 676).

5 REFERENCES

ALLEN E.J. & NELSON E.W. (1910) On the artificial culture of marine plankton organisms. *J. mar. biol. Ass. U.K.* **8,** 421–74.

BAATZ I. (1941) Die Bedeutung der Lichtqualität für Wachstum und Stoffproduktion planktontischer Meeresdiatomeen. *Planta* **31,** 727–66.

BERGON P. (1907) Biologie des diatomées. Les processus de division, de rajeunissement de la cellule et de sporulation chez le *Biddulphia mobiliensis* Bailey. *Bull. Soc. bot. Fr.* **54,** 327–57.

BETHGE H. (1925) *Melosira* und ihre Planktonbegleiter. *Pflanzenforschung* **3,** 1–78.

BRAARUD T. (1945) Experimental studies on marine plankton diatoms. *Avh. norske Vidensk. Akad. Oslo, Matematisk naturvidenskapelig Klasse* **10,** 1–16.

BRUCKMAYER-BERKENBUSCH H. (1954) Die Beeinflussung der Auxosporenbildung von *Melosira nummuloides* durch Außenfaktoren. *Arch. Protistenk.* **100,** 183–211.

BUFFHAM T.H. (1885) Newly observed phenomena in the conjugation of the diatom *Rhabdonema arcuatum. J. Quekett microsc. Club Ser.* **2, 2,** 131–37.

CHOLNOKY B. (1928) Über die Auxosporenbildung der *Anomoeoneis sculpta* E. Cl. *Arch. Protistenk.* **63,** 23–57.

DREBES G. (1965) Zur Entwicklungsgeschichte und Entwicklungsphysiologie der zentrischen Diatomee *Stephanopyxis turris* (Grev. at Arn.) Ralfs. Dissertation, Universität Marburg.

DREBES G. (1966) On the life history of the marine plankton diatom *Stephanopyxis palmeriana* (Grev.) Grunow. *Helgoländer wiss. Meeresunters.* **13,** 101–14.

DREBES G. (1967) *Bacteriastrum solitarium* Mangin, a stage in the life history of the centric diatom *Bacteriastrum hyalinum. Marine Biol.* **1,** 40–2.

DREBES G. (1968) Subdiözie bei der zentrischen Diatomee *Coscinodiscus granii. Naturwissenschaften* **55,** 236.

DREBES G. (1969a) Ungeschlechtliche Fortpflanzung der Kieselalge *Stephanopyxis turris* (Centrales). Begleitveröff. zum Film C 982, 14 pp. Inst. Wiss. Film, Göttingen.

DREBES G. (1969b) Geschlechtliche Fortpflanzung der Kieselalge *Stephanopyxis turris* (Centrales). Begleitveröff. zum Film C 983, 16 pp. Inst. Wiss. Film. Göttingen.

DREBES G. (1972) The life history of the centric diatom *Bacteriastrum hyalinum* Lauder. *Beih. Nova Hedwigia* **39,** 95–110.

DREBES G. (1974) *Marines Phytoplankton—Eine Auswahl der Helgoländer Planktonalgen (Diatomeen, Peridineen).* 186 pp. Thieme, Stuttgart.

DRING M.J. (1974) Reproduction. In *Algal Physiology and Biochemistry* (ed. W.D.P. Stewart) pp. 814–37, Blackwell Scientific Publications, Oxford.

ERBEN K. (1959) Untersuchungen über Auxosporenentwicklung und Meioseauslösung an *Melosira nummuloides* (Dillw.) C.A. Agardh. *Arch. Protistenk.* **104,** 165–210.

ETTL H., MÜLLER D.G., NEUMANN K., STOSCH V. H.A. & WEBER W. (1967) Vegetative Fortpflanzung, Parthenogenese und Apogamie bei Algen. In *Handb. Pfl. Physiol.* (ed. W. Ruhland) Vol. 18, pp. 597–776. Springer, Berlin-Heidelberg-New York.

GEITLER L. (1932) Der Formwechsel der pennaten Diatomeen (Kieselalgen). *Arch. Protistenk.* **78,** 1–226.

GEITLER L. (1952a) Oogamie, Mitose, Meiosis und metagame Teilung bei der zentrischen Diatomee *Cyclotella. Öst. bot. Z.* **99,** 507–20.

GEITLER L. (1952b) Untersuchungen über Kopulation und Auxosporenbildung pennater Diatomeen. III. Gleichartigkeit der Gonenkerne und Verhalten des Heterochromatins bei *Navicula radiosa. Öst. bot. Z.* **99,** 469–82.

GEITLER L. (1956) Automixis, Geschlechtsbestimmung und Pyknose von Gonenkernen bei *Cymbella aspera. Planta* **47,** 359–73.

GEITLER L. (1957) Die sexuelle Fortpflanzung der Diatomeen. *Biol. Rev.* **32,** 261–95.

GEITLER L. (1963) Alle Schalenbildungen der Diatomeen treten als Folge von Zell- oder Kernteilungen auf. *Ber. dt. bot. Ges.* **75,** 393–96.

GEITLER L. (1969) Comparative studies on the behaviour of allogamous pennate diatoms in auxospore formation. *Amer. J. Bot.* **56,** 718–22.

GEITLER L. (1973) Auxosporenbildung und Systematik bei pennaten Diatomeen und die Cytologie von *Cocconeis*-Sippen. *Öst. bot. Z.* **122,** 299–321.

GRAN H.H. (1915) The plankton production in the north European waters in the spring of 1912. *Bull. plankt. p. l'année* 1912, Kopenhagen, 1–146.

GROSS F. (1937) The life history of some marine plankton diatoms. *Phil. Trans. R. Soc. London* **228B,** 1–47.

GROSS F. (1940) The development of isolated resting spores into auxospores in *Ditylum brightwellii* (West). *J. mar. biol. Ass. U.K.* **24,** 375–80.

HEATH I.B. & DARLEY W.M. (1972) Observations on the ultra-structure of the male gametes of *Biddulphia levis* Ehr. *J. Phycol.* **8,** 51–9.

HEURCK H. VAN (1896) *A treatise on the Diatomaceae.* 558 pp. Wesley & Son. London.

HOFKER J. (1928) Die Teilung, Mikrosporen- und Auxosporenbildung von *Coscinodiscus biconicus* v. Breemen. *Annls. Protist.* **1,** 167–94.

HOLMES R.W. (1966) Short-term temperature and light conditions associated with auxospore formation in the marine centric diatom, *Coscinodiscus concinnus* W. Smith. *Nature, Lond.* **209,** 217–8.

HUSTEDT F. (1930) Die Kieselalgen, Teil 1. In *Rabenhorst's Kryptogamenflora* (ed. L. Rabenhorst). 920 pp. Akad. Verlagsges., Leipzig.

IKARI J. (1926) On some *Chaetoceros* of Japan. I. *Bot. Mag., Tokyo* **40,** 517–34.

IYENGAR M.O.P. & SUBRAHMANYAN R. (1944) On reduction division and auxospore formation in *Cyclotella meneghiniana. J. Indian bot. Soc.* **23,** 125–52.

KARSTEN G. (1899) Die Diatomeen der Kieler Bucht. *Wiss. Meeresunters. N.F. Kiel* **4,** 17–205.

KARSTEN G. (1912) Über die Reduktionsteilung bei der Auxosporenbildung von *Surirella saxonica. Z. Bot.* **4,** 417–26.

KARSTEN G. (1928) Bacillariophyta (Diatomeae). In *Natürl. Pfl. Familien* (eds. A. Engler & K. Prantl), 2nd edt. pp. 105–203. W. Engelmann, Leipzig.

KLEBAHN H. (1896) Beiträge zur Kenntnis der Auxosporenbildung. I. *Rhopalodia gibba* (Ehrenb.) O. Müller. *Jb. wiss. Bot.* **29,** 595–654.

KRIEGER W. (1927) Zur Biologie des Flussplanktons. Untersuchungen über das Potamoplankton des Havelgebietes. *Pflanzenforschung* (ed. Kolckwitz) **10,** 1–66.

LOCKER F. (1950) Beiträge zur Kenntnis des Formwechsels der Diatomeen an Hand von Kulturversuchen. *Öst. bot. Z.* **97,** 322–32.

LÜNING K. & DRING M.J. (1972) Reproduction induced by blue light in female gametophytes of *Laminaria saccharina. Planta (Berl.)* **104,** 252–6.

MAGNE-SIMON M.-F. (1960) Note sur le processus de l'auxosporulation chez une diatomée marine, *Grammatophora marina* (Lyngb.) Kütz. *C. r. hebd. Séanc. Acad. Sci., Paris* **251,** 3 040–2.

MAGNE-SIMON M.-F. (1962) L'auxosporulation chez une Tabellariacée marine, *Grammatophora marina* (Lyngb.) Kütz. *Cah. Biol. mar.* **3,** 79–89.

MANTON I. & STOSCH V. H.A. (1966) Observations on the fine structure of the marine centric diatom *Lithodesmium undulatum. Jl. R. microsc. Soc.* **85,** 119–34.

MANTON I., KOWALLIK K. & STOSCH V. H.A. (1969a) Observations on the fine structure and development of the spindle at mitosis and meiosis in a marine centric diatom (*Lithodesmium undulatum*). I. Preliminary survey of mitosis in spermatogonia. *J. Microscop.* **89,** 295–320.

MANTON I., KOWALLIK K. & STOSCH V. H.A. (1969b) Observations on the fine structure and development of the spindle at mitosis and meiosis in a marine centric diatom (*Lithodesmium undulatum*). II. The early meiotic prophases in male gametogenesis. *J. Cell Sci.* **5,** 271–98.

MANTON I., KOWALLIK K. & STOSCH V. H.A. (1970a) Observations on the fine structure and development of the spindle at mitosis and meiosis in a marine centric diatom (*Lithodesmium undulatum*). III. The later stages of meiosis I in male gametogenesis. *J. Cell Sci.* **6,** 131–57.

MANTON I., KOWALLIK K. & STOSCH V. H.A. (1970b) Observations on the fine structure and development of the spindle at mitosis and meiosis in a marine centric diatom (*Lithodesmium undulatum*). IV. The second meiotic division and conclusion. *J. Cell Sci.* **7,** 407–43.

MEUNIER A. (1915) Microplancton de la Mer Flamande. Pt. 2. Les Diatomacées (suite). *Mém. Mus. Roy. Hist. Nat. Belgique* **7**(3), 1–120.

OSORIO TAFALL B.F. (1935) La auxosporulación en *Bacteriastrum hyalinum* Lauder. *Boln. Soc. esp. Hist. Nat.* **35,** 111–24.

PASCHER A. (1932) Über das Vorkommen von kontraktilen Vakuolen bei pennaten Diatomeen. *Beih. bot. Zbl. Abt. A.,* **49,** 703–9.

PAVILLARD J. (1921) Sur la reproduction du *Chaetoceros Eibenii* Meunier. *C. r. hebd. Séanc. Acad. Sci., Paris,* **172,** 469–71.

PERSIDSKY B.M. (1929) *The development of auxospores in the group of the Centricae (Bacillariaceae).* 16 pp. Published by the author, Moscow.

PFITZER E. (1871) Untersuchungen über Bau und Entwicklung der Bacillariaceen (Diatomaceen). *(J. Hanstein's) Bot. Abh. Morphol. Physiol. (Bonn)* **1,** 1–189.

REIMANN B. (1960) Bildung, Bau und Zusammenhang der Bacillariophyceenschalen. *Nova Hedwigia* **2,** 349–73.

RIETH A. (1953) Zur Auxosporenbildung bei *Melosira nummuloides. Flora (Jena)* **140,** 205–8.

ROZUMEK K.-E. (1968) Der Einfluss der Umweltfaktoren Licht und Temperatur auf die Ausbildung der Sexualstadien bei der pennaten Diatomee *Rhabdonema adriaticum* Kütz. *Beitr. Biol. Pfl.* **44,** 365–88.

SCHMIDT P. (1927) Weiteres über die Fortpflanzung der Diatomee *Biddulphia sinensis,* etcs., *Int. Revue ges. Hydrobiol.* **18,** 400–14.

SCHREIBER E. (1931) Über Reinkulturversuche und experimentelle Auxosporenbildung bei *Melosira nummuloides. Arch. Protistenk.* **73,** 331–45.

SCHULTZ M.E. & TRAINOR F.R. (1968) Production of male gametes and auxospores in the centric diatoms *Cyclotella meneghiniana* and *C. cryptica. J. Phycol.* **4,** 85–8.

SCHULTZ M.E. & TRAINOR F.R. (1970) Production of male gametes and auxospores in a polymorphic clone of the centric diatom *Cyclotella. Can. J. Bot.* **48,** 947–51.

SIMONSEN R. (1972) Ideas for a more natural system of the centric diatoms. *Beih. Nova Hedwigia* **39,** 37–54.

STEELE R.L. (1965) Induction of sexuality in two centric diatoms. *Bioscience* **15,** 298.

STOSCH V. H.A. (1950) Oogamy in a centric diatom. *Nature* **165,** 531–2.

STOSCH V. H.A. (1951a) Zur Entwicklungsgeschichte zentrischer Meeresdiatomeen. *Naturwissenschaften* **38,** 191–2.

STOSCH V. H.A. (1951b) Entwicklungsgeschichtliche Untersuchungen an zentrischen Diatomeen. I. Die Auxosporenbildung von *Melosira varians. Arch. Mikrobiol.* **16,** 101–35.

STOSCH V. H.A. (1953) Einige Bemerkungen zur Phylogenie der Diatomeen. *Naturwissenschaften* **6,** 141.

STOSCH V. H.A. (1954) Die Oogamie von *Biddulphia mobiliensis* und die bisher bekannten Auxosporenbildungen bei den Centrales. *Com. VIII. Congr. intern. Bot. Sect.* **17,** 58–68.

STOSCH V. H.A. (1955) Pennate Diatomeen. *Z. Bot.* **43,** 89–99.

STOSCH V. H.A. (1956) Entwicklungsgeschichtliche Untersuchungen an zentrischen Diatomeen. II. Geschlechtszellenreifung, Befruchtung und Auxosporenbildung einiger grundbewohnender Biddulphiaceen der Nordsee. *Arch. Mikrobiol.* **23,** 327–65.

STOSCH V. H.A. (1958a) Entwicklungsgeschichtliche Untersuchungen an zentrischen Diatomeen. III. Die Spermatogenese von *Melosira moniliformis* Agardh. *Arch. Mikrobiol.* **31,** 274–82.

STOSCH V. H.A. (1958b) Kann die oogame Araphidee *Rhabdonema adriaticum* als Bindeglied zwischen den beiden grossen Diatomeengruppen angesehen werden? *Ber. dt. bot. Ges.* **71,** 241–9.

STOSCH V. H.A. (1962) Über das Perizonium der Diatomeen. *Vortr. Gesgeb. Bot. Nat.* **1,** 43–52.

STOSCH V. H.A. (1965) Manipulierung der Zellgrösse von Diatomeen im Experiment. *Phycologia* **5,** 21–44.

STOSCH V. H.A. & DREBES G. (1964) Entwicklungsgeschichtliche Untersuchungen an zentrischen Diatomeen. IV. Die Planktondiatomee *Stephanopyxis turris*—ihre Behandlung und Entwicklungsgeschichte. *Hegoländer wiss. Meeresunters.* **11,** 209–57.

STOSCH V. H.A. & KOWALLIK K. (1969) Der von Geitler aufgestellte Satz über die Notwendigkeit einer Mitose für jede Schalenbildung von Diatomeen. Beobachtungen über die Reichweite und Überlegungen zu seiner zellmechanischen Bedeutung. *Öst. bot. Z.* **116,** 454–74.

Stosch v. H.A., Theil G. & Kowallik K. (1973) Entwicklungsgeschichtliche Untersuchungen an zentrischen Diatomeen. V. Bau und Lebenszyklus von *Chaetoceros didymum*, mit Beobachtungen über einige andere Arten der Gattung. *Helgoländer wiss. Meeresunters.* **25,** 384–445.

Thaler F. (1972) Beitrag zur Entwicklungsgeschichte und zum Zellbau einiger Diatomeen. *Öst. bot. Z.* **120,** 313–47.

Went F.W. (1925) Microsporenvorming bij *Chaetoceros. Ned. kruidk. Archf.* 1 924, 75.

Werner D. (1971a) Der Entwicklungscyclus mit Sexualphase bei der marinen Diatomee *Coscinodiscus asteromphalus.* I. Kultur und Synchronisation von Entwicklungsstadien. *Arch. Mikrobiol.* **80,** 43–9.

Werner D. (1971b) Der Entwicklungscyclus mit Sexualphase bei der marinen Diatomee *Coscinodiscus asteromphalus.* II. Oberflächenabhängige Differenzierung während der vegetativen Zellverkleinerung. *Arch. Mikrobiol.* **80,** 115–33.

Werner D. (1971c) Der Entwicklungscyclus mit Sexualphase bei der marinen Diatomee *Coscinodiscus asteromphalus.* III. Differenzierung und Spermatogenese. *Arch. Mikrobiol.* **80,** 134–46.

Wiedling S. (1948) Beiträge zur Kenntnis der vegetativen Vermehrung der Diatomeen. *Bot. Notiser* **3,** 322–54.

Wiese L. (1969) Algae. In *Fertilization. Comparative Morphology, Biochemistry and Immunology,* Vol. 2, eds. Metz C.B. & Monroy A. pp. 135–88. Academic Press, New York & London.

CHAPTER 10

ECOLOGY OF FRESHWATER DIATOMS AND DIATOM COMMUNITIES

RUTH PATRICK

The Academy of Natural Sciences,
Nineteenth and the Parkway,
Philadelphia, Pennsylvania 19103, USA

1 DIATOM COMMUNITIES*

Diatom communities may be found in aerial habitats, in soils, and in water. The communities that live in water under natural conditions usually consist of a great many more species than those communities found in aerial habitats or in the soil.

1.1 Types of habitats

1.2 *Aerial habitats*

Species which can live in aerial habitats must be able to endure more rigorous environmental changes than those that live in the water. They must be able to withstand flooding as well as extreme drought and sudden changes in temperature. Thus, as would be expected, the diatoms which can survive such changes represent a small specialized group. For the most part they are small forms. Very few large diatoms are found under aerial conditions, particularly dry conditions. Several different classifications of aerial habitats have been proposed (Beger, 1927; Kolbe, 1932; Petersen, 1935, 1943). Perhaps the most convenient way to divide these are those that live under moist aerial habitats and those that live under dry aerial habitats.

Under the group of moist aerial habitats are included wet rocks and moss, caves, snow and ice, and the spray or surf zones of lakes and rivers. In such habitats occur aerophilous forms which live in or out of the water. However, it should be pointed out that the aerophilous forms typically do not occur as frequently in water or develop populations as large as may occur under aerial conditions.

On rocks or moss kept wet by seeping springs or bog water one often finds *Pinnularia borealis, Melosira roeseana, Navicula fragilarioides,* and *N. confervacea*; also certain species of *Cymbella, Gomphonema, Synedra, Achnanthes,* and *Epithemia.* In very wet moss such as the tips of *Sphagnum,* species of the genera *Eunotia* and *Frustulia* are of common occurrence.

Caves also furnish a suitable habitat for diatoms if the light is not a limiting factor. The very moist atmosphere permits a great variety of diatoms to flourish, including *Fragilaria construens* var. *venter, Melosira dickiei, M. roeseana, Navicula kotschyi, N. perpusilla,* and *Pinnularia borealis* (Schröder, 1916; Hustedt, 1922; Kolbe, 1932).

Another interesting habitat for diatoms is the spray zone of lakes, rivers, etc. In the very moist areas, Lowe & Collins (1973) found that diatoms growing on a bluff of black sandstone in Hocking County, Ohio, formed a community dominated by *Melosira roeseana.* Other aerial diatoms were

* The author has used extensively Chapter 23, Volume II of G.E. Hutchinson's *A Treatise of Limnology.*

Achnanthes coarctica and *Navicula contenta* f. *biceps* Intermixed with these were several species which are characteristic of water habitats.

Species that live in the dryer spray zones or those subject to wider fluctuations in moisture often belong to genera which produce a well-developed jelly sheath, such as certain species of *Navicula* (Schizonema) and *Cymbella* (Encyonema). Closer to the water's edge or in areas that are more continually moist are found *Achnanthes, Cymbella, Gomphonema, Epithemia,* and *Denticula.*

Diatoms that can live on dry rocks are *Eunotia fallax* var. *gracillima, Melosira dickiei, Navicula contenta, N. krasskei,* and *N. sohrensis* living among *Protococcus* (Krasske, 1929). Species which can live in such habitats are rather limited in number.

The extent to which diatoms will occur on dry moss seems to be limited for the most part by the drying of the moss, the amount of light, and the temperature of the substrate. Cool, damp, humus-rich moss on kaolinic rock will have more species than warm, dry, humus-poor moss on limestone rocks (Beger, 1927; Kolbe, 1932). Diatoms which can live in such habitats must be able to withstand extreme variation in moisture conditions and, thus, great variation in salt concentration—which may result in great variation in the osmotic pressure of their cells. Usually the smaller varieties of species are found in dry moss. For example, *Navicula contenta* var. *parallela* and var. *elliptica* instead of *N. contenta; N. mutica* var. *cohnii* instead of *N. mutica*; smaller forms of *N. fragilarioides* and *Pinnularia borealis*; also the smaller forms of *Melosira roeseana* rather than its long filamentous varieties, *spiralis* and *epidendron* (Krasske, 1936).

Diatoms have been found to develop various methods for withstanding severe desiccation. Some aerophils have been found to store large amounts of oil. Often diatoms living in such habitats build inner plates, as found most commonly in *Melosira dickiei* and *M. roeseana.* The reduction in size, increase in oil storage, and the building of inner plates may well be special adaptations of these species to withstand drying out and increased salt concentration (Kolbe, 1932; Hustedt, 1938). The diatoms occurring in such habitats are for the most part geographically widely distributed species, although a few endemics have been found.

1.3 *Soil habitats*

Our knowledge of diatoms that live in the soil is quite limited and, no doubt, many new species will be found in this specialized habitat. For the most part they are small species belonging to the Naviculales and Bacillariales. It is believed that they are able to live in this more or less unfavourable habitat because they have a raphe and are able to move to deeper regions when the soil dries out. Diatoms are less often encountered in woods soil than in field or garden soil, and are most plentiful in the top few centimetres. *Hantzschia amphioxys, Navicula atomus, N. nitrophila, N. mutica, N. contenta* f. *biceps, Pinnularia balfouriana, P. brebissonii,* and *P. borealis* are the more common soil species (Bristol, 1920; Petersen, 1928; Hedlund, 1913; Moore & Carter, 1926; Francé, 1912; Hayek & Hulbary, 1956).

1.4 *Aquatic habitats*

(a) *Lake, pond, and reservoir habitats*

The aquatic habitats in which diatoms may live are lakes and ponds, reservoirs, free-flowing streams and rivers, and estuaries.

Lakes are generally characterized by having a slow exchange of water—that is, the outflow is very small in comparison to the volume of the lake. Currents are usually relatively slow unless they are generated by wind. Typically lakes are deep, and the bed of the lake lies beyond the photosynthetic zone. Most lakes have a wide expansion of open water. During the summer a thermocline typically develops in deep lakes, and those that are mesotrophic to eutrophic become stratified. When the oxygen becomes exceedingly low in the waters, particularly those waters lying close to the bed of the lake, an unoxidized microzone develops. As a result, minerals and chemicals will enter the lake waters from the sediments. In the spring and the autumn in temperate zone lakes and more frequently in tropical lakes, changes in temperature bring about an overturn of these waters resulting in increased nutrient concentration near the surface waters, often producing a diatom bloom in the plankton.

Ponds are characterized by being much more shallow than lakes. Typically they have a well-developed benthos across the entire bed of the pond and the plankton contains many of these benthic forms. Often a true plankton does not develop.

Reservoirs are more-or-less mixed current habitats depending upon the structure of the reservoir. Typically there is a river that flows into the reservoir and there is a release of water from the reservoir. This exchange of water may be much more rapid if the reservoir is used for power generation than if it is not; therefore the diatom communities of some reservoirs may have many of the characteristics of river or lake floras.

Free-flowing streams and rivers are characterized by a current and a rapid inflow and outflow of water; thus the types of species developed are quite often different than in a lake. Leopold (1964) has classified streams according to orders; the first order stream being the short stream without any tributaries; as they increase in size and number of tributaries, the order increases. In these headwater streams, which are typically first to third order streams, the water is typically shallow and the benthic diatom flora extends over the bed of the stream. There are also epipelic and epiphytic species as classified by Round

(1965), as rocks are very common. Often there are filamentous algae to which diatoms are attached.

As these streams increase in size, rooted and floating aquatic plants are more common, and the epiphytic flora of diatoms typically increases. In large rivers when the bed of the river is typically beyond the photosynthetic zone, the typical habitats for diatom growth are on the shoaling edges of meanders and on floating debris. Also large beds of floating aquatics often accumulate if the stream waters are rich in nutrients. These support large epiphytic diatom floras. Such large rivers typically form large meanders and develop oxbows and sloughs. These areas are excellent for the development of diatom floras, because the water is shallow and less turbid. Diatoms living in these areas often are carried by the current into the river and thus supplement the food chain. They are typically epipelic, epiphytic, or benthic species.

In the estuary we find a combination of different types of diatom growth. In the shallow parts of the estuary where the bed is in the photosynthetic zone many benthic species are present. Species are often found which live in the mud and migrate up and down through the top few centimetres. In open waters of the estuary one often finds a plankton flora which may be a mixture of brackish-water and marine species. In some areas typically freshwater plankton species may be found. Epiphytic forms are very common and live attached to macroscopic algae, floating masses of aquatic plants or rooted plants, and sometimes grow attached to the base of stems of emergent aquatics. This is a particularly common habitat in the marshlands.

In the marshes the large tidal flats exposed during low tides are excellent habitats for many species. Also the spray zones from waves are habitats for other semi-aerial forms.

(b) *Plankton*

The plankton algae of lakes are typically those that spend their vegetative lives afloat and are not the plankton of the open ocean in which the reproductive stages, resting stages, and vegetative stages are maintained afloat. The plankton species find their best development where there are large expanses of open water and often develop large populations referred to as blooms. Just what factors are involved in initiating a bloom seems to vary depending on the lake. In Lake Michigan (Eddy, 1927a; Chandler, 1944) and Lake Erie, turbidity of the water seems to greatly influence the amount and duration of diatom blooms. In Lake Erie, Chandler (1944) found that wind, precipitation, and solar radiation are also very important. Nutrients have also been found to be important in instigating such blooms. According to Pearsall (1930, 1932) and West (1912) nitrogen and phosphorus may not only be limiting factors. Some studies on English lakes emphasize the importance of dissolved salts in relation to diatom productivity. Pearsall (1930) states that most diatoms must have calcium carbonate hardness of about 3 mg/litre and Lund (1950) has estimated that the amounts of silica present should vary between 0·4 mg/litre and 0·5 mg/litre for some diatoms to occur in abundance. More recently the work of Patrick *et al.* (1975) has indicated that metals may be important in determining the species of diatoms which are present and whether diatoms are able to out-compete other forms of algae in large blooms. Thus, it would seem that any one of several factors may limit diatom production.

As a rule, the spring diatom bloom is much larger than the autumn bloom, although some exceptions may occur (Petersen, 1943; Chandler, 1940, 1942; Gotschall & Jennings, 1933; Budde & Burkholder, 1929; Damann, 1945; Eddy, 1927b). Although some diatoms may exhibit both a spring and an autumn pulse, usually the dominant species are different. As noted by Chandler (1940) those genera which are usually dominant in spring blooms are *Asterionella, Fragilaria* and *Synedra,* which are Pennate forms. Those that dominate the autumn pulse are *Melosira, Cyclotella,* and *Stephanodiscus,* all of which belong to the Coscinodiscales. Of course, all genera may occur at any time, but their time of dominance varies—for instance, many workers have noted that *Asterionella* usually occurs earliest in the spring and it is then succeeded by other genera. Maeda & Ichimura (1973) found in Lake Haruna that *Synedra ulna* and *Asterionella gracillima* were the most common algae under the ice. *Asterionella gracillima* continued to be the dominant plankter until April and May, and also dominant in April and May was *Melosira italica. Melosira italica* was very common in August and November, in addition, *Cyclotella* sp. was also one of the common species. Pearsall (1932) and others believed that succession in species is because some diatoms such as *Asterionella formosa* and *Fragilaria crotonensis* demand higher nutritive requirements than other species. Akehurst (1931) however, believed that this succession is due to the fact that one group produces substances which are toxic to itself but stimulating to other groups. For example, *Asterionella* produces substances toxic to itself but would stimulate *Synedra,* and as a result a *Synedra* pulse would follow an *Asterionella* one.

Talling (1957a) working with *Asterionella formosa* and *Fragilaria crotonensis* found no evidence that either species produced an extracellular substance that appreciably modified the growth of the other.

The theory that diatoms belonging to the Fragilariales are most common in early spring, and those belonging to the Coscinodiscales are most common in the autumn is not borne out by the studies of Kozarov (in Stankovic, 1960) who found that *Cyclotella ocellata* and *Stephanodiscus hantzschia* compose the spring maximum. Likewise, it has been shown in the Ossiachersee that the first diatom maximum in April and May is largely due to increases in *Cyclotella* (Hutchinson, 1967). Likewise, in Lake Mendota (Birge & Juday, 1922), the spring maximum is usually developed by enormous populations of *Stephanodiscus astrea.* Williams (1972) in his study of rivers and the Great

Lakes (USA) found that this ratio of dominance for spring and fall diatom blooms was not true.

It would appear that not only temperature and light, but nutrients in the water are important in determining what species develop most abundantly. Lund (1950) has shown that silica is extremely important for the bloom of *Asterionella formosa*. Furthermore, according to Hutchinson (1967), natural populations of diatoms in the English lake district seem to be able to support division when the phosphorus is less than 1 mg/m^3. The greatest population recorded was $1{\cdot}2 \times 10^7$/litre; the minimum phosphorus that would have been required to produce such a population would be about 0·7 μg/litre. Furthermore, there is little evidence that nitrogen becomes limiting in the English lake district. An extensive discussion of the effect of nutrients on diatoms is set forth by Hutchinson (1967).

The genus *Melosira* is composed of many species of diatoms, some of which seem to have their best development under low light intensity and in cooler waters such as *Melosira italica,* particularly *M. italica* var. *subarctica, M. islandica* var. *helvetica*; whereas such species as *M. granulata, M. varians, M. nyassensis* and *M. agassizii* have their best development in warmer waters with higher light intensity. *M. varians* becomes dominant in the plankton of many lakes and streams in late summer in the eastern part of the USA if they are nutrient rich. However, if the water is warm and nutrient rich it will remain very common throughout almost any season of the year. Usually it is most common when the day length is fairly long.

Richardson (1968) found in African lakes that the plankton species of tropical lakes were very different from those of temperate zone lakes. For example, *Nitzschia* sp. may dominate the plankton in nutrient rich lakes. Furthermore, various species of *Melosira* dominated under various pH conditions, e.g. *M. ikapoensis* dominated in acid water and *M. goetzeana* in alkaline water.

(c) *Benthic and epiphytic species*

Besides the plankton, the benthic and epiphytic diatoms are also important in lakes. The benthic diatoms are those that live attached to hard substrates or live in and on fine sediments such as silt and fine sand. Round (1965) has referred to those communities that develop on hard surfaces as epilithic communities. Such communities in lakes are formed of species which attach by either gelatinous pads or stalks. Some of the more common species that dominate such habitats are *Diatoma vulgare, Diatoma hiemale, Melosira varians, Meridion circulare,* various species of *Synedra,* some species of *Gomphonema* and *Cymbella,* and some species of *Fragilaria.* Species belonging to other genera may also be found in these habitats, if they are able to attach either by gelatinous pads or stalks. Those species which live in or on the sediments Round has called epipelic communities. Genera often found in these habitats are some of the species of *Caloneis, Diploneis, Fragilaria, Frustulia, Gyrosigma, Cymatopleura, Navicula, Neidium, Pinnularia, Nitzschia, Stauroneis,* and *Surirella.* These are the most common genera found in or associated with lake sediments. Various species of *Tabellaria* may also be found in such habitats.

Epiphytic diatoms are typically associated with floating masses of aquatic plants or grow on the surfaces of various types of filamentous algae or on the stems and undersides of leaves of aquatic plants such as water lilies, etc. Some of the more common genera found in these habitats are *Achnanthes, Cocconeis, Epithemia, Navicula, Rhoicosphenia,* and *Synedra.* It is believed that many of the species that live on the surface of sediments or attached to various kinds of aquatic plants are facultative heterotrophs and utilize organic matter found in the substrate.

(d) *Stream communities*

Stream communities of diatoms are usually composed of the benthic and epiphytic genera found in lakes and less often have truly planktonic species present. In contrast to lakes which have many of their nutrients recycled, streams or flowing waters depend mainly upon the watershed for nutrients. Because the river is turbulent, an oxidized microzone usually exists across the sediments and as a result they contribute rather little in the way of nutrients to the water. Some exceptions to these are low redox potential areas found in some pools.

The diatom flora differs considerably in various parts of a stream. In the headwaters which often derive most of their water from springs, we find those species which like cool, rather constant temperatures. Such waters are often rich in CO_2 and bicarbonates. Among the species found under these conditions are many species of *Diatoma, Fragilaria* and also *Meridion circulare.* One also may find many species of *Navicula* in such habitats. Further downstream where the temperature of the water is more variable and where the current is usually stronger, quite different types of diatom communities are formed. In the fast-flowing waters on rocks and other hard surfaces one typically finds those kinds of diatoms that can grow attached to substrates such as *Achnanthes* and *Cocconeis.* In slower flowing water on hard substrates one often finds *Melosira varians,* various species of *Synedra, Gomphonema,* and *Cymbella,* whereas in or on the sediments one typically finds, as in lakes, many species of *Campylodiscus, Cyclotella, Cymatopleura, Diploneis, Fragilaria, Gyrosigma, Navicula, Nitzschia, Surirella,* etc. Hustedt (1938) has classified those species which live in strong-flowing water as rheophils. These species are often endemic and are more characteristic of specific habitats than are those that live in slower-flowing water.

In the estuaries on the mud flats we have the largest development of what Round (1965) refers to as epipelic communities. Besides those genera listed above for this type of habitat in lakes and rivers, one often finds brackish-water forms such as various species of the genus *Pleurosigma, Amphora,* and, of course under marine conditions, typically marine species.

(e) *Bog floras*

Bog diatoms are often different species from those one finds in lakes or streams. They are often associated with the mosses that make up the marshy areas such as *Sphagnum*. It is in the *Sphagnum* tips one often finds many species of diatoms, particularly those belonging to the genus *Eunotia*. One also finds in bogs benthic species such as *Tabellaria flocculosa, Actinella* sp., *Stenopterobia* sp., some of the species of *Pinnularia,* and a few species of *Navicula*.

2 STRUCTURE OF DIATOM COMMUNITIES

The structure of diatom communities, as communities of many other kinds of aquatic organisms, seems to be largely determined by available species pool, invasion rate, density-independent factors, density-dependent factors, predator pressure, parasitism, and competition.

2.1 *Factors important in establishing diatom communities*

In order to evaluate the importance of these various forces, Patrick (1967a) devised a series of experiments to test the importance of size of area, available species pool, and invasion rate in establishing a complex diatom community. In order to determine the effect of the size of area a series of small islands were created by mounting various sizes of glass substrates onto small pedicels attached to glass slides which were then anchored in various kinds of containers. The results of these experiments are shown in Table 10.1. These results indicate that the size of the area has considerable effect on the numbers of species that one may expect in a given area.

The diversity of this area is also important. For example, we have found that when glass slides are placed into a stream the community which first develops is a two-dimensional community consisting of colonies of various species of diatoms. As these species grow and there is more competition for space, those that can stand on one end, such as many of the *Synedras* which attach by jelly pads will build vertical colonies. Once this is done a three-dimensional community is established and new habitats are available. Under these conditions the diversity becomes much greater. It is undoubtedly true

Table 10.1. Effects of size of area and species pool on number of species established in an area. A, Experiments in September–October, 1964.

	Roxborough Spring (Pennsylvania)						*Ridley Creek (Pennsylvania)*
	No. of species						*No. of species*
	625 mm^2		36 mm^2		9 mm^2		36 mm^2
	Box 1	Box 2	Box 3	Box 4	Box 5	Box 6	Box 7
4 days	46	37	23	23	1	3	—
1 week	40	32	28	24	7	—	—
2 weeks	54	35	—	22	10	10	—
8 weeks	—	—	29	14	19	14	160

B, Experiments in Roxborough Spring during Summer, 1964.

	1 week 144 mm^2 slide		2 weeks 144 mm^2 slide		1 week 625 mm^2 slide		2 weeks 625 mm^2 slide	
	Box 1	Box 2	Box 3	Box 4	Box 5	Box 6	Box 7	Box 8
No. of species	32	28	23	22	47	44	29	28

that very different current patterns exist on a glass slide after it has been well colonized than when the first invaders reach it. It is interesting to note that *Gomphonema* which form dendritic colonies do not become established until the three-dimensional community is well developed.

The importance of an available species pool was tested by putting similar size glass slides in a small stream (Roxborough Spring Creek) where we knew that the total number of species present at any one time was about 100 and in a much larger stream wherein previous studies had shown that the total number of species present in any one area at a given point in time was about 350 (Ridley Creek). We see from Table 10.1 that there is a great difference in numbers of species that establish themselves on the same size area. The numbers seem to be closely related to the numbers of species available for species occupancy.

These studies further show that those species which had developed fairly large communities never became extinct. It was those species with extremely small populations that would appear and disappear on the slides. A further series of experiments was conducted to determine the effects of reinvasion of species on a given slide upon the structure of the community. In these experiments we set up a series of boxes containing glass slides. They were developed after much experimentation so that roughly the same species would

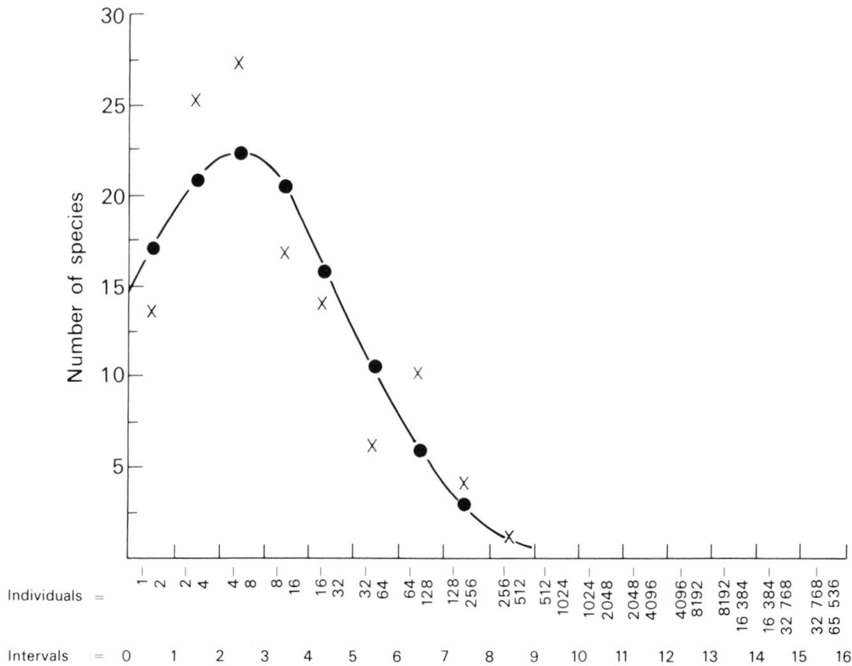

Fig. 10.1. Studies on invasion rate. Structure of diatom community. Invasion rate 550–600 litre/hour, October to November, 1964. Height of mode, 22·4 species; observed species, 123; σ^2 6·2; intervals covered by the curve, 9; diversity index, 3·805.

be present in all of the experimental boxes. We are able to establish, according to Patrick (1968), that 75–85% of the species were the same in all boxes and of those that were not the same, 75% had less than 4 specimens when a total of 8 000 to 10 000 specimens were counted. All of these boxes had the same species pool capable of invasion. We then altered the rate of invasion. The results of these studies are given in Figs 10.1 and 10.2. Those boxes that had a high rate of invasion of approximately 650 litres/hour of new water with a current speed of approximately $\frac{1}{2}$ cu. ft/second developed very diverse communities with relatively small variance in population sizes or shown by σ^2. In contrast, those boxes in which the invasion rate of diatoms had been lowered to 1·5 litres/hour, although the current need was maintained at $\frac{1}{2}$ cu. ft/second by filtered water and recycling, showed much more variable sizes in the populations of diatoms producing a much larger σ^2. Also, there was considerable reduction in the numbers of species present in the community.

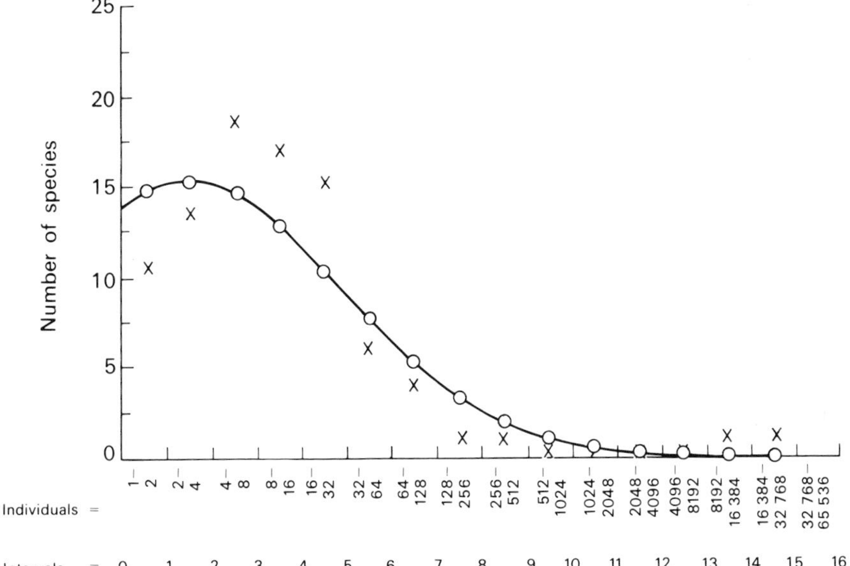

Fig. 10.2. Studies on invasion rate. Structure of diatom community. Invasion rate 1·5 litre/hour, October to November, 1964. Height of mode, 15·3 species; observed species, 97; σ^2, 12; intervals covered by the curve, 15; diversity index, 0·972.

2.2 *Factors important in the maintenance of diatom communities*

2.3 *Density-independent factors*

The density-independent factors are many and have considerable effect on diatom communities. Among the most important of these are light, temperature, and rate of flow of the water. The nutrient levels also seemed to have a considerable effect upon the kinds of species of diatoms that compose a community. However, as pointed out by Gerloff & Skoog (1954) many species of algae can accumulate large amounts of various nutrients under very favourable nutrient-rich conditions and thus are not dependent upon the external medium for some period of time after a nutrient has been reduced to a suboptimum level. Thus we often do not find as direct a correlation of diatom communities with nutrients as we do with temperature.

(a) *Current*

The strength of current has been recognized by Ruttner (1940), Kolbe (1932), and others as one of the most important factors influencing the kinds of species which one tends to find in a stream. Some of the important effects of current are that it continually renews nutrients and takes away excretion products from the area in which they are produced. As a result, the chance that any nutrient will become limiting is much more reduced and autotoxic effects are also greatly mitigated.

Schumacher & Whitford (1965) studied the effect of current on respiration rate and $^{32}PO_4^{3-}$ uptake on the diatom *Eunotia pectinalis*. They found that O_2 consumed at a current speed of 18 cm/second at 22°C was about twice as much as the amounts consumed in still water. Furthermore, they found that the $^{32}PO_4^{3-}$ uptake at 18 cm/second when compared with that in still water was about 7 times greater. Hustedt (1939) and others have noted that diatoms will change their shape due to the speed of current. *Desmogonium rabenhorstianum* was less capitate and much more streamlined in its morphology in fast-flowing water than in slow-flowing water.

(b) *Light*

Light is a very important density-independent factor influencing diatom growth. The amount of light and duration of light necessary for optimum growth vary greatly depending on the species under consideration. Just as in higher plants, there are some species which prefer abundant light while others live in regions of low light intensity. Those diatoms that prefer abundant light are usually found in the plankton or in shallow littoral zones (Schroeder, 1939). Sunshine seems to favour the development of *Cyclotella meneghiniana, Fragilaria capucina,* and *Navicula cryptocephala* (Rice, 1938). Diatoms which seem to tolerate the lowest light conditions live in caves. An example of this is *Melosira roeseana,* and also those that live on the beds of lakes in deep water such as species of the genera *Campylodiscus* and *Surirella* (Koorders, 1901). The fact that in winter diatoms seem to grow in more shallow water than in summer may be due to small amounts of light penetration under the ice (Godward, 1937). Chandler (1940, 1942) found that the amount of light penetration affected the magnitude and duration of phytoplankton pulses. It is difficult to separate light effects from nutrient effects. Lund (1954) noted that *Melosira italica* var. *subarctica* grew at low light intensities and low temperatures. On the other hand, *Asterionella formosa* was a high-light and low-temperature species. In the Blue Nile the strong absorption of light did not necessarily inhibit the amount of phytoplankton production (Prouse & Talling, 1958). Talling (1975b) reported that *Melosira nyassensis* nar. *victoriae* developed its greatest growth at 30 to 60 metres depth in Lake Victoria. The amount of light together with the other characteristics of the water were probably responsible for the growth at these depths. Verduin (1952) found that the *Asterionella-Cyclotella* community in western Lake Erie occurred at a depth of approximately half that of the *Stephanodiscus* community. Light may have been one of the factors influencing the depth to which these various species grew. Rodhe (1948) has reported evidence to show that the periodicity of *Melosira islandica* var. *helvetica* was caused by temperature and light. Patrick (1971) has shown that increasing day length by artificial light was not as effective in increasing the diversity of diatom communities and the biomass as was natural light. These results indicate that the kind of light which penetrates a great deal of the atmosphere when day lengths are prolonged may be quite different in quality from the kind of light added by artificial extension of the daylight period.

(c) *Temperature*

Temperature effects on diatom growth may be both direct and indirect. To many temperate zones species the reduction in temperature seems to be the most profound density-independent environmental effect in the production of biomass. It is well known that certain species have a narrow temperature range and others are able to withstand rather large temperature ranges. In the studies by Patrick (1971) in White Clay Creek it was evident that most of the species had wide ranges of tolerance and that the species composition in winter and summer as well as in the autumn and spring were quite similar. However, the dominance or the development of large populations of given species seemed to be quite clearly correlated with temperature; some species growing best under cool water conditions and others under warm. General diatom diversity seemed to increase as one approached the optimum range of temperature within the range of tolerance for most of the species of a community. When the temperature regime moved away from the optimum either by becoming colder or hotter the diversity was decreased and the biomass was also affected.

Some diatoms seem to be able to stand fairly wide temperature ranges and maintain fairly good growth rates, whereas other diatoms seem to have small ranges of temperature tolerance. Hustedt has classified diatoms as in Table 10.2. For example, Stockner (1967) has found that some diatoms can live in very high temperatures. However, most diatoms living in the temperate zone seem to prefer temperatures less than 30°C. There are, however, exceptions to this—for example, Barker (1935) found that *Nitzschia palea* reached its maximum rate of photosynthesis at 33°C and at 40°C the rate of photosynthesis was irreversibly lowered. Wallace (1955) found that the growth of *Nitzschia linearis* was greatly reduced or inhibited at 30°C and *Nitzschia filiformis* at 34°C. In contrast, *Gomphonema parvulum* could grow fairly well at 34°C, although the rate of cell division was not as great as at 20°C. Rodhe (1948) has found that *Melosira islandica* var. *helvetica* could be reared at 5°C, 10°C, and 20°C, although the best growth was produced at 20°C, but it did not maintain this growth for a very long period of time. The best temperature for growth over time was at 5°C. Phinney & MacIntire (1965) found that the increase in temperature from 11·9°C to 20°C produced

Table 10.2. Thermal tolerance classification of diatoms.

Stenotherms	
Cold-water stenotherms	15°C
Temperate stenotherms	15°–25°C
Warm-water stenotherms	above 25°C
Meso-stenotherms: those forms that can withstand 10°C variation in temperature	
Tropical cold-water forms	10°–20°C
Temperate forms	15°–25°C and 20°–30°C
Warm-water forms	25°–35°C and 30°–40°C
Meso-eurytherms: those forms that can withstand 15°C variation in temperature	
Cold-water to temperate forms	10°–25°C
Temperate forms	15°–30°C
Temperate to warm-water forms	30°–45°C
Eu-eurytherms: those forms that can live in 20°C or more variation in temperature.	

an increase in oxygen evolution from 3·37 to 4·4 mg $O_2.m^{-2}.h^{-1}$ when the intensity of illumination was approximately 22 000 lux. However, if the temperature was increased and the light was below 11 100 lux the increased production of oxygen by photosynthesis did not occur. Patrick (1969) has found that a considerable shift in diatom species occurred with increase of temperature. In White Clay Creek the maintenance of an average temperature of 30·4°C caused a shift from a diatom-dominated community to one dominated by a mixture of green and blue–green algae and diatoms. In September, when the temperature was maintained at 34°C, a shift from the diatom community to a blue–green community occurred, whereas in August this shift occurred between 37·8 and 38·8°C. This would indicate that during August the diatom species seemed to be able to tolerate higher temperatures than in September. Since many other factors such as day length and intensity of sunlight were variables, it is difficult to know why this shift occurred. However, repeated experiments have shown that maintaining the temperature around 34°C does cause a shift in the structure of diatom communities and that most species are not able to survive, and the community is dominated by blue–green algae.

Many studies have been made concerning the effect of sudden and short exposures of diatoms to high temperatures. Patrick (1974) has summarized some of these studies. In general, one can conclude that if the exposure time at much higher temperatures is very short then the harm to diatoms is minimum. Another important factor is whether the high temperature during the time of exposure is near the maximum which the species can tolerate. For example, if one raises the temperature from 5°C to 9°C above the ambient temperature and the ambient temperature is near the lower range of tolerance for diatoms, then a stimulatory effect might possibly occur. Certainly it has been shown not to be deleterious. However, if this same degree rise in temperature is near the upper limit of temperature tolerance for diatom species and the maximum tolerance temperature is reached or exceeded, then more-or-less severe harm may occur. Such temperature rises typically occur in the passage of diatoms by entrainment through power plants.

Temperature may have many indirect effects upon diatom communities. For example, it may lower the viscosity of the water and thus increase the sinking rate of plankton diatoms. It is well known that those diatoms living in warm water often have less silica in their cell walls than those living in very cold water. It is also known that increases in temperature affect the diffusion rates of chemicals and lower the amount of oxygen that water may maintain. These changes in the environmental conditions in which the diatoms live may affect their reproductive rates and metabolism.

(d) *Turbidity*

One of the more important density-independent factors affecting diatom growth is turbidity. It may have the effect of clogging up habitats and homogenizing sediments, and thus reducing the diversity of current patterns for various species of diatoms, or it may cut down light penetration. The type of suspended solids may make a great deal of difference in the amount of light which penetrates. For example, two bodies of water may appear to have the same amount of suspended solids, but in one case it is fairly large particulate matter whereas in another case it is colloidal in nature. The one that is colloidal in nature will eliminate more light than one composed of larger suspended particles. As a result, a river may be very high in nutrients and capable of supporting a considerable amount of diatom growth, but if the solids are present this growth will not occur. An example of this was in the Kansas River below Topeka, Kansas in the USA. In this case, the river was high in nutrients and capable of supporting a very rich diatom flora, but due to the suspended solids, a well developed diatom flora did not occur until the solids dropped out in slower-flowing water many miles downstream (Patrick, unpublished data).

Chandler (1942) has found that in western Lake Erie turbidity is one of the more important factors affecting the development of large populations of diatoms. High turbidity prevented spring 'blooms' of plankton and low turbidity enabled a 'bloom' to develop under the ice.

(e) *Oxygen*

Oxygen is required for respiration as is characteristic of most forms of aquatic life. It is produced by the process of photosynthesis. Light and temperature as well as the nutrient levels of the water and other environmental conditions seem to be important in determining the rates of photosynthesis and therefore oxygen production. Cholnoky (1968) has emphasized that *Achnanthes minutissima* requires high oxygen, and he has developed various schemes by which one can judge the reoxygenation of water after pollution by the increase in abundance of *Achnanthes minutissima*. Schoeman (1973) has stated that *Navicula seminulum* and *Nitzschia amphibia* can grow successfully in fairly low amounts of oxygen. He also stated that *Nitzschia formalis* can live in oxygen-deficient waters. It is probable that many species of diatoms that are found in highly organic-polluted waters are able to tolerate lower levels of oxygen in the water than those that cannot live under such circumstances. Kolbe (1932) found that *Caloneis permagna* grew abundantly in the sediments of deep-water lakes that were low in oxygen.

In general, dystrophic or brown waters contain less oxygen than those that have a circumneutral pH. Lake waters that are deep also contain less oxygen. More study is needed to determine whether the species of diatoms forming the floras in these various situations are characterized by a need for less oxygen in the surrounding medium or if the generation of oxygen by photosynthesis mitigates the effect of the oxygen concentration in the external medium. Certainly in water environments which have very low oxygen concentration the redox potential is lower, and the solubility of many trace metals and other chemicals is higher. Thus the external environment may be altered in many ways by extremely low amounts of oxygen. These effects may be more important than that of a small amount of oxygen on the growth of species of diatoms.

(f) *Hydrogen ion concentration*

In considering the effects of pH or the hydrogen ion concentration on diatom communities, one should think not only of its direct effect upon the organisms but what is even more important is its indirect effect on the solubility of various substances. It is well known that very acid lakes or streams often support smaller numbers of species than circumneutral ones. One of the most important effects of pH is its effect upon the carbonate–bicarbonate buffering system. Over evolutionary time most species of diatoms have evolved to live within the range of pH in which this system is operative. At pHs above and below the carbonate–bicarbonate buffering system the numbers of species are relatively few. When the pH is low, the form of carbon available for diatom growth is mostly in the form of CO_2 or HCO_3^-, whereas at high pHs it is in the carbonate (+bicarbonate) forms. Some species have evolved to be able to live successfully in low pH waters rich in humates which often occur in bogs or in streams draining wetlands. In such waters one typically finds a characteristic flora consisting of many species of *Eunotia, Actinella,* and some species of *Frustulia, Stenopterobia,* and *Pinnularia.* Also, *Tabellaria flocculosa* is of common occurrence (Schroeder, 1939; Hustedt, 1939; Patrick, 1945). Likewise, waters which are rich in sodium and potassium, such as some of the very alkaline lakes in the midwestern and western parts of the USA, have a restricted characteristic diatom flora often consisting mainly of species of *Denticula,* some species of *Epithemia,* and species of other genera that typically occur in brackish-waters, i.e. euryhaline forms.

In contrast to these naturally occurring extremes in pH, acid waters produced by man's activities such as acid mine waters do not have a characteristic diatom flora. This is because these waters would typically support a circumneutral flora, and the very low pHs are very deleterious to the growth of such species. As stated above, in the acid humate waters one typically finds a few species some of which develop large populations and others have relatively small ones. The structure of this type of community as found in Egg Harbor River, New Jersey is illustrated in Fig. 10.3.

The effect of pH on circumneutral diatom floras seems to vary depending upon the accumulated stresses due to other environmental factors. Patrick *et al.* (1968) found that exposing a circumneutral diatom flora to a pH around 5 did not have as much effect at a temperature average of 22°C and fairly long

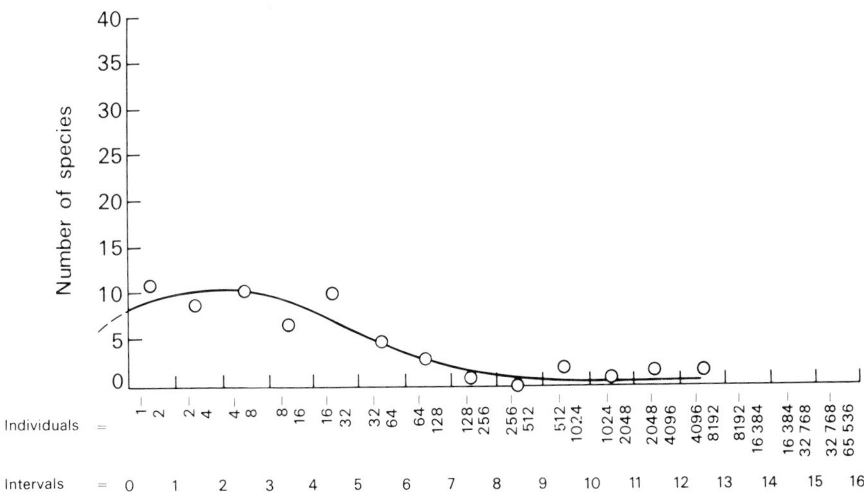

Fig. 10.3. Structure of diatom community in Egg Harbor River, New Jersey.

Table 10.3. Effects of various pH ranges on diatom communities. Data developed from truncated normal curves and diversity index ($\Sigma p_1 \log_e p_1$).

Test	σ^2	*Position of mode*	*Height of mode*	*Species observed*	*Theoretical no. species*	*Int.*	*Rows counted*	*Diversity index*	H_1-H
Series I									
October–November, 1965									
Control Box 8-3	8·9	2·2	17·7	109	132	9	10	3·462	
pH 9·66 Box 5-2	8·2	2·9	9·5	62	68	9	145	2·727	0·1451 = 1·13
pH 5·26 Box 6-2	4·9	2·1	11·1	57	61	8	230	2·872	0·3102 = 1·36
pH 6·42 Box 7-2	9·2	2·4	14·2	92	103	11	30	2·689	0·1425 = 1·13
Series II									
May–June, 1966									
Control Box 8-2	13·3	2·1	18·3	128	167	13	10	2·742	
pH 9·63 Box 5-2	13·5	2·5	15·7	115	144	12	14	2·894	0·0345 = 1·01
pH 5·15 Box 6-2	10·1	2·3	19·7	128	157	13	30	2·230	0·3968 = 1·48
pH 6·38 Box 7-2	8·7	2·3	19·3	122	142	10	10	3·228	0·3239 = 1·39

day length (14 hours 47 minutes average) on the reproductive rates of diatoms as if this had occurred under short day length (10 hours 29 minutes average) with much cooler water (9°C) conditions. The effect of lowering the pH on such communities seemed to be to restrict the division rate; thus the biomass was much less than in the control and the diatom community represented by many species had smaller populations. The structure of the community under such conditions is shown in Table 10.3.

Hustedt (1956) has classified the diatoms as alkalibionte forms—those that prefer pHs above 7; as alkaliphile forms—those that prefer pH around 7; as indifferent forms—those that live in a fairly wide range of pHs above and below 7; as acidophile forms—those that prefer a pH below 7; and acidobionte forms—those that prefer a pH of 5 or lower.

(g) *Alkaline ions*

Calcium, magnesium, sodium, and potassium are the main alkaline ions commonly found in surface waters. The calcium and magnesium contribute to the hardness of the water, whereas sodium and potassium, which are much more soluble, contribute more to the alkalinity of the water.

Calcium is extremely important in the carbonate–bicarbonate buffering system, and because it brings about the precipitation of many heavy metals and excessive phosphate it is often beneficial to diatom growth. It is known to antagonize toxic effects of various metals, probably by precipitation, but this is not clearly understood. Bacterial activity is often not as great in waters with very low calcium concentration, probably due to the lower pH which usually accompanies such conditions. The amount of calcium in solution is somewhat dependent on the type of water—for example, whether oligotrophic or eutrophic conditions exist. The soluble calcium bicarbonate is often changed to insoluble calcium carbonate by pH change due to photosynthesis in the epilimnion of lakes and as such settles to the bottom and out of solution in oligotrophic lakes. However, in eutrophic lakes where the bacterial activity is relatively high remain as calcium bicarbonate in the epilimnion and a higher calcium content of the water is maintained (Kolbe, 1932). Pearsall (1922) states that those lakes having a low sodium–potassium to calcium–magnesium ratio are more prolific than those that have high ratio. For example, diatom phytoplankton tends to dominate when the sodium–potassium divided by calcium–magnesium concentration is less than 1·5. Furthermore, Pearsall states that when the ratio is low, water is usually rich in nitrates, carbonates, bicarbonates, and silicates. Pearsall (1924) states that in calcium-rich waters fat-producing species such as diatoms dominate.

Some species of diatoms seem to definitely prefer calcium. Most species of *Synedra* seem to like a considerable amount of calcium as does *Achnanthes minutissima, Gomphonema olivaceum,* and certain species of *Cymbella, Diploneis, Navicula,* etc.

In calcium-poor waters one typically finds lower pHs and often fairly large amounts of iron and other trace metals. Species that seem to dislike calcium are often referred to as calciophobes. These include many species of the genera *Eunotia, Stenopterobia, Actinella, Pinnularia, Frustulia,* and some species of *Fragilaria.* As stated above, in oligotrophic and eutrophic lakes calcium is very important in the carbonate–bicarbonate buffering system. In dystrophic or brownwater lakes, the pH is controlled by a humate buffering system which is typically not as strong as the carbonate–bicarbonate buffering system. Whereas calcium and magnesium are present in fairly large quantities in both hard waters and brackish-waters, in brackish-water the sodium and potassium are present in much higher amounts than in hard water. This is particularly true for sodium. Alkali waters of the midwest and western part of the USA are often very rich in potassium; in contrast, the more brackish waters are richer in sodium. The sodium and potassium increase the conductivity of the water, which produces a shift in the kinds of species present. Cholnoky (1960) points out that total conductivity may be more important than chlorides in determining the kinds of species present. Kolbe (1932) and Hustedt (1956) have classified diatoms according to their ability to withstand various concentrations of chlorides which, in turn, usually relates to their ability to withstand sodium.

Kolbe divides diatoms into four groups—'polyhalobiens,' 'euhalobiens,' 'mesohalobiens,' and 'oligohalobiens'. The polyhalobiens are species which can stand a salt concentration greater than that of the sea. To this group belongs *Navicula longirostris.* The euhalobien species develop best in water with total salt concentration of 3–4% (NaCl 1·7–2%). To this group belong the marine and brine-water species. The mesohalobien species have their optimum in a total salt concentration of 0·5–2% (NaCl 0·2–1·5%). To this group belong such brackish-water species as *Achnanthes brevipes* var. *intermedia, Amphora coffeaeformis, Nitzschia hungarica, Stauroneis salina* var. *latior, Navicula salinarum, N. integra, N. pygmaea,* and *Diploneis interrupta.* The oligohalobien species have their optimum condition in water with a very low salt concentration. This group is subdivided into three subgroups. First, those species which have their best development in water with a small amount of salt, such as *Navicula cincta, Anomoeoneis sphaerophora, Caloneis amphisbaena, Cyclotella meneghiniana, Diatoma elongatum,* and *Navicula hungarica.* These are known as halophilic species. Second, those freshwater species, such as *Diploneis elliptica, Cymbella lacustris, Gyrosigma attenuatum, Melosira arenaria,* and *Hantzschia elongata* which are not sensitive to a little salt are known as indifferent species. Third, those which live in very pure water and dislike high concentrations of salts, are halophobe. To this group belong most species of *Eunotia, Actinella, Stenopterobia, Tabellaria flocculosa,* and

Asterionella ralfsii and some of the species of *Pinnularia* and *Frustulia*. These species are often referred to as sphagnophils, because they are found in waters which are usually rich in humic material. It is hard to be sure whether the lack of salt, the accumulation of humic materials, or other unknown factors determine their preference for this type of habitat.

Sodium chloride is certainly one of the most important salts which limits the distribution of diatom species. Indeed diatoms may be generally classified into those which are specific for certain salt concentrations and those which are euryhaline or indifferent. Richter (1906) has noted that with *Nitzschia putrida* it is the sodium which causes the influence of NaCl on diatom development. In other diatoms it would seem that it is the chloride which is the important ion. Petersen (1943) found in his study of the lakes and bogs of Denmark that the threshold of the effect of the chloride factor was at about 100 mg Cl/litre. Where the chloride content was below this amount, indifferent species formed 80–95% of the flora. Above 100 mg Cl/litre these species dropped off to 56–70% of the flora and halophilic and mesohalobic species increased correspondingly. If the chloride content was between 16–42 mg/litre it was of no significance, since other factors were more important.

The so-called brackish-water habitats are mainly of three types: those which are formed by the mixing of fresh- and salt-water; inland salt lakes; and wastewater of various industries. The first type formed by the mixing of fresh- and salt-water contains mainly euhalobien and oligohalobien species. The littoral flora, frequently very rich, is composed of many characteristic species which are often fairly restricted geographically. The inland salt lakes if old, contain many mesohalobiens. If new, there are fewer species. The brine or wastewaters contain very few species. A salt content of 10% (Krasske, 1932) was found to support a very large population of *Amphora coffeaeformis, A. delicatissima, Nitzschia closterium, N. frustulum, Navicula cincta,* and *N. salinarum.*

In polluted waters of the USA Patrick (unpublished data) has found that diatom communities in waters with fairly high conductivity can typically withstand weakly toxic pollution that increases the conductivity of the water much better than can communities with water with a low conductivity. For example, brine from oil wells can be introduced into hard waters or slightly brackish waters with much less effect on diatom communities than if it is introduced into water with very low conductivity.

(h) *Sulphur*

Sulphur in surface waters that are well oxygenated is typically found as sulphates, whereas in waters that are anaerobic or with very low redox potential H_2S or sulphides may be found. Whereas sulphates are utilized in the metabolism of diatoms and are one of the required nutrients, sulphides are often toxic. Sulphates are commonly found abundantly in hard waters and in brackish-waters, whereas oligotrophic waters are typically fairly low in sulphur compounds. There are a number of diatoms that can tolerate fairly large amounts of sulphides. Schroeder (1939) found that *Hantzschia amphioxys* and *Nitzschia palea* can tolerate 3·9 mg/litre of H_2S, whereas *Cyclotella meneghiniana, Neidium bisulcatum, Navicula minima, Nitzschia ignorata, Nitzschia tribionella* var. *deblis,* and *Surirella ovalis* var. *sabina* can tolerate concentrations of 1·5 to 3·7 mg/litre.

(i) *Nitrogen and Phosphorus*

Ammonia, nitrates, and phosphates which are commonly referred to as nutrient chemicals are utilized by diatoms in varying amounts. Some diatoms typically known as oligotrophic diatoms seem to prefer very low amounts of nitrates or ammonia and phosphates, whereas those that typically live in eutrophic and polysaprobic conditions seem to prefer high amounts of these substances. The preference of various species of diatoms for varying amounts of these chemicals has led to the development of the saprobic system of Fjerdingstad (Table 10.4) which was mainly devised to classify various conditions of pollution due to sewage. In waters in which decomposition of organic wastes is rapid, ammonia is often the form of nitrogen that is present. Some species of diatoms can tolerate or prefer these conditions; other species of diatoms seem to prefer nitrogen in the form of nitrates. Patrick (1967b) studied the effects on diatom communities of varying concentrations of nitrogen in the form of nitrates and ammonia, and of phosphates. These were diatoms that prefer circumneutral pH, in White Clay Creek Pennsylvania, which is a water of low hardness. The results of these studies showed that whereas varying the concentrations of nitrogen and phosphorus affected the abundance of various species and some species disappeared under varying concentrations and forms of these substances, there was not a shift to green or blue–green algal communities by the manipulation of just these elements when all other nutrients were in sufficient quantities. It is true that nitrogen (NO_3, NH_4) and phosphorus (PO_4) were not tested in concentrations that would be so low that they would be limiting. The purpose of the experiments was to test the effects of high concentrations of these elements.

It is evident from the literature that diatom succession is often due to the change in concentrations of nutrients in the water due to uptake of various previous species. Once these concentrations are too low to support the existing species, species which prefer low concentrations then produce large populations.

Seasonal succession of diatoms is often attributed to shifts in the concentrations of nutrients in the water together with shifts in other ecological factors. For example, the shifts in dominant species from *Fragilaria*

Table 10.4. Aquatic communities representing various zones of pollution. Survey of the saprobic zones and the corresponding communities (Fjerdingstadt, 1965).

Zone I.	**Coprozoic zone** a. the bacterium community b. the *Bodo* community c. both communities
Zone II.	**α-polysaprobic zone** 1. *Euglena* community 2. Rhodo-Thio bacterium community 3. Pure Chlorobacterium community
Zone III.	**β-polysaprobic zone** 1. *Beggiatoa* community 2. *Thiothrix nivea* community 3. *Euglena* community
Zone IV.	**γ-polysaprobic zone** 1. *Oscillatoria chlorina* community 2. *Sphaerotilus natans* communities
Zone V.	**δ-mesosaprobic zone** a. *Ulothrix zonata* community b. *Oscillatoria benthonicum* community (*Oscillatoria brevis, O. limnosa, O. splendida* with *O. subtilissima, O. princeps,* and *O. tenuis* present as associate species) c. *Stigeoclonium tenue* community
Zone VI.	**β-mesosaprobic zone** a. *Cladophora fracta* community b. *Phormidium* community
Zone VII.	**γ-mesosaprobic zone** a. Rhodophyce community (*Batrachospermum moniliforme* or *Lemanea fluviatilis*) b. Chlorophyce community (*Cladophora glomerata* or *Ulothrix zonata* (clean-water type)
Zone VIII.	**Oligosaprobic zone** a. Chlorophyce community (*Draparnaldia glomerata*) b. Pure *Meridion circulare* community c. Rhodophyce community (*Lemanea annulata, Batrachospermum vagum* or *Hildenbrandia rivularis*) d. *Vaucheria sessiis* community e. *Phormidium inundatum* community
Zone IX.	**Katharobic zone** a. Chlorophyce community (*Chlorotylium cataractum* and *Draparnaldia plumosa*) b. Rhodophyce community (*Hildenbrandia rivularis*) c. Lime-incrusting algal communities (*Chamaesiphon polonius* and various *Calothrix* species)

a, b, c, d, e = as alternatives.
1, 2, 3 = as differences in degree.

crotonensis and *Asterionella formosa* in the early spring of the year to *Stephanodiscus astrea* later on in the year, and finally in late summer to *Melosira granulata* after blooms of other algae, are believed to be due to the lower requirements of these successive species for nitrogen and phosphorus. In the winter the main decline of *Asterionella formosa* in an English lake occurred when the phosphate–phosphorus content reached 0·2 to 0·25 mg/litre. In contrast, the main decline in *Fragilaria* occurred at about 0·15 to 0·17 mg/litre. It has also been noted that the decline in *Asterionella formosa* took place when the phosphate decreased from 85 to 60 μg/litre; whereas, the maximum of *Stephanodiscus astrea* occurred when the phosphate concentration was in the order of 0·7 mg/litre (Hutchinson, 1967).

It is often very difficult to say that the decline of a species and the subsequent maximum of another species is due to one or several chemicals unless all chemical and physical characteristics of the water are taken into consideration, because at the same time nitrogen or phosphorus are declining, some other limiting factor, not measured, might have increased in value due to some input into the lake or stream, and therefore no longer being minimum may have caused a shift in species.

Whereas orthophosphate is the common form of phosporus and nitrate that of nitrogen used by most diatoms, organic and reduced forms of these chemicals may also be nutrients. Chu (1946) found that *Nitzschia palea* could utilize phytin and glycerophosphoric acid as phosphorus sources. It has also been noted that ammonia, nitrates, and organic forms of nitrogen may be utilized by diatoms in growth. Typically with a pH of 7 or below, ammonia and/or nitrogen as nitrates may be utilized; whereas at a pH above 7 the ammonia often forms ammonium hydroxide which is not readily utilized by diatoms because it is toxic. Therefore, diatoms can grow optimally at higher pH if nitrate rather than ammonia is the source of nitrogen.

Certain species of diatoms in natural habitats seem to grow best in the presence of nitrates. In streams in eastern parts of the USA we have found, for example, that *Melosira varians, Synedra ulna, Navicula viridula, Navicula mutica, Cocconeis placentula,* and *Cyclotella pseudostelliger,* become very common in the presence of high nitrate concentrations in water (2–3 mg/litre), Bahls (1973) has found that the optimum growth of *Nitzschia epiphytica* and *Navicula minima* are positively correlated with ammonia concentrations. Schoeman (1973) has shown that certain species of diatom prefer organic compounds as a source of nitrogen. Some of these are facultative nitrogen heterotrophic species such as *Nitzschia epiphytica, Navicula seminulum, Cyclotella meneghiniana, Nitzschia amphibia, Nitzschia formalis, Nitzschia frustulum, Nitzschia intermedia, Gomphonema parvulum,* and *Navicula confervacea.* Other species seem to be obligate nitrogen heterotrophic forms. These are *Navicula muralis, Navicula perparva, Nitzschia fonticola, Nitzschia palea,* and *Nitzschia formalis.*

As pointed out by many workers, the concentrations that seem to be optimum for growth or limit growth in cultures are often quite different from those found in natural waters. Furthermore, the condition of the diatom—that is, whether it has stored nitrogen or phosphorus previous to the time of observation—is also important. It is well known that many species can go for considerable lengths of time in the presence of very little nitrogen and phosphorus if sufficient amounts are stored within the cells. Many species of diatoms have been classified as oligotrophic, mesotrophic, or eutrophic species. Others have been classified according to their saprobic index. Both such considerations of the occurrence of diatoms involved the presence of nitrogen and phosphorus, but do not define other characteristics of the water as to concentrations. Therefore, much of the eutrophic literature is difficult to interpret in terms of the requirements of the species of diatoms for either nitrogen or phosphorus.

Seenayya (1972) found relatively few species of diatoms (six in number) to be common in Golkanda Pond which was abnormally high in oxidizable and nitrogenous matter, inorganic carbon, nitrogen, and phosphorus. But in Hydrilla Pond, India, were these compounds were in much lower concentrations 50 species were common. Only a few diatoms, such as *Navicula cryptocephala* and *Nitzschia palea,* multiplied profusely in the polluted waters—that is, those high in nitrogen, phosphorus, and carbon. He states that *Nitzschia amphibia* and *Nitzschia gracilis* could be regarded as avoiding pollution, and found that *Nitzschia gracilis* only utilized nitrates as a source of nitrogen.

(j) *Carbon*

In most waters there is sufficient carbon present in the form of CO_2 or carbonates or bicarbonates for diatom growth. It should be further pointed out that pH determines what form of carbon is the most common soluble form. Furthermore, at higher pHs when carbonates are prevalent one often finds precipitation of phosphorus in association with such carbonates, and as a result it is difficult to tell whether a shift in species may be associated with low phosphorus or with a shift in the form of carbon.

(k) *Silicon*

Silicon is a required element for the formation of the cell walls of diatoms. Its uptake as silicic acid or silicates by diatom cells seems to be greatly influenced by sulphydryl groups. In the cell membrane are enzymes containing sulphydryl groups. It is also known to be important in cell division. The recent work of Darley & Volcani (1969) indicates that it is related to DNA formation in *Cylindrotheca fusiformis*. Various species seem to require varying amounts of silicate. As a result when the concentration of silicate in the water varies, diatom succession seems to take place, i.e. a shift in the silicate concentration is one of the factors that brings about diatom succession. For example, populations of *Asterionella formosa* (Lund, 1950) have been found to decrease greatly when the silicate concentrations reach 6·5–5 mg/litre. This decrease was accompanied by a decrease in phosphorus from 0·085 to 0·06 mg/litre. In contrast, *Stephanodiscus astrea* developed maximum populations when the phosphate was in the order of 5 μg/litre and silicate was in the order of 0·7 mg/litre (Hutchinson, 1967).

Schleiske & Stoermer (1972) believe that the shift from a diatom dominated community to a blue–green algae dominated one occurs when phosphorus increases and silicate decreases. It is well known that silicate is much more available in alkaline waters than in acid waters. If it were not the high concentrations of sand in acid waters which is almost pure quartz, it is doubtful if a large number of individuals of diatoms such as *Eunotia* could live. However, recent work has shown that when diatoms grow attached to sand particles they may obtain silicate from the quartz. Jørgensen (1957) found that this diatom decreased when the silicon content was 30–35 μg/litre. Recent work of Kilham (1971) has also emphasized the low silicon requirements of *Stephanodiscus astrea*. In Lyngby Sø (Denmark), maximum growth of *Stephanodiscus hantzschia* occurred when the silicon content was 1–3 mg/litre. Jørgensen found that in Furesø there was a spring pulse of *Stephanodiscus hantzschia,* but when the silicate content was reduced in the lake large populations of epiphytic diatoms occurred on the stems of *Phragmites* which are known to contain silicate. Thus a shift of habitat occurred for diatom growth and the kinds of species that were most common changed, depending upon the availability of silicate. The silicate metabolism is considered in detail in Chapter 4.

(l) *Trace elements*

Only a few aspects of the function of trace elements and the toxic effects of higher concentrations will be mentioned here. Trace elements are particularly important in determining the kinds of algae present in a surface body of water. The solubility of these elements varies greatly according to the redox potential and also according to the pH of the medium. Most of them tend to be more soluble under low pHs and low redox potentials. Metals often recognized as trace metals essential for algal growth are molybdenum (Mo), copper (Cu), vanadium (V), cobalt (Co), manganese (Mn), zinc (Zn), boron (B), and iron (Fe).

It has been shown by Goldberg (1952) and others that these various trace metals can be utilized by diatoms in both a precipitated and soluble form, provided that the diatom comes in close contact with the precipitate.

Iron is one of the most important trace metals as it is a constituent of many enzymes and cytochromes and certain other porphyrins. Thus it is intimately related to the general metabolism of the cell. It is also important in chlorophyll biosynthesis. Typically we find that acid waters and waters rich in humates usually contain more iron than alkaline waters. Some species of diatoms can tolerate large concentrations of iron. Typical diatoms belonging to this group are species of the genus *Eunotia* and some of the larger forms of the genus *Pinnularia, Stauroneis phoenicenteron, Anomoeoneis serians* var. *brachysira, Navicula subtilissima, Pinnularia subcapitata* var. *hilseana, Stenopterobia intermedia, Surirella linearis, Gomphonema acuminatum,* and *Pinnularia microstauron* (Schroeder, 1939; Kolbe, 1932). Iron oxides, which form a yellow floc, do not seem to be a desirable substrate for diatom growth. As a result, acid mine streams that have been neutralized often form a yellow floc over the stream bed and this floc prevents to a large extent, diatom growth.

Molybdenum is known to be important in nitrogen fixation which is carried on by some species of blue–green algae. It is also an essential component of the nitrate reductase system.

Copper in very small amounts is necessary for the metabolic processes of many algae. The concentrations at which it seems to be advantageous to some algae are in parts per billion. In larger concentrations it is toxic to algae. Various experiments have shown that species of *Navicula* are adversely affected by 0·07 mg/litre.

Schroeder (1939) has shown that *Fragilaria virescens, Synedra ulna, Achnanthes affinis, Neidium bisulcatum, Navicula viridula, Cymbella naviculaformis, Cymbella ventricosa, Gomphonema parvulum,* and *Nitzschia palea* could live in concentrations of 1·5 mg/litre of copper. Two mg/litre of copper was tolerated by *Achnanthes affinis, Cymbella ventricosa,* and *Nitzschia palea.* Corbella *et al.* (1958) found that *Achnanthes nodosa* was very tolerant to copper and ammonia.

The role of zinc in diatoms has apparently not been well studied. Since lactate metabolism plays an important role in heterotrophic nutrition (see Chapter 6) and diatoms contain unusually high activities of lactate dehydrogenase (Werner, 1967) with zinc as essential constituent, a special requirement for this element can be postulated. In various laboratory tests using *Navicula seminulum* var. *hustedtii* it has been found that 50% reduction in division rate in this diatom occurred in 5 days if it was exposed to concentrations between 1·3 and 4·5 mg/litre of zinc (Patrick, 1965).

Vanadium seems to be associated with the process of photosynthesis. In experiments carried out by Patrick *et al.* (1975) on the effects of various trace metals on the structure of diatom communities and their replacement by other algae, it was found that in very small amounts (20 µg/litre) vanadium was stimulating to diatom growth. At 0·2–2 mg/litre the diatom diversity was good. Green algae and blue–green algae were also present. At 4·8 mg/litre blue–green algae completely dominated the community, and the diatoms were out-competed by these algae. Thus we see at very low concentrations of 20 μg/litre, vanadium seems to be stimulating to diatom communities, at higher concentrations it is more deleterious and the tolerant blue–greens replace the diatoms. The reaction of the diatom community to a given trace metal such as vanadium is greatly influenced by other environmental conditions such as light and temperature and the division rates of other algae occurring in the community.

The response of algae to a given concentration of a trace metal seems to be more closely correlated with the amount accumulated within the cells—that is, the μg/g of dry weight accumulated. We found that a concentration of 20 μg/litre of vanadium in the medium resulted in 230 to 260 μg/g being accumulated in the biomass which was mainly diatoms with very few other algae or sediments being present. At concentrations of approximately 4 mg/litre in the water, the biomass which was mainly blue–green algae accumulated about 26·4 mg/g.

Cobalt which is often present in water as cyanocobalamin, vitamin B_{12}, is shown to be necessary for several diatoms. Diatom species that have been shown to require B_{12} by Lewin & Lewin (1960) are *Amphora coffeaeformis, Amphora lineolata, Nitzschia frustulum, Nitzschia ovalis, Opephora* sp. and *Cyclotella* sp.

Manganese has been found to be important for photosynthesis in many algae. It has also been found to activate a number of enzymes that are concerned with carbohydrate breakdown. Constantopoulos (1970) has shown that manganese is necessary for fat formation in *Euglena*. Studies carried out by Patrick *et al.* (1969) have shown that if the manganese concentration is maintained below 40 μg/litre, and optimally at 15–20 μg/litre, blue–green algae will take over a diatom community; whereas if the manganese content is maintained above 40 μg/litre and between 0·04 and 0·4 mg/litre a diatom community will be maintained. Higher concentrations of manganese seem to be somewhat toxic to diatoms as they are to other algae. Other studies by Patrick *et al.* (1975) show that other environmental factors which can cause stress may influence the effect of manganese upon diatom communities. Likewise, the accumulation of manganese within the cells varies with environmental conditions such as temperature and light.

The role of *boron* in diatom metabolism has been studied by Lewin (1966) and Werner (1969). Patrick *et al.* (1975) have found that, depending on the concentration of boron, a diatom community will dominate or will be supplanted by communities of green and blue–green algae. At boron concentrations of 0·03 to 0·12 mg/litre diatoms dominate. At concentrations between 0·15 to 2 mg/litre blue–greens increased and diatoms decreased; thus showing that blue–greens were favoured by the highest concentrations and seemed to be more tolerant to boron than diatoms.

Other trace metals that are not known to be necessary as trace elements for diatom growth have been found to distinctly affect the ability of diatom communities to compete with other algae (Patrick *et al.* 1975). For example, diatom communities dominate at concentrations of 49 to 50 μg/litre hexavalent chromium in the medium. However, if this was increased to 95–97 μg/litre *Stigeoclonium lubricium* and diatoms were very common. As with other similar experiments, the shift of a diatom community to a green algal community, in this case, *Stigeoclonium lubricum*—seems to be closely correlated with the amount that accumulates within the cells. Also, we find a variation according to the season of the year as in other similar studies.

Selenium, depending upon the form that it is in, seems to stimulate diatom growth or to be very toxic to them. Patrick *et al.* (1975) have found that at concentrations of 0·1 to 10 mg/litre of selenite, diatom communities maintained a very high diversity and developed a high biomass, indicating that cell division was at an optimum condition. At 20–40 mg/litre we found that the diatom diversity was much less, and at 40 mg/litre the division rate was somewhat lower and the biomass was not as high as in the lower concentrations. Again, we see a difference in the accumulation at various seasons of the year. From September to October diatom communities exposed to 10 mg/litre of selenium as selenite accumulated 5 600 μg/g of selenium; whereas in November and December the accumulation was about 3 200 μg/g. In contrast, selenate was very toxic to diatoms at 0·1 mg/litre and greater concentrations.

Nickel, another trace metal that is not known to be important in the metabolism of diatoms, has been shown to greatly influence the ability of a diatom community to maintain itself. In March–April at 2 μg/litre, diatom diversity was poor, and they were mixed with green algae in considerable amounts and some blue–green algae; whereas at 8–10 μg/litre the diatom community was replaced by blue–green algae. In May–July experiments, the marked increase in green and blue–green algae was seen at 2 μg/litre as well as at 8–9 μg/litre. The effect of the season upon diatoms and their response to varying concentrations was evident.

(m) *Organic compounds*

Besides vitamin B_{12}, thiamine has been shown to stimulate diatom growth for a few species, whereas other species seem to have no vitamin requirements. Those that are stimulated by thiamine are *Amphora coffeaeformis, Amphora paludosa* var. *duplex,* and *Nitzschia closterium.* Those diatoms that seem to require both thiamine and vitamin B_{12} are *Amphipleura rutilans* and *Nitzschia closterium* (Lewin & Lewin, 1960). These species which Lewin believes to not require vitamins are *Navicula corymbosa, Navicula insecta, Navicula meniscula, Nitzschia affinis, Nitzschia filiformis, Nitzschia laevis, Nitzschia lanceolata, Nitzschia marginata,* and *Nitzschia obtusa.*

2.4 *Species interaction*

Density-dependent factors are produced by biological organisms or are caused by the interaction of biological organisms either by direct interface or by the production of chemicals which will favour one species over another.

(a) *Reaction to space limitation*

Competition in diatoms differs very much from competition as commonly interpreted for other groups of organisms. To date, studies have not been made that indicate how the movement of one diatom will affect the growth of another. We have, however, studied the development of diatom communities on substrates and know that at first the community is a two-dimensional one and the numbers of habitats as defined by variation in current structure, nutrients, light effects, etc., are relatively few. However, as these diatoms develop populations on the substrate an irregular current pattern develops; and more niches are available for species occupancy. When the growth becomes relatively heavy many of the species that formerly laid flat on the substrate now will stand upright because of jelly pads or their ability to produce jelly stalks. Thus a three-dimensional community will be produced which will have varying light effects, current patterns, varying effects of interaction of species by secretions, and varying nutrient effects. This micro-three-dimensional community was able to maintain a high diversity of species.

For example, *Achnanthes lanceolata* and *Cocconeis placentula* are both diatoms that form a flat 'pavement-like' growth over the surface of the substrate. It has been noted by several workers that these two diatoms never equally dominate a substrate—in other words, if *Cocconeis placentula* first establishes itself in a habitat, it will divide at such a rapid rate that *A. lanceolata* will not be able to become established, and *C. placentula* will be able to out-compete *A. lanceolata*. In contrast, if *A. lanceolata* establishes large populations in a habitat before the invasion of *C. placentula, C. placentula* will not become established in large colonies as a dominant (personal observation). Brown & Austin (1973) have noted that *A. minutissima* and *C. placentula* were not dominant at the same time, but that *C. placentula* was dominant in August and September and *A. minutissima* in late September and October.

(b) *Organic excretions*

The excretions of organic compounds which either are autotoxic or stimulatory or heterotoxic or stimulatory have been observed. By this type of species interaction a species may either encourage or discourage the development of its population or the association with another species. Another stimulation has been observed by Fogg (1971) in several species of diatoms. In such cases the excretion of glycolic acid into the external medium seemed to stimulate cell division; however, more research is needed to clearly define this effect.

Autoantibiosis has been observed by von Denffer (1948) in *Nitzschia palea*. The autoantibiotic effect was to block mitosis. Antibiosis between various species of algae have been recorded by several workers. Rice (1954) found that in mixed cultures of *Chlorella vulgaris* and *Nitzschia frustulum* mutual antibiosis occurred. Not as much substance needed to be excreted for *Nitzschia frustulum* to inhibit *C. vulgaris* as was necessary for *C. vulgaris* to inhibit the growth rate of *N. frustulum*. Jørgensen (1956) found that *Asterionella formosa* and *Nitzschia palea* grown together produced antibiotics against each other. However, sometimes in old cultures a stimulatory effect of *A. formosa* on *N. palea* was observed. It is possible that *Asterionella formosa* may on occasion produce a substance promoting its own growth. This type of phenomenon has given rise to many controversial papers. This phenomenon also has been found in Protozoa.

Nalewajko & Lean (1972) have found that *Navicula pelliculosa* and *Asterionella* could excrete organic compounds. Small molecular compounds such as glycolates were being reutilized whereas the larger molecular compounds were accumulated in the medium. They also state that it has been shown that bacteria in ordinary lake waters also produce large molecular compounds and utilize the small molecular compounds. Thus the bacteria enhance the affect of the algae, both of them utilizing the small molecular compounds and allowing the large molecular compounds to accumulate in the medium. Brown & Austin (1973) found that *Achnanthes minutissima* and *Cocconeis placentula* adversely affect each other. As stated above, this may be a matter of competition for space or it may be the production of some biotic factor that needs further investigation.

The interaction of *Anabaena circinalis* and *Fragilaria crotonensis* is another interesting example of species interaction (Hutchinson, 1967). In the interaction of these species it would appear that *F. crotonensis* requires an external source of inorganic nitrogen whereas *Anabaena circinalis* can probably fix molecular nitrogen. Thus *F. crotonensis* can bloom when the compounds of inorganic nitrogen are present. When it reduces these down to a level not suitable for its growth *A. circinalis* can fix nitrogen and can take advantage of the existing nutrient ratios. Furthermore, when *A. circinalis* uses up other nutrients and finally dies, N-compounds are liberated or formed by bacterial action, and the growth of *F. crotonensis* is possible. As Hutchinson points out, this *Fragilaria–Anabaena* system alters the environment itself and does so reversibly and rapidly in comparison with the time taken for the development and decline of populations.

(c) *Parasitism*

Parasitism is another type of species interaction which may in turn lead to seasonal succession. Canter (1949) found that populations of *Asterionella formosa* could be severely limited by the chytrid *Rhizophidium planktonicum,* particularly if the populations of *Asterionella* were very large. She found that *Asterionella formosa* suddenly declined in late March due to this fungus. However, the nutrient level in the water was such that *Fragilaria crotonensis* and *Tabellaria fenestriata* var. *asterionelloides* could come in and develop large populations. In this case, the fungus seems to have played a complex role in determining seasonal successions in nutrient-rich lakes. It has been noted by Canter that *F. crotonensis* can also be attacked by this fungus. Thus a combination of the effect of the fungus at a time when the nutrients were not exhausted allowed a substantial growth of other species of diatoms to occur.

(d) *Predation*

Predation of phytoplankton by various types of Crustacea and rotifers has been shown to greatly limit population sizes of the species involved. For example, Anderson (1958) found in Lake Lenore that the spring and late summer phytoplankton maxima are due to small species of *Amphora* and *Chaetoceros elmorei.* The interval between the two major phytoplankton pulses was characterized by rapid development of populations of *Leptodiaptomus sicilis* which declined as the *Chaetoceros* pulse developed. During the latter part of the phytoplankton minima *Moina hutchinsoni* was also present in moderate numbers. Also present as *Chaetoceros* declined was the rotifer *Hexarthra fenica,* resulting in a small secondary maximum in zooplankton after the *Chaetoceros* pulse.

Another example of predator pressure affecting the diversity of diatom communities is the results of studies by Roop (reported by Patrick, 1970). In these studies she found that the snail *Physa heterostropha* selected in feeding against the diatom *Achnanthes lanceolata* and less against *Cocconeis placentula.* These species were allowed to divide and develop large populations, whereas other species were grazed in a random manner. Thus the diversity of the community went down because of the very large populations of these two species. Many observations have been made by Patrick (unpublished data) indicating, that protozoans, particularly ciliates, select diatoms of a given size and shape as their food, thus these species are preyed upon and their populations reduced in relation to the other species present. Other observations made by Patrick and others have shown that certain insect larvae will select certain species of diatoms such as *Rhoicosphenia curvata* in greater proportions than they occur in the natural flora, thus bringing about a reduction in the population of this species that would not occur if predator pressure was absent.

3 METHODS OF MEASURING VARIATION IN THE STRUCTURE OF DIATOM COMMUNITIES

Natural diatom communities typically consist of many species representing many different genera of diatoms. The number of species characteristic for a given type of natural condition varies depending upon the invasion rate and species pool capable of invading the area. It is also influenced by other environmental factors. For example, in acid brownwater streams the numbers of species forming a diatom community will not be as large as those in a circumneutral stream. Also in springs where the environment is usually fairly constant there typically will not be as many species as in a free-flowing stream not under the influence of the more-or-less constant environmental characteristics of springs. For example, Patrick (1967a) found that a stream near the bowl of a spring in Montgomery County, Pennsylvania, supported a total diatom flora of about 100 species whereas a free-flowing stream removed from this source receiving drainage from the watershed of a type similar to the spring stream supported a community of well over 300 species of diatoms at any one time. This difference in species numbers has also been verified by many studies of dystrophic streams as contrasted with circumneutral streams (Patrick, 1964).

3.1 *Measurements based on numbers of species and size of populations*

3.2 *Model of community*

Patrick *et al.* (1954) devised a model which represents the structure of natural diatom communities (Fig. 10.4). This was a truncated normal curve, the curve being truncated because it did not represent the whole community of diatoms that might be present. This was the first time a model had been made for diatom communities. It was found that if the sample size of the diatom community studied was sufficient to always place the mode in the second or third interval the shape of the curve remained very constant over time if there was no change in the ecological characteristics of an area due to pollution. It was necessary in developing such a system to make sure that the communities developed on the slides were very similar to the benthic communities. They found that the communities on the slides also included the more common species in the plankton or floating communities in the stream. It was found that if one carefully hand-collected the stream area in which the slides were placed, 75–85% of the species found in the collections were found on the slides. Furthermore, 95% of the species represented by eight or more specimens when 8 000 specimens were counted on the slides were found also in the benthic communities. This showed that it was only the rare species that were different. These studies have been repeated many times. The community structure stays

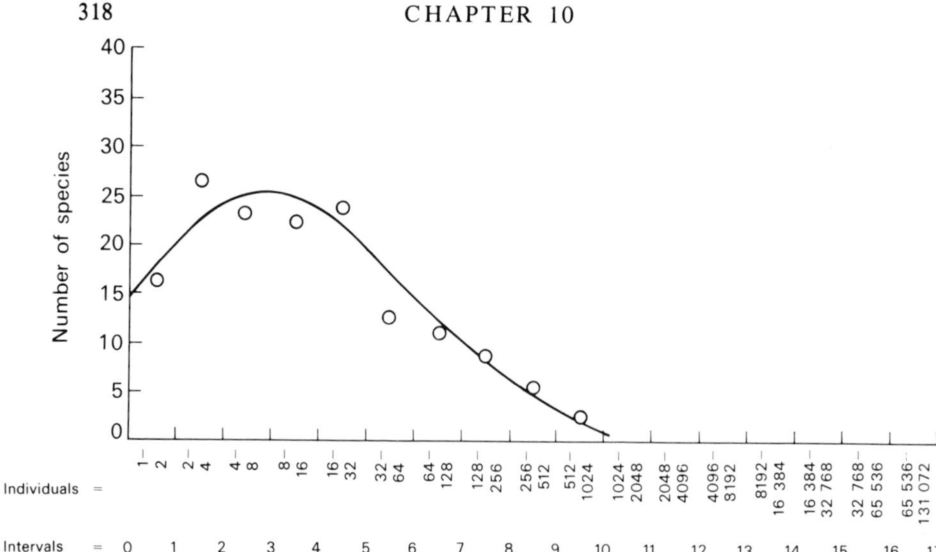

Fig. 10.4. Structure of natural diatom communities in Ridley Creek, Chester County, Pa.

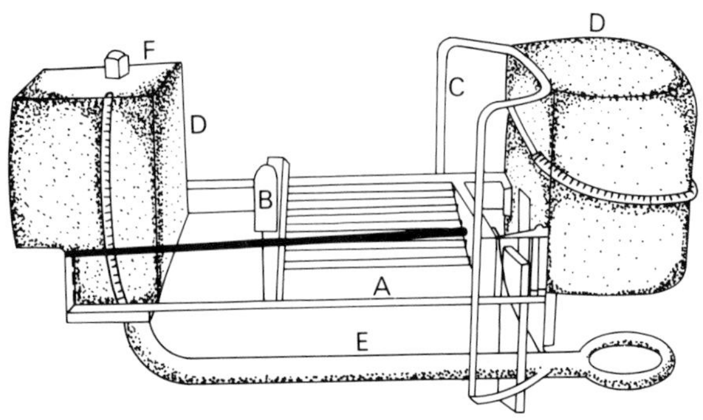

Fig. 10.5. Catherwood diatometer. A, slide holder; B, retaining bar; C, deflector; D, styrofoam float; E, brass rod; and F, identification tag.

relatively constant over time as indicated by Table 10.5. The instrument (the diatometer) used for collecting such communities is shown in Fig. 10.5.

Butcher (1947) had previously used artificial substrates to indicate the condition of a stream. His interpretations were based on indicator species present or the dominance of certain species.

Table 10.5. Summary of Catherwood diatometer readings at Station 1 October 1953 to January 1958.

Date	*Specimen number of modal interval*	*Species in mode*	*Species observed*	*Species in theoretical universe*
Oct 1953	4–8	22	150	178
Jan 1954	4–8	19	151	181
Apr 1954	2–4	24	169	200
July 1954	2–4	23	153	193
Oct 1954	4–8	21	142	168
Jan 1955	4–8	19	132	166
Apr 1955	2–4	25	165	221
July 1955	2–4	20	132	180
Oct 1955	2–4	27	171	253
Jan 1956	2–4	30	185	229
Apr 1956	4–8	35	215	252
July 1956	2–4	24	147	185
Oct 1956	2–4	23	149	206
Jan 1957	2–4	29	177	233
Apr 1957	2–4	21	132	185
July 1957	4–8	29	181	203
Oct 1957	2–4	25	157	232
Jan 1958	2–4	27	152	212
(Apr 1954– 1958 averages)		24	151	194

3.3 *Diversity index*

Margalef (1957) was the first to use diversity indices to indicate variations in the structure of phytoplankton communities, particularly diatoms. He used the following types of indices:

$$I = kN \sum_{i=1}^{i=s} Pi \log_e Pi \log_e Pi \quad \text{or} \quad I = k \log_e \frac{N}{n_1 n_2 \ldots n_s}$$

I = information
N = total specimens in sample
n_i = number of specimens in a species
$Pi = n_i/N$ or portion of the sample a given species represents
$K = 1$
$\log_e$ = natural logarithm

Patten (1962) in his study of Chesapeake Bay used the Shannon-Weiner index to measure H or community diversity:

$$H = -k \sum_{i=1}^{i=s} \frac{n_i}{N} \log_e \frac{n_i}{N}$$

where

H = community diversity
N = total number of organisms
n = number of individuals per taxon
s = total number of species in a unit area

Since Margalef's introduction of the Shannon-Weiner information theory into the study of algae, it has been followed by many researchers. Its limitation is that it measures the evenness of distribution of specimens in species more accurately than it measures changes in number of species. For example, the diversity index of a community which has fifty species in it with three representing 70% and 47 species evenly distributed between 30% is not very different from one with three species representing 70% and 10 species evenly distributed between the remaining 30%.

3.4 *Effects of pollution*

(a) *Saprobien system*

Kolkwitz & Marsson (1908) published the first paper that classified species of algae as to their tolerance to various kinds of pollution and stated that by the presence of certain species of algae one could define various zones of degradation in a river. They described the most severe of these zones as the polysaprobic zone, characterized by a wealth of high molecular decomposable organic matter. Chemicals were usually present in a reduced state, and little, if any, dissolved oxygen was present. A gamma-mesosaprobic zone represented the stage in the recovery of the river from heavy pollution in which complex matter was present but oxidation was proceeding. The beta-mesosaprobic zone was that area in which most of the organic matter had been mineralized. The oligosaprobic zone was the zone in which cleaner water mineralization had taken place. The katharobic zone was characterized as the clean, unpolluted water often found in mountain streams.

Each of these zones or conditions contain species belonging to a given condition. Thus by knowing what a given zone indicated according to this chart one could tell the condition of the water.

Many workers such as Hentschel (1925), Naumann (1925, 1932), Butcher (1947), and Liebmann (1951) used these systems and became aware of the fact that some of the species classified as characteristic of a zone of pollution often did not occur when such a condition was present or might be found under very different river conditions. This lack of applicability was due to several factors. One was the realization, as Butcher stated, that species of algae are often resistant to types of pollution other than organic and grow rather well in these substances. Thus these species may be found in many other river conditions besides those to which they were assigned by Kolkwitz and Marsson.

Another important reason why this system failed was that the types of pollution had changed greatly since the early part of the century. In the first part of the 20th century, pollution was mainly organic in nature. Today, pollution is a collective noun referring to conditions resulting from (1) the inflow of many kinds of chemicals, some of which are very toxic; (2) various physical conditions such as warm water; (3) various kinds of organic materials coming from industrial sources, farms, and homes; and (4) various kinds of sediment loads caused by the development of the watershed or industrial uses.

Owing to the complexity of pollution, it has become increasingly difficult to state that any one species characteristically might be found in all kinds of pollution or be generally indicative of pollution. Wuhrmann (1951), Sramek-Husek (1956), and Fjerdingstad (1960) increased the number of zones characteristic of stages of pollution from the five recognized by Kolkwitz & Marsson to a total of fifteen recognized by Wuhrmann.

(b) *An index of pollution*

Pantle & Buck (1955) tried to improve the system by developing an index of pollution based upon saprobic zones. The occurrence ratings were as follows:

Saprobic zone rating:
oligosaprobic rates as 1
beta-mesosaprobic rates as 2
gamma-mesosaprobic rates as 3
polysaprobic rates as 4

1. Occurring incidentally
2. Occurring frequently
3. Occurring abundantly.

To determine the degree of pollution, a mean saprobic index S is calculated for each locality.

$$S = \frac{\Sigma rh}{\Sigma h}$$

where

S = saprobic index
r = polysaprobic zone rating
h = the occurrence rating.

They concluded that a saprobic index equal to 1·0–1·5 denoted oligosaprobic; 1·5–2·5 denoted beta-mesosaprobic; 2·5–3·5 denoted gamma-mesosaprobic; and 3·5–4·0 denoted polysaprobic.

Wantanabe (1962) developed an index of the relative degree of water pollution based upon the types of diatoms.

$$\frac{2A + B - 2C}{A + B - C} \times 100$$

where

A = number of intolerant species
B = number of indifferent species
C = number of pollution tolerant species.

Fjerdingstad (1964, 1965), realizing the importance of recognizing that species may or may not occur only in a given kind of pollution, revised the system of Kolkwitz and Marsson. He recognized saprobiontic species as those that occurred only in heavily polluted waters in large numbers; saprophilous, organisms that occur generally in polluted waters may also occur in other communities, namely, organisms that to a certain extent are indifferent; saproxenous, organisms that occur generally in biotopes other than polluted ones, but may survive even in the presence of pollution; and saprophobous, organisms that will not survive in polluted waters (Table 10.4).

Patrick (1961) in various studies of diatoms has realized that the kinds of species change greatly over time with no change in the quality of water, and that such changes are due to other environmental conditions. It seemed important, therefore, that means other than the species indicator system should be developed. In 1949, Patrick set forth the principle that in natural or healthy streams the algal flora was represented by a high number of species, most of them with relatively small populations. Furthermore, the species were largely diatoms with a few green and blue–green algae present. The effects of pollution are to reduce species numbers; to cause a greater unevenness in sizes of populations of species, with some becoming extremely common; and to cause a shift of kinds of species composing the algal community from one being dominated by diatoms to one being dominated by various kinds of filamentous green or blue–green algae, or, in a few cases, unicellular green and blue–green algae. Smaller shifts were noted by the kinds of diatoms changing from narrowly tolerant to broadly tolerant species. The types of shifts were dependent upon the effects of various kinds of pollution.

(c) *Phytoplankton quotient*

Nygaard (1949) following Thunmark (1945) recognized the following phytoplankton quotients based on the number of species in each group: Myxophyceae/Desmidieae; Chlorococcales/Desmidieae; Centrales/Pennales; Euglenineae/(Myxophyceae + Chlorococcales).

The compound quotient was (Myxophyceae + Chlorococcales + Centrales + Euglenineae)/Desmidieae. This kind of a formula only applies in habitats where desmids are quite common.

If the compound quotient is below 1, the water is probably oligotrophic. If the compound quotient is above 1, the water is probably eutrophic. The true saprorophy is revealed only by the Euglenineae quotient; the border between eutrophy and saprotrophy being 2 to 3 as saprotrophy is above 3. Stockner & Benson (1967) used the Centrales/Pennales index to follow the enrichment of Lake Washington over time. This index, though useful in Lake Washington according to Stockner & Benson, has not proved useful in streams according to Williams (1972) and according to our findings. The reason why this does not hold true in many rivers and estuaries studied at the Academy of Natural Sciences is that there are a great many Centrales that are indicative of fairly high nutrient levels in the streams, and in contrast there are many Centrales and many species belonging to what he would classify as Pennales that are indicative of clean water conditions.

Cholnoky developed a method for measuring changes in the amount of pollution by changes in the dominant species of diatoms (Fig. 10.6). This system is only applicable if ecological changes other than pollution do not produce a similar shift in the abundance of species. Patrick *et al.* (1954, 1963)

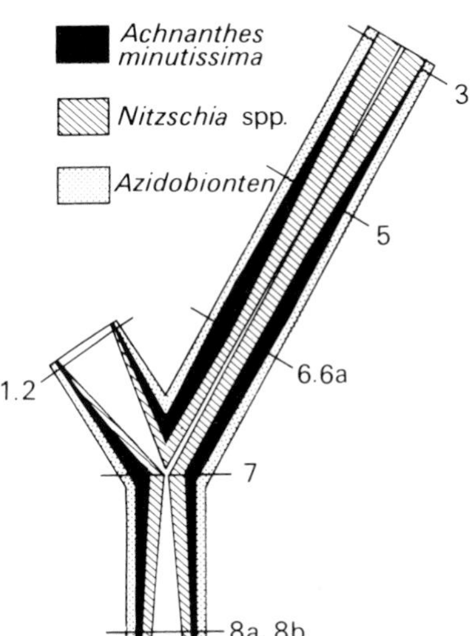

Fig. 10.6. Changes in dominance of certain diatoms because of pollution. From Cholnoky, 1968.

show that the structure of diatom communities change with the increase of pollution and that these changes are very characteristic of the type of pollution which occurred. The effect of organic pollution was to cause certain species to become excessively common and thus the curve covered more intervals, usually 13–15. More severe organic pollution would bring about a reduction in the height of the mode, σ^2 would increase, and the curve would cover 15–16 intervals. The effect of toxic pollution, however, was to greatly decrease the height of the mode, and usually the intervals covered by the curve did not increase very much. However, under some types of pollution one or two species were able to tolerate the pollutant and grow successfully, and the curve would cover many more intervals. Under these conditions σ^2 greatly increased.

Patrick *et al.* (1968) showed that toxic pollutants might produce another type of curve. This was the result of studies on the effect of the low pH (5·15 in the autumn of the year when conditions were unfavourable) on diatom communities. Under such conditions the low pH limited diatom reproduction but did not kill the diatoms. As a result, the height of the mode was high, σ^2 was very small, and the truncated curve covered only a few intervals. Indeed, the curve very much resembled that of a natural community. This type of pollution might be confused with natural conditions, but if one considers the amount of biomass, it is very evident that it is much less when the pH is low than under natural conditions. These studies clearly show that one should consider not only the structure of the diatom community but also the kinds of species and the total biomass. Furthermore, one is able to predict what kinds of pollution are occurring by the types of shifts in the structure of the curve.

Patrick & Hohn (1956) applied this method of studying the structure of diatom communities to brackish-waters in Maryland and Texas. They found that typically in brackish-water the height of the mode was somewhat less as were the numbers of observed species. The curve typically covers more intervals. Based upon studies in brackish- and freshwater, Patrick & Strawbridge (1963) developed a method whereby one could determine from a single analysis whether a community fell within a 95 or 99% confidence interval for a natural stream. This method used a formula of Bennett & Franklin (1954) for a normal bivariate distribution function. Such an equation using two variables can be applied to the structure of these curves if the numbers of the species counted always result in placing the mode in the same interval or approximately the same interval.

Wilhm & Doris (1966) have used the Shannon-Weiner diversity index first developed by Margalef to indicate varying degrees of water pollution. Usually, the ordinates indicate the diversity index and the abscissae indicate the position of the station in a river, thus one is able to show changes in the diversity index along the course of the river. They mainly have applied this method to invertebrates and not to diatoms; however, it might be applied to diatoms. Patrick and others have found that a more reliable indication of the kind of community is obtained by plotting the number of species as the ordinate and the diversity index as the abscissa (unpublished data).

The redundancy may be measured by comparing the numbers of species common to any two communities with the total number of species in each of the two communities. This is usually done by the Jaccard coefficient:

$$\frac{C}{A + B - C}$$

where

A = total species in Sample 1
B = total species in Sample 2
C = species common to both samples

By this formula the percent under the same two communities can be estimated in terms of the total number found in the two communities.

A more sophisticated method has been developed by Patten (1962). He measures the redundancy for comparing the diversity of a given community with the maximum and minimum possible diversity for that community.

$$H_{max} = k[\log_e N - m \log_e (N/m)]$$

$$H_{min} = k\{\log_e N - \log_e [N - (m - 1)]\}$$

$$R = \frac{H_{max} - H}{H_{max} - H_{min}}$$

m = number of species.

3.5 COMPARISON OF COMMUNITIES

MacIntire (1969) has used a distance index to compare the structure of two diatom communities:

$$Djh = \sqrt{\sum_{i=1}^{i=s} (Xij - Xih)^2}$$

where

Xij = the proportion of the i species in the j community,
Xih = the proportion of the i species in the h community, and
D = the degree of difference between the j and h communities.

Patrick, following MacArthur (1965), has used another method for measuring differences between two communities of diatoms:

$$\sum_{i=1}^{i=s} \frac{Pi^1 + Pi^2}{2} \log_e \frac{Pi^1 + Pi^2}{2} - \frac{\sum_{i=1}^{i=s} Pi^1 \log_e Pi^1 + \sum_{i=1}^{i=s} Pi^2 \log_e Pi^2}{2}$$

$$Pi = \frac{n_i}{N} \quad \text{where} \quad \begin{array}{l} n_i = \text{specimens/species} \\ N = \text{total specimens in sample.} \end{array}$$

3.6 *Measurements based on pigments*

Margalef (1957) has tried to measure the diversity of a community by the diversity of pigments present. Other workers have studied the relation of the amount of chlorophyll *a* to the amount of primary production. Others have used primary production as a measure of the amount of algae present. Still other workers have used chlorophyll $a:c$ ratios to determine relative types of algae present. Others have added to this measures of the phycoerythin and phycocyanin pigments. These measures are based on physiological activity which is correlated with the amount of living algae present. Much more study needs to be done to work out the reliability of such an index.

3.7 *Conclusions*

It should be pointed out that the using of any kind of mathematical model or index only gives the worker part of the information concerning these diatom communities. One of the most important items of information that can be obtained by species identification is the recognition of small changes or trends of shifts from sensitive to tolerant species.

4 REFERENCES

AKEHURST S.C. (1931) Observations on pond life with special reference to the possible causation of swarming of phytoplankton. *J. R. microsc. Soc.* London Ser. 3, **51,** 237–65.

ANDERSON G.C. (1958) Seasonal characteristics of two saline lakes in Washington. *Limnol. Oceanogr.* **3,** 51–68.

BAHLS L.L. (1973) Diatom community response to primary waste-water effluent. *J. Wat. Poll. Control Fed.* **45**(1), 134–44.

BARKER H.A. (1935) Photosynthesis in diatoms. *Arch. Mikrobiol.* **6,** 141–56.

BEGER H. (1927) Beiträge zur Ökologie und Soziologie der luftlebigen (atmosphytischen) Kieselalgen. *Ber. dt. bot. Ges.* **45**(6), 285–407.

BENNETT C. & FRANKLIN N. (1954) *Statistical Analyses in Chemistry and Chemical Industry*. John Wiley and Sons, New York.

BIRGE E.A. & JUDAY C. (1922) The inland lakes of Wisconsin. The plankton. I. Its quantity and chemical composition. *Bull. Wis. geol. nat. Hist. Surv.* **64** (Sic. ser. 13), 222 pp.

BRISTOL B.M. (1920) On the alga-flora of some desiccated English soil: an important factor in soil biology. *Ann. Bot.* **34,** 35–80.

BROWN S.D. & AUSTIN A.P. (1973) Diatom succession and interaction in littoral periphyton and plankton. *Hydrobiologia* **43** (3–4), 333–56.

BUDDE H. & BURKHOLDER P. (1929) Microplankton studies of Lake Erie. *Bull. Buffalo Soc. nat. Sci.* **14,** 73–93.

BUTCHER R.W. (1947) Studies in the ecology of rivers. VII. The algae of organically enriched waters. *Jour. Ecol.* **35** (1/2), 186–91.

CANTER H.M. (1949) The importance of fungal parasitism in limnology. *Verh. int. Verein. theor. angew. Limnol.* **10,** 107–8.

CHANDLER D.C. (1940) Limnological studies of western Lake Erie. I. Plankton and certain physical–chemical data of the Bass Islands region, from September, 1938, to November, 1939. *Ohio J. Sci.* **40,** 291–336.

CHANDLER D.C. (1942) Limnological studies of western Lake Erie. III. Phytoplankton and physical–chemical data from November, 1939 to November, 1940. *Ohio J. Sci.* **42,** 24–44.

CHANDLER D.C. (1944) Limnological studies of western Lake Erie. IV. Relation of limnological and climatic factors to the phytoplankton of 1941. *Trans. Am. microsc. Soc.* **63,** 203–36.

CHOLNOKY B.J. (1958) Beitrag zu den Diatomeenassoziationen des Sumpfes Olifantsvlei südwestlich Johannisburg. *Ber. dt. bot. Ges.* **71,** 177–87.

CHOLNOKY B. J. (1960) The relationship between algae and the chemistry of natural waters. *Coun. sci. industr. Res.* Reprint R.W. No. 129, pp. 215–25.

CHOLNOKY B.J. (1968) *Die Ökologie der Diatomeen in Binnengewässern* (The ecology of diatoms in inland waters). 699 pp. J. Cramer, Lehre.

CHU S. (1946) The utilization of organic phosphate by phytoplankton. *J. mar. biol. Ass. U.K.* **26,** 285–95.

CONSTANTOPOULOS G. (1970) Lipid metabolism of manganese deficient algae. I. Effect of manganese deficiency on the greening and the lipid composition of *Euglena gracilis Z. Pl. Physiol.* **45**(1), 76–80.

CORBELLA C., TONOLLI V. & TONOLLI L. (1958) I sedimenti del lago d Orta testimoni di una disastorsa polluzione cuproammoniacale. *Memorie Ist. ital. Idrobiol.* **10,** 9–50.

DAMANN K.E. (1945) Plankton studies of Lake Michigan. *Am. Midl. Nat.* **34,** 796–7.

DARLEY W.M. & VOLCANI B.E. (1969) A silicon requirement for deoxyribonucleic acid synthesis in the diatom *Cylindrotheca fusiformis* Reimann & Lewin. *Expl. Cell Res.* **58,** 334–42.

DENFFER V. D. (1948) Über einen Wachstumshemmstoff in alternden Diatomeenkulturen. *Biol. Zbl.* **67**, 7–13, 464.

EDDY S. (1927a) The plankton of Lake Michigan. *Bull. Ill. St. nat. Hist. Surv.* **17,** 203–32.

EDDY S. (1927b) Growth of diatoms in relation to dissolved gases. *Trans. Ill. St. Acad. Sci.* **20,** 63–66.

FJERDINGSTAD E. (1960) Forurening af vandlb biologisk bedmt. *Nord. hyg. Tidskr.* **Vol. 41,** Nos. 7 and 8, 149–196.

FJERDINGSTADT E. (1962) Some remarks on a new saprobic system. In *Biological Problems in Water Pollution,* 3rd seminar. Public Health Service Publication 999-WP-25, 1965, pp. 232–5.

FJERDINGSTAD E. (1964) Pollution of streams estimated by benthal physomicro-organisms. I. A saprobic system based on communities of organisms and ecological factors. *Int. Revue ges. Hydrobiol. Hydrogr.* **49**(1), 63–131.

FJERDINGSTAD E. (1965) Some remarks on a new saprobic system. In *Biological Problems in Water Pollution,* 3rd Seminar, *USPHS Publ.* 999-WP-25, pp. 232–5.

FOGG G.E. (1971) Extra cellular products of algae in fresh water. *Arch. Hydrobiol. Beih. Ergebn. Limnol.* **5,** 1–25.

FRANCE R.H. (1912) Studien über edaphische Organismen. *Zentbl. Bakt. Parasitkde.* Jena **32**(2), 1–7.

GERLOFF J.C. & SKOOG J.K. (1954) Cell contents of nitrogen and phosphorus as a measure of their availability for growth of *Microcystis aeruginosa. Ecology,* **35**, 348–53.

GODWARD M. (1937) An ecological and taxonomic investigation of the littoral algal flora of Lake Windermere. *J. Ecol.* **25,** 496–568.

GOLDBERG E.D. (1952) Iron assimilation of marine diatoms. *Biol. Bull. Wash. Dep. Game.* **102,** 243–8.

GOTTSCHALL R.Y. & JENNINGS O.E. (1933) Limnological studies at Erie, Pennsylvania. *Trans. Am. microsc. Soc.* **52,** 181–91.

HAYEK J.M. & HULBARY R.L. (1956). A survey of soil diatoms. *Proc. Iowa Acad. Sci.* **63,** 327–38.

HEDLUND T. (1913) Till frågan om växternas frosthärdighet. I–II. *Bot. Notiser* for 1913, 65–78, 153–74.

HENTSCHEL E. (1925) Abwasserbiologie. *Abderhalden's Handb. Biol. Arbeitsmethod.,* Abt. 9, Teil 2, 1 Hälfte, 233–80.

HUSTEDT F. (1922) Bacillariales aus Schlesien. I. *Ber. dt. bot. Ges.* **40,** 98–103.

HUSTEDT F. (1939) Systematische und ökologische Untersuchungen über die Diatomeenflora von Java, Bali und Sumatra (Systematic and ecological investigations of the diatom flora of Java, Bali, and Sumatra). *Arch. Hydrobiol., Suppl.* **15,** 131–77; **16,** 1–155; **16,** 187–295; **16,** 393–506.

HUSTEDT F. (1956) *Kieselalgen* (Diatomeen). Kosmos-Verlag Franckh, Stuttgart.

HUTCHINSON G.E. (1967). *A Treatise on Limnology,* Vol. II. 1115 pp. John Wiley & Sons, New York, London, Sydney.

JØRGENSEN E.G. (1956) Growth inhibiting substances formed by algae. *Physiologia Pl.* **9,** 712–26.

JØRGENSEN E.G. (1957) Diatom periodicity and silicon assimilation. *Dansk. bot. Ark.* **18**(1), 54 pp.

KILHAM P. (1971) A hypothesis concerning silica and the freshwater planktonic diatoms. *Limnol. Oceanogr.* **16,** 10–18.

KOLBE R.W. (1932) Grundlinien einer allgemeinen Ökologie der Diatomeen. *Ergebn. Biol. Berlin,* **8,** 221–348.

KOLKWITZ R. & MARSSON M. (1908) Ökologie der pflanzlichen Saprobien. *Ber. dt. bot. Ges.* **26,** 505–19.

KOORDERS S.H. (1901) Notiz über die dysphotische Flora eines Süsswassersees in Java. *Natuurk. Tijdschr. Ned.-Indië.* **41,** 119–26.

KRASSKE G. (1929) Beiträge zur Kenntnis der Diatomeenflora Sach sens. *Bot. Arch. Leipzig,* **27,** 348–80.

KRASSKE G. (1932) Diatomeen deutscher Solquellen und Gradierwerke. II. Die Diatomeen von Bad Nauheim, Wisselheim und Bad Salzungen a.d. Werra. *Hedwigia,* **74**(4/5), 135–43.

KRASSKE G. (1936) *Die Diatomeenflora der Moosrasen des Wilhelmshöher Parkes.* Festschrift des Vereins für Naturkunde zu Kassel zur Feier seines hunderjährigen Bestehens, 151–64.

LEWIN J.C. (1966) Boron as a growth requirement for diatoms. *J. Phycol.* **2,** 160–3.

LEWIN J.C. & LEWIN R.A. (1960) Auxotrophy and heterotrophy in marine littoral diatoms. *Can. J. Microbiol.* **6,** 127–34.

LEOPOLD L.B. (1964) *Fluvial Processes in Geomorphology.* W.H. Freeman & Co., San Francisco.

LIEBMANN H. (1951) *Handbuch der Frischwasser und Abwasserbiologie,* Vol. I. Verlag R. Oldenbourg, München.

LOWE R.L. & COLLINS G.B. (1973) An aerophilous diatom community from Hocking County, Ohio. *Trans. Am. microsc. Soc.* **92**(3), 492–6.

LUND J.W.G. (1950) Studies on *Asterionella formosa* Hass. II. Nutrient depletion and the spring maximum. Parts I–II. *J. Ecol.* **38,** 1–35.

LUND J.W.G. (1954) The seasonal cycle of the plankton diatom *Melosira* (Ehr.) Kütz. subsp. *subarctica* O. Mull. *J. Ecol.* **42,** 151–79.

MACARTHUR ROBERT H. (1965) Patterns of species diversity. *Biol. Rev.* **40**(4), 510–33.

MAEDA O. & ICHIMURA S. (1973) On the high density of a phytoplankton population found in a lake under ice. *Int. Revue ges. Hydrobiol. Hydrogr.* **58**(5), 673–85.

MARGALEF D.R. (1957) La teoria de la informacion en ecologia. *Mems. R. Acad. Cienc. Artes Barcelona* **23,** 373–449.

MCINTIRE C.D. (1969) Ecological–physiological investigations of littoral diatom communities of the Yaquina River Estuary, Oregon. *Prog. Rep. natn. Sci. Found. Res. Grant GB-7203.*

MOORE G.T. & CARTER N. (1926) Further studies on the subterranean algal flora of the Missouri Botanical Garden. *Ann. Mo. bot. Gdn.* **13,** 101–40.

NALEWAJKO C. & LEAN D.R.S. (1972) Growth and excretion in planktonic algae and bacteria. *J. Phycol.* **8,** 361–6.

NAUMANN E. (1925) Die Arbeitsmethoden der regionalen Limnologie. *Abderhalden's Handb. biol. Arbeitsmeth.*

NAUMANN E. (1932) *Grundzüge der regionalen Limnologie.* Binnengewässer, Band 11. 176 pp.

NYGAARD G. (1949) Hydrobiological studies of some Danish ponds and lakes. II. The quotient hypothesis and some new or little unknown phytoplankton organisms. *K. danske Vidensk. Selsk. Skr.* **7,** No. 1, 1–293 pp.

PANTLE R. & BUCK H. (1955) Die biologische Überwachung der Gewässer und die Darstellung der Ergebnisse. *Gas. Wassfach.* 96–604.

PATRICK R. (1945) A taxonomic and ecological study of some diatoms from the Pocono Plateau and adjacent regions. *Farlowia,* **2**(2), 143–221.

PATRICK R. (1949) A proposed biological measure of stream conditions based on a survey of Conestoga Basin, Lancaster County, Pennsylvania. *Proc. Acad. nat. Sci. Philad.* **101,** 277–341.

PATRICK R., HOHN M.H. & WALLACE J.H. (1954) A new method for determining the pattern of the diatom flora. *Not. natn. Acad. natn. Sci. Philad.* **259,** 12 pp.

PATRICK R. & HOHN M.H. (1956) The diatometer—a method for indicating the conditions of aquatic life. *Proc. Am. Petrol. Inst.* Sect. III, Refining **36**(3), 332–9.

PATRICK R. (1961) A study of the numbers and kinds of species found in rivers in eastern United States. *Proc. Acad. Nat. Sci. Philadelphia,* **113,** 215–58.

PATRICK R. (1963) The structure of diatom communities under varying ecological conditions. Conf. on the problems of environmental control on the morphology of fossil and recent protobionta. *N.Y. Acad. Sci.* **108**(2), 359–65.

PATRICK R. & STRAWBRIDGE D. (1963) Variations in the structure of natural diatom populations. *Am. Nat.* **97**(892), 51–7.

PATRICK R. (1964) A discussion of natural and abnormal diatom communities. In *Algae and man* (ed. D.F. Jackson) pp. 185–204, Plenum Press, New York.

PATRICK R. (1965) Algae as indicators of pollution. In *Biological problems in water pollution,* pp. 225–31. US Dept. Health, Education & Welfare, PHS Publ. 999-WP-25, Cincinnatti, Ohio.

PATRICK R. (1967a) The effect of invasion rate, species pool, and size of area on the structure of the diatom community. *Proc. natn. Acad. Sci. U.S.A.* **58**(4), 1 335–42.

PATRICK R. (1967b) The effect of varying amounts and ratios of nitrogen and phosphate on algae blooms. *Proc. 21st Ann. industr. Waste Conf.,* Purdue U., Purdue, Indiana, 41–51.

PATRICK R. (1968) The structure of diatom communities in similar ecological conditions. *Am. Nat.* **102**(924), 173–83.

PATRICK R., ROBERTS N.A. & DAVIS B. (1968) The effect of changes in pH on the structure of diatom communities. *Not. natn. Acad. Nat. Sci. Philadelphia* No. 416, 16 pp.

PATRICK R. (1969) Some effects of temperature on freshwater algae. In: *Biological Aspects of Thermal Pollution* (ed. P.A. Krenkel and F.L. Parker). 161–85 pp. Vanderbilt Univ. Press.

PATRICK R., CRUM B. & COLES J. (1969) Temperature and manganese as determining factors in the presence of diatom or blue–green algal floras in streams. *Proc. natn. Acad. Sci. U.S.A.* **64**(2), 472–8.

PATRICK R. (1970) Benthic stream communities. *Am. Scient.* **58**(5), 546–9.

PATRICK R. (1971) The effects of increasing light and temperature on the structure of diatom communities. *Limnol. Oceanogr.* **16**(2), 405–21.

PATRICK R. (1974) Effects of abnormal temperatures on algal communities. In *Thermal Ecology* (ed. J.W. Gibbons & R.R. Sharitz). *Proc. Symp. held at Augusta, Ga.,* May 3–5, 1973. U.S. Atomic Energy Commission.

PATRICK R., BOTT T. & LARSON R. (1975) The role of trace elements in management of nuisance growths. *U.S. Env. Protection Agency,* Corvallis, Oregon.

PATTEN B.C. (1962) Species diversity in net plankton of Raritan Bay. *J. mar. Res.* **20**(1), 57–75.

PEARSALL W.H. (1922) A suggestion as to factors influencing the distribution of free-floating vegetation. *J. Ecol.* **9**(2), 241–53.

PEARSALL W.H. (1924) Phytoplankton and environment in the English Lake District. *Revue algol.* **1,** 53–67.

PEARSALL W.H. (1930) Phytoplankton in the English Lakes. I. The proportions in the water of some dissolved substances of biological importance. *J. Ecol.* **18**(2), 306–20.

PEARSALL W.H. (1932) Phytoplankton in the English Lakes. II. The composition of the phytoplankton in relation to dissolved substances. *J. Ecol.* **20**(2), 241–62.

PETERSEN J.B. (1928) The aerial algae of Iceland. *Botany Icel.* Vol. 2, part 2 (8), 325–447.

PETERSEN J.B. (1935) Studies on the biology and taxonomy of soil algae. *Dansk. bot. Ark.* **8,** 1–180.

PETERSEN J.B. (1943) Some halobion spectra (Diatoms). *K. danske Vidensk. Selsk. Meddel. Copenhagen* **17**(9), 3–95.

PHINNEY H.K. & MCINTIRE C.D. (1965) Effect of temperature on metabolism of periphyton communities in laboratory streams. *Limnol. Oceanogr.* **10**(3), 341–4.

PROUSE G. & TALLING J. (1958) The seasonal growth and succession of plankton algae in the White Nile. *Limnol. Oceanogr.* **3**(2), 222–38.

RICE C.H. (1938) Studies in the phytoplankton of the River Thames (1928–1932). I and II. *Ann. Bot.,* New Series **2**(7), 539–57, 559–81.

RICE T.R. (1954) Biotic influences affecting population growth of planktonic algae. *Fish. Bull. Fish. Wildl. Serv., U.S.* **87,** 227–45.

RICHARDSON J.L. (1968) Diatoms and lake typology in East and Central Africa. *Int. Revue ges. Hydrobiol. Hydrogr.* **53**(2), 299–338.

RICHTER O. (1906) Zur Physiologie der Diatomeen. J. Mitteilung. *Sitzungsberichte der Mathematisch-Naturwissenschaftlichen Klasse der Kaiserlichen Akademie der Wissenschaften.* Wien, **115**(1), 27–119.

RODHE W. (1948) Environmental requirements of freshwater plankton algae. Experimental studies in the ecology of phytoplankton. *Symb. bot. upsal.* **10**(1), 149.

ROUND F.E. (1965) *The Biology of Algae.* Arnold, London.

RUTTNER F. (1940) *Grundriss der Limnologie.* 167 pp. W. de Gruyter, Berlin.

SCHLEISKE C.L. & STOERMER E.F. (1972) Phosphorus, silica and eutrophication of Lake Michigan. *Am. Soc. Limnol. Oceanogr.* **1,** 157–71.

SCHOEMANN F.R. (1973) A systematical and ecological study of the diatom flora of Lesotho with special reference to the water quality. *National Institute for Water Research.* Pretoria, South Africa. 355 pp.

SCHRÖDER B. (1916) *Melosira roeseana* Rabenh., eine 'leuchtende' Bacillariaceae. *Ber. dt. bot. Ges.* **34**(9), 796–800.

SCHROEDER H. (1939) Die Algenflora der Mulde. *Pflanzenforschung,* **21,** 1–88.

SCHUMACHER G.J. & WHITFORD L.A. (1965) Respiration and P^{32} uptake in various species of freshwater algae as affected by a current. *J. Phycol.* **1**(2), 78–80.

SEENAYYA G. (1972) Ecological studies in the plankton of certain freshwater ponds of Hydraabad, India, II. Phytoplankton 2. *Hydrobiologia* **39**(2), 247–71.

SRAMEK-HUSEK R. (1956) Zur biologischen Charakteristik der höheren Saprobitätsstufen. *Arch. Hydrobiol.* **51,** 376–90.

STANKOVIC S. (1960) The Balkan Lake Ohrid and its living world. *Monographiae biol.* **9,** 357 pp.

STOCKNER J.G. (1967) Observations of thermophilic algal communities in Mount Rainier and Yellowstone National Parks. *Limnol. Oceanogr.* **12**(1), 13–17.

STOCKNER J.G. & BENSON W.W. (1967) The succession of diatoms assemblages in the recent sediment of Lake Washington. *Limnol. Oceanogr.* Vol. 12, No. 3, 513–32.

TALLING J.E. (1957a) The growth of two plankton diatoms in mixed culture. *Physiologia Pl.* **10,** 215–23.

TALLING J.E. (1975b) The phytoplankton population as a compound photosynthetic system. *New Phytol.* **56,** 133–49.

THUNMARK S. (1945) Zur Soziologie des Süsswasserplanktons. *Folia Limnol. Scand.* Vol. 3.

VERDUIN J. (1952) The volume-based photosynthetic rates of aquatic plants. *Am. J. Bot.* **39,** 157–9.

WALLACE N.C. (1955) The effect of temperature on the growth of some freshwater diatoms. *Not. Nat., Acad. Nat. Sci. Phila.*, No. 280, 11 pp.

WATANABE T. (1962) On the biotic index of water pollution based upon the species number of Bacillariophyceae in the Tokoro River in Hokkaido. *Jap. J. Ecol.* **12**(6), 216–22.

WERNER D. (1967) Hohe Aktivitäten von Laktatdehydrogenase bei aerob wachsenden Diatomeen. *Naturwissenschaften* **54,** 474–5.

WERNER D. (1969) Silicoborate als erste nicht C-haltige Wachstumsfaktoren. *Arch. Mikrobiol.* **65,** 258–74.

WEST W. & WEST G.S. (1912) On the periodicity of the phytoplankton of some British lakes. *J. Linn. Soc. of London, Botany* **40,** 393–432.

WILHM J.L. & DORIS T.C. (1966) Species diversity of benthic macroinvertebrates in a stream receiving domestic and oil refinery effluents. *Am. Midl. Nat.* **76**(2), 427–49.

WILLIAMS L.G. (1972) Plankton diatom species biomasses and the quality of American rivers and the Great Lakes. *Ecology* **53**(6), 1 038–50.

WUHRMANN K. (1951) 'Über die Biologische Prüfung von Abwasser-reinigungsanlagen'. *Gesundheitsingenieur,* **72,** 253–61.

CHAPTER 11

MARINE LITTORAL DIATOMS: ECOLOGICAL CONSIDERATIONS

C. DAVID McINTIRE

and

WENDY W. MOORE

Department of Botany and Plant Pathology
Oregon State University,
Corvallis, Oregon 97331, USA

1 INTRODUCTION

The non-planktonic diatoms found in estuaries and marine coastal waters include taxa that grow attached to other plants (*epiphyton*), particularly marine angiosperms and macroalgae; on relatively large rocks (*epilithon*) and other non-living surfaces (e.g. wood, plastic, metal, and glass); and on sediments. Diatoms associated with sediments are sometimes subdivided further into those taxa that grow attached to sand grains (*epipsammon*) and taxa, usually motile forms, that live on silty sediment but are not firmly attached to the particles (*epipelon*). The terms *endolithon, endophyton, endopelon,* and *epizoon* are sometimes used to refer to assemblages of organisms growing within rock cavities, plant tissues, sediment, and attached to animals, respectively. We refer to a collection of cooccurring diatom populations in an area of interest as an *assemblage*. The term *community* is avoided here, as virtually nothing is usually known about the dynamic interrelationships and dependencies among closely associated diatom populations and between the diatoms and other groups of organisms.

The ecological properties of diatom assemblages can be examined from essentially two points of view: (1) their distributional patterns and community structure; and (2) their bioenergetic or functional attributes. The former deals primarily with the distribution and diversity of genetic information along environmental gradients, while the latter is concerned with primary production and energy relationships between diatom assemblages and other groups of organisms. Ideally, both the functional and structural attributes of diatom assemblages should be investigated simultaneously, although there are very few examples of such studies in the literature. Allen & Koonce (1973), using principal components ordinations and a cluster analysis, found that most phytoplankton species in Lake Wingra (Wisconsin, USA) could be classified in one of three categories: (1) ungrazed, slow-growing and very persistent; (2) ungrazed, fast-growing and of intermediate duration; and (3) grazed, fast-growing and ephemeral. Riznyk (1969, 1973) and Riznyk & Phinney (1972a,b) described some functional and structural attributes of diatom assemblages associated with the sediment of two tidal flats in Yaquina Estuary (Oregon, USA).

This chapter is concerned with some ecological aspects of marine and estuarine benthic diatom assemblages with emphasis placed on the structure of epilithic, epiphytic, epipelic and epipsammic intertidal assemblages and the general distributional patterns in these assemblages along chemical and physical environmental gradients. A complete review of literature related to the structure and function of marine and brackish-water benthic diatom assemblages is not attempted here, as an excellent introduction to the subject is available in a comprehensive review article by Round (1971). Instead, we first present a historical review of selected floristic studies followed by a brief review of investigations concerned with measurements of primary production, biomass, and pigment concentrations in assemblages of intertidal diatoms and with nutritional relationships between such assemblages and other groups of marine and estuarine organisms. The final part of the chapter presents a summary of a quantitative approach for the analysis of distributional patterns and a simple example of the use of this approach.

2 REVIEW OF SELECTED STUDIES

2.1 *Floristic studies*

A list of selected floristic studies of non-planktonic (littoral) diatoms is presented in Table 11.1. This list is by no means complete, but does represent some of the better studies that are readily accessible in the literature to a beginning student. Moreover, the list is obviously biased towards particular geographical regions, namely South Africa, the south coast of England, coastal regions of northern Europe, and coastal regions of the northwestern United States. This bias reflects, in part, a disproportionate local interest in littoral diatoms as well as difficulties in obtaining certain relevant publications. Other studies, particularly those dealing with distributional patterns in the subtidal zone and in the sediments of the deep sea, are cited by Round (1971).

One interesting feature of marine and estuarine littoral diatoms is that many of the same taxa, i.e. species or varieties of species, have been found in samples from different parts of the world. Table 11.2 represents examples of such cosmopolitan taxa and indicates references to floristic studies in which a particular taxon is reported (Table 11.1). The list includes taxa collected from a variety of substrates, but excludes taxa that are usually considered planktonic in growth form. It is important to emphasize that the failure of a particular taxon to be included in distributional records for a certain region does not necessarily mean that it will not occur in samples obtained in the future from that location. In fact, the total effort expended in sampling marine and estuarine littoral diatoms has been minimal considering their relative importance in the bioenergetics of intertidal communities. Indeed, most ecological studies of marine and estuarine littoral diatoms are still in the descriptive, qualitative stage of scientific investigation, and the continuation of such studies in unsurveyed parts of the world is crucial for an understanding of general, worldwide distributional patterns. Unfortunately, financial support for survey studies is often difficult to obtain.

The fact that many of the same morphological taxa of marine and estuarine diatoms are found in different parts of the world does not mean that they are randomly distributed over various substrates in the intertidal region. The diatoms listed in Table 11.2 include taxa that are characteristically abundant at times on one or more of the substrates listed in Table 11.1, but are relatively rare on others. Furthermore, ecologists have found that intertidal marine and estuarine diatoms exhibit local responses to the spatial and temporal dynamics of chemical and physical environmental gradients. However, very little progress has been made toward an understanding of biological interactions among diatom taxa associated together in an assemblage and between the littoral diatom flora and other taxonomic groups

Table 11.1. A list of references to selected floristic studies of marine diatoms that grow attached to (or associated with) various living and nonliving substrates. Substrates investigated during each study are designated as S (sand), M (mud), P (macroalgae or angiosperms), W (wood), R (rock, concrete, glass, or plastic), and N (not indicated).

Geographical region	*Reference*	*Substrate(s) investigated*
Africa	1. Cholnoky, B.J. (1963)	N
	2. Cholnoky, B.J. (1968)	M, P, R
	3. Giffen, M.H. (1963)	N
	4. Giffen, M.H. (1967)	N
	5. Giffen, M.H. (1970a)	P
	6. Giffen, M.H. (1970b)	M, P, S
	7. Giffen, M.H. (1971)	P
	8. Giffen, M.H. (1973)	P, S
Arctic	1. Cleve, P.T. & Grunow, A. (1880)	M
Asia	1. Meister, F. (1932)	S, M
British Isles	1. Aleem, A.A. (1949)	M, R
	2. Aleem, A.A. (1950a)	P, R
	3. Aleem, A.A. (1950b)	M
	4. Ghazzawi, F.M. (1933)	S
	5. Hendey, N.I. (1951)	R
	*6. Hendey, N.I. (1974)	N
	7. Hustedt, F. & Aleem, A.A. (1951)	M
	8. Smyth, J.C. (1955)	M, R
Caribbean	1. Hagelstein, R. (1938)	W, R, P, M
Europe	1. Aleem, A.A. (1973)	P, M
	2. Brockmann (1950)	S, M
	3. Castenholz, R.W. (1967)	P, R
	4. Edsbagge, H. (1965)	S, M, P, R
	5. Edsbagge, H. (1966a)	N
	6. Edsbagge, H. (1966b)	P
	7. Gemeinhardt, K. (1935)	M, P
	8. Hustedt, F. (1939)	S, M, P
	9. Simonsen, R. (1962)	S, M
	10. Taasen, J.P. (1973)	P
Galapagos	1. Hendey, N.I. (1971)	†
North America	1. Castenholz, R.W. (1963)	R
	2. Drum, R.W. & Webber, E. (1966)	M, P, W, R
	3. Hustedt, F. (1955)	M
	4. Main, S.P. & McIntire, C.D. (1974)	P
	5. McIntire, C.D. & Overton, W.S. (1971)	R
	6. Phifer, L.D. (1929)	S, R
	7. Riznyk, R.Z. (1973)	S, M
	8. Wood, E.J.F. (1963)	M

* Check list of British Marine diatoms.

† Collected from foam washed up by surf at Punta Espinosa.

of plants and animals. The principal mission of field-oriented diatom research in the littoral zone of coastal and estuarine regions therefore is to determine the extent to which various environmental factors influence the vertical, horizontal, and seasonal distribution and abundance of the intertidal flora, usually with an emphasis on the epilithic, epiphytic, epipelic, and epipsammic growth forms.

(a) *Epilithic assemblages**

Investigations of littoral diatom assemblages associated with non-living substrates other than sediment are exemplified by the work of Aleem (1950a) at Swanage (Dorset). In this study, substrates under investigation included concrete surfaces, large boulders, and the pebbles in intertidal pools. The assemblages growing on concrete exhibited a vertical zonation that was subject to some seasonal modification by climatic conditions. *Achnanthes brevipes* and associated blue-green algae were prominent during winter and early summer in the upper intertidal region including the spray zone. This taxon and other closely related species of *Achnanthes* are relatively tolerant of intertidal exposure to desiccation (e.g. see Castenholz, 1963; McIntire & Reimer, 1974). Below the mean high water level, the diatom flora consisted primarily of *Fragilaria striatula, F. hyalina, Licmophora gracilis, Melosira moniliformis, Opephora marina,* and *Synedra tabulata.* Maximum growth occurred during the early spring, but the assemblage as a whole was adversely affected by periods of hot weather, disappearing from the littoral region during the summer. In the tide-pools, taxa tolerant of brackish-water included *Amphipleura rutilans, Navicula cryptocephala,* varieties of *Nitzschia frustulum,* and *Stauroneis constricta*; while the common marine littoral taxa were *Achnanthes brevipes, Fragilaria striatula, Melosira moniliformis, Opephora marina, Navicula (Schizonema) ramosissima, Navicula (Schizonema) grevillei,* and *Synedra tabulata.* On reefs and boulders, assemblages dominated by *Grammatophora marina* and *Navicula (Schizonema) ramosissima* were found in addition to the *Achnanthes*–blue-green and *Melosira–Fragilaria* assemblages. *Grammatophora marina* was associated with *Cladophora rupestris,* particularly during the spring, and was restricted to the lower littoral region and regions protected from high light intensities; this assemblage also included other diatoms, namely *Cocconeis scutellum, Grammatophora serpentina, G. oceanica* var. *macilenta, Fragilaria hyalina, Opephora marina, Rhoicosphenia marina, Synedra tabulata,* and *Striatella unipunctata.* Assemblages dominated by *Navicula ramosissima* extended over a wide tidal range and were found on small rocks or on the vertical sides of reefs during the winter months.

* This discussion includes diatom assemblages associated with nonliving substrates other than sediment.

Table 11:2. A list of selected cosmopolitan non-planktonic diatoms that have been reported for samples from marine and brackish-water. The numbers under each geographical location correspond to the references to floristic studies listed in Table 11.1.

	Africa	*Arctic*	*Asia*	*British Isles*	*Caribbean*	*Galapagos*	*Europe*	*North America*
Achnanthes brevipes Ag.	5, 8, 1, 6, 7			6, 2, 1, 5, 4, 7	1	1	1, 3, 6, 2, 8, 7	1, 8, 6, 3, 7
Achnanthes brevipes var. *intermedia* Kütz.	5, 8, 1, 6, 7			6, 1			5, 1, 6, 9, 7	5, 4
Achnanthes groenlandica Cl.	8			6		1	5, 1, 6	
Achnanthes hauckiana Grun.	2			6, 7		1	5, 1, 6, 2, 8	8, 2, 3, 5
Achnanthes longipes Ag.	1		1	6, 7, 1, 5, 4	1		5, 9, 1, 3, 6, 10, 2, 8, 7	8, 6, 2, 3
Achnanthes parvula Kütz.					1	1	5, 6	5, 4
Actinoptychus senarius Ehr.	5, 8, 2, 1, 6		1	6, 7, 1, 5, 4	1		1, 8, 7	8, 6, 2, 7
Amphipleura rutilans (Trent.) Cl.	3, 8, 2, 1			6, 2, 7, 1, 5, 4	1		5, 8, 1, 4, 3, 6, 10, 9, 2	1, 2, 3, 7, 4
Amphiprora alata (Ehr.) Kütz.			1	6, 4	1		1, 2, 8	6, 2, 3, 7
Amphora angusta Greg.	8, 2, 1, 6, 7			6, 7	1		5, 1, 6, 8	8, 7, 5
Amphora coffaeiformis (Ag.) Kütz.	2	1		6, 7, 5, 8, 4	1		1, 6, 8	8, 2, 3
Amphora exigua Greg.	3, 8, 2, 1, 6, 7			6, 1, 5	1		5, 1, 6, 2, 8	2
Amphora proteus Greg.	8, 6, 7	1		6, 7, 8	1	1	5, 1, 6, 2, 8, 7	8, 3, 7
Bacillaria paxillifer (Müll.) Hendey	2, 1			6			3, 2, 8	8, 2, 3, 7, 5
Biddulphia aurita (Lyngb.) Bréb.	5, 3, 2, 6	1	1	7, 6, 2, 1, 3, 5, 4	1	1	5, 1, 4, 3, 6, 8, 7	8, 6, 2, 3, 7, 5
Caloneis brevis (Greg.) Cl.	4			6, 5			1, 2, 8, 7	2, 3, 7
Campylosira cymbelliformis (Schm.) Grun. *ex* V.H.	7			6, 2, 1, 5				8, 3, 7
Cocconeis scutellum Ehr.	5, 3, 8, 1, 7	1		6, 2, 7, 1, 5, 8, 4	1	1	5, 2, 1, 4, 10, 3, 6, 8, 9, 7	8, 6, 2, 3, 7, 4

Cocconeis scutellum var. *parva* Grun. *ex* Cl.	3, 8, 2, 1,7			6, 1	1	1	5, 7	5, 4
Cocconeis scutellum var. *Stauroneiformis* V.H.	5, 8, 7			6, 7		1	5, 1, 4, 7	8
Coscinodiscus nitidus Greg.	8, 6		1	6, 4	1		8	3, 7
Cymbella pusilla Grun.	2, 1, 6		1				8	
Dimerogramma minor (Greg.) Ralfs *ex* Pritch.				6, 7, 5	1		5, 1, 6	8, 7
Diploneis bombus Ehr.	5, 4, 2, 1, 6			6, 1, 5, 7	1		1, 2, 8, 7	8, 3, 7
Eunotogramma marinum (W. Sm.) Perag.	1			6				3
Fragilaria hyalina Kütz.) Grun.				6, 2, 1	1		5, 1, 4, 3, 6, 10	
Fragilaria striatula Lyngb.				6, 2, 1, 4			3, 10, 7	6
Fragilaria striatula var. *californica* Grun.								1, 5, 4
Grammatophora angulosa Ehr.	5, 2, 1, 7			6, 4		1	5, 6, 8, 7	7
Grammatophora arcuata Ehr.	7	1					5, 6, 7	
Grammatophora marina (Lyngb.) Kütz.	8, 2, 1, 6, 7			6, 2, 7, 1, 4	1	1	5, 1, 4, 3, 6, 10, 2, 8, 7	8, 2, 3, 7
Grammatophora oceanica Ehr.	2		1	6, 2, 1		1	5, 4, 6, 2	8
Grammatophora oceanica var. *macilenta* (W. Sm.) Grun.	2			6, 2, 1	1		5, 1, 4, 7	3
Grammatophora serpentina Ehr.	5, 1			6, 2, 1			5, 1, 4, 6	2
Gyrosigma balticum (Ehr.) Rabh.	6		1	6, 2, 7, 1, 3, 5, 4	1	1	1, 2, 8, 7	8, 6, 3, 7
Gyrosigma fasciola (Ehr.) Cl.	6	1	1	6, 7, 1, 4	1		1, 2, 8, 7	8, 2, 3, 7, 5
Gyrosigma spencerii (W. Sm.) Cl.	6, 7		1	6, 7, 4	1		1, 2, 8, 7	2, 3
Hantzschia marina (Donk.) Grun. *ex* Cl. & Grun.	3, 7			6	1		2, 8	8, 7
Hantzschia virgata (Roper) Grun.	4, 2, 1, 6			6, 4	1		2, 8	7
Hyalodiscus radiatus (O'Meara) Grun.	5			6, 2, 1		1	1	

Table 11.2—(cont.)

	Africa	Arctic	Asia	British Isles	Caribbean	Galapagos	Europe	North America
Hyalodiscus scoticus (Kütz.) Grun.				6, 7	1		5, 1, 6, 9, 7	
Hyalodiscus subtilis Bail.				6	1		5, 6, 7	
Isthmia enervis Ehr.	7			6	1		5, 4, 6	
Isthmia nervosa Kütz.				6	1			7
Licmophora ehrenbergi (Kütz.) Grun.	8, 7			6, 2, 7	1	1	5, 4, 6, 10	1, 3, 7
Licmorphora flabellata (Carm.) Ag.			1	6, 2, 7, 1, 5, 4	1	1	5, 3, 6, 10, 7	1, 8
Licmophora gracilis (Ehr.) Grun.	5, 7			6, 2, 1, 5, 4, 7	1		5, 1, 6, 2, 7	1, 2, 7, 5, 4
Licmophora juergensii Ag.	5, 7	1		6, 2, 7, 1, 4			5, 1, 4, 6	
Licmophora paradoxa (Lyngb.) Ag.				6, 7, 5	1		5, 1, 3, 6, 7	1
Mastogloia exigua Lewis	2, 1, 6			6	1		5, 1, 6, 2, 8	2, 7
Melosira jurgensii Ag.				6	1		5, 6, 8, 7	2
Melosira moniliformis (Müll.) Ag.	2, 1, 6			6, 2, 1, 5, 7			5, 7, 1, 3, 6, 8, 9	6, 7, 5, 4
Melosira nummuloides (Dillw.) Ag.	5, 8, 2, 1, 6, 7	1		6, 1, 5, 4	1		5, 7, 1, 4, 3, 6, 8, 9	1, 8, 6, 2, 3, 5, 4
Melosira sulcata (Ehr.) Kütz.	2, 6, 7		1	2, 7, 1, 5, 8, 4	1		5, 3, 6, 8, 7	8, 3, 7, 5
Navicula arenaria Donk.	2			7	1		1, 2, 8	2, 3
Navicula cancellata Donk.	5, 8, 6, 7	1		6, 7, 1, 3, 7	1	1	1, 2, 8, 7	6, 3, 7
Navicula cincta Ehr.	2			6	1		5, 1, 6, 2, 8	9
Navicula comoides (Ag.) Perag.	8, 6, 7			6			5, 4, 6	3, 7, 5
Navicula complanatula Grun. *ex* Cl. & Grun.	5, 8, 2, 6	1		6			5, 3, 6	3, 5
Navicula crucigera (W. Sm.) Cl.				6, 7, 5, 4			5, 7, 1, 8, 4, 3, 6, 2	2
Navicula cryptocephala Kütz.	2, 1, 6			6, 2, 1, 5, 7	1		1, 2, 8, 7	2, 5
Navicula cryptocephala var. *veneta* (Kütz.) Cl.				6, 7, 1, 7	1		1, 7	5

Navicula digito-radiata (Greg.) Ralfs	1	1		6, 7, 1, 3		1	1, 2, 8, 7	6, 3, 7
Navicula directa (W. Sm.) Cl.	1, 6, 7	1		6	1	1	5, 3, 6, 9, 7	1, 8, 5, 4
Navicula diserta Hust.	2			6			2, 8	3, 5, 4
Navicula finmarchica (Cl. & Grun.) Cl.		1		6			1, 2, 8	2, 7
Navicula forcipata Grev.	2, 6			6, 7, 5		1	1, 2, 8, 7	8, 2, 3
Navicula forcipata var. *densestriata* Schm.			1	6, 5	1		1	7
Navicula gracilis Ehr.			1		1		1, 2, 8	
Navicula gregaria Donk.	2, 1, 6			6, 7, 8			1, 6, 2, 8	3, 5, 4
Navicula grevillei (Ag.) Cl.	3, 1, 6, 7		1	6, 2, 1, 5, 4, 7	1, 9		5, 1, 4, 3, 6	1, 8, 3, 7
Navicula lyra Ehr.	6	1		6, 1, 5, 4, 7	1	1	1, 2, 7	3
Navicula mutica Kütz.	2, 1, 6, 7			6, 5	1		1, 2, 8	2, 3, 5, 4
Navicula palpebralis Bréb. *ex* W. Sm.				6, 7, 1			1, 7	3, 7
Navicula ramosissima (Ag.) Cl.	2			6, 2, 7, 1, 5, 8	1		1, 3, 10, 9	1
Navicula ramossima var. *mollis* (W. Smith) Aleem	8			7, 1			1	
Navicula salinarum Grun.	6	1		6			1, 2, 8	3, 5
Navicula salinicola Hust.	3, 1, 6, 7			6, 7			1, 8	
Navicula scopulorum Bréb. *ex* Kütz.	6, 7		1	6, 5	1		2, 8	2, 3
Navicula viridula var. *avenacea* (Bréb.) V.H.				1				5
Nitzschia acuminata (W. Sm.) Cl. *ex* Cl. & Grun.	8, 6			6, 7, 1, 5	1		1, 2, 8	2, 3, 7
Nitzschia angularis W. Sm.	5, 2, 1	1		6, 7, 1, 4	1		1, 7	6, 7, 5
Nitzschia apiculata (Greg.) Grun. *ex* Cl. & Grun.	2, 1, 6			6, 7, 1, 5	1	1	1, 2, 8	2, 3, 7
Nitzschia closterium (Ehr.) W. Sm.	8, 6	1	1	2, 7, 1, 5, 3, 8, 4	1		1, 3, 10, 2, 8, 7	6, 2, 3
Nitzschia frustulum (Kütz.) Grun. *ex* Cl. & Grun. (and varieties)	8, 2, 1, 6, 7		1	6, 2, 1	1		1, 2, 8	1, 2, 5, 4

Table 11.2—(*cont.*)

	Africa	*Arctic*	*Asia*	*British Isles*	*Caribbean*	*Galapagos*	*Europe*	*North America*
Nitzschia longissima (Bréb.) Ralfs *ex* Pritch.	8, 1, 6		1	6, 7, 1, 5	1	1	1, 3, 7	8, 3
Nitzschia punctata (W. Sm.) Grun.	2, 6			6, 7, 5	1		1, 2, 7	8, 7
Nitzschia sigma (Kütz.) W. Sm.	5, 8, 2, 1, 6, 7		1	6, 4, 2, 7, 1, 3, 5, 8	1		1, 2, 8, 7	8, 2, 3, 7, 5
Nitzschia socialis Ralfs	8, 6, 7	1		6, 7	1	1	1, 8, 7	3, 7, 5
Opephora marina (Greg.) Petit				6, 2, 7, 1			8, 5, 1, 4, 10, 3, 6, 2	3, 7
Opehora pacifica (Grun.) Petit	7, 5, 8, 4, 2, 1, 6			6	1	1	5, 6, 2, 8, 7	8, 3
Plagiogramma staurophorum (Greg.) Heib.				6, 1	1		5, 1, 6, 8	7
Plagiogramma vanheurckii Grun.	7			6			1, 8	5, 4
Pleurosigma aestuarii (Bréb. *ex* Kütz.) W. Sm.	6		1	6, 7, 1, 3			1, 2, 8	3
Pleurosigma angulatum (Quekett) W. Sm.			1	6, 1, 5, 8, 4, 7	1		2, 8, 7	8, 2
Pleurosigma normanii Ralfs	8, 2, 1, 6			6, 7	1	1	1	3
Pleusosigma strigosum W. Sm.	8	1		6, 5	1		2	2, 3
Rhabdonema arcuatum (Lyngb.) Kütz.	5, 8	1		6, 2, 1, 5, 4			7, 5, 1, 4, 3, 6, 10	3, 7
Rhaphoneis amphiceros Ehr.	2, 1		1	6, 2, 7, 1, 5	1		5, 6, 2, 8	8, 3, 7
Rhaphoneis surirella Ehr. (Grun.)	1, 6, 7			7, 1	1		5, 6, 2, 8	3
Rhoicosphenia curvata (Kütz.) Grun.	1		1	6			5, 1, 10, 3, 6, 8, 7	7, 5
Rhoicosphenia marina (W. Sm.) Schm.				6, 2, 1			5, 1, 4, 10, 6	

Rhopalodia gibberula Kütz.	2, 1, 6		1	6	1		5, 6, 8	8, 6, 2, 3
Rhopalodia musculus (Kütz.) Mull.	2, 1, 6			6, 8, 7	1		5, 1, 4, 6, 2, 8	8, 3, 7
Scoliopleura latestriata Bréb.				1, 3	1		1, 2	
Scoliopleura tumida (Bréb. *ex* Kütz.) Rab.		1	1	6, 7, 1, 5	1		1, 2, 8, 7	
Stauroneis constricta (W. Sm.) Cl.	5, 6			6, 2, 1, 5	1		1	2
Striatella delicatula (Kütz.) Grun. *ex* V.H.	2, 1, 7			6, 5	1		5, 3, 6, 7	
Surirella gemma (Ehr.) Kütz.	1, 6, 7		1	6, 7, 1, 5, 4	1		1	7
Surirella ovalis Bréb.	7, 5, 2, 6			6, 7			1, 2, 8	3
Synedra gailloni (Bory) Ehr.	7			6, 2, 7, 1, 5			5, 8, 1, 4, 3, 6, 10, 7	2, 3
Synedra tabulata (Ag.) Kütz.	5, 8, 1, 6, 7	1	1	2, 7, 1, 5, 4	1		2, 8, 9, 5, 1, 4, 10, 3, 6	1, 2, 3, 7, 5, 4
Thalassionema nitzschioides Hust.	1			6, 7			1, 8	3, 7, 5, 4
Trachyneis aspera (Ehr.) Cl.	1, 5, 3, 8, 2, 6, 7			6, 2, 1, 5, 7	1	1	1, 3, 2, 7	8, 3, 7
Triceratium alternans Bail.				1, 7			5, 6, 8, 7	3
Triceratium favus Ehr.	8		1	6		1	6	3
Tropidoneis vitrae (W. Sm.) Cl.	6		1	6, 7, 1, 3	1		1, 2, 8	3, 7

Epilithic diatom assemblages similar in species composition to those found at Swanage, Dorset, UK have been reported by other investigators. For example, Castenholz (1963) described diatom assemblages on the concrete base of a footbridge near Coos Bay, Oregon (USA). These assemblages also incuded *Achnanthes brevipes, Amphipleura rutilans, Fragilaria striatula* (var. *californica*), species of *Licmophora, Navicula grevillei, N. ramosissima,* and *Synedra tabulata.* In this study, *Achnanthes brevipes* again appeared in greatest numbers near the upper limit of the diatom cover, while *Fragilaria striatula* often was associated with another species of *Melosira (M. nummuloides)* in the upper intertidal region at levels exposed to desiccation for periods up to about 2·5 hours. A similar study by Castenholz (1967) on the western coast of Norway revealed epilithic assemblages consisting of these same species as well as certain other taxa also reported by Aleem for Swanage (e.g. *Grammatophora marina, Cocconeis scutellum, Opephora marina, Rhabdonema arcuatum,* and *Synedra gaillonii*).

(b) *Epiphytic assemblages*

Visual examination of diatom assemblages associated with large seaweeds and marine angiosperms reveals a great diversity of life form. Diatoms attached to their hosts by mucilaginous pads or stalks often include various species and varieties of *Cocconeis, Licmophora, Achnanthes, Melosira,* and *Synedra* (Fig. 11.1). In addition, tube-forming species of *Navicula* and *Amphipleura rutilans* and species of *Navicula, Nitzschia,* and *Amphora* that form masses of cells within a gelatinous matrix can also be associated with host macrophytes (Main & McIntire, 1974). Other diatoms in these assemblages include large motile taxa and planktonic species that are entrapped in the surface film or entangled among the attached filaments.

One of the most fascinating ecological problems relating to distributional patterns in the littoral diatom flora is the question of host–epiphyte (or endophyte) specificity. While it is intuitively tempting to accept the hypothesis that at least some epiphytic diatoms have coevolved with specific host plants, there is no convincing quantitative evidence that there are any obligate epiphytic diatoms or that a particular epiphytic taxon can survive only when associated with a particular species of host plant. Perhaps, the most interesting case of diatom–host plant specificity is *Navicula endophytica* Halse, an endophyte found in the receptacles of *Ascophyllum nodosum* (Hasle, 1968), *Fucus vesiculosus* and *F. serratus* (Tassen, 1972), and *Fucus evanescens* (Main & McIntire, 1974). In this case, specificity seems to be for a taxon above the species level—the order Fucales. Apparently, *Navicula endophytica* is an obligate epiphyte–endophyte and may be unable to survive without an interaction with its host. Another endophytic diatom, *Navicula dumontiae* Baardseth & Taasen, has been reported from the internal mucilage of *Dumontia incrassata* (Baardseth & Taasen, 1973).

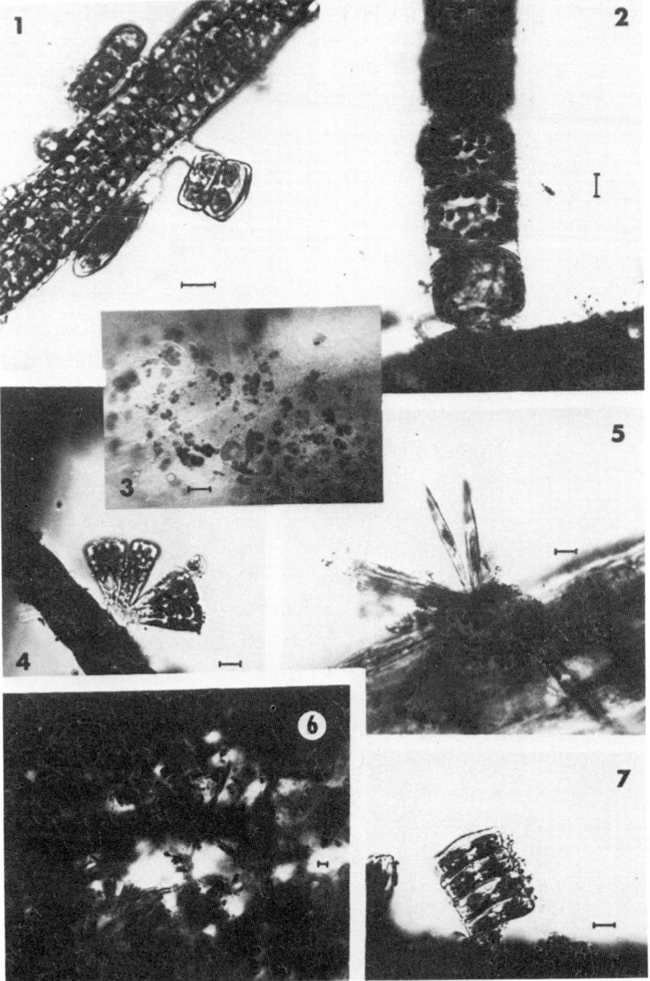

Fig. 11.1. Life forms of epiphytic diatoms. The line is equivalent to 10 μm. 1. *Achnanthes* sp. on *Enteromorpha* sp.; 2. *Melosira moniliformis* on *Polysiphonia paniculata.*; 3. *Cocconeis* sp. on *Polysiphonia* sp.; 4. *Licmophora* sp. on an epiphytic brown alga on *Fucus evanescens*; 5. *Synedra fasculata* on *Polysiphonia paniculata*; 6. Assemblage on *Zostera marina*; 7. *Achnanthes* sp. on *Enteromorpha intestinalis.*

While the existence of host–epiphyte specificity in the marine littoral diatom flora is questionable, there is no doubt that many diatoms occur in greater frequencies when associated with macroalgae and marine angiosperms than with non-living substrates (Edsbagge, 1966b; Aleem, 1950a; Hopkins, 1964a; Round, 1971; Main & McIntire, 1974). Furthermore, many so-called epilithic diatoms are, in fact, growing epiphytically on other diatoms which are attached to rock. Round (1971), from observations around British coastal regions, concluded that some algal macrophytes often exhibit a unialgal epiphytic diatom flora, frequently of the genera *Licmophora* and *Achnanthes*. Moreover, he suggested that large mucilaginous algae of the genera *Pelvetia, Fucus, Ascophyllum, Chorda, Laminaria, Himanthalia, Sargassum, Rhodymenia, Chondrus,* and *Gigartina* have relatively few epiphytic diatoms, whereas such filamentous genera as *Cladophora, Polysiphonia, Ceramium,* and *Ectocarpus* support much denser assemblage of epiphytes. Carpenter (1970) found only 13 species (6 genera) of epiphytic diatoms on *Sargassum* from the western Sargasso Sea and concluded that there are fewer species of epiphytic diatoms on this floating alga than on many littoral marine plants. On the Oregon coast, *Rhodomela larix* often supports near unialgal populations of *Isthmia nervosa* Kütz., a large diatom that usually forms branching colonies on its host. However, the distribution of *Isthmia nervosa* is by no means confined to *Rhodomela*. Aleem (1950b) and Edsbagge (1966c) ranked various seaweeds according to their apparent relative importance as host plants for epiphytic diatoms. Aleem's ranking considers members of the Chlorophyceae and Rhodophyceae as more important hosts than representives of the Phaeophyceae, while Edsbagge ranks the Phaeophyceae and Rhodophyceae ahead of the Chlorophyceae. The conclusions of Aleem correspond more to Round's observations than those of Edsbagge, differences are likely to be related to geographical location and sampling strategy.

Epiphytic assemblages of diatoms have been considered from essentially two points of view: (1) as a mosaic of more or less discrete, well-defined associations that are readily amenable to various sociological classification systems (e.g. see Braun-Blanquet, 1928); or (2) as a system of overlapping populations distributed along physical and chemical environment gradients, but modified—sometimes considerably—by local differences or discontinuities in the substrate quality and by species interactions with other marine organisms. While we essentially subscribe to the latter point of view (McIntire & Overton, 1971; McIntire, 1973b; Main & McIntire, 1974), it is recognized that environmental heterogeneity and discontinuities in chemical and physical gradients sometimes can justify an organismic or sociological approach to the analysis of distributional patterns. Hopefully, diatomists are not evolving toward the same dichotomy of thinking that has plagued the field of terrestrial plant ecology during the past 50 years.

Edsbagge (1966b,c), a leading proponent of the Braun-Blanquet sociological system for the classification of epiphytic diatom assemblages, classified the epiphytic flora on *Zostera marina* as the *'Cocconeis-Typus'*, citing *C. pediculus, C. placentula, C. placentula* var. *euglypta,* and *C. scutellum* as dominant taxa on this host; other assemblages also were classified according to the dominant genus (e.g. the *Licmophora*-type association on *Chordaria flagelliformis*). Main & McIntire (1974) also found *Cocconeis scutellum* associated with *Zostera marina* but concluded that the relationship was probably the result of similar responses to the physical environment rather than a specific, obligate biochemical interaction between epiphyte and host. This study also demonstrated that the structure of diatom assemblages on adjacent host macrophytes of the same species often can differ as greatly as the community structures of assemblages from nearby host macrophytes of different species. A comparison of this work with an earlier study of epilithic diatoms by McIntire & Overton (1971) indicated that the epiphytic and epilithic floras were similar in species composition, although *Cocconeis scutellum, C. scutellum* var. *parva, Achnanthes parvula, A. brevipes* var. *intermedia,* and *Licmophora gracilis* occurred in greater relative frequencies as epiphytes.

The relationship between the epiphytic diatom flora and such factors as temperature, salinity, light, tidal cycle, and geographical position on the western coast of Sweden has been discussed by Edsbagge (1966d). In this region, the majority of non-planktonic diatoms grow at temperatures below 12°C, but certain taxa apparently exhibit growth maxima that correspond to temperature changes. Taxa exhibiting their maximum abundances during the winter season include *Biddulphia aurita, Melosira moniliformis, Rhabdonema arcuatum,* and *Cocconeis scutellum* var. *stauroneiformis*. However, these taxa have been collected in Yaquina Estuary (Oregon) during the summer at temperatures as high as 15°C (McIntire & Overton, 1971). Edsbagge also found that the majority of epiphytic diatoms on the coast of Sweden could be classified as Meioeuryhaline Polyhalobous (salinity: 35 to ca. 30–17‰) according to the system of Simonsen (1962). Species requiring salinities greater than 30‰ (Oligoeuryhaline Polyhalobous) are *Isthmia enervis* and *Synedra hennedyana* Greg., while freshwater taxa that are able to tolerate salinities up to 30‰ (Holoeuryhaline Oligohalobous) include *Cocconeis placentula* Ehr. and *Rhoicosphenia curvata.* Taxa tolerant of high light intensities are *Achnanthes brevipes* var. *intermedia* and certain species of *Licmophora,* and *Achnanthes brevipes* and *Melosira moniliformis* are taxa apparently requiring periods of intertidal exposure to the air for optimum growth. Edsbagge classifies the epiphytic flora according to geographical position as: (1) arctic species (e.g. *Grammatophora arcuata* and *Achnanthes groenlandica*); (2) northern species, found from the North Sea to the Arctic Ocean (e.g. *Melosira moniliformis, Rhabdonema arcuatum, R. minutum, Cocconeis scutellum* var. *stauroneiformis,* and *Navicula grevillei*); and (3) southern species occurring below the Baltic current (e.g. *Rhabdonema adriaticum* Kütz., *Rhaphoneis surirella* var. *australis* (Ehr.) Grun., *Synedra baculus* Greg., *S. hennedyana, Campyloneis grevillei* (W. Sm.) Grun., and *Cocconeis quarnerensis* Grun.).

(c) *Epipsammic and epipelic assemblages*

The diatom floras associated with intertidal sediments usually consist of a diversity of taxa, many of which, because of their small size, present the investigator with troublesome taxonomic problems. Some difficulties associated with the analysis of sediment-associated assemblages include (1) the separation of diatoms from the sediment for slide preparation; (2) the determination of the degree of contamination from deposits of epilithic, epiphytic, and planktonic diatoms; and (3) the problem of separating living and dead cells. Furthermore, sometimes it is desirable to estimate the relative importance of epipsammic and epipelic forms in the flora, although it is not always easy to clearly discriminate between these growth forms. In actual practice, separation is arbitrarily based on the fractions derived from a particular sampling procedure. Examples of such procedures have been described by Hickman (1969) and Hickman & Round (1970). The selection of a sampling strategy for the analysis of epipelic and epipsammic diatom floras usually involves the examination of the trade-offs between complex procedures for a detailed analysis at a few selected locations and the relatively simple procedures that are more compatible with distributional studies over broad geographical regions.

In a study of the epipsammon in freshwater habitats, Round (1965) found that the diatom flora was composed of non-motile taxa usually less than 20 μm in length, and often belonging to the Araphidineae (Fragilariales) or Monoraphidineae (Achnanthales). These generalizations also tend to hold for marine habitats. For example, Hustedt (1939) studied the diatom flora associated with sediments along coastal regions of the North Sea, and taxa considered to be typically associated with sand ('Sandwatt-Diatomeen') included species of the genera *Rhaphoneis, Opephora, Plagiogramma, Dimerogramma, Cymatosira, Campylosira, Achnanthes, Cocconeis, Diploneis, Amphora,* and *Catenula*. Amspoker (1973), while investigating the diatom assemblages associated with beach sand near La Jolla, California (USA), found that the most abundant epipsammic diatoms were *Eunotogramma leve* Grun., *Dimerogramma hyalinum* Hust., *Navicula hummi* Hust., *Amphora sabyii* Salah, *Plagiogramma vanheurckii, Rhaphoneis grossepunctata* Hust., and unidentified species of *Achnanthes* (1), *Amphora* (1), *Cocconeis* (5), *Navicula* (2), and *Rhaphoneis* (2). Most of the taxa found in the lower intertidal zone also were collected in the upper intertidal zone. Moreover, a quantitative analysis of species–numbers data clearly indicated that there was no significant change in the relative abundance of these taxa along transects from the lower through the upper intertidal zones. The failure to find a conspicuous vertical zonation during this study is interesting and warrants further investigation. Perhaps, the homogeniety of the flora, in this case, was related to the instability of the beach sand.

Ghazzawi (1933) suggested that a minute sand grain can provide a surface area for the attachment of 'not less than 20 or 30 diatoms of various genera and species'. Meadows & Anderson (1966, 1968) have examined the sand grain as a microhabitat for diatoms and other microorganisms. These studies showed that microorganisms and organic material are always present on the surfaces of sand grains and that the organisms tend to be localized in patches between which there are large areas of bare surface. In general, these patches are found in the depressions and cracks of the grains. The study at Etterick Bay indicated that diatoms were very abundant on grains from the lower littoral zone—up to 2 600 cells/mm^2 of sand grain surface—but were less common on grains in samples collected from the upper littoral and the deep-water sublittoral zones. Furthermore, there is evidence that most of the living diatoms occur in the uppermost centimetre of sand (Phifer, 1929; Harper, 1969; Meadows & Anderson, 1966; Riznyk & Phinney, 1972a). However, viable diatom cells are often found at sediment depths of 10 cm or more (Meadows & Anderson, 1968; Riznyk & Phinney, 1972a,b), a phenomenon apparently related to the magnitude of sediment disturbance and perhaps to the ability of some diatoms to adopt a heterotrophic mode of nutrition (Lewin & Lewin, 1960; Lewin, 1953, 1963; Chapter 6) or lower their metabolic rates for extended periods (McIntyre *et al.* 1970). An experiment designed to determine the effect of abrasion on the colonization of sand grains indicated that the development of diatom colonies on the open surfaces of the grains was inhibited by shaking the sediment in flasks (Meadows & Anderson, 1968).

Harper (1969) has investigated the movement and migration of diatoms on sand grains and concluded that epipsammic diatoms tend to cease movement and remain in cracks and depressions of the grains. Again, there is the problem of clearly defining the difference between epipsammic and epipelic forms. Round (1965) and Harper (1969) have considered epipsammic diatoms to be frustules that remain on the sand grains after repeated washings. Along with the non-motile forms, this flora usually includes some small forms that exhibit limited movements and have relatively strong frustules. Harper (1969) also found that diatoms on sand grains in a freshwater lake migrated to and from the grains at the surface, and that such movements continued during the normal photoperiod or in continuous darkness. However, the diatoms did not leave the surface when exposed to a constant illumination intensity of 750 lux.

While epipsammic diatoms exhibit relatively little or no movement, the epipelic forms are usually larger motile diatoms with a well-developed raphe system. Round (1971) reviewed selected studies of intertidal epipelic diatoms (viz. Carter, 1932, 1933; Hustedt, 1939; Brockmann, 1935, 1950, Drum & Weber, 1966; Round, 1960) and concluded that the diversity of epipelic assemblages on sand is usually relatively low as compared to such assemblages associated with silt. However, epipelic species on sand tend to be larger in size than those found on silt. Brockmann (1950) identified distinct diatom associations on the Schleswig-Holstein coast and classified them according to their affinity for three sediment types: (1) clean sand (Sand), (2) sandy silt (Schlicksand), and (3) silt (Schlick). The assemblages on sand ('Das *Pinnularietum cruciformis*') were dominated by *Pinnularia cruciformis* Donk. and *Amphora laevissima* Greg., whereas *Navicula arenaria* Donk. was the most common diatom found on sandy silt ('Das *Naviculetum arenariae*'); the assemblages on fine sediment ('Das *Scoliopleuretum tumidae*') consisted of primarily *Scoliopleura tumida, Pleurosigma angulatum* var. *quadrata* (W. Sm.) V. H., *Nitzschia sigma, Gyrosigma spenceri,* and *Navicula cincta*. Studies in Yaquina Estuary (Oregon) have indicated that a plant sociological approach to community analysis may be more meaningful when applied to sediment-associated floras than when used in the attempt to classify epiphytic and epilithic diatoms into discrete assemblages. In this particular estuary, populations of epiphytic and epilithic diatoms exhibit overlapping distributions along the horizontal salinity and temperature gradients and along the vertical desiccation and insolation gradients (McIntire & Overton, 1971; McIntire, 1973b; Main & McIntire, 1974), but effects of such gradients on patterns in the epipsammic and epipelic floras apparently are greatly modified by the chemical and physical properties of the sediment (Amspoker, personal communication). Therefore, it is not unreasonable to consider the sediment-associated flora in this estuary as a patchy mosaic of assemblages, the species composition of which is determined in part by discontinuities in sediment type and in part by the more continuous gradients associated with the aquatic milieu. Of course, such generalizations are only interpretable within the context of a particular sampling strategy, i.e. the times and places of a series of collections.

Examples of floristic studies that included collections of epipelic diatoms are represented in Table 11.1 by substrate type M (mud). Aleem (1950b) described an epipelic assemblage from the mud-flats at Whitstable, Kent that consisted largely of *Pleurosigma balticum* W. Sm., *P. aestuarii, Stauroneis salina* W. Sm., *Scoliopleura latestriata, Navicula cancellata, N. digito-radiata, N. ammophila* Grun., *Tropidoneis vitrae, Nitzschia sigma, N. closterium,* and *Biddulphia aurita*. Epipelic assemblages consisting of many of these same or closely related taxa have been reported from Sweden (Aleem, 1973), Canvey Island in Essex (Carter, 1933), Oregon (Riznyk, 1973; Amspoker, personal communication), Germany (Brockmann, 1950; Hustedt, 1939), North Carolina, USA (Hustedt, 1955), and the Ouse Estuary in Sussex (Hopkins, 1964b). The last investigation is particularly interesting, as it included quantitative data (cell counts) obtained from summer and winter collections. In the Ouse Estuary, ecological factors considered to be important in determining the distribution of eleven species of mud-flat diatoms included: (1) the ability to tolerate desiccation near the mean high water level; (2) the ability to live under short periods of illumination near the mean low water level; and (3) the ability to tolerate organic materials present in the black sulphurous layer below the diatom cover. In this study, it was found that the top 2 mm of mud supported 85% of the diatom assemblage in the light and 80% in the dark. *Navicula ammophila* var. *flanatica* Grun. and *Pleurosigma balticum* were tolerant of sulphurous materials, and *N. ammophila* var. *flanatica, N. cyprinus* (Ehr.) Kütz., *N. cancellata, Stauroneis salina, Nitzschia closterium,* and *Triceratium favus* were able to survive at low illumination intensities; *Navicula cancellata, Tropidoneis vitrae, Pleurosigma angulatum, P. aestuarii,* and species of *Campylodiscus* were relatively tolerant of desiccation.

One of the most fascinating ecological properties of diatom assemblages on mud-flats is the periodicity associated with vertical movements of the epipelic forms. Physiological aspects of diatom movement have been reviewed by Hendey (1964), Round (1971), Nultsch (1974), and in Chapter 8 of this monograph. Aleem (1950b) observed a diurnal periodicity in the diatom assemblage on the mudflats at Whitstable. The assemblage appeared at the surface of the sediments at low tide if the light intensity was high enough; but if low tide occurred near sunset or during the night or if the day was exceptionally cloudy, the assemblage remained below the surface. In addition, it was found that the diatoms were concentrated in the upper 2 mm of sediment during both light and dark periods; a similar vertical distribution was reported by Hopkins (1963) and Palmer & Round (1965). Perkins (1960) concluded that the littoral diatoms of the River Eden estuary have a diurnal rhythm dependent only on light intensity and that these organisms remain at the surface as long as the incident light exceeds some minimum intensity. Therefore, it appears certain from the work cited above and from other field and laboratory studies (e.g. Palmer, 1960; Round & Palmer, 1966; Round, 1966; Palmer & Round, 1967) that light rather than tidal cycle is the most important factor controlling the vertical movements of epipelic diatoms. However, the process apparently is a complex combination of what Hopkins (1965, 1966) calls transient motility and repeated motility which involves both diurnal rhythms phased by day length conditioning and tidal rhythms that affect the moisture content of the sediment. Hopkins (1965) also found that cytological changes take place during the diurnal rhythm that are correlated in timing with the entrainment of either positive geotaxy with the negative phototaxy or negative geotaxy with the positive phototaxy.

2.2 *Trophic considerations*

(a) *Primary production*

Relatively little information is available on rates of primary production and respiration in assemblages of marine and estuarine benthic diatoms.

Methodology employed in the estimation of these rates usually involves either one of two approaches: (1) isolation of a sample in a bell jar or chamber and the subsequent monitoring of carbon-14 uptake or changes in the concentration of metabolic gases in the medium (e.g. see Pomeroy, 1959; Pamatmat, 1965, 1968, 1971; Pamatmat & Fenton, 1968; Pamatmat & Banse, 1969; McIntire & Wulff, 1969); and (2) the measurement of the rate of biomass accumulation on artificial substrates, usually glass microscope slides (Castenholz, 1960, 1961a, 1967; Kevern *et al.* 1966; King & Ball, 1966; Wetzel, 1964, 1965; Slâdecková, 1962). Both of these approaches attempt to measure the productivity of the entire assemblage—including microorganisms other than diatoms—and require certain assumptions that are usually violated to some degree depending on the particular situation.

The difficulty in partitioning primary production and respiration among the constituents of a periphyton assemblage is one of the most complex problems in productivity studies. Bair & Wetzel (1968) and Steele & Baird (1968) described a method for measuring carbon assimilation by epipsammic organisms, and the method of Hickmann (1969) is designed to separate epipelic and epipsammic algae for productivity measurements. Patten & Van Dyne (1968) presented a non-linear programming method—a statistical approach—for partitioning productivity among constituents of a phytoplankton assemblage that could also be applied to benthic algal assemblages. Another promising approach involves the use of carbon-14 autoradiography to determine relative productivity of individual cells in mixed algal assemblages (Maguire & Neill, 1971). This method is designed to estimate the relative contribution of each component taxon to primary production in the entire assemblage and to show size-specific production by cells in each of these taxa. Autoradiography also has been used to determine the site of nutrient sorption in periphyton assemblages (Rose & Cushing, 1970). To our knowledge, there is still no satisfactory method for partitioning community respiration into bacterial and producer respiration in assemblages of microorganisms.

The study by Steele & Baird (1968) indicated that rates of primary production on beach sand were relatively low, in the range of 4–9 g $C.m^{-2}.yr^{-1}$. There was an increase in chlorophyll and organic carbon with depth below the low-water level to 13 m, but decreases in light intensity with depth apparently accounted for a corresponding decrease in the ratio of ^{14}C uptake to the concentration of chlorophyll *a*. In any case, organic carbon content under 1 m^2 of beach sand to a depth of 20 cm was about 50 g. Furthermore, viable populations of diatoms were found to a depth of 20 cm at the low-water stations. This distribution of living organisms was attributed to mixing of the sand by wave action and stimulated speculation concerning metabolic rates of diatoms buried below the zone of effective light penetration for extended periods.

Other estimates of total primary production for assemblages of benthic microorganisms associated with sandy silt or mixtures of silt and clay are higher than those reported above for epipsammic assemblages alone. Grøntved (1960) investigated the productivity of the microbenthos in some Danish fjords and estimated that the average carbon fixation was 116 g $C.m^{-2}.yr^{-1}$. It also was found that average fixation was 142 mg $C.m^{-2}.(2\ hr)^{-1}$ for 87 samples from sand (mean water depth of 0·85 m) and 139 mg $C.m^{-2}.(2\ hr)^{-1}$ for 42 samples from sandy silt (mean depth of 1·11 m). After correcting the data for depth effects, it appeared that productivity was greater when the bottom material contained silt and clay than when the substrate was 'pure' sand. Moreover, maximum productivity was at water depths between 0·5 and 0·7 m, and rates at one locality (Naeraa Strand) for the period from November to February were about one-third of the rates found for the rest of the year. In another study, Grøntved (1962) found that photosynthetic potential was about four times higher on the exposed tidal flats in the Danish Wadden Sea than in the Danish fjords.

Gross primary production of microalgae in the intertidal marshes on the coast of Georgia (USA) was measured by the oxygen method with bell jars (Pomeroy, 1959). The annual rate of gross production was estimated to be 200 g $C.m^{-2}$; efficiency of photosynthesis ranged from 1 to 3% at light intensities less than 100 $kcal.m^{-2}.hr^{-1}$ and was 0·1% or less at intensities in excess of 300 $kcal.m^{-2}.hr^{-1}$. Pamatmat (1968) using similar methods found that the productivity on an intertidal sandflat on False Bay, San Juan Island (USA) was comparable to the Danish Wadden Sea and the salt marshes of Georgia. In this case, photosynthetic efficiency over the year averaged 0·10, 0·11, and 0·12% of total incident radiation at the three stations under investigation. Also, rates of photosynthesis exhibited an endogenous rhythm apparently related to tidal cycle; rates were relatively high during flood and ebb tide and depressed during low and high water.

Taylor & Gebelein (1966) investigated the vertical distribution of plant pigments in intertidal sediments at Barnstable Harbor, Massachusetts (USA). Highest concentrations of all pigments occurred in the upper 1 mm. Chlorophyll *a* and *c* and fucoxanthin concentrations decreased with depth and were 20 to 50% of surface values at 5 cm; diatoxanthin, diadinoxanthin, and carotene concentrations did not decrease with depth. In a related study, Taylor (1964) showed that 10% of the solar radiation penetrated to a depth of 1·5 mm in sand with a grain size between 63 and 177 μm in diameter; 1% reached a depth of 3 mm. Microalgae living on sediment in this area required only 12 g $cal.cm^{-2}.hr^{-1}$ to obtain their maximum photosynthetic rate and were able to photosynthesize at 90% of their maximum rate at a depth of 2 mm at noon on a clear day. Riznyk & Phinney (1972a,b) also investigated the vertical distribution of chlorophyll *a* and primary production in the intertidal sediments of Southbeach and Sally's Bend in Yaquina Estuary. The sandy silt of

Southbeach had an estimated annual gross primary production of 275–325 g $C.m^{-2}.yr^{-1}$, while the finer silt of Sally's Bend had estimated values of 0–125 g $C.m^{-2}.yr^{-1}$. These differences were attributed to the presence of large populations of bacteria and meiofauna in the fine, detritus-rich sediment of Sally's Bend. The greatest biomass of microalgae on both tidal flats was found in the upper 1 cm of sediment, but viable diatoms were found throughout the length of the piston core sample (9·1 cm in length). Chlorophyll *a* at Southbeach ranged from a mean concentration of 9·3 μg/cm^3 at a depth of 7·8–9·1 cm to 20·7 μg/cm^3 in the upper 1·3 cm; corresponding mean concentrations at Sally's Bend were 2·9 and 4·7 μg/cm^3, respectively.

Summarizing the productivity studies cited above, we can say that annual rates of gross primary production by microalgae on tidal flats frequently fall between 100 and 400 g $C.m^{-2}.yr^{-1}$, and values for the epipsammic assemblages associated with shifting beach sands may be as low as 10 g $C.m^{-2}.yr^{-1}$ or less. Organisms and plant pigments usually are concentrated in the upper few millimetres of sediment, but living cells are often found at depths of 10 cm or more. Vertical distribution apparently is related to the extent to which the sediment is mixed by water movements, although epipelic diatoms can exhibit a vertical migration in the upper few millimetres as noted earlier. There is some indirect evidence that primary productivity and the development of assemblages of autotrophic organisms may be inhibited to some extent when there is a large amount of organic detritus in the sediments. Bacterial activity in such sediments reduces the oxygen concentration, alters the pH, and perhaps generates compounds that are toxic to some diatoms.

(b) *Grazing*

Most data relating to grazing by marine and estuarine animals on benthic diatoms are simply records of occurrences of various taxa in the stomach contents of these animals. Literature concerned with such records has been reviewed by Round (1971) and includes the work of Baddeley (1858), Mare (1942), Crosby & Wood (1959), Takano (1960), and Hendey (1964). Hendey (1964) provides a list of 80 diatom taxa that were found in the stomach contents of sea urchins or representatives of the genera *Ostrea, Ciona, Crepidula, Mytilus,* and *Pectin*; many of these diatom taxa are attached or benthic forms.

Another approach to the investigation of grazing is to compare the accumulation of plant biomass in the absence of grazing with that at different grazing pressures. Castenholz (1961b) introduced known numbers and volumes of *Littorina scutulata* into bare, sterilized pools to measure their ability to maintain diatom-free areas. Results of these experiments indicated that a littorine volume greater than 0·2 cm^3/dm^2 (about 3 animals/dm^2) was enough to keep an area nearly free of diatoms for 2 weeks, while volumes less than this allowed diatom patches to develop. It was concluded that diatoms are the principal food of *Littorina scutulata* and most epilithic species of *Acmaea* in the littoral zone along the southern Oregon coast.

In order to understand the dynamics of grazing relationships, the bioenergetics of grazing as a process must be investigated. Unfortunately, in this respect information concerning marine littoral diatom assemblages and associated animals is lacking. The kind of information that is needed includes a knowledge of periphyton dynamics, e.g. rates of primary production and community respiration as a function of relevant environmental variables as well as estimates of food consumption, assimilation, and respiration for the grazer. Rates of food consumption are particularly difficult to measure directly, and Warren (1971) has suggested that the growth rate of a consumer can sometimes be expressed as a simple function of the density of its food organisms. This approach assumes a dynamic equilibrium between the consumer and its food. It is not particularly useful for the investigation of short-term seasonal dynamics when it is essential to monitor the effect of the consumer on its food supply.

Paine (1971) investigated energy flow in a natural population of the herbivorous gastropod *Tegula funebralis*. This animal is known to graze on diatoms, other microalgae, and species of macroalgae. However, in this particular case the dominant plants were the seaweeds *Endocladia muricata* and *Hedophyllum sessile*, and the impact of grazing on diatoms was not reported. In any case, *Tegula* consumed 1 071 kcal.m^{-2}.yr^{-1} and had an assimilation efficiency of about 70%. Literature relating to the grazing process in marine plankton has been reviewed by Parsons & Takahashi (1973) and in Chapter 12.

McIntire (1973a) developed a mathematical model of periphyton dynamics based on data obtained from laboratory streams and in the field. Output from the model indicated that it is bioenergetically feasible for a periphyton biomass of about 10 g/m^2 to support a consumer biomass of 150 g/m^2 or more if the productive capacity of the system is sufficient. Although parameter estimation was based on experimental work with freshwater organisms, many structural and behavioural attributes of this model are applicable to analogous marine systems. A modelling approach provides an analytical way of synthesizing the results of field and laboratory studies and helps gain insights into the dynamics of ecological systems that can not be obtained by intuition alone or by examining the results of individual experiments.

2.3 *A laboratory approach to ecological studies*

McIntire & Wulff (1969), Wulff & McIntire (1972), and Bergland (1973) studied the effects of illumination intensity, exposure period, salinity, and temperature on the vertical distribution and growth of estuarine benthic diatoms in a laboratory model ecosystem. This general approach was originally employed to investigate lotic communities in laboratory streams (McIntire *et al.* 1964; McIntire & Phinney, 1965). Examples of other laboratory marine ecosystems were described by Odum *et al.* (1963).

The laboratory system described by McIntire & Wulff (1969) consists of a fibreglassed wooden trough, 3 m long, 76 cm wide, and 80 cm deep, with the bottom graduated in a 'stair-step' manner. Tidal cycles are simulated by periodically pumping seawater into the system, and the illumination intensity is regulated by adjusting the height of a large lamp fixture over the trough. A respirometer chamber was designed to monitor changes in dissolved oxygen concentration in water surrounding a sample community. Such samples usually consist of assemblages of diatoms that develop in the laboratory ecosystem on acrylic plastic plates or other substrates of interest.

The diatom flora that developed in the laboratory ecosystem was similar to that found at field stations in lower Yaquina Bay. The vertical distribution of many taxa was closely related to illumination intensity and period of exposure to desiccation. For example, *Navicula directa, N. socialis* Greg., *Melosira nummuloides, Synedra fasciculata* (Ag.) Kütz. (=*S. tabulata*), and *Nitzschia subhybrida* Hust. were more abundant on the substrate with no exposure to desiccation than on substrates with exposure periods of 1 or 8 hr/day (McIntire & Wulff, 1969). *Nitzschia lanceolata* var. *minor* V. H. and *N. sigma* had highest relative abundances on the substrate exposed for 1 hr/day, whereas *Achnanthes brevipes* (=*A. yaquinensis* McIntire & Reimer) and *Thalassionema nitzschioides* were tolerant of an exposure period of 8 hr/day. In another experiment, vertical distribution also was closely related to illumination intensity in the absence of exposure to desiccation. *Cocconeis scutellum* var. *parva* represented 32·6% of the total number of cells on the lowest substrate (1 290 lux), but was less prominent on the higher steps. *Fragilaria striatula* var. *californica, Amphipleura rutilans,* and *Melosira nummuloides* had their highest relative abundances on the uppermost substrate at 11 300 lux, and *Bacillaria paxillifer* did best at 5 490 lux. A sudden, unseasonable decrease in salinity or increase in water temperature in one of the laboratory systems had a much greater effect on diatom assemblages exposed to 12 270 lux than on those that developed at either 1 030 or 4 710 lux; the most conspicuous changes in community structure were a decrease in species diversity and a rapid growth of a population of *Melosira nummuloides* (Wulff & McIntire, 1972).

The laboratory ecosystems also provide the opportunity to investigate rates of primary production and community respiration. Biomass accumulated most rapidly on plastic substrates subjected to relatively high illumination intensities without exposure to desiccation (McIntire & Wulff, 1969; Wulff & McIntire, 1972). In the absence of grazing, biomass ranged from 17·2 g/m^2 on a substrate exposed to the air for 8 hr/day to 128 g/m^2 on a substrate with no exposure; corresponding concentrations of chlorophyll *a* were 0·037 and 0·837 g/m^2, respectively. Primary productivity in diatom assemblages exposed to periods of desiccation was less under winter conditions than under corresponding conditions in the summer; and productivity in assemblages not exposed to desiccation was strongly affected by illumination intensity during both the summer and winter experiments. Rates of primary production at 18 490 lux ranged from about 0·08 to 1·00 g O_2.m^{-2}.hr^{-1} depending on the biomass and chlorophyll concentration; the assimilation ratio was greater during winter than summer.

Laboratory ecosystems provide the ecologist with a useful tool for examining simplified communities under controlled conditions, while retaining many of the salient features of natural ecosystems. In the work cited above, they represent an experimental approach to the examination of hypotheses related to productivity and structure of diatom assemblages in an estuary. Results of these experiments were consistent with patterns of the diatom flora observed during a concurrent field sampling program (McIntire & Overton, 1971) and with the outcome of culture studies of individual taxa by Castenholz (1964). The laboratory systems also can be used to investigate plant–substrate relationships and grazing.

3 QUANTITATIVE ANALYSIS OF DISTRIBUTIONAL PATTERNS

Notwithstanding the excellent floristic studies cited in Table 11.1, there is still much to be learned about the ecological properties of non-planktonic marine and estuarine diatoms. Obviously, physiological and biochemical studies of individual taxa and of host–epiphyte (or endophyte) interactions are needed. In the field, we recommend a more rigorous, quantitative approach to the study of distributional patterns. In this section, we present one such approach and provide a simple example from our work in Yaquina Estuary.

3.1 *The approach*

Our general approach to the quantitative analysis of distributional patterns involves: (1) estimation of community composition parameters; (2) computations of a similarity (or difference) measure for comparing the composition of two assemblages; (3) multiple correlation of similarity values with selected environmental (predictor) variables; and (4) standard multivariate statistical methods, i.e. clustering, principal components, canonical correlation, and discriminant analysis, for examining the distribution of selected taxa of interest.

Community composition parameters include a variety of diversity statistics and numerical expressions of evenness, equitability, redundancy, and dominance (see Pielou, 1969; Lloyd *et al.* 1968; Lloyd & Ghelardi, 1964; Simpson, 1949; Hurlbert, 1971; McIntosh, 1967). These statistics provide a convenient way of summarizing species-numbers relationships within an assemblage. However, we have not found estimates of such parameters particularly useful in our effort to understand spatial and temporal patterns in the diatom flora, although some investigators consider them valuable for monitoring water quality (e.g. Patrick, 1954; Wilhm & Dorris, 1968).

Similarity, difference, or distance measures are useful for comparing the structures of diatom assemblages distributed along environmental gradients. One measure which is related to Simpson's measure of concentration and analogous to a correlation coefficient is

$$SIMI\,(a,\,b) = \frac{\sum_{i=1}^{S} P_{ai}P_{bi}}{\sqrt{\sum_{i=1}^{S} P_{ai}^2} \cdot \sqrt{\sum_{i=1}^{S} P_{bi}^2}}$$

SIMI is a degree of similarity between assemblages *a* and *b*; P_{ai} and P_{bi} are the proportions of individuals represented by the *i*-th taxon in assemblages *a* and *b*, respectively; and *S* is the total number of taxa in the assemblages (Stander, 1970). Other measures of similarity, difference, and distance have been described by MacArthur (1965), McIntosh (1967), and Sokal & Sneath (1963). In some of our studies, we have constructed matrices of difference values comparing the composition of epilithic and epiphytic diatom assemblages along salinity and desiccation gradients in an estuary (McIntire & Overton, 1971; Main & McIntire, 1974). If these values are ordered in the proper sequence corresponding to the location of the assemblages along the gradient, the longest diagonal row of values can be examined for discontinuities. Furthermore, these values can be matched with corresponding differences in values for environmental properties for multiple correlation analysis. This approach provides a heuristic indication of groupings and relationships that can be examined in greater detail by experimentation in the field and laboratory. In our work, we were able to relate between 50 and 70% of the variability in a difference measure to corresponding differences in mean salinity, mean salinity range, and period of intertidal exposure.

After the structures of whole diatom assemblages have been investigated, it is sometimes desirable to analyze the distributional patterns of a few selected taxa. For this purpose we recommend a sequence of standard multivariate statistical methods. Theoretical aspects of these methods are presented in various textbooks (e.g. Anderson, 1958; Seal, 1966; Morrison, 1967; Cooley & Lohnes, 1971). Briefly, these procedures involve a linear transformation of a set of variables into a set of new uncorrelated variables. If there is a high degree of correlation within the original data set, the transformation concentrates a large proportion of the information (variance) in relatively few of the new variables (principal components or canonical variables), hopefully reducing the number of dimensions necessary for a meaningful interpretation of the data set. Although the mathematical theory of these methods is well known, their practical application in ecology, and particularly in phycology, is in an early stage of development and evaluation (see Allen, 1971; Allen & Skagen, 1973; Allen & Koonce, 1973; Levandowsky, 1972; Hughes *et al.* 1972; McIntire, 1973b; Cassie & Michael, 1968; Cassie, 1972a,b). The principal difficulty comes with trying to attach biological meaning to linear combinations of the original data.

A useful series of multivariate and correlation analyses was outlined by Cassie (1972a). This approach involves a sequence of statistical analyses on the composite matrix [**YX**], in which the elements of **Y** are measures of species abundance, and the elements of **X** are measures of environmental properties. A principal components analysis is performed on **Y** and **X** individually, followed by a correlation analysis between **X** and the principal components of **Y**. Next, a canonical correlation analysis of [**YX**] is performed along with a computation of the relative information content of the various environmental properties as they relate to the principal components of **Y**. The data can be examined further by considering each sample from the different stations as a point in p-dimensional space, where p is the number of species and the location of a point (sample) with respect to p orthogonal axes depends on the relative abundance of the p species at the different stations. A clustering algorithm (McIntire, 1973b) is then used to partition this cloud of points into discrete groups. If these groups can be given a meaningful interpretation in terms of what is known about the properties of the environment, the matrix **Y** is rearranged to correspond to the desired clusters. A stepwise discriminant analysis is performed on the reordered matrix to evaluate the integrity of the clusters. In biological terms, the discriminant analysis sometimes can provide insights into the nature of environmental gradients relative to a particular sampling strategy and group of species.

In general, we are pleased with the results of this approach and feel that it is a powerful tool for analyzing the distributional patterns in the diatom flora when the observations are species-numbers data. Moreover, we consider this sequence of multivariate analyses a valuable supplement to the total community structure analysis outlined above. The principal constraint is the number of variables and cases that the computing facility will accommodate.

3.2 *An example*

Examples of total community structure analysis were given by McIntire (1974). Here, we present a simple example of a multivariate analysis of the distributional patterns of selected taxa. The data files were derived from the work of McIntire & Reimer (1974) who studied the distribution, cell wall morphology, and taxonomy of eight marine and estuarine *Achnanthes* in Yaquina Estuary. For this analysis, we also included three freshwater taxa of *Achnanthes* that occur in the upper estuary and six environmental measures: mean period of intertidal exposure, light energy, mean salinity, mean salinity range, mean temperature, and mean temperature range. Three sets of samples were obtained along gradients from marine to freshwater during February, May, and August, 1969. Samples were removed from PVC plastic sampling devices (McIntire & Overton, 1971), and microscope slides of the material were prepared in the usual way (Hohn & Hellerman, 1963). A count of 500 diatoms was made for each sample, and the total number of samples was 54. The description of Yaquina Estuary and the locations of the sampling stations were summarized by McIntire & Overton (1971) and McIntire & Reimer (1974).

The species data matrix **Y** consists of 11 variables or taxa (columns), and the environmental matrix **X** has 6 variables (columns); both matrices have 54 samples or cases (rows). The variable names and corresponding acronyms are

Table 11.3. Variables and corresponding acronyms of **Y*** and **X**.

Variables of **Y**	*Acronym*	*Variables of* **X**	*Acronym*
Achnanthes cocconeioides	ACHCOC	Exposure period (h/day)	EXPO
Achnanthes kuwaitensis	ACHKUW	Light energy (ly/day)	LITE
Achnanthes groenlandica var. *phinneyi*	ACHGP	Mean salinity (‰)	SBAR
Achnanthes groenlandica var. *phinneyi* f. *jaydei*	ACHGPJ	Mean salinity range (‰)	SALR
Achnanthes yaquinensis	ACHYAQ	Mean temperature (°C)	TBAR
Achnanthes brevipes var. *intermedia*	ACHBRE	Mean temperature range (°C)	TEMR
Achnanthes javanica f. *subconstricta*	ACHJAV		
Achnanthes parvula	ACHPAR		
Achnanthes lanceolata	ACHLAN		
Achnanthes minutissima	ACHMIN		
Achnanthes deflexa	ACHDEF		

* Authorities for the variables of **Y** were reported by McIntire & Overton (1971) and McIntire & Reimer (1974).

presented in Table 11.3. The elements of **Y** and **X** (y_{ij} and x_{ik}, respectively) were found from transformations of raw data, where

$$Y_{ij} = \ln (y^*_{ij} + 1)$$

$$y_{ij} = (Y_{ij} - \bar{Y}_j)/s_j$$

$$x_{ik} = (X_{ik} - \bar{X}_k)/s_k$$

y^*_{ij} is the count of the j-th taxon in the i-th sample; X_{ik} is the measurement of the k-the environmental variable corresponding to the i-th sample; and $\bar{Y}_j$ and $\bar{X}_k$ are the means and s_j and s_k are the standard deviations corresponding to Y_{ij} and X_{ik}, respectively. Cassie & Michael (1968) have discussed the first transformation, while the last two standardize the variables of **Y** and **X**.

Correlations between the first two principal components of **Y** and each of the original variables (taxa) are presented in Table 11.4. These correlations, sometimes called factor loadings, help in the interpretation of linear combinations of the original data set. In this case, the first two principal components account for about 60% of the total variability within **Y**. In the first component, the freshwater taxa (*Achnanthes lanceolata, A. minutissima,* and *A. deflexa*) have high negative loadings, and the marine taxa (*A. cocconeioides, A. groenlandica* var. *phinneyi* and f. *jaydei, A. yaquinensis,* and *A. parvula*) have relatively high positive loadings. The brackish-water taxa (*A. kuwaitensis, A. brevipes* var. *intermedia,* and *A. javenica* f. *subconstricta*) have low correlations with this component. The first component therefore constrasts the

Table 11.4. Factor loading coefficients for the first two principal components, and multiple correlation coefficients relating each taxon to the six environmental variables (R_{yx}) and the first two principal components to the six environmental variables (R_{zx}).

	Factor loadings		
Taxon	*1*	*2*	R_{yx}
ACHCOC	0·7598	−0·3768	0·7793
ACHKUW	0·3215	0·5563	0·4823
ACHGP	0·4997	−0·3318	0·6352
ACHGPJ	0·7983	−0·2176	0·7637
ACHYAQ	0·4643	0·4061	0·6027
ACHBRE	0·1527	0·7626	0·4696
ACHJAV	0·1693	0·8620	0·3925
ACHPAR	0·6038	0·3496	0·6429
ACHLAN	−0·8200	0·0311	0·7784
ACHMIN	−0·7697	0·1141	0·7339
ACHDEF	−0·8452	0·1204	0·7774
R_{zx}	0·8497	0·5172	

freshwater taxa with the marine taxa. This component has a relatively high correlation with mean salinity ($r = 0{\cdot}81$) and a weak correlation ($r = 0{\cdot}55$) with mean temperature; correlations with other environmental variables are low ($r < 0{\cdot}26$). The second principal component identifies the brackish-water taxa with high positive loadings, but correlations with environmental variables are low ($r < 0{\cdot}48$). Multiple correlation coefficients R_{zx} relating principal components 1 and 2 of **Y** to the six variables of **X** were 0·85 and 0·52, respectively (Table 11.4).

It is sometimes instructive to regress each taxon individually on the environmental variables. In our example, the multiple correlation coefficients R_{yx} derived from these regressions are relatively high for the marine and freshwater taxa ($R > 0{\cdot}60$) and low ($R < 0{\cdot}50$) for the brackish-water forms (Table 11.4). This makes sense biologically, as brackish-water taxa are

Table 11.5 Canonical correlations relating **Y** and **X** and the results of Bartlett's test for canonical correlations.

	χ^2	d.f.	*Significance*
0·8894	68·89	16	P < 0·01
0·8397	53·73	14	P < 0·01
0·7144	31·42	12	P < 0·01
0·5701	17·29	10	0·05 < P < 0·10
0·5308	14·56	8	0·05 < P < 0·10
0·2664	3·24	6	0·75 < P < 0·90
Total	189·13	66	P < 0·01

Table 11.6. Simple correlation coefficients relating the first two canonical variables of **Y** and **X** to the original variables of **Y** and **X**.

Variables of **Y**	*CV-1*	*CV-2*	*Variables of* **X**	*CV-1*	*CV-2*
ACHCOC	0·5349	0·6223	EXPO	0·3974	0·5341
ACHKUW	0·4219	−0·2401	LITE	0·5577	0·4842
ACHGP	0·1312	0·2292	SBAR	0·9524	−0·1823
ACHGPJ	0·6741	0·5039	SALR	0·2296	0·8708
ACHYAQ	0·6146	0·1945	TBAR	0·3061	0·7175
ACHBRE	0·3040	−0·3106	TEMR	0·6925	−0·0755
ACHJAV	0·2506	−0·3126			
ACHPAR	0·6584	−0·0583			
ACHLAN	−0·7836	0·1951			
ACHMIN	−0·7221	0·2444			
ACHDEF	−0·7550	0·1087			

presumably more indifferent to environmental variations than marine or freshwater taxa.

Another way of examining the correlation structure of the data set is by canonical correlation analysis (Tables 11.5 and 11.6). This algorithm maximizes the covariances between linear combinations of **Y** and **X**, given that each pair of functions is uncorrelated with all previously located functions in both domains. Bartlett's test of canonical correlations (Cooley & Lohnes, 1971) suggests that most of the information in [**YX**] can be collapsed into three dimensions (Table 11.5). Interpretation of these dimensions is aided by correlating the canonical variables of **Y** and **X** with the original variables of those matrices (Table 11.6). The first canonical variable of **Y** (CV-1) seems to contrast marine and freshwater taxa, and its associated canonical variable of **X** has a very high correlation with mean salinity. In CV-2 of **Y**, brackish-water taxa with negative correlations are contrasted with the other taxa, and the corresponding canonical variable of **X** has high correlations ($r > 0{\cdot}70$) with mean salinity range and mean temperature. This analysis therefore provides a numerical indication that the brackish-water *Achnanthes* are relatively indifferent to fluctuating salinities, an insight that was not manifested from the principal components analysis. The interpretation of the first canonical axis is consistent with our interpretation of the first principal component and its correlation with the variables of **X**.

It is interesting to examine the degree to which cases (samples) of **Y** exhibit a multiple cluster structure as opposed to a more continuous structure. As a preliminary indication of groups, we clustered the 54 cases of **Y** over 11 dimensions (taxa). Cassie (1972b) did essentially the same thing by examining plots of the principal components of **Y**. With the aid of a clustering algorithm, the *Achnanthes* data were grouped into three clusters, roughly corresponding to samples exposed to marine, brackish-water, and freshwater conditions. Next, we ask whether these clusters are discrete or contiguous. Ecologically, we are interested in whether our distributional pattern is a continuum or is made up of discrete assemblages. In reality, environmental gradients usually are continuous within some spatial level of resolution, so we are evaluating the steepness of such gradients relative to the taxa under consideration and the location of sampling stations.

By imposing a 3-cluster structure on **Y**, a discriminant analysis allows the cases to be projected onto two dimensions (Fig. 11.2). The algorithm finds linear combinations of the data that maximize the ratio of among-groups to within-groups sum of squares, subject to the restriction of orthogonality. In our example, the clusters are more or less contiguous, indicating that some populations of *Achnanthes* exhibit overlapping distributions in the estuary and that the distributional pattern is a continuum relative to these taxa and the sampling strategy. We also have examined 19 taxa of other diatom genera in this manner (unpublished) and found a continuum of marine and brackish-

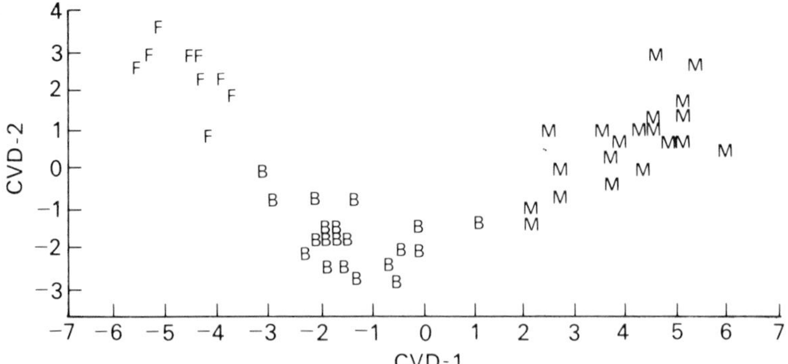

Fig. 11.2. A plot of the first and second canonical variables of **Y** from the discriminant analysis. The letters F, B, and M refer to samples from the freshwater, brackish-water, and marine clusters, respectively.

water samples and three relatively discrete clusters. Relative to these taxa, the latter suggested discontinuities between freshwater and brackish-water and between the lower and upper intertidal region in marine and freshwater.

The discriminant functions of **Y** often can be interpreted in terms of the variables of **X**. if we project the points in Fig. 11.2 down on each axis, the first canonical variable (CVD-1) has its highest correlation with mean salinity ($r = 0{\cdot}72$) and the second canonical variable (CVD-2) with mean salinity range ($r = 0{\cdot}63$). Also, it is sometimes instructive to perform a discriminant analysis on **X** after it has been reordered to correspond to biologically meaningful clusters of **Y**.

3.3 *Conclusions*

The sequence of multivariate procedures provides different ways of looking at the same data. In the example, we have illustrated only part of the details of this approach. We did not, for example, regress each taxon against the principal components of **X** or present measures of information content (Cassie, 1972a). These and other ramifications should be included in a routine procedure for processing a data set. Computer programs similar to the multivariate package described by Cassie (1972a) are useful for establishing this standard procedure. In any case, after the analyses are performed, the investigator must decide what aspects of the results are worth reporting. Such decisions will vary depending on the properties of each data set and the objectives of the research.

Rigorous statistical demonstrations of field data are often criticized on the grounds that they do not tell you anything that can not be understood by an intuitive interpretation of the raw data. While we are fully aware of the value of intuitive ecology, it has been our experience that it is often difficult to extract all the information out of a large set by intuition alone. In particular, it is difficult to discriminate between strong and weak relationships without some kind of numerical approach. Furthermore, intuition often dictates or eliminates relationships that are inconsistent with the correlation structure of the data. This is not necessarily unacceptable scientifically, unless some portion of the raw data is offered as supporting evidence. Unfortunately, this is sometimes done, and the reader is not always in the position to detect this deception. In summary, a multivariate analysis of distributional patterns in the diatom flora provides a numerical method of indicating possible groupings and relationships that, along with good ecological intuition, helps increase the probability of obtaining maximum information from field data. At the very least, the approach offers a convenient way to condense and summarize such information for publication.

ACKNOWLEDGMENTS

The senior author gratefully acknowledges the professional assistance of two long-time friends, Dr Charles W. Reimer and Dr W.S. Overton. We also would like to express our gratitude to Dr Stephen P. Main for contributing Fig. 11.1. The intellectual interaction with Michael C. Amspoker was particularly valuable in the review of work with epipsammic and epipelic diatoms.

Some of this work was supported in part by Research Grant No. DES72-01412 from the National Science Foundation.

4 REFERENCES

ALEEM A.A. (1949) Distribution and ecology of marine littoral diatoms. *Bot. Notiser* **4,** 414–40.

ALEEM A. A. (1950a) Distribution and ecology of British marine littoral diatoms. *J. Ecol.* **38,** 75–106.

ALEEM A. A. (1950b) The diatom community inhabiting the mud-flats at Whitstable. *New Phytol.* **9,** 174–88.

ALEEM A.A. (1973) Contribution to the study of littoral diatoms on the west coast of Sweden. *Botanica mar.* **16,** 193–200.

ALLEN T.F.H. (1971) Multivariate approaches to the ecology of algae on terrestrial rock surfaces in North Wales. *J. Ecol.* **59,** 803–26.

ALLEN T.F.H. & KOONCE J.F. (1973) Multivariate approaches to algal stratagems and tactics in systems analysis of phytoplankton. *Ecology* **54,** 1 234–46.

ALLEN T.F.H. & SKAGEN S. (1973) Multivariate geometry as an approach to algal community analysis. *Br. Phycol. J.* **8,** 267–87.

AMSPOKER M.C. (1973) The distribution of intertidal episammic diatoms on Scripps Beach, La Jolla. MS Thesis. California State Univ. San Diego, California. 54 pp.

ANDERSON T.W. (1958) *An Introduction to Multivariate Statistical Analysis.* 374 pp. Wiley, New York.

BAARDSETH E. & TAASEN J.P. (1973) *Navicula dumontia* sp. nov., an endophytic diatom inhabiting the mucilage of *Dumontia incrassata* (Rhodophyceae). *Norw. J. Bot.* **20,** 79–87.

BADDELEY W.H.C. (1858) On some diatomaceae that are found in *Notiluca milaris,* and the best means of obtaining them. *Q. Jl. microscop. Sci.* **6,** 79–80.

BAIRD I.E. & WETZEL R.G. (1968) A method for the determination of zero thickness activity of C^{14} labeled benthic diatoms in sand. *Limnol. Oceanogr.* **13,** 379–81.

BERGLAND L.A. (1973) Laboratory studies of successional patterns in assemblages of attached estuarine diatoms. MS Thesis. Oregon State Univ., Corvallis, Oregon. 71 pp.

BRAUN-BLANQUET J. (1928) *Pflanzensoziologie.* J. Springer, Berlin.

BROCKMANN C. (1935) Diatomeen an Schlick im Jadegebiet. *Abh. senckenberg. naturforsch. Ges.* **430,** 1–64.

BROCKMANN C. (1950) Die Watt-Diatomeen der schleswig-holsteinischen Westküste. *Abh. senckenburg. naturforsch. Ges.* **478,** 1–26.

CARPENTIER E.J. (1970) Diatoms attached to floating *Sargassum* in the Western Sargasso Sea. *Phycologia* **9,** 269–74.

CARTER N. (1932) A comparative study of the algal flora of two salt marshes. Part I. *J. Ecol.* **20,** 341–70.

CARTER N. (1933) A comparative study of the algal flora of two salt marshes. Part III. *J. Ecol.* **21,** 385–403.

CASSIE R.M. (1972a) A computer programme for multivariate statistical analysis of ecological data. *J. expl. mar. Biol. Ecol.* **10,** 207–41.

CASSIE R.M. (1972b) Fauna and sediments of an intertidal mud-flat. An alternative multivariate analysis. *J. expl. mar. Biol. Ecol.* **9,** 55–64.

CASSIE R.M. & MICHAEL A.D. (1968) Fauna and sediments of an intertidal mud-flat: a multivariate analysis. *J. expl. mar. Biol. Ecol.* **2,** 1–23.

CASTENHOLZ R.W. (1960) Seasonal changes in the attached algae of freshwater and saline lakes in the Lower Grand Coulee, Washington. *Limnol. Oceanogr.* **5,** 1–28.

CASTENHOLZ R.W. (1961a) An evaluation of a submerged glass method of estimating productivity of attached algae. *Verh. int. Verein. theor. angew. Limnol.*

CASTENHOLZ R.W. (1961b) The effect of grazing on marine littoral diatom populations. *Ecology* **42,** 783–94.

CASTENHOLZ R.W. (1963) An experimental study of the vertical distribution of littoral marine diatoms. *Limnol. Oceanogr.* **8,** 450–62.

CASTENHOLZ R.W. (1964) The effect of daylight and light intensity on the growth of littoral marine diatoms in culture. *Physiol. Pl.* **17,** 951–63.

CASTENHOLZ R.W. (1967) Seasonal ecology of non-planktonic marine diatoms on the Western Coast of Norway. *Sarsia* **29,** 237–56.

CHOLNOKY B.J. (1963) Beiträge zur Kenntnis des marinen Littorals von Südafrika. *Botanica mar.* **5,** 38–83.

CHOLNOKY B.J. (1968) Die Diatomeenassoziationen der Santa-Lucia-Lugune in Natal (Südafrika). *Botanica mar. Suppl.* **11,** 7–121.

CLEVE P.T. & GRUNOW A. (1880) Beiträge zur Kenntnis der arktischen Diatomeen. *Kong. Sven. Vet. Handl.* **17,** 1–121.

COOLEY W.W. & LOHNES P.R. (1971) *Multivariate Data Analysis*. 364 pp. Wiley, New York.

CROSBY L.H. & WOOD E.J.F. (1959) Studies on Australian and New Zealand diatoms. II. Normally epontic and benthic genera. *Trans. R. Soc. N.Z.* **86,** 1–58.

DRUM R.W. & WEBBER E. (1966) Diatoms from a Massachusetts salt marsh. *Botanica mar.* **9,** 70–7.

EDSBAGGE H. (1965) Vertical distribution of diatoms. *Svensk. bot. Tidskr.* **59,** 463–8.

EDSBAGGE H. (1966a) Distribution notes on some diatoms not earlier recorded from the Swedish West Coast. *Botanica mar.* **11,** 54–63.

EDSBAGGE H. (1966b) Zur Ökologie der marinen angehefteten Diatomeen. *Bot. Gothoburg.* **6,** 9–139.

EDSBAGGE H. (1966c) Some problems in the relationship between diatoms and seaweeds. *Botanica mar.* **11,** 64–7.

EDSBAGGE H. (1966d) The composition of the epiphytic diatom flora on the Swedish West Coast. *Botanica mar.* **11,** 68–71.

GEMEINHARDT K. (1935) Diatomeen von der Westküste Norwegens. *Ber. dt. bot. Ges.* **53,** 42–142.

GHAZZAWI F.M. (1933) The littoral diatoms of the Liverpool and Port Erin shores. *J. mar. biol. Ass. U.K.* **19,** 165–76.

GIFFEN M.H. (1963) Contributions to the diatom flora of a South African Province. I. Diatoms of the estuaries of the Eastern Cape. *Hydrobiologia* **21,** 201–65.

GIFFEN M.H. (1967) Contributions to the diatom flora of South Africa. III. Diatoms of the marine littoral regions at Kidd's Beach near East London, Cape Province, South Africa. *Nova Hedwigia* **13,** 245–92.

GIFFEN M.H. (1970a) Contributions to the diatom flora of South Africa. IV. The marine littoral diatoms of the estuary of the Kowie River, Port Alfred, Cape Province. *Nova Hedwigia* **13,** 259–307.

GIFFEN M.H. (1970b) New and interesting marine and littoral diatoms from Sea Point near Cape Town, South Africa. *Botanica mar.* **13,** 87–99.

GIFFEN M.H. (1971) Marine littoral diatoms from Gordon's Bay region of False Bay, Cape Province, South Africa. *Botanica mar.* **14,** 1–16.

GIFFEN M.H. (1973) Diatoms of the marine littoral of Steenberg's Cove in St. Helena Bay, Cape Province, South Africa. *Botanica mar.* **16,** 32–48.

GRØNTVED J. (1960) On the productivity of microbenthos and phytoplankton in some Danish Fjords. *Meddr. Danm. Fisk.-og Havunders. N.S.* **3,** 55–92.

GRØNTVED J. (1962) Preliminary report on the productivity of microbenthos and phytoplankton in the Danish Wadden Sea. *Meddr. Damn. Fisk.-og Havunders.* **3,** 347–78.

HAGELSTEIN R. (1938) *Scientific Survey of Porto Rico and the Virgin Islands*. Vol. 8, Pt. 3, 450 pp. New York Academy of Sciences, New York.

HARPER M. (1969) Movement and migration of diatoms on sand grains. *Br. phycol. J.* **4,** 97–103.

HARPER M. & HARPER J. (1967) Measurements of diatom adhesion and their relationship with movement. *Br. phycol. Bull.* **3,** 195–207.

HASLE G.R. (1968) *Navicula endophytica* sp. nov., A. pennate diatom with an unusual mode of existence. *Br. phycol. Bull.* **3,** 475–80.

HENDEY N.I. (1951) Littoral diatoms of Chichester Harbour with special reference to fouling. *J. royal microscop. Soc.* **71,** 1–86.

HENDEY N.I. (1964) An introductory account of the smaller algae of British Coastal waters. Part V. Bacillariophyceae (Diatoms). *Min. Agr. Fisheries and Food, Fisheries Invest. Series IV.* 317 pp.

HENDEY N.I. (1971) Some marine diatoms from the Galapagos Islands. *Nova Hedwigia* **22,** 371–422.

HENDEY N.I. (1974) A revised check-list of British marine diatoms. *J. mar. biol. Ass. U.K.* **54,** 277–300.

HICKMAN M. (1969) Methods for determining the primary productivity of epipelic and episammic algal associations. *Limnol. Oceanogr.* **14,** 936–41.

HICKMAN M. & ROUND F.E. (1970) Primary production and standing crops of episammic and epipelic algae. *Br. phycol. J.* **5,** 247–55.

HOHN M.H. & HELLERMAN J. (1963) The taxonomy and structure of diatom populations from three eastern North American rivers using three sampling methods. *Trans. Am. microsc. Soc.* **82,** 250–329.

HOPKINS J.T. (1963) A study of the diatoms of the Ouse Estuary, Sussex. I. The movement of the mud-flat diatoms in response to some chemical and physical changes. *J. mar. biol. Ass. U.K.* **43,** 653–63.

HOPKINS J.T. (1964a) A study of the diatoms of the Ouse Estuary, Sussex. III. The seasonal variation in the littoral epiphyte flora and the shore plankton. *J. mar. biol. Ass. U.K.* **44,** 613–44.

HOPKINS J.T. (1964b) A study of the diatoms of the Ouse Estuary in Sussex. II. The ecology of the mud-flat diatom flora. *J. mar. biol. Ass. U.K.* **44,** 333–41.

HOPKINS J.T. (1965) Some light-induced changes in behaviour and cytology of an estuarine mud-flat diatom. In *Light as an Ecological Factor.* 205–29 pp. (ed. R. Bainbridge). Blackwell Scientific Publications, Oxford.

HOPKINS J.T. (1966) The role of water in the behaviour of an estuarine mud-flat diatom. *J. mar. biol. Ass. U.K.* **46,** 617–26.

HUGHES R.N., PEER D.L. & MANN K.H. (1972) Use of multivariate analysis to identify functional components of the benthos in St. Margaret's Bay, Nova Scotia. *Limnol. Oceanogr.* **17,** 111–21.

HURLBERT S.H. (1971) The non-concept of species diversity: a critique and alternative parameters. *Ecology* **52,** 577–586.

HUSTEDT F. (1939) Die Diatomeenflora des Küstengebietes der Nordsee vom Dollart bis zur Elbemündung. *Naturwissenschaftlicher Verein zu Bremen.* **31,** 572–677.

HUSTEDT F. (1955) Marine littoral diatoms of Beaufort, North Carolina. *Duke Univ. Mar. Stat. Bull.* **6,** 1–67.

HUSTEDT F. & ALEEM A.A. (1951) Littoral diatoms from the Salstone near Plymouth. *J. mar. biol. Ass. U.K.* **30,** 177–97.

KEVERN N.R., WILHM J.L. & VAN DYNE G.M. (1966) Use of artificial substrata to estimate the productivity of periphyton. *Limnol. Oceanogr.* **11,** 499–502.

KING D.L. & BALL R.C. (1966) A qualitative and quantitative measure of Aufwuchs production. *Trans. Am. microsc. Soc.* **85,** 232–40.

LEVANDOWSKY M. (1972) An ordination of phytoplankton populations in ponds of varying salinity and temperature. *Ecology* **53,** 398–407.

LEWIN J.C. (1953) Heterotrophy in diatoms. *J. gen. Microbiol.* **9,** 305–13.

LEWIN J.C. (1963) Heterotrophy in marine diatoms. In *Symposium on Marine Microbiology.* pp. 229–35. (ed. C.H. Oppenheimer) C.C. Thomas, Springfield, Illinois.

LEWIN J.C. & LEWIN R.A. (1960) Auxotrophy and heterotrophy in marine littoral diatoms. *Can. J. Microbiol.* **6,** 127–34.

LLOYD M. & GHELARDI R.J. (1964) A table for calculating the 'equitability' component of species diversity. *J. Anim. Ecol.* **33,** 217–25.

LLOYD M., INGER R.F. & KING F.W. (1968) On the diversity of reptile and amphibian species in a Bornean rain forest. *Am. Nat.* **102,** 497–515.

MACARTHUR R.H. (1965) Patterns of species diversity. *Biol. Rev.* **40,** 510–33.

MAGUIRE B. & NEILL W.E. (1971) Species and individual productivity in phytoplankton communities. *Ecology* **52,** 903–7.

MAIN S.P. & MCINTIRE C.D. (1974) The distribution of epiphytic diatoms in Yaquina Estuary. Oregon (USA). *Botanica mar.* **17,** 88–99.

MARE M.F. (1942) Study of a marine benthic community with special reference to microorganisms. *J. mar. biol. Ass. U.K.* **25,** 517–54.

MCINTIRE C.D. (1973a) Periphyton dynamics in laboratory streams: a simulation model and its implications. *Ecol. Monogr.* **43,** 399–420.

MCINTIRE C.D. (1973b) Diatom associations in Yaquina Estuary, Oregon: A multivariate analysis. *J. Phycol.* **9,** 254–9.

MCINTIRE C.D., GARRISON R.L., PHINNEY H.K. & WARREN C.E. (1964) Primary production in laboratory streams. *Limnol. Oceanogr.* **9,** 92–102.

MCINTIRE C.D. & PHINNEY H.K. (1965) Laboratory studies of periphyton production and community metabolism in lotic environments. *Ecol. Monogr.* **35,** 237–58.

MCINTIRE C.D. & WULFF B.L. (1969) A laboratory method for the study of marine benthic diatoms. *Limnol. Oceanogr.* **14,** 667–78.

MCINTIRE C.D. & OVERTON W.S. (1971) Distributional patterns in assemblages of attached diatoms from Yaquina Estuary, Oregon. *Ecology* **52,** 758–77.

MCINTIRE C.D. & REIMER C.W. (1974) Some marine and brackish-water *Achnanthes* from Yaquina Estuary. Oregon (USA) *Botanica mar.* **17,** 164–75.

MCINTOSH R.P. (1967) An index of diversity and the relation of certain concepts to diversity. *Ecology* **48,** 392–404.

MCINTYRE A.D., MUNRO A.L.S. & STEELE J.H. (1970) Energy flow in a sand ecosystem. In *Marine Food Chains.* (ed. J.H. Steele) pp. 19–31. Univ. of Calif. Press, Berkeley, California.

MEADOWS P.S. & ANDERSON J.G. (1966) Microorganisms attached to marine and freshwater sand grains. *Nature* **212,** 1 059–60.

MEADOWS P.S. & ANDERSON J.G. (1968) Microorganisms attached to marine sand grains. *J. mar. biol. Ass. U.K.* **48,** 161–75.

MEISTER F. (1932) *Kieselalgen aus Asien.* Verlag von Gebrüder Borntraeger, Berlin. 56 pp.

MORRISON D.F. (1967) *Multivariate Statistical Methods.* 338 pp. McGraw-Hill, New York.

NULTSCH W. (1974) Movements. In *Algal Physiology and Biochemistry.* (ed. W.D.P. Stewart) pp. 864–93. Blackwell Scientific Publications, Oxford.

ODUM H.T., SILER W.L., BEYERS R.J. & ARMSTRONG N. (1963) Experiments with engineering of marine ecosystems. *Publs. Inst. mar. Sci. Univ. Tex.* **9,** 373–403.

PAINE R.T. (1971) Energy flow in a natural population of the herbivorous gastropod *Tegula funebralis. Limnol. Oceanogr.* **16,** 86–98.

PALMER J.D. (1960) The role of moisture and illumination on the expression of the rhythmic behavior of the diatom *Hantzschia amphioxys. Biol. Bull. mar. biol. Lab., Woods Hole* **119,** 330.

PALMER J.D. & ROUND F.E. (1965) Persistent vertical-migration rhythms in benthic microflora. I. The effect of light and temperature on the rhythmic behavior of *Euglena obtusa. J. mar. biol. Ass. U.K.* **45,** 567–82.

PALMER J.D. & ROUND F.E. (1967) Persistent vertical-migration rhythms in benthic microflora. VI. The tidal and dirunal nature of the rhythm in the diatom *Hantzschia virgata. Biol. Bull. mar. biol. Lab., Woods Hole* **132,** 44–55.

PAMATMAT M.M. (1965) A continuous-flow apparatus for measuring metabolism of benthic communities. *Limnol. Oceanogr.* **10,** 486–89.

PAMATMAT M.M. (1968) Ecology and metabolism of a benthic community on an intertidal sandflat. *Int. Rev. ges. Hydrobiol.*, **53**, 211–98.

PAMATMAT M.M. (1971) Oxygen consumption by the seabed. IV. Shipboard and laboratory experiments. *Limnol. Oceanogr.* **16**, 536–50.

PAMATMAT M.M. & FENTON D. (1968) An instrument for measuring subtidal benthic metabolism *in situ*. *Limnol. Oceanogr.* **13**, 537–40.

PAMATMAT M.M. & BANSKE K. (1969) Oxygen consumption by the seabed. II. *In situ* measurements to a depth of 180 m. *Limnol. Oceanogr.* **14**, 250–9.

PARSONS T. TAKAHASHI M. (1973) *Biological Oceanographic Processes*. 186 pp. Pergamon Press, Oxford.

PARTICK R. (1954) A new method for determining the pattern of the diatom flora. *Notul. Nat., Philadelphia* **259**, 1–12.

PATTEN B.C. & VAN DYNE G.M. (1968) Factorial productivity experiments in a shallow estuary: Energetics of individual plankton species in mixed populations. *Limnol. Oceanogr.* **13**, 309–14.

PERKINS E.J. (1960) The diurnal rhythm of the littoral diatoms of the River Eden estuary, Fife. *J. Ecol.* **48**, 725–8.

PHIFER L.D. (1929) Littoral diatoms of Argyle Lagoon. *Publs. Puget Sound mar. biol. Sta.* **7**, 137–49.

PIELOU E.C. (1969) *An Introduction to Mathematical Ecology*. 286 pp. Wiley-Interscience, New York.

POMEROY L.R. (1959) Algal productivity in salt marshes of Georgia. *Limnol. Oceanogr.* **4**, 386–97.

RIZNYK R.Z. (1969) Ecology of benthic microalgae of estuarine intertidal sediments. Ph.D. Thesis. Oregon State Univ., Corvallis, Oregon. 196 pp.

RIZNYK R.Z. & PHINNEY H.K. (1972a) Distribution of intertidal phytosammon in an Oregon estuary. *Mar. Biol.* **13**, 318–24.

RIZNYK R.Z. & PHINNEY H.K. (1972b) Manometric assessment of interstitial microalgae production in two estuarine sediments. *Oecologia* **10**, 193–203.

RIZNYK R.Z. (1973) Interstital diatoms from two tidal flats in Yaquina Estuary, Oregon, U.S.A. *Botanica mar.* **16**, 113–38.

ROSE F.L. & CUSHING C.E. (1970) Periphyton: Autoradiography of zinc-65 adsorption. *Science* **168**, 576–7.

ROUND F.E. (1960) The diatom flora of a salt marsh on the River Dee. *New Phytol.* **59**, 332–48.

ROUND F.E. (1965) *The Biology of Algae*. 269 pp. St. Martin's Press, New York.

ROUND F.E. (1966) Persistent vertical-migration rhythms in benthic microflora. V. The effect of artificially imposed light and dark cycles. *5th Int. Seaweed Symp.* pp. 197–203. Pergamon Press, Oxford.

ROUND F.E. & PALMER J.D. (1966) Persistent, vertical-migration rhythms in benthic microflora. II. Field and laboratory studies of diatoms from the banks of the River Avon. *J. mar. biol. Ass. U.K.* **46**, 191–214.

ROUND F.E. (1971) Benthic marine diatoms. *Oceanogr. Mar. Biol. Ann. Rev.* **9**, 83–139.

SEAL H. (1966) *Multivariate Statistical Analysis for Biologists*. 209 pp. Methuen, London.

SIMONSEN R. (1962) Untersuchungen zur Systematik und Ökologie der Bodendiatomeen der westlichen Ostsee. *Int. Rev. ges. Hydrobiol. (Syst. Beih.)* **1**, 9–144.

SIMPSON E.H. (1949) Measurement of diversity. *Nature* **163**, 688.

SLÁDECKOVÁ A. (1962) Limnological investigation methods for the periphyton ('Aufwuchs') Community. *Bot. Rev.* **28**, 286–350.

SOKAL R.R. & SNEATH P.H.A. (1963) *Principles of Numerical Taxonomy*. 359 pp. Freeman, San Francisco.

STANDER J.M. (1970) Diversity and similarity of benthic fauna off Oregon. M.S. Thesis. Oregon State Univ., Corvallis, Oregon. 72 pp.

STEELE J.H. & BAIRD I.E. (1968) Production ecology of a sandy beach. *Limnol. Oceanogr.* **13,** 14–25.

SMYTH J.C. (1955) A study of the benthic diatoms of Loch Sween (Argyll). *J. Ecol.* **43,** 149–71.

TAASEN J.P. (1972) Observations on *Navicula endophytica* Hasle (Bacillariophyceae). *Sarsia* **51,** 67–82.

TAASEN J.P. (1973) Remarks on th epiphytic diatom flora of *Dumontia incrassata* (Mull.) LaMour. (Rhodophyceae). *Sarsia* **55,** 129–32.

TAKANO H. (1960) Diatoms in pearl shell fishing grounds in the Arafura Sea. *Bull. Tokai reg. Fish. Res. Lab.* **27,** 1–6.

TAYLOR W.R. & GEBELEIN C.D. (1966) Plant pigments and light penetration in intertidal sediments. *Helgoländer wiss. Meeresunters* **13,** 229–37.

TAYLOR W.R. (1964) Light and photosynthesis in intertidal benthic diatoms. *Helgoländer wiss. Meereseunters* **10,** 29–37.

WETZEL R.G. (1964) A comparative study of the primary productivity of higher aquatic plants, periphyton and phytoplankton in a large, shallow lake. *Int. Rev. ges. Hydrobiol.* **49,** 1–61.

WETZEL R.G. (1965) Techniques and problems of primary productivity measurements in higher aquatic plants and periphyton. *Memorie Ist. ital. Idrobiol.,* 18 Suppl. pp 249–67.

WARREN C.E. (1971) *Biology and Water Pollution Control.* 434 pp. W.B. Saunders, Philadelphia.

WILHM J.L. & DORRIS T.C. (1968) Biological parameters for water quality criteria. *Bioscience* **18,** 477–81.

WOOD E.J.F. (1963) A study of the diatom flora of fresh sediments of the South Texas Bays and adjacent waters. *Publs. Inst. mar. Sci. Univ. Tex.* **9,** 237–310.

WULFF B.L. & MCINTIRE C.D. (1972) Laboratory studies of assemblages of attached estuarine diatoms. *Limnol. Oceanogr.* **17,** 200–14.

CHAPTER 12

THE ECOLOGY OF MARINE PLANKTONIC DIATOMS

ROBERT R.L. GUILLARD

Woods Hole Oceanographic Institution,
Woods Hole, Massachusetts 02543, USA

and

PETER KILHAM

Department of Ecology and Evolutionary Biology,
University of Michigan,
Ann Arbor, Michigan 48104, USA

1 INTRODUCTION

This review of marine planktonic diatom ecology is restricted almost entirely to considerations of geographical distribution and seasonal succession. Studies of autecological relationships by means of culture experiments are the subject of another chapter. There has been no general survey of diatom ecology, and of distribution in particular, since the review by Patrick (1948), the title of which—*Factors effecting the distribution of diatoms*—reveals the autecological approach that persists to the present time. The intuitions of earlier workers, elaborated and transmitted by others e.g. H.H. Gran, continue to guide current research to a remarkable degree, even though methods have changed radically in some cases and concepts derived from other scientific fields have been applied to marine ecology. Thus, concerning the effects of plant nutrients, Gran (1912, p. 367, p. 377) wrote in his splendid paper on pelagic plant life in the sea:

> 'We cannot get away from the view, which was first confidently put forward by Brandt, that certain indispensable nutritive substances occur so sparsely that, according to Liebig's minimum law, they act as factors which limit production',

and:

> 'Experiments with pure cultures of different plankton-diatoms, made by Allen and Nelson at Plymouth, show that they do not thrive without a regular supply of nitrogenous compounds. The plan of working which they adopted may also be employed with advantage when we wish to ascertain what concentration of dissolved nitrogenous compounds induced the plankton-algae to augment most rapidly. This is the first thing to find out if we desire to know whether a want of dissolved nutritive substances is the limiting factor of production. It is quite possible that augmentation diminishes from lack of nitrogen long before the total amount of this essential has been fully consumed; yet augmentation must not fall below a certain minimum if the species is to hold its own, because of the larger or smaller number of individuals that are constantly perishing. Questions like these can only be settled by experiment, so that the cultivation method of Allen and Nelson is bound to be of great assistance to us eventually.'

The recent studies of nutrient kinetics *in vitro* and observations of nutrient limitation *in situ* are presaged in these words, just as there is a glimpse of the generally unstated postulate that phytoplankton distribution is influenced to an important extent by the now-familiar μ, the growth rate (Gran's 'augmentation'). Yet, important as growth rate is, it seems quite probable that a high value of μ is neither necessary nor sufficient for survival of a plant species in a given region, and thus is not really uniquely responsible for its distribution! Growth rate is merely easy to measure. This point was quite obvious to the early plankton naturalists, who considered the effects of such other phenomena as spore or zygote formation, which are discontinuous functions of environmental circumstances (much like seed production by land plants), and thus different from essentially continuous variables such as the growth rate. Study of the ecological consequences of these phenomena has barely begun, at least for diatoms. Certainly the culture technique has important application here, as it has had for study of the growth rate.

Treatment of phytoplankton distribution in recent books is somewhat unsatisfactory, mostly because the authors have chosen, perhaps wisely, to generalize overmuch rather than become involved in a tangle of detailed and interrelated concepts (and a huge body of literature). Even for the best-studied marine phytoplankters, diatoms, there are no recent summaries covering world populations, and, especially for diatoms, the literature is enormous. Several thousand papers were processed for this review. We have felt it advisable to present a general summary of planktonic diatom distribution as it is seen from the literature available now, because of the lack mentioned above and because distribution does, after all, represent the basic body of data that phytoplankton studies seek to explain. A survey can hardly fail to bring into focus those important problems that can be managed now, whether the authors of the survey perceive them or not.

2 DISTRIBUTION OF MARINE PLANKTONIC DIATOMS

2.1 *Introductory remarks*

The fundamental question is whether or not there is a non-uniform distribution of species or of communities of diatoms. The affirmative answer, which is developed in this chapter, leads to the other large question of plankton ecology—why are the organisms non-uniformly distributed? There are some entangled conceptual and procedural complications best dealt with before reviewing the literature on diatom distributions.

At the spatial scale of thousands of kilometres, differences in populations are considered relative to topographic features of the earth, such as the positions of the continents, or to latitude, which, overall, determine temperature and light regimes. On the scale of hundreds of kilometres, differences in populations of bodies of water separated by geographic or hydrographic features are of interest. These two scales cover what are generally called 'biogeographical considerations' and correspond to the provincial level of organization. Valentine (1973, p. 337), describes provinces 'as regions in which communities maintain characteristic composition. . . . there is no reason to employ any particular arbitrary level of endemism, and it is in fact theoretically possible that a province could possess no endemic species at all and yet have distinctive communities'. His definition of community (p. 270) is a 'recurrent association' of species, 'with the proviso that the association is flexibly defined and need only include a certain percentage of the species in some definitive set of species in order to qualify'. Source references for diatoms at the biogeographic scale are by Gran (1912), Patrick (1948), Smayda (1958) and Braarud (1962). On the smaller scale of tens to about one kilometre, the populations of open water 'patches' are of interest, as well as the inhabitants of salt ponds, estuaries, fjords, etc. (Bainbridge, 1957). Even the 'microdistribution' of plankton algae, at the metre to decimetre scale has been studied, though generally in context of the statistics of sampling. The general problem of sampling merits consideration as the first of several involved with the facts of distribution over all scales.

2.2 *Technical and other considerations*

(a) *Sampling criteria*

In regard to sampling, the ultimate question concerns the zero abundance category, i.e. determining that a given species is absent from a given region. This is critical to the concept of endemism and is important also for establishing floras. Establishing that a diatom does not occur in some region usually reduces to its not having been recognized among several thousand individual specimens examined by a few workers. In many regions sampling has been local, sporadic, or by techniques unsuited to the capture of certain species. Endemic or cosmopolitan forms of low abundance could easily be missed. Confusion is also introduced by the transport of species ('expatriates') from their normal habitats to regions in which they do not thrive (where they are 'guest species' or 'allochthonous species'); this is often discussed in plankton studies, e.g. by Beklemishev (1964) in connection with zonation in the Antarctic oceanic regions.

Sometimes there is merely a question of establishing quantitative differences in the species composition of adjacent water bodies, or of showing non-random distribution of some species in a given region. Though less critical, the sampling problems remain; they concern collection of samples, preservation, subsampling, and enumeration.

(b) *Collection and identification techniques*

First, as to collection, no one technique will sample all diatoms adequately. Diatoms vary in size from about 2 μm to over 1 000 μm, and in concentration in nature, from less than 1/litre in the 'oceanic deserts' to the order of 10^6/ml in polluted estuaries. While the finest nylon nets used (about 35 μm apertures) do not adequately sample the nannoplankton or ultraplankton, they certainly capture the larger species. The 153 μm mesh Phinque net used by Venrick (1969) retained only 3% of the diatom crop in the North Pacific (see also Steyaert, 1973a, p. 315). Netting can in principle be used to sample for specific size ranges of phytoplankton by using either different nets or nets arranged in series according to mesh size. Quantitative estimates of large species are possible by using calibrated tow nets (Clarke & Bumpus, 1950) or by pumping water in measured volumes through netting arrays on board ship; however, this is seldom done. In principle, net techniques are adequate for large species that are abundant to moderately rare in eutrophic waters, and even for very rare species in oligotrophic regions, where there are few organisms *in toto* to obscure observation of the rarest forms.

The Hardy continuous plankton recorder captures on netting both zooplankton and the larger phytoplankton. It has been used to study plankton distribution patterns, chiefly in waters of the North Atlantic region and around

Japan; a bibliography of work with this instrument and a series of distribution charts in the North Atlantic arranged by species is available (Edinburgh Oceanographic Laboratory, 1973). At least one new diatom species has been discovered from catches made in the Hardy recorder (Hendey, 1964, p. 188).

For phytoplankters not adequately sampled by nets, only settling techniques will suffice. The standard device is the separable settling chamber, which, with the inverted microscope, permits use of high resolution lenses and phase contrast illumination (Utermöhl, 1958; Lund *et al.* 1958). (High resolution water immersion objectives can be used with the separable chambers, but one cannot then count cells around the periphery of the chamber.) Individual plankton cells can even be removed from the settling chamber for further examination in the electron microscope. The volume of the original sample is usually about 300 ml, and the volume searched no more than 100 ml. Thus, species rarer than 10/litre are likely to be missed. Sedimentation techniques can be used to concentrate the specimens in a larger volume of sample, and the concentrate can then be examined in the chamber of the inverted microscope (Hasle, 1959) or in some other appropriate chamber used with a standard microscope (see discussion of chambers in Guillard, 1973). Hasle (1959) estimated a mere 10% loss of phytoplankton in the concentrating process in conjunction with the inverted microscope, and recommended this technique for sparse phytoplankton when debris is not so abundant as to interfere.

The membrane filter technique allows phytoplankton from samples up to about 1 litre in volume to be concentrated on the surface of a filter about 3 cm in diameter. The dried filter is rendered transparent by an appropriate medium and the phytoplankton is examined at high magnification with the light microscope. The technique as used extensively by Soviet oceanographers is described by Kozlova (1964). Thorrington-Smith (1970, 1971) recorded 237 phytoplankton taxa by Holmes' version of the technique.

Combinations of the techniques mentioned work well for large diatoms that are abundant to rare, and for small species that are abundant in any waters or rare in oligotrophic waters. However, no technique easily exposes rare small diatoms in waters rich in other species, largely because individual specimens must be studied with the electron microscope for identification. A technique is needed like that used by Okada & Honjo (1973) to observe coccolithophores using the transmission or scanning electron microscopes directly on the surface of membrane filters on which species were collected. Up to the present, many of the small diatoms that have been studied have been cultured first (see Hargraves & Guillard, 1974).

Industrial or large laboratory continuous centrifuges ('separators') have been used on research vessels under way to collect large masses of phytoplankton—and all suspended material—from 10 to over 100 tonnes of water (about 10–100 m^3). This represents an integrated surface sample from a cruise track of 50 to perhaps 400 miles (Kozlova, 1964). The diatom shells are freed of organic matter and mounted on slides for qualitative and quantitative study. Kozlova (1964, pp. 42–44) points out that estimates of diatom abundance from examination of continuous centrifuge processed samples always exceed the estimates made by the membrane filter technique; the factors in different regions around Antarctica varied from roughly 2 to 10. They agreed best in the subantarctic zone, lying between the Antarctic Divergence and the Antarctic Convergence. Presumably the numbers of species found by the two methods did not differ significantly. Grant & Kerr (1970) examined phytoplankton samples collected by continuous centrifugation of 4·5 litre samples. From comparison of cell counts with pigment analyses made on filtered water samples, they concluded that significant numbers of small naked flagellates and possibly small diatom species were lost.

We have no knowledge of direct comparisons of oceanic phytoplankton (or diatom) abundance estimates made simultaneously with calibrated nets or the 'separator' technique and with the inverted microscope. Published results of different investigators in the same bodies of water show in most instances at least an order of magnitude higher counts with the inverted microscope, agreement being best where the abundant species are large. Very rare species are not detected by the inverted microscope. In examining published estimates made by net tows, one should note that vertical hauls, which are often used, integrate the whole water column's population and thus do not show the maximum abundance occurring at the station.

The 'centrifuge method' in which phytoplankton is concentrated, and immediately observed alive, will work with diatoms though its advantage is for species that cannot be preserved for subsequent examination.

In summary of collection techniques, the remark by Gran (1912, p. 361) is as true as ever: 'It follows, therefore, that we must abandon all thought of a universal method.'

Relative abundances of species are usually reported in terms of cell numbers and for floristic purposes this is best. However, because the sizes of diatoms vary by almost three orders of magnitude the relative contributions to the biomass differ considerably from the numerical abundances. Following Paasche (1960) and Smayda (1965), many workers have estimated and reported the surface area of phytoplankton, which is a better index to productivity than either cell numbers or cell volume, presumably because cell area best approximates the active protoplasmic volume. A clear example of differences between the two estimates is shown in Saunders & Glenn (1969, Table 5) who compared the rankings of the 58 diatom species contributing most to the total cell area with their ranking according to numerical abundance (in coastal Gulf of Mexico waters off Florida). Their data for the 10 highest ranks in each category are given (Table 12.1). The most abundant species by numbers (*Skeletomena costatum*) had an average abundance of

Table 12.1. Comparison of ranks of planktonic diatoms, arranged according to total cell area or to numerical abundance. From Saunders & Glenn, 1969.

Species	*Rank according to cell area*	*Rank according to numerical abundance*
Rhizosolenia alata	1	17
Rhizosolenia setigera	2	9
Rhizosolenia stolterfothii	3	11
Skeletonema costatum	4	1
Leptocylindrus (minimus and *danicus)*	5	2
Rhizosolenia fragillissima	6	3
Hemidiscus hardmanianus	7	45
Guinardia flaccida	8	27
Bellerochea malleus	9	18
Cerataulina pelagica	10	16
Nitzschia pungens	12	4
Chaetoceros pelagicus	20	5
Chaetoceros compressus	14	6
Chaetoceros didymus	21	7
Asterionella glacialis (japonica)	18	8
Nitzschia sp. *(closterium?)*	27	10

$1{\cdot}8 \times 10^5$ cells/litre, while the forty-fifth in abundance (*Hemidiscus hardmanianus*) numbered 239 cells/litre. Their respective areas were 177 μm^2 and 88 376 μm^2, thus their contributions to the biomass were about the same. The plant contributing most to the biomass, *Rhizosolenia alata,* was seventeenth in abundance. Note that the small cells, which are usually numerically abundant, may not necessarily be important contributors to the biomass, but they are usually much faster growing species than the larger plants and hence may contribute more to the daily production than is suggested by their relatively small biomass.

(c) *Small-scale non-random distribution*

Two recent studies have revealed significant non-random distribution of planktonic diatoms in both oceanic and estuarine environments. These papers refer to virtually all the previous literature on phytoplankton 'patchiness' as well as to statistical treatments used to analyze sampling variability and to detect aggregation or contagion (statistically described as 'over-dispersion'). In the estuarine Narragansett Bay (Rhode Island, USA), McAlice (1970) observed 12 species of diatoms in samples collected simultaneously at intervals of about 6 m, 2 m, and 10 cm, the last mentioned samples distributed vertically as well as horizontally. Over the 6 m scale, all 10 species counted were significantly aggregated; six of these species were probably grouped, indicating that they had been influenced similarly by whatever determined their distribution. However, one species seemed to be negatively correlated with most of the others. Platt (1972), who measured phytoplankton abundance fluorometrically over the spatial scale range 10–1 000 m, saw that the *variance* of the total abundance followed a minus 5/3 power relationship with distance, which parallels the spectral distribution of certain physical properties of the environment. If, as Platt suggests, turbulence largely controls the local concentration of total phytoplankton, it is still quite possible that the distribution of individual species on the contrary reflects in addition biotic factors such as differential grazing, sinking, or growth rates. Bernhard & Rampi (1965) found that various species were differently correlated in abundance, for example in their study of patchiness in the Ligurian Sea (Mediterranean). To return to McAlice's results, one study of *Skeletonema costatum* over a 2 m scale showed that the species was randomly distributed in only one of three series of observations. Another set of observations, presumably at the scale of metres, showed that 9 of 12 species observed (8 diatoms) were aggregated. *Striatella unipunctata* and *Thalassionema nitzschioides* were two of the randomly distributed species. Finally, there was little evidence of aggregation at the 10 cm scale. This last study was made during a bloom of *Asterionella glacialis* (ex-*japonica*; Körner, 1970), and only one vertical series of samples showed any evidence of contagion. McAlice judged that the patches he observed during his studies had the dimensions of 1–12 metres, which agrees with estimates made by previous observers. He found no correlation of phytoplankton abundance with small-scale variation in temperature or salinity.

Venrick (1969, 1971, 1972) studied the distribution of diatoms at two oceanic stations, one in the Central North Pacific, the other in the Central Pacific. Stations were occupied at intervals of 0·5, 1, or 2 nautical miles over a 10·5 nautical mile stretch. Her preliminary study (Venrick, 1969, 1971) showed how the variability of population estimates (made by counting specimens) depended upon the volumes of successive subsamples (sample bottles and counting chamber volumes) relative to the size of the initial sample taken (water bottle volume). Her paper (1971) should be consulted for design of efficient sampling and subsampling programs for specific purposes. The point here is that for her particular technique she could compute the departure from randomness. As might be expected, the degree of contagion (patchiness) was considerably less in the oceanic locations than in the estuarine ones previously considered. Within the relatively rich northern region (about 5×10^4 diatom cells/litre), nine species were counted in samples taken from three depths at several stations. Of 24 distributions studied, 10 were significantly aggregated; the species involved were always the most abundant ones (Venrick, 1972). There was significant concordance (agreement of distribution pattern) between all species at each of the depths, suggesting that the environment had influenced all species in the same way. Because fewest aggregations occurred in the 10 m samples, which were in the mixed layer, the wind-driven turbulence was apparently enough to keep most species randomly distributed.

In the Central Pacific location, in contrast, where the diatom populations were lower (3×10^2 cells/litre), only three of 20 distributions were aggregated at the same level of statistical significance seen in the Subarctic region. However, a further test based on presence or absence showed that another five were also aggregated. The patchiness was not related to species abundance, nor was there any concordance between species. Venrick estimated the precision of her technique; for any single count x of a species (one depth), the expression:

$$0{\cdot}3x \leqslant X \leqslant 3{\cdot}2x$$

included the observed population mean X at least 90% of the time. Single counts can vary by an order of magnitude even in a relatively homogeneous area with good counting techniques.

The statistical techniques of multivariate analysis are available to quantify studies of diatom associations in adjoining water bodies or geographical regions; in these cases the differences between populations lie mostly in the abundances of the various species, rather than in presence or absence of species. Principal components analysis is the technique of multivariate analysis most commonly employed by phytoplankton ecologists. Its use in phycological studies has been explained by Allen & Skagen (1973). Principal components analysis has been used to identify groups of associated phytoplankton species in the Caribbean Sea (Margalef & González-Bernáldez, 1969), in the North Sea (Ibanez, 1972), off the Ivory Coast in the Atlantic Ocean (Dandonneau, 1971; Reyssac & Roux, 1972) and in the upwelling region off the coast of Peru (Blasco, 1971). Using an association matrix and principal components analysis, Dandonneau (1971) identified eight groups of species (I–VIII) off the Ivory Coast. Groups I and II were present in all types of water, groups III and VII were predominant in recently upwelled water, and group IV was present in less recently upwelled water. Groups V and VI bloomed when there was land drainage. Group VIII was found specifically in oligotrophic tropical waters. In the relatively complicated hydrography of the Caribbean Sea (see p. 426), the analysis by Margalef & González (1972), Bernáldez (1969) could account for only 16% of the original variance, and yielded only two components that could be correlated in any way with factors of ecological significance; these factors, not surprisingly, were the oceanic influence and the coastal influence (not salinity).

Thorrington-Smith (1971) adapted the methods of cluster analysis, used in numerical taxonomy, to identify phytoplankton population groups in the Western Indian Ocean. This technique reduces (or changes) the subjectivity involved in forming the groups; its purpose remains the same as that of many previous studies, to identify specific bodies of water by their populations. Her analysis divided the area of study into 'phytohydrographic regions' that corresponded to water masses identifiable by characteristic values of the temperature, salinity, and phosphate concentrations. Grant & Kerr (1970) attempted a study of the same sort, using two different statistical techniques. Their samples were taken over a season in shallow water 7 miles offshore at Port Hacking, eastern Australia. In this much more difficult hydrographic situation their methods also revealed population groupings, but only one grouping of any obvious ecological significance—a population below the thermocline.

Two points emerge, as concluded by McGowan (1971, p. 64) in reference to zooplankton communities; first, that the more stable the hydrographic situation, the more definitely the community is defined, and second, the species composition of the community, though not its trophic structure, is strongly influenced by the history of the water body—this includes an 'inoculum effect'.

The distributions of different phytoplankton species may be dissimilar in depth, just as they may differ horizontally. A particular species may be abundant at one depth and undetectable (by the counting technique used) at other depths. For floristic purposes this could be critical but it is relatively easily accounted for by examining vertical net tows or pooled bottle samples. Most studies that have considered vertical distribution have viewed it in relation to such factors as stratification of the water column, depth of the euphotic zone, or nutrient status of the crop. A good example is Ohwada's study (1972) of the vertical diatom distribution in the Sea of Japan, which revealed also that turbulence can circulate living diatoms to at least 300 m in coastal areas. In some instances non-uniform vertical distribution has been attributed to grazing (Sakshaug & Myklestad, 1973). These authors observed an inverse relationship between diatom and zooplankton abundances over the water column. In what may be a special case, Hulburt & Corwin (1972), station D found an inhomogeneous vertical distribution of diatom species at or near a loop in the current flowing into the Gulf of Mexico from Yucatan Strait. Species ubiquitous in the region (*Coccolithus huxleyi, Nitzschia closterium*) decreased in abundance with depth much as usual for nearby waters while certain other species were found regularly only near the surface. However, two diatoms, *Thalassiosira subtilis* and *Rhizosolenia stolterfothii,* essentially occurred only below 60 m, attaining maximum abundance at 80–90 m, near the bottom of the euphotic zone where nutrients were somewhat more abundant. Apparently both nutrient enrichment and some other factor associated with the current (and possibly the loop) were necessary for both the increased plankton population which occurred at all depths and for maintenance of the local deep-water maximum of the two diatoms. Relatively few studies other than that of Venrick (see above) have dealt

Table 12.2. Frequency of species occurring in the sediments of the Equatorial Region of the Pacific, Indian and Atlantic Oceans (arranged in decreasing order of their frequency). From Kolbe, 1957.

1	*Coscinodiscus*	*Coscinodiscus nodulifer*	*Nitzschia marina*
2	*Fragments (Ethmodiscus)*	*Fragments (Ethmodiscus)*	*Coscinodiscus nodulifer*
3	*Nitzschia marina*	*Nitzschia marina*	*Fragments (Ethmodiscus)*
4	*Hemidiscus cuneiformis*	*Hemidiscus cuneiformis*	*Pseudoeunotia doliolus*
5	*Thalassionema + -thrix*	*Coscinodiscus excentricus**	*Coscinodiscus africanus*
6	*Asteromphalus elegans*	*Thalassionema + -thrix*	*Coscinodiscus excentricus**
7	*Coscinodiscus lineatus*	*Coscinodiscus lineatus*	*Coscinodiscus lineatus*
8	*Asteromphalus hiltonianus*	*Coscinodiscus africanus*	*Thalassionema + -thrix*
9	*Coscinodiscus excentricus**	*Actinocyclus subtilis*	*Hemidiscus cuneiformis*
10	*Coscinodiscus africanus*	*Rhizosolenia bergonii*	*Roperia tesselata*
11	*Pseudoeunotia doliolus*	*Coscinodiscus obscurus*	*Asteromphalus arachne*
12	*Actinocyclus elongatus*	*Thalassiosira decipiens*	*Actinocyclus subtilis*

The abundance (frequency in numbers) of fragments (*Ethmodiscus*) from cores taken at a region of *Ethmodiscus* ooze is not taken into consideration.
* Now *Thalassiosira excentrica*.

directly with lack of concordance between species abundances vertically. A spectacular example was found by Ryther & Hulburt (1960), who remarked that despite the existence of a deep well-mixed layer as judged by physical and chemical criteria, *Asterionella bleakeleyi* and an unidentified centric species had developed sizeable maxima in the upper 10 m, while *Leptocylindrus danicus* and *Lauderia borealis* had apparently flowered at some time then settled out of the euphotic zone. Similar phenomena have been recorded from weatherships near Japan (Marumo, 1954) and from a buoy laboratory in the Mediterranean (Léger, 1972, and earlier papers in his series). In the Red Sea the two dominant diatoms were found at different depths; *Climacodium frauenfeldianum* in the upper 20 m, and *Ditylum sol* from 25–65 m (Sukhanova, 1969).

Strickland (1965) found a remarkable concentrated layer of *Skeletonema costatum* only about two metres thick; it was discovered in the process of comparing vertical chemical profiles obtained from bottle samples with profiles obtained from an automated system pumping continuously over the range of depths studied.

Catches of apparently living diatoms (and other phytoplankters) at depths far exceeding the compensation depth (about 100 m in clear waters) have led to speculation on the origin, activity, and fate of these populations. Passive sinking of cells and of whole populations certainly occurs in the sea, but the interesting finds involve displacements through relatively large distances at velocities higher than expected. Smayda, in his reviews (1970, 1971), marshals convincing evidence that some mechanism other than passive sinking is necessary to account for the presence of the living algae, and, for that matter, to account also for the deposition of diatom shells in many sediments, because of the relatively rapid rate of solution of the silica of which shells are formed. Vertical circulation in Langmuir convection cells can exceed normal diatom sinking rates by at least two orders of magnitude, but this circulation seldom goes below about 100 m, which is scarcely enough (Smayda, 1971, p. 110). While downwelling currents may be significant in sòme cases, the spectacular events are storm-induced vertical currents that can move water through 500–800 m in less than a day (see Smayda, 1971, p. 116). F. Bernard (1967) reports his observation that immediately after a storm near the Straits of Gibraltar, Atlantic surface water, normally confined to the upper 80 m, was at 500 m, carrying its attendant characteristic diatoms.

In the Indian Ocean, Nel (1968) found living oceanic diatoms (of 48 species) regularly in samples collected in bottles or in closing nets (with meshes too small to permit significant contamination) from 500–3 000 m; species found at depth were sometimes uncommon or undetected above. His observations are at least consistent with the mechanism proposed. Malone *et al.* (1973) have cultured diatoms, belonging to littoral and coastal species, from samples of North Atlantic red clay sediments collected from 6 150 m. They suggested that specimens were transported to the depths in faecal material or by turbidity currents, or that both mechanisms were involved.

The fate of algae in deep water, whether on the bottom or planktonic, remains unknown. Recent studies with bacteria (Jannasch & Wirsen, 1973) suggest that life processes in general may proceed much more slowly in deep waters than heretofore suspected. If this is true of algae, it supports the suggestion (e.g. Kimball *et al.* 1963) that populations in deep water may, if brought to the surface by upwelling or vertical currents, serve as 'seed' for populations in the euphotic zone.

(d) *Identification and taxonomy*

The number of species of diatoms is enormous; Helmcke (1961) cites Hustedt's opinion that of some 10^5 described species, about 10^4 may be recognized as valid. The number known from the marine plankton is about an order of magnitude less (than 10^4) but it is not possible to decide which species should be considered truly planktonic; many are casual or accidental. Regional floristic works purporting to deal with the plankton typically list 100–200 species:

Hendey (1974): 212 planktonic species (916 total)
Trégouboff & Rose (1957): 121 species
Cupp (1943): 160 species
Lebour (1930): 160 species
Gran (1912): 298 species
Taylor (1966): 220 species.

A list compiled from the works of Wood and co-workers in the waters around Australia, Indonesia, and New Zealand contains about 400 species estimated to occur in the plankton at least part of the time (Crosby & Wood, 1958; Wood & Crosby, 1959; Wood, 1961, 1963a, 1965a; Cassie, 1961). The number is no doubt large because in the widespread shallow Indonesian seas, and in estuaries an unusually large number of tychopelagic forms occur.

It is difficult to estimate the number of planktonic species found in the Antarctic. Hasle (1969) identified 96 species from offshore plankton collections of the Brategg Expedition, which represents a minimum number. Kozlova (1964) reported 92 species (10 new) from extensive collections in the Indian and Pacific sectors of the Southern Ocean. Van der Werff (1954) estimated that there were 250 valid Antarctic diatom species, including benthic types. At least 100 new species have been added since then (Frenquelli & Orlando, 1958; Hustedt, 1958; Manguin, 1960; Kozlova, 1964). Kozlova (1964) attributes the extensive synonomy in Antarctic diatoms to the 'great polymorphism' peculiar to them. *Charcotia actinophilus* with 15 synonyms and *Schimperiella antarctica* with 12 are given as examples.

Many planktonic species are cosmopolitan, so that a world list is far smaller than the sum of the regional floras. Hendey is credited with the estimate that about 600 species are reasonably characteristic of the plankton, including pennate species that are usually benthic (Wimpenny, 1966a; p. 53).

Studies of distribution are hampered by difficulties in describing and recognizing diatoms. Wood (1967, p. 95) complained '—on occasions when I have asked my colleagues to draw the same organism, the differences in the resulting illustrations would warrant specific rank'. Frequent use of high quality light micrographs in conjunction with schematic line drawings has done much in the last decades to improve descriptions. Observation of thin-shelled species has been aided by the availability of phase contrast and Nomarsky (differential interference contrast) microscopy (Simonsen, 1974). However, it is to the electron microscope that much of the recent progress is due, particularly in respect to small species. A brief review of studies with the transmission electron microscope is given by Lewin & Guillard (1963); its use is now routine. The scanning electron microscope has proved especially useful for observing diagnostic features on a number of planktonic diatoms (Hasle, 1968a,b; Fryxell & Hasle, 1971; Round, 1970, 1972). The study of *Thalassiosira tumida* (Hasle *et al.* 1971) is a model of the combined use of light photomicroscopy and the two kinds of electron microscopy. The new degree of structural resolution available makes discrimination at the specific and varietal level much better, permitting, in some cases, the reassignment of taxa in ways that clarify distributional observations. Thus, *Nitzschia pungiformis* could be distinguished by a central nodulus from the similar *N. pungens* (Hasle, 1971), and finds of what appear to be the latter species in open sea samples are likely to be of *N. pungiformis*. There are examples in the genus

Thalassiosira also (Hasle, 1972a; Fryxell & Hasle, 1972). The necessary nomenclatural changes that accompany taxonomic revision remain a source of bibliographic confusion. Important or characteristic species are moved to other genera, which sometimes results in a new species name also (e.g. *Cyclotella nana* Hustedt = *Thalassiosira pseudonana* Hasle & Heimdal). The former genus *Fragilariopsis* Hustedt, which contains species important in cold oceanic waters, has been reduced to the rank of a section of the genus *Nitzschia* by Hasle (1972b); nine species names could be preserved, but four new ones were necessary. Species of the major planktonic genus *Thalassiosira* were transferred to *Coscinodiscus* by Cleve-Euler (1951) but later separated again and made the basis of a new family, now including some of the species of *Coscinodiscus* (Hasle, 1973a; Fryxell & Hasle, 1972). The useful revision of 'Mills' Index', an enormous undertaking, is not yet complete (Van Landingham, 1967, 1968, 1969, 1971, 1975), and just the recognition of names in the literature remains a problem to non-specialists. Mann (1907) pointed out that *Actinocyclus ehrenbergii* Ralfs. has 'somewhat over 120 synonyms'!

(e) *Species concept in diatoms*

In speaking of the distribution of species, it is implied that one knows what a species is. The idea of Hendey (1964, p. 54) seems equivalent to that of others dealing with diploid organisms capable of sexual reproduction; the 'species' is a population of individuals with significant actual or potential genetic exchange. In diatoms, the taxonomic species are defined by structural features of the silica shell, and it is not known to what extent these features reflect the genetic make-up of the populations (see also Lewin & Guillard, 1963, pp. 374–5). There are some rather obvious incongruities. In three important planktonic genera it has been discovered that the cells of one species, studied in clonal culture, produced the characteristic valve structure of more than one species (Holmes & Reimann, 1966; Holmes, 1967; Drebes, 1967; Hasle *et al.* 1971; Hasle, 1973a). (See Simonsen, 1974, for an alternate taxonomic explanation of the phenomena observed by Holmes & Reimann, 1966.) Clonal cultures of a marine saprophytic *Nitzschia* produced shells with the structures of both the genera *Nitzschia* and *Hantzschia*—not a difficult geometrical transformation (Lauritis *et al.* 1967). Wood (1959, 1963c, p. 238) observed, in natural collections, shells of large marine centric diatoms having the two valves quite dissimilar, each with the character of a different genus. (In a few planktonic species the two valves differ characteristically.) Also, in some species, taxonomic 'forms' are established for variant morphologies but cells can be found with one valve of each type (see Hendey, 1964, p. 55). Differences between the two valves of *Nitzschia (Fragilariopsis) kerguelensis* and *Asteromphalus hookeri* are shown and discussed by Van der Spoel *et al.* (1973).

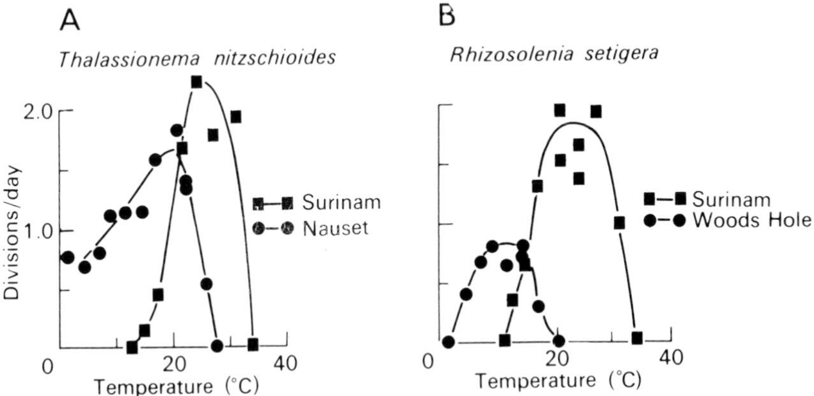

Fig. 12.1 Growth rates of *Thalassionema nitzschioides* and *Rhizosolenia setigera* clones, isolated from tropical waters ■—■ or temperate waters ●—●, when cultured at various temperatures after brief periods of adaptation to each temperature. Each point represents a period of several days of growth at essentially constant growth rate, in enriched seawater medium.

The use and significance of taxonomic characters other than frustule structure are discussed by Hasle (1973b, p. 134–5) in connection with *Skeletonema costatum* and *S. tropicum,* which are distinguished morphologically only by chloroplast number. The extent to which these populations are separated genetically is not known. The same uncertainty applies to taxonomic varieties of many species, especially when these are geographically well separated, and even to taxonomic 'forms', which are often found in circumscribed local populations (see later discussion). The existence of physiological (ecological) races of marine plankton diatoms has been suggested (with reservations) by Braarud (1962, p. 644), by Smayda (1958, p. 171), and, somewhat more forcefully, by Guillard & Ryther (1962), Carpenter & Guillard (1971), Fisher *et al.* (1973), and by Hargraves & Guillard (1974). Guillard & Hulburt (unpublished) have observed that strains of the same species of some large and easily identified (sic!) diatoms from tropical coastal and northern coastal waters of the Atlantic had very different temperature optima and ranges for growth as determined in culture; two are illustrated in Fig. 12.1 and a summary shown in Fig. 12.2. (See Guillard & Hellebust, 1971, for details on *Phaeocystis,* and Guillard *et al.* 1974, for similar studies of three *Skeletonema* species. Hutchinson, 1967, pp. 433–4, discusses thermal races of freshwater planktonic diatoms.) Jensen & Rystad (1974) have found that a clone of the neritic species *Skeletonema costatum* isolated from the Sørfjord, which is heavily contaminated with zinc, tolerates

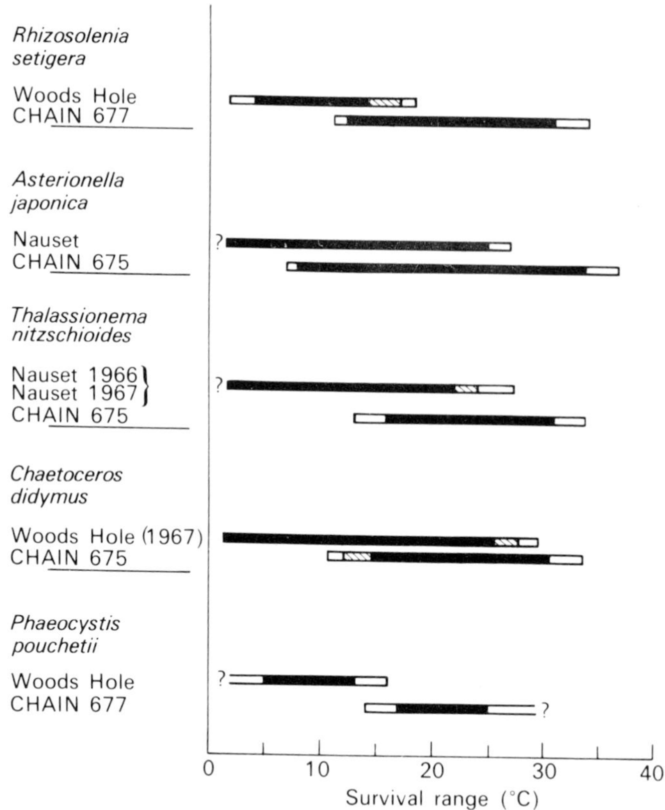

Fig. 12.2. Survival ranges of pairs of clones of several phytoplankton species isolated from temperate waters (upper bar) or tropical waters (lower bar). Culture conditions as in Fig. 12.1. *Black portion of bar*: growth relatively reliable though temperature dependent. *Hatched portion of bar*: variable growth. *White portion of bar*: temperature interval in which growth fell to zero in repeated experiments. The black segment is an estimate of the temperature range in which the species could survive, as deduced from the properties of the clones studied.

an order of magnitude more zinc (in dialysis culture without added chelator) than a clone from the relatively unpolluted Trondheimfjord. Presumably a strain has been selected that resists high levels of the metal under natural chelation levels.

A search for ecological races should be rewarding in the Black and Azov Seas and especially in the isolated Caspian and Aral Seas. All these not only have a considerably lower salinity than that of seawater—the Mediterranean particularly—but have significantly different ionic composition from seawater diluted to the same total solids content (or to the same osmotic pressure). Zenkevitch (1963, pp. 366, 551, 652) points out that all these seas have more Ca, Mg, and SO_4 than seawater of the same salinity and relatively less Na and Cl; K content varies somewhat. The effects of individual anions and cations and of ionic ratios are not negligible (see p. 536 of review by Guillard, 1962). The stenohaline marine clone of *Thalassiosira pseudonana* (clone 13–1) was sensitive to K concentration (Guillard & Myklestad, 1970); unpublished results by the same authors showed that the Mg content of natural seawater was above the optimum for this clone. It would be extremely interesting to determine experimentally if the diatom species abundant in isolated seas differ from their counterparts in neighbouring waters. Proshkina-Lavrenko (1961) has noticed that in the Black Sea, certain widespread diatom species, among them *Skeletonema costatum, Leptocylindrus danicus, Rhizosolenia calcar-avis,* and *Coscinodiscus gigas* occur in populations of small-sized cells that often show variations from the usual morphology. Certain of the variants have been assigned rank as forms or varieties. Size differences of diatoms in the other seas have been remarked on as well, and inasmuch as the ionic contents of these seas differ still more from that of seawater, the effect should be more pronounced; in fact, the Mediterranean flora is reduced in them.

It is not difficult to envisage selection of physiological strains in the confines cf a fjord or a confined sea, but it might be considered questionable that discrete populations would remain intact and thus detectable in more open waters. Plankton recorder studies have revealed that in the northern North Atlantic three species of planktonic diatoms had morphologically recognizable, geographically circumscribed populations that persisted for some time (Robinson, 1961; Robinson & Colburn, 1970). There were populations of three different forms of *Rhizosolenia alata* and of two varieties of *Thalassionema nitzschioides.* In the best documented case, there were eight populations in three varieties of *Rhizosolenia styliformis,* occupying regions from the Bay of Biscay to Newfoundland. Taxonomic questions surrounding this species and *R. hebetata* (Seaton, 1970; Takano, 1972) do not obscure the point that populations obviously closely related, at least by morphological criteria, have become dominant and maintained themselves in regions of somewhat different chemical and physical characteristics. Recognizable populations of *Nitzschia (Fragilariopsis) kerguelensis, Coscinodiscus lentiginosus, C. minor,* and *Asteromphalus hookerii* were found in the South Atlantic and Southern Ocean (Van der Spoel *et al.* 1973). It should come as no surprise to find physiological varieties, which might or might not show morphological distinctions also.

Speciation in higher terrestrial plants and the contribution of ecological races to their distribution patterns have been studied extensively (see Stebbins, 1950, Ch. II). Comparable progress in knowledge of recent diatom distributions is not possible without genetic data. Genetic comparisons through the analysis of enzymes by gel electrophoresis offer some promise, especially when coupled with morphological and physiological data. The techniques are being applied to species for which there are many isolates in culture (Murphy & Guillard, 1976) and would be of obvious application to species such as those in the *Nitzschia seriata* group, or the section *Fragilariopsis* of *Nitzschia*.

(f) *Concepts in distribution patterns*

Examples of endemism and of bipolarity of species distribution will be discussed in connection with specific regions, mainly the Antarctic and Australian waters. The origins and development of the terms 'neritic', 'oceanic', 'meroplanktonic', and 'holoplanktonic' are discussed in detail by Smayda (1958). He points out that some confusion has developed in application to phytoplankton because of presumed causal connections between factors; for example, that a species found near the coasts has a requirement for shallow waters because it forms resting stages which must not sink too far. He proposed a binary classification for phytoplankton. This uses prefixes to specify the innate life cycle; *mero* signifies that the species has a stage that is dormant, whether a spore or not, and *holo* signifies that the species lacks such a stage. The stems *paractic, anoictic,* and *adiaphoric* denote, respectively, that the species lives in the coastal influence, in the open sea, or is indifferent (lives in both regions). Recent literature gives the impression that the words 'neritic' and 'oceanic' are now used essentially in the sense of Smayda's 'paractic' and 'anoictic', without further implication. 'Adiaphoric' appears to be equivalent to Hart & Currie's (1960) 'panthalassic'. (The terms 'holoplankton', 'meroplankton', and 'tychoplankton' are still used in the sense of species that are, respectively, obligately planktonic, planktonic during part of their life cycle, and only accidently planktonic, being normally benthic.) 'Cosmopolitan' refers in the widest sense to species found everywhere, but occurs most frequently in the diatom literature in reference to species found everywhere up to the polar regions. Terms such as 'pantropical' or 'circumtropical', and 'circumpolar' are encountered, and compound terms such as 'neritic pantropical' or 'boreal oceanic'. Hart (1942, p. 283) found that none of these concepts fit adequately in the Antarctic and established 5 categories that are fortunately unnamed save for the 'neritic/ice edge' one that includes with neritic forms certain presumably meroplanktonic ones that use the ice-edge as a coast and persist briefly in the open waters after the pack breaks up.

Cleve (1897), recognizing certain loose associations of phytoplankton species in waters of the North Atlantic, named a series of 'plankton types', characterized by particular species, that were felt to reflect the selective influence of the environment on the populations (see discussions in Patrick, 1948, p. 483; Wimpenny, 1966a, p. 205). The descriptive terms coined by Cleve ('sira-plankton', etc.) are now rarely encountered although descriptive terms such as '*Chaetoceros-plankton*' or '*Skeletonema plankton*' occur. Gran (1912, p. 340) pointed out that Cleve's 'plankton-types' have no species in common while in reality species distributions overlap significantly. Hart & Currie (1960) remarked that Cleve's 'types' can be recognized wherever environmental gradients are abrupt, as is usual around Scandinavia. Experience has generally confirmed the low fidelity of marine phytoplankton species.

It may be expected that the broad distribution of planktonic diatoms will, in general, correspond to the main patterns of distribution of marine animals, for which the information is much more complete (Ekman, 1953). McGowan (1971) made an extensive review of the biogeography of the oceanic Pacific, utilizing newer quantitative data as well as the older qualitative studies. In pointing out that similar distribution patterns are shown by groups from protozoa to vertebrates, he comments that there is as yet fragmentary but still strongly suggestive evidence that phytoplankton species follow the same scheme. Figure 12.3 shows the main regions for the Pacific (from McGowan, 1974). (Note that the Antarctic region, which is quite distinct, is not shown separately.) Oceanic coccolithophores also fit into the general pattern (Okada & Honjo, 1973), and so do the assemblages of diatom shells that sink to the ocean floor, even though they but imperfectly represent the populations in the waters immediately above (Kanaya & Koizumi, 1966; Jousé *et al.* 1971; Kozlova & Mukhina, 1967).

The historical development of knowledge of marine diatom distribution cannot be traced here, for lack of space. Some is given by Patrick (1948) from a botanical point of view, and more by Gran (1912), who summarized the contributions to phytoplankton research made by the great oceanographic expeditions up to that time. There seems to be no complete general account of recent oceanographic explorations.

There are some well-documented instances of 'invasions' of regions by planktonic species. The classical cases (citations in Smayda, 1958, and Braarud, 1962) are the rapid establishment of *Biddulphia sinensis,* an Indo-Pacific form, in the North Sea beginning in 1903, and the spread of *Asterionella glacialis* (ex-*japonica*) in Norwegian coastal waters, presumably from British coastal waters in 1927. *Thalassiothrix mediterranea* is apparently an addition of the last decade to the Black Sea flora (see p. 432) and *Rhizosolenia calcar-avis* invaded the Caspian Sea in 1934, where it became a dominant for several decades (see p. 434). Lewin (1974) has documented the changes in dominance of the diatom species comprising the surf-zone blooms along the Olympic Peninsula coast (Washington, USA) from 1925 to the present. The three true surf-loving species *Aulacodiscus kittonii, Chaetoceros armatum,* and *Asterionella socialis* Lewin and Norris occur all along the West Coast of the United States. Lewin has been unable to correlate changes in the dominance of these species (which is extreme) with any observed changes in the physical or chemical environment.

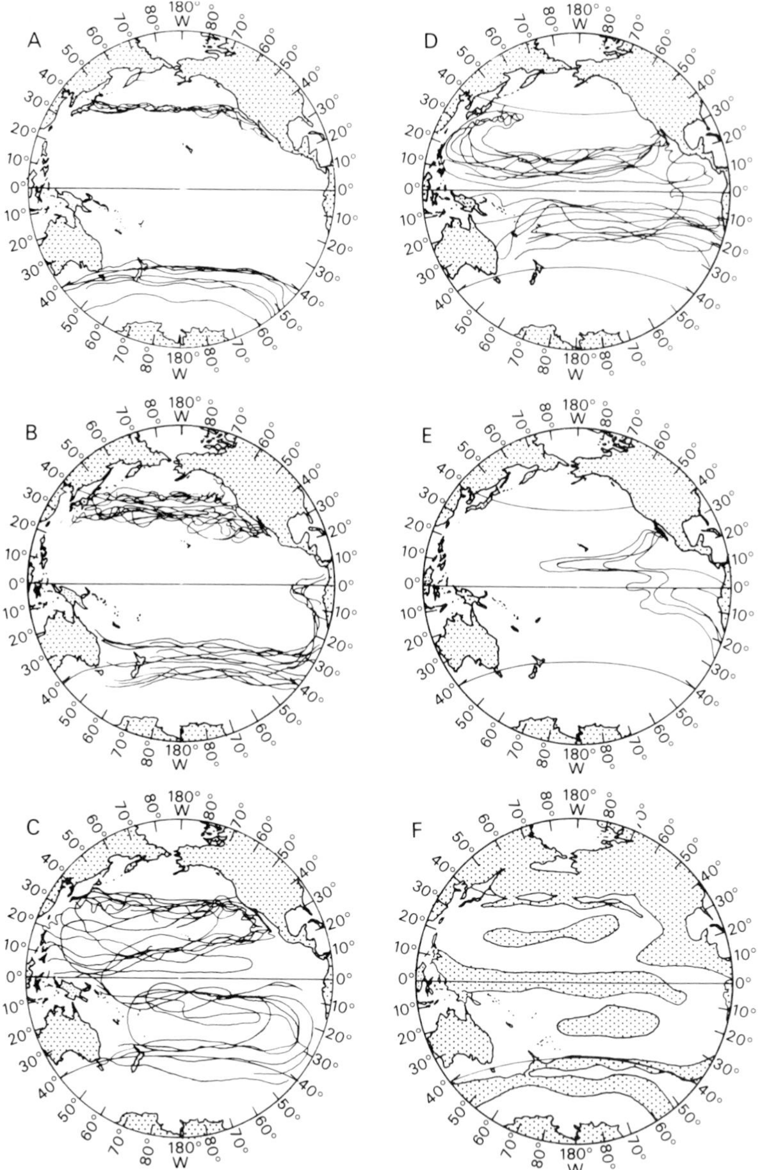

Fig. 12.3. Distributional boundaries of zooplankton species from regions of the Pacific: A, Subarctic species; B, Transition zone species; C, Central species; D, Equatorial species; E, Eastern tropical species; F, The patterns of the basic biotic provinces of the oceanic Pacific from McGowan (1974).

3 BIOGEOGRAPHY OF MARINE PLANKTON DIATOMS, BY REGIONS

The following summary of regional floristic and ecological studies of diatoms can list only some of the more complete recent studies. An effort has been made to include those that we feel best summarize the floristic or ecological advances made since publication of the sources mentioned before. Citations are given at least once in the text for recent changes of species names.

3.1 *Antarctic and the Southern Ocean*

The Antarctic flora, of which diatoms comprise the major portion, has long been of interest; between 1844 and 1969 some 62 accounts of it have been published. Hasle (1969) studied the phytoplankton of the Pacific sector of the Southern Ocean (Brategg Expedition) using the inverted microscope method. She summarizes the biological and biogeographic information in the earlier papers and also presents pertinent hydrographical information and productivity data from still other sources. A condensed account, together with an excellent series of maps of diatom distributions is given by Hasle (in Balech *et al.* 1968). To Hasle's bibliography should be added the papers of Wood (1960), El-Sayed (1971), and Steyaert (1973a, 1973b, 1974). The number of planktonic species found in Antarctic waters was discussed before, in the section *Identification and taxonomy.*

Special oceanographic conditions occur in the Southern Ocean (Hasle, 1969; Deacon, 1963). In brief, the Antarctic region is relatively isolated by the Subtropical and Antarctic convergences, which are regions of sinking surface water, at roughly latitudes 40°S and 55° ± 5°S respectively. The easterly flowing West Wind Drift, with its northerly component of flow, is a further barrier. The Drift and to a lesser extent the oppositely flowing Antarctic Circumpolar Current contribute to the relatively uniform circumpolar distribution of the biota. Surface waters are enriched by upwelling at the Antarctic Divergence and elsewhere.

Endemism is high in the Antarctic biota, as might be expected. Among fishes, 65% of the genera and 90% of the species are endemic; corresponding figures for echinoderms are 27% and 73% (Ekman, 1953). Moreover, there is endemism at taxonomically higher levels. About 35% of the macroscopic benthic algal species are endemic (Neushal, in Balech *et al.* 1968), from 80–85% of the dinoflagellates are likewise, and virtually all the tintinnid species (Balech, *ibid.*). Endemism is not great in taxa above the species level.

Hasle (1969) discusses the difficulties in estimating the degree of endemism in the Antarctic diatom flora; she quotes estimates by various authors based on specific collections, which generally include both benthic and planktonic species. Recent estimates are 60–65% of the species. (An additional estimate can be made from the tables of Heiden & Kolbe, 1928, p. 676 *et seq.*, by comparing the total number of species and varieties found from Kerguelen Island south with the corresponding number found *only* from Kerguelen Island south; it is 51%.) Hasle (1969) does not give an estimate of endemism, but one can be made (37%) from the total number of diatom species and varieties she observed (100) and the numbers found restricted to the Southern Ocean (in her distributional groups 4–7, p. 117, plus those so designated in Table 41, p. 135—these totalled 37). Kozlova (1964) cited seven of the 92 species seen as endemic, but made no comment about the others. We note that another of the supposedly characteristic Antarctic forms, *Nitzschia (Fragilariopsis) kerguelensis,* has been found in continuous distribution to 30°N (Van der Spoel *et al.* 1973). At the generic level, most of the important taxa in the Antarctic plankton are cosmopolitan, but the two monospecific genera *Charcotia* and *Micropodiscus* are apparently endemic. *Asteromphalus* and *Tropidoneis* are better represented in the Antarctic flora than elsewhere, as is the section *Fragilariopsis* of the genus *Nitzschia* (Hasle, 1969, 1972b; members of this section were formerly placed in the genus *Fragilariopsis*).

The genera *Chaetoceros, Coscinodiscus, Nitzschia, Thalassiosira* and *Rhizosolenia* account for more than half the common planktonic species in the Southern Ocean. The species most important numerically (as well as for production) are in the size class of the nannoplankton (10–50 μm; terminology of Strickland, 1965, p. 480). The most important diatom species numerically in different latitudinal zones, are given by Hasle (1969; in Balech *et al.* 1968) and Steyaert (1974):

Subantarctic: *Chaetoceros neglectus, Nitzschia barkleyi, N. turgiduloides;*
North Antarctic Zone: *Chaetoceros dichaeta, C. neglectus, Nitzschia pseudonana, N. prolongatoides, N. subcurvata;*
South Antarctic Zone: *N. curta, N. cylindrus, N. subcurvata.*

Some of these species were found in concentrations as high as 10^6 cells/litre. Large species are not uncommon; species such as *Thalassiothrix antarctica, Synedra reinholdii, Corethron criophilum* and *Chaetoceros criophilus* occur up to perhaps 5×10^4 cells/litre, and *Thalassioria tumida* attained $4{\cdot}4 \times 10^5$ cells/litre in patches in the Weddell Sea (El-Sayed, 1971). *Nitzschia* (ex-*Fragilaria*) *kerguelensis* is widespread.

Hasle's summary of the zonation probably applies essentially to the whole of the Southern Ocean, though it was written of the Pacific sector. Briefly the Subantarctic region is distinguished from the more southerly regions by high or diatom diversity and by the addition of allochthonous (visitor) species from the north to the characteristic subantarctic flora, which occurs only occasionally either to the north or the south. The South Antarctic Zone has a group of characteristic species seldom found to the North; many of them are found on or near the ice pack.

The bibliography of Steyaert (1973a) should be consulted for references to regional studies of the Antarctic phytoplankton, and diatoms in particular. She documented for one region large differences in phytoplankton abundance in two years because of hydrographical conditions (Steyaert, 1973b).

The concept of bipolarity arises in context of the Antarctic diatom flora. A general discussion of bipolarity and its occurrence among animal taxa is given by Ekman (1953, Chapter XI), who points out that the critical fact is discontinuous distribution across the tropics (i.e. the taxa involved need not have ranges extending all the way to the poles). Organisms that need not photosynthesize can be transported across (under) the warm tropics in the colder waters below the euphotic zone; this is possible for photosynthetic diatoms only if long-lived resting spores can be transported.

Smayda (1958) concluded that for a species to thrive at both poles it must also become cosmopolitan, which is to say, able to maintain itself within the tropics as well as in colder waters. As with animals, the number of diatoms considered to be bipolar has diminished as knowledge accumulates. Based on the admittedly meagre evidence of the *Challenger* diatom collection, Murray (1895, p. 1 456) concluded that many of the pelagic diatom species of the Arctic and Antarctic were identical; Karsten (1905) regarded 31 species as bipolar, Mangin (1915) thought there were six, Hustedt (1958) mentions four, but three were considered different at the varietal level. Frenguelli (1960) listed seven, with four varieties. Hasle & Heimdal (1968) have shown that the supposedly characteristic Antarctic species *Thalassiosira antarctica* Comber occurs in the Northern hemisphere also and is distinguishable only with great difficulty from at least two other species of the genus. However, *T. antarctica* has not as yet been recorded from latitudes lower than 58° in either hemisphere, hence it remains an example of a supposedly bipolar distribution (as does *Porosira glacialis,* Hasle, 1973a). Populations may yet be discovered in intermediate locations, or, as an alternative hypothesis to explain the distribution, the several similar species that are found at various latitudes may not be genetically isolated.

A large fraction of the primary production in both Antarctic and Arctic seas results from the photosynthesis of algae, mostly diatoms, frozen in the ice (Meguro *et al.* 1967; Burkholder & Mandelli, 1965; Bunt, 1963). The concentrations of plants in the ice (and of plant pigment) can be two orders of magnitude greater than that in the surrounding water, and the production correspondingly higher. The communities of some 10 or 20 species present in the ice contain a few of the surrounding planktonic species, but most species are ones normally considered benthic—usually relatively small cylindrical pennate ones. There are in fact two different communities. One is in ice crystals and the interstitial water of the ice matrix at the bottom surface of ice one or more metres thick. The other is found in the layer of ice and slush between old pack ice and an overlying snow layer. The weight of the snow sinks the old ice and allows seawater to invade the zone between them and provide a habitat. Not many samples have as yet been examined from either polar region, but so far no species common to ice of both polar regions has been reported and in fact, the species reported by various authors and even by the same authors from different sampling locations differ markedly (compare Horner & Alexander, 1972; Meguro *et al.* 1967). There are surely differences in identification, but note that Meguro *et al.* (1967) examined ice algae from both polar regions, and their identifications would be consistent. Also, Bunt & Wood (1963) point out that investigators had, years before, found some of the same species that they did in Antarctic ice. The number of species seen up to now cannot be determined from the literature; we estimate it to be about 60. The important genera are common to northern and southern ice floras. They are *Amphiprora, Gomphonema, Navicula, Nitzschia, Pinnularia, Fragilaria, Pleurosigma,* and *Stenoneis*. In addition, in the Antarctic at least, there occur genera normally planktonic, such as *Chaetoceros, Rhizosolenia, Coscinodiscus,* and *Synedra*. Bunt (1967) has cultured several Antarctic species all of which are cryophilic and capable of extreme shade adaptation. *Fragilaria sublinearis* and *Chaetoceros fragile* can grow after 3 months exposure to total darkness in the absence of added organic matter (Bunt & Lee, 1972), and the first species adapts to low temperatures by reduced respiration, as determined using oxygen isotopes (Bunt *et al.* 1966). All properties examined demonstrate adaptation to the Antarctic environment. Horner & Alexander (1972), using labelled substrates and microradioautographic techniques, found no convincing evidence for heterotrophic growth of Arctic ice algae. Rapid photosynthesis at low temperature and light intensity in the presence of abundant nutrients, plus durability and low respiration in the dark and cold, must account for survival of these cryophilic diatoms.

3.2 *Arctic seas and adjoining waters*

Here are considered the ring of seas lying above the Arctic Circle (about 67°N) together with certain intimately connected regions at slightly lower latitudes. The area is dominated by the ice pack, and even Nansen remarked on the bleakness of some of it. About half the shoreline of the Arctic Seas borders the Soviet Union. Zenkevitch (1963) treats regions individually and it is a measure of the inaccessibility of the East Siberian Sea that there is no chapter on its biota, and no studies of its plankton were found for this review. The closest stations were taken by Nansen at about 78°N, 136°W, at the border of the Laptev and East Siberian Seas (Gran, 1904a). Gran (1908) is the source taxonomic reference for this whole region.

The current systems in the Arctic are more complex than in the Antarctic, even in the relatively land-free waters of the Beaufort Sea (see Johnson, 1956) and in the Arctic Mediterranean (Norwegian, Greenland, and Barents Seas; see Sverdrup *et al.* 1942). There is a general movement of ice from Eastern Siberia across the Polar region to Greenland, as was suggested by the similarity of ice-floe diatoms found near Greenland to those at Cape Wankarema (Chukchi Sea). On the other hand, the characteristic ice diatoms of the intervening seas along the West Siberian and north European coasts were not found around Greenland (see Patrick, 1948, pp. 482–3 for discussion of the work of Cleve and others). Gran (1904a) had remarked the richness of the ice flora compared to that in the plankton beneath the ice; at some stations he could find no phytoplankton—probably in part because only nets were used. The drift of the ice, which exchanges its flora with the water to some extent, as well as the drift of current systems, act to distribute the species, and there is a characteristic flora that is pan-Arctic at least to genus level. If the species are not circumpolar, then most are found over many degrees of longitude to judge from species lists in the literature. There are also contributions to the Arctic plankton by intrusions of Atlantic water into the Norwegian Sea and Davis Strait, and by Pacific water *via* the Bering Sea, in each case adding boreal or temperate species at certain seasons. Salinities of the Siberian seas and other regions are lowered significantly by river flow, especially in shallow areas, where the salinity may not exceed a seventh that of oceanic seawater (Zenkevitch, 1963). These regions have an estuarine or even freshwater phytoplankton.

An excellent bibliography of research on the Arctic has been published as the *Arctic Bibliography*; the index to earlier research on diatoms is given in Vol. III, (1953, p. 3 329). Each volume thereafter is indexed individually. Allen (1971) has written a short review on high latitude phytoplankton.

Beginning at the 0° meridian, which lies east of Greenland, and proceeding to the east, the Polar Seas are: the Greenland Sea and the Norwegian Sea south of it; the Barents Sea with the White Sea below it; the Kara Sea; the Laptev Sea; the East Siberian Sea; the Chukchi (or Chukhotsk) Sea with Bering Sea to its south, connected only by the shallow Bering Strait; the Beaufort Sea; various sounds and straits through the Canadian Polar islands, including Fox Basin and Hudson Bay below it; and last, Baffin Bay and Davis Strait, which lie to the west of Greenland. The series of paragraphs that follow contain at least one reference to each region except, as noted before, the East Siberian Sea, together with remarks on the flora.

(a) *Norwegian Sea*

The Norwegian Sea has been much studied. The papers of Ramsfjell (1960), Paasche (1960), and Paasche & Rom (1962) contain references to the previous literature and Paasche (1961) adds taxonomic and nomenclatural notes on some important species. Paasche (1960) recorded 46 diatom species in the plankton, as did Sakshaug (1972) in the Trondheimsfjord (at 64°N); only 15 species are not listed in common. The Denmark Strait had many of the same forms (Braarud, 1935). The complex hydrography of the Norwegian Sea involves mixtures of Polar, Arctic, Atlantic, and Coastal waters; eight different vegetation areas have been recognized (Paasche, 1960). The following diatoms are important in the various vegetation areas:

I North Norwegian coast: *Nitzschia* (ex-*Fragilariopsis*) *atlantica, N. pseudonana* (ex-*F. nana*) (the dominant), *N. delicatissima, N. closterium, Thalassiosira bioculata* var. *raripora.*

II Atlantic: The species in Area I plus *Chaetoceros decipiens* (a dominant).

III Atlantic–Arctic: The species in Area II plus *Chaetoceros atlanticus, C. borealis, C. debilis* (a dominant), *C. densus, C. teres, Corethron hystrix, Coscinodiscus centralis, Eucampia zoodiacus, Rhizosolenia alata, R. styliformis* (a dominant), *Thalassiosira gravida, Thalassionema nitzschioides, Thalassiothrix longissima.* (A 'styliformis' community forms an important part of the total population.)

IV A narrow Atlantic–Arctic zone with some Polar influence: Most of the species in Area III, with *Rhizosolenia hebetata,* f. *semispina* and *Thalassiothrix longissima* replacing *Rhizosolenia alata* and *R. styliformis* as dominants, and with the addition of *Asteromphalus robustus, Coscinodiscus curvatulus, C. kützingii, C. lineatus, Thalassiosira nordenskioldii, T. hyalina, T. poroseriata* (ex-*Coscinosira poroseriata*; Hasle, 1972d), *T. excentrica* (ex-*Coscinodiscus excentricus*; Fryxell & Hasle, 1972). (A 'semispina' community is important in this region.)

V, VI Polar: Populations are very variable from year to year; the dominants during the period of study were those of Area II plus *Thalassiosira gravida.* The 'arctic-neritic' species expected, such as *Achnanthes taeniata, Nitzschia grunowii* (ex-*Fragilaria oceanica*; Hasle, 1972b) or *Chaetoceros furcellatus,* were not found.

VII Polar, near Jan Mayen: Variable, mostly larger species of Areas III or IV; the species most important quantitatively were *Chaetoceros decipiens, Nitzschia delicatissima,* and *Thalassiosira gravida.* Species characteristic of Area VII were *Chaetoceros convolutus, C. borealis* f. *concavicornis, Thalassiosira hyalina, T. nordenskioldii,* and *Rhizosolenia alata.*

VIII Arctic or Polar: The characteristic components were *Chaetoceros furcellatus, C. septentrionalis* (?), *Nitzschia grunowii, N. pseudonana, N. closterium* (?), *Rhizosolenia hebetata* f. *semispenia, Thalassiosira gravida,* and *T. nordenskioldii.*

(b) *Barents Sea*

The Barents Sea plankton was examined by Gran (1904b), Meunier (1910), and Kisselew (quoted in Zenkevitch, 1963, p. 91). Kisselev reports 92 diatom species, of which the dominant were *Chaetoceros diadema, Coscinodiscus subbulliens, Corethron criophilum, Skeletonema costatum, Rhizosolenia styliformis,* and *R. semispina*. To these, and to the list of species reported for the Norwegian Sea, Gran (1904b) added the oceanic species *Coscinodiscus oculus-iridis* and *Chaetoceros criophilum* as well as the neritic *Melosira hyperborea* and *Eucampia groenlandica*. Meunier's list (1910) is the largest and serves as a convenient check-list for all Arctic species—it also covers the Kara Sea.

(c) *White, Kara and Laptev Sea*

The White Sea phytoplankton is neritic in character and dominated chiefly by species also found to the West—*Skeletonema costatum* and *Chaetoceros* species in summer, but by *Nitzschia grunowii* Hasle (ex-*Fragilaria oceanica*) in spring (Konoplya, 1973). Fedorov (1969) lists 46 species, including *Asterionella bleakeleyi*.

Samples from the Kara Sea were examined by Meunier (1910). Usatchev (1947, quoted in Zenkevitch, 1963) reports 52 species of diatoms. The Kara Sea has very low surface salinities in summer.

The Laptev Sea is even more influenced by river run-off than the Kara Sea. The low salinities may extend 500 km from shore. Kisselew (1932, quoted in Zenkevitch, 1963) estimated that only 5% of the phytoplankton was of marine origin. He listed the important diatom species as *Thalassiosira baltica, Coscinodiscus marginatus, Chaetoceros gracile, C. wighami, Caloneis brevis,* and species of *Navicula*. Gran (1904a) as mentioned before, had examined samples taken by the *Fram* expedition near the New Siberian Islands. The phytoplankton in the water was not at all abundant, and consisted of *Chaetoceros boreale, C. decipiens, C. contortum, C. criophilum, Coscinodiscus oculis-iridis,* and *Nitzschia grunowii* Hasle (ex-*Fragilaria oceanica*). The ice had a larger population and more species, most of which were, as remarked earlier, known from Greenland waters. (Many are also listed from the Norwegian Sea and even the fjords.)

(d) *Chukchi, Beaufort Seas and Hudson Bay*

The shallow Chukchi Sea is also subject to wide fluctuations in salinity. Shirshov (1936, quoted in Zenkevitch, 1963) found the important species to be *Thalassiosira baltica, Coscinodiscus marginatus, Chaetoceros gracile, C.*

wighami, Caloneis brevis, and a *Navicula.* The distribution of animals, which has been studied more, follows the complex current patterns, with Pacific boreal species from Bering Sea being mixed with high Arctic species and even Atlantic species. The phytoplankton has apparently not been observed in such detail.

The stations occupied by Bursa (1963) at Point Barrow are at the juncture of the Chukchi and Beaufort Seas. The station with highest salinity was offshore and contained the most marine diatom species, some 46 in number, found in and under pack ice. There were eight species of *Chaetoceros, Nitzschia closterium,* species of *Navicula* and *Thalassiosira,* and the typical neritic *Skeletonema costatum* and *Leptocylindrus danicus.* No additional studies are recorded.

Bursa (1961a, 1961b) also studied the phytoplankton in Hudson Bay, recording 88 species of diatoms. The communities resembled those on the two sides of the boreal Atlantic. The influence of the Labrador Sea was detected.

(e) *Baffin Bay and Bering Sea*

The circle about the Arctic is completed by Baffin Bay and Hudson Strait. Seidenfaden (1947) has reviewed the literature on phytoplankton of the region. He lists some 80 diatom species, grouped as typical arctic, boreal, transition, temperate (autumn visitors), and ice or littoral forms. Most of the species in his first three categories are also in Meunier's (1910) list of Barents Sea diatoms, showing the generally pan-Arctic nature of the flora. Interestingly, Seidenfaden does not list *Skeletonema costatum.*

Bering Sea, which is connected to the Arctic Ocean by the shallow Bering Strait, is conveniently considered at this point. The paper by Karohji (1972) refers to most of the previous studies (by Maruno, Takano, Karawada, and others). Recent studies have mostly used either the settling method or the centrifuge technique, which Phifer (1934) had used here much earlier. The list of species is familiar, with typical boreal oceanic and neritic species and the arctic neritic species *Chaetoceros furcellatus, Nitzschia grunowii* (ex-*Fragilaria oceanica*), and *Thalassiosira hyalina.* The pan-Arctic community thus extends at least partly into Bering Sea. Cosmopolitan temperate oceanic and neritic forms also occur, and *Denticula seminae* and *Stephanopyxis nipponica,* reflecting Pacific influence (Phifer, 1934; Marumo, 1956). *Denticula seminae* (ex-*D. marina* Semina) is so far known only from the northwestern Pacific (Simonsen & Kanaya, 1961). Karohji (1972) distinguished five major regions containing different phytoplankton communities. He found that overall the dominant net-caught diatoms in Bering Sea were:

> *D. seminae, Thalassiothrix longissima, Chaetoceros atlanticus, C. convolutus, Coscinodiscus curvatulus, C. oculus-iridis, Nitzschia seriata,* and *Rhizosolenia hebetata* f. *hiemalis.*

There were neritic populations too, with *Thalassiosira nordenskioldii* important.

3.3 *Pacific Ocean and adjacent seas*

(a) *Biocoenoses and thanatocoenoses*

An overall picture of the abundance and distribution of diatoms of the North Pacific and Southwestern Pacific Basins (roughly Bering Sea to the Antarctic, and Australia to 120°W longitude) has been provided by Soviet biological oceanographers using standardized techniques (Kozlova & Mukhina, 1967; Jousé *et al.* 1971). Phytoplankton samples (of about litre volumes) were collected by membrane filtration, and seston from the euphotic zone and from deep waters was collected by centrifugation. Bottom sediment samples were taken for examination of diatom shells and silicoflagellate skeletons in contemporary deposits (the 'thanatocoenoses'). The species dominating (or even presnt) in the sediments are by no means necessarily those most abundant in the euphotic zone whence they came, but regional differences in thanatocoenoses usefully reflect relatively stable differences in the corresponding living plankton populations ('biocoenoses') both in regard to abundance and species composition. Seven thanatocoenoses are defined by Jousé *et al.* (1971):

Arctoboreal thanatocoenosis: *Thalassiosira nordenskioldii; T. gravida; Bacteriosira fragilis; Chaetoceros furcellatus; Biddulphia aurita.*

Northboreal thanatocoenosis: *Thalassiosira excentrica; Coscinodiscus curvatulus; C. marginatus; Actinocyclus divisus; Rhizosolenia hebetata* Gran. (f. *hiemalis*); *Thalassiothrix longissima; Denticula seminae.*

Subtropical thanatocoenosis: *Thalassiosira decipiens; T. lineata; Coscinodiscus radiatus; Thalassionema nitzschioides* (a complex of forms); *Pseudoeunotia doliolus; Roperia tesselata; Nitzschia bicapitata; N. interrupta; N. sicula; C. wailesii.*

Tropical thanatocoenosis: *Coscinodiscus crenulatus; C. nodulifer; Hemidiscus cuneiformis; Thalassiosira oestrupii; Rhizosolenia bergonii; Nitzschia marina; Ethmodiscus rex; Planktoniella sol.*

Equatorial thanatocoenosis: *Asteromphalus imbricatus; Coscinodiscus africanus; Triceratium cinnamoneum; Asterolampra marylandica.*

Subantarctic thanatocoenosis: *Fragilariopsis antarctica* (now *Nitzschia kerguelensis*, see Hasle, 1972b); *Coscinodiscus lentiginosus; Thalassiothrix antarctica; Schimperiella antarctica.*

Antarctic thanatocoenosis: *Nitzschia* (ex-*Fragilariopsis*) *curta; Nitzschia* (ex-*Fragilariopsis*) *cylindrus; Eucampia balaustium; Thalassiosira gracilis; Charcotia actinochilus; C. oculoides; C. symbolophorus; Biddulphia weissflogii.*

Distinction between neritic and oceanic planktonic populations can be detected in the thanatocoenoses, as can the influences of major currents such as the cold Oyashio and the warm Kuroshio. The influence of the Equatorial upwelling is seen. The regions of maximum diatom deposition are the belt just south of the Antarctic Convergence, and the boreal Okhotsk and Bering Seas. In the greatest depths of the sea, notably of the Philippine and Marianas trenches, there occur deposits of shells of *Ethmodiscus rex,* a huge diatom that can attain a diameter of 1·8 mm. This species and others of the Tropical thanatocoenoses listed above are never abundant in the plankton but have shells that resist dissolution during sinking or upon sedimentation, at least in certain places. (Diatoms of recent origin that fall onto red abyssal clays may dissolve; Jousé *et al.* 1971). It is estimated that in some regions of relatively heavy deposition e.g., on the Antarctic Shelf, or in the Bering Sea, the proportion of the planktonic shells that survive in the sediments is in the range 0·06–6%. In sediments underlying oceanic regions of low productivity the proportion is doubtless smaller, though no estimate is given (essentially no diatoms occur in sediments). Just the opposite occurs in oceanic regions of Antarctic Seas. There, virtually the entire diatom crop, largely *Nitzschia kerguelensis* (ex-*Fragilariopsis antarctica*) is deposited (Kozlova & Mukhina, 1967). The diatoms act as a mechanism for depositing upwelled silicic acid.

The zonation for biocoenoses chosen by Kozlova & Mukhina (1967) is slightly more elaborate than that for thanatocoenoses cited above (Jousé *et al.* 1971) and the boundaries for the zones do not match exactly. The biocoenoses, determined from examination of seston collections, are given as follows:

Subarctic: 13 species, with *Thalassiosira nordenskioldii, Nitzschia grunowii* (ex-*Fragilaria oceanica*), *N.* (ex-*Fragilaria*) *cylindrus,* and *Chaetoceros subsecundus* as dominant (samples in one location only).

Boreal: 34 species which include *Denticula seminae, Asteromphalus robustus, Thalassiothrix longissima, Coscinodiscus radiatus, Chaetoceros messanensis, C. concavicornis.*

North Tropical: 19 species containing *Skeletonema costatum* (this may include *S. tropicum,* see Hulburt & Guillard, 1968), *Plantoniella sol, Nitzschia bicapitata, N. delicatissima,* and *Pseudoeunotia doliolus* among them.

Equatorial: 23 species having *Thalassiothrix frauenfeldii, Coscinodiscus nodulifer, Asteromphalus diminutus, Rhizosolenia calcar-avis, R. bergonii, Thalassionema nitzschioides, Nitzschia bicapitata, N. oceanica,* and *N. delicatissima* as characteristic.

South Tropical: No species given (flora too poor).

Intermediate zone (mixed Subantarctic and Subtropical waters): 54 species;

Nitzschia kerguelensis (ex-*Fragilariopsis antarctica*) and *Thalassiothrix antarctica* are the principal species.

Subantarctic: 41 species including *Nitzschia kerguelensis* (ex-*Fragilariopsis antarctica*), *Coscinodiscus lentiginosus, Thalassiothrix antarctica,* and *Dactyliosolen antarcticus.*

Antarctic: 34 neritic species with *Nitzschia* (ex-*Fragilariopsis*) *curta, N.* (ex-*Fragilariopsis*) *cylindrus, N. angulata* (ex-*F. rhombica*) and *Thalassiosira gracilis* as dominants, and a small unspecified number of oceanic species which include *Nitzschia kerguelensis* (ex-*Fragilariopsis antarctica*) and *Thalassiothrix antarctica.*

(b) *North Pacific and Sea of Okhotsk*

The phytoplankton of Bering Sea, which was considered as part of the Arctic Seas, is very similar to that of Northern Pacific waters (Karohji, 1972) and indeed to that of the boreal Sea of Okhotsk, which lies west of the Kamchatka Peninsula and the Kuril Islands (Smirnova, 1959; Karawada, 1960; Zenkevitch, 1963, Chapter 15). According to Zenkevitch (1963), 290 species of diatoms (unlisted) have been identified from the Okhotsk Sea. Oceanic and neritic species cited as abundant in one location or another by Zenkevitch or Karawada (who examined sedimented plankton samples) include the following:

Chaetoceros concavicornis, C. convolutus, C. constrictus, C. debilis, C. furcellatus, C. atlanticus, C. borealis, Thalassiosira nordenskioldii, T. gravida, Detonula confervacea, Denticula seminae (ex-*marina*), *Nitzschia grunowii,* (ex-*Fragilaria oceanica), Nitzschia closterium, Rhizosolenia fragilissima, Bacterosira fragilis, Leptocylindrus danicus, Skeletonema costatum, Coscinodiscus marginatus,* and *Thalassiothrix longissima.*

On the surface of sediments from great depths of the Okhotsk Sea, the durable shells of *Coscinodiscus oculus-iridis, C. marginatus, Thalassiothrix longissima* and *Rhizosolenia hebetata* dominate (Zenkevitch, 1963); in shallow waters, thinner shelled and neritic species are also represented in the deposits.

Karohji (1972) recognized six regions across the subarctic Pacific from the waters east of the Kuril Islands to those south of the Alaska Peninsula. The regions were characterized by hydrographic features, overall phytoplankton abundance, and relative abundances of species. There was no evidence for localization (regional endemism) of any of the planktonic diatom flora. Assemblages differed largely in species abundance, the species being those just listed in connection with the Bering and Okhotsk Seas.

(c) *Waters around Japan*

The waters around Japan have been studied extensively by a variety of techniques. Karawada *et al.* (1968b) distinguished essentially seven regional phytoplankton populations (with subdivisions). To the east of Japan the populations reflect the neritic influence and the presence of the Kuroshio current and its associated currents. Motoda & Marumo (1963) summarized knowledge of the plankton, including diatoms, of the Kuroshio; they consider the tropical oceanic diatoms *Gossleriella tropica, Hemiaulus hauckii* and *Chaetoceros coarctatus* to indicate relatively undiluted water of Kuroshio origin, though they list 17 other species, including eight of *Rhizosolenia,* as characteristic inhabitants of the Kuroshio. The Kuroshio water divides, and, as pointed out by Karawada *et al.* (1968b), the diatom communities in it develop differently in the various regions invaded.

The phytoplankton to the west of Japan, in the Sea of Japan, has been studied by Kisselew (1959), who recorded 137 species of diatoms, most of which seem cosmopolitan (all but about 20 occur in European waters). Zenkevitch (1963) says 306 diatom species had been recorded from the Sea of Japan. Konovalova (1972) observed seasonal changes of communities in Amur Bay. Choe (1966) and Uhm & Yoo (1967) have observed that diatoms in the Korea Strait show the mixing of neritic and boreal populations with Kuroshio populations. This also occurs in the shallow East China Sea (Karawada *et al.* 1968b) and to some extent at least in the Yellow Sea (Aikawa, 1936; pp. 130–136), which was characterized by *Coscinodiscus* species, by *Biddulphia sinensis* and *B. longicruris,* and by the tychopelagic *Licmophora lyngbyei,* presumably originating from the shallower waters around the Korean Archipelago. Aikawa attributes the presence of characteristic Kuroshio diatoms in the Gulf of Chihli, far to the north in the enclosed Yellow Sea, to upwelling of subsurface water near the coast. Choe (1967) has surveyed the neritic phytoplankton along the Korean coast of the Yellow Sea.

There are no recent reports of plankton studies from the Yellow Sea, the South China Sea, or the Gulf of Siam. The Philippine Sea has been crossed by sections south from Japan towards New Guinea along meridians 137°E and 130°E by Karawada *et al.* (1968a) and along 135°E and 130°E by Belyaeva (1970). Belyaeva's net samples are discussed later in connection with Equatorial waters. Rich neritic communities prevail near Japan and New Guinea, but oceanic diatom species prevail in the several currents or water bodies lying between the islands. These waters include, besides the Kuroshio Current and its Countercurrent, the diatom-poor North Equatorial Current, the Equatorial Countercurrent, and the Equatorial Undercurrent, containing New Guinea Coastal Current water (Karawada *et al.* 1968a). The 12 most abundant oceanic diatoms seen in the settled litre samples of Karawada *et al.* (1968a) are, with two exceptions, on Belyaeva's list of net-caught diatoms (see section following, on the Equatorial Pacific; one of the names missing, *Nitzschia seriata,* may be on Belyaeva's list under another species name).

(d) *Indonesian Seas*

There are few studies in Indonesian seas. Mann (1925) described planktonic species only incidently as they occurred in his dredging samples, and, in any event, the locales of most samples are unfortunately not known. Allen & Cupp (1935) compiled a flora of 117 species and 10 varieties of net-caught diatoms from the Java Sea (110°–113°E, 2°–10°N). Wood (1963a) listed many species from 'Indonesian waters' but gave no station locations or abundance estimates. He stresses the large number of essentially benthic species that occur in plankton catches in these shallow seas. We know of no distributional studies in this region.

Diatoms at four locations in the pearl oyster fishery grounds of the shallow Arafura and eastern Timor Seas were examined by Takano (1960a), who lists some 61 planktonic and 36 benthic species taken in net hauls or bottom samples. Though the waters are shallow, pan-tropical oceanic species were common. The cyanophyte *Trichodesmium* dominated in some regions, while diatoms did in others. Dominant forms were *Chaetoceros* species, *Bacteriastrum* species, *Coscinodiscus* species, *Rhizosolenia* species, and, in a few places, *Guinardia flaccida.* Takano did not specifically mention the interesting *Rhizosolenia arafurensis,* though it has been recorded there recently (Wood, 1963b). It has been reported from the Atlantic (see Mills' Index) but not to our knowledge from waters between. Phytoplankton communities of the Arafura, Coral, and Tasman Seas are discussed by Wood (1964) but details are not given.

(e) *Australian and New Zealand waters*

The papers of Wood and co-workers describing the flora of the Australian–Indonesian–New Zealand region were cited in the section 'Identification and taxonomy'. The early valuable study of the coastal phytoplankton off New South Wales (Port Jackson) by Dakin & Colfax (1940) listed 84 diatom species caught by nets (approximately 125 μm mesh). Inasmuch as the authors acknowledge the help of Allen and Cupp, identifications made for many diatoms over a wide area—the Java Sea, Australia, and the Pacific coast of North America—are consistent. Virtually all the species recorded are found also along the East African coast (F.J.R. Taylor, 1966; Sournia, 1970a). Grant & Kerr (1970) identified 241 phytoplankton species (not listed) off Port Hacking (south of Port Jackson). Diatoms dominated except during the periods of lowest phytoplankton abundance, when dinoflagellates were more numerous. The most important species throughout the year were the diatoms *Leptocylindrus danicus, Navicula* sp., *Nitzschia longissima, Rhizosolenia alata* and *R. stolterfothii.*

It is interesting that there appears to be essentially no endemism in the planktonic diatom flora of the Australian region. On the other hand, the southern Australian attached algal flora has an endemism greater than that of any other part of the world (Womersley, 1959), being as high as 30% of the genera and 75% of the species among members of the Rhodophyceae. One might on this basis expect some degree of endemism among the purely benthic diatom species. An indication of this is seen by comparing the species lists in papers by Wood and co-workers (see above) for the Pacific, with those in Wood's studies of the Atlantic flora (Wood, 1965b, 1966, 1968). Most of the planktonic species are the same, but the species of benthic or tychopelagic genera differ noticeably. Of course, this may only reflect a tendency to use finer taxonomic criteria in these genera (e.g., *Navicula*).

F.J. Taylor (1970) has compiled a useful check-list of diatoms from New Zealand coastal waters. Of some 420 taxa about 115 species are considered to be essentially planktonic. Cassie (1961) studied the net phytoplankton of the current systems around New Zealand; about a dozen diatom species were restricted to only one or two current zones and hence were useful indicator species. Four of these species, *Pleurosigma naviculaceum, P. balticum, Synedra fulgens,* and *Amphora chinensis,* which are confined to the East Auckland Current (north of New Zealand), have not been reported from the Western Indian Ocean, as have most other diatoms of the region.

(f) *Equatorial waters*

Diatoms of equatorial and immediately adjacent Pacific waters from roughly New Guinea to Panama have been included in recent studies by several workers using different techniques, as follows: Belyaeva (1970, 1971), nets; Desrosieres (1969), Hardy recorder; Hasle (1959), inverted microscope; Lapshina (1971), nets; Marshall (1972), inverted microscope; Smayda (1966), inverted microscope. (The study by Karawada *et al.* 1968a, by sedimentation, was discussed in connection with the Philippine Sea, south of Japan, p. 403.)

Belyaeva's (1970, 1971) nets caught 110 species, mainly pantropical, in the western region and 97 generally similar species in the central and eastern region (there were a few unidentified species and two forms that could not be identified to the genus level). Of the 31 most common species caught in the western region, which are listed below, all but four (indicated by *) were also reported from the central and eastern region:

*Ethmodiscus rex; Coscinodiscus gigas**; C. janischii**; C. concinnus; C. nodulifer; C. crenulatus; Asterolampra marylandica; Planktoniella sol; Roperia tessellata; Rhizosolenia bergonii; R. styliformis; R. calcar-avis; R.*

*alata; Thalassiosira subtilis; T. lineata**; Chaetoceros peruvianus; C. lorenzianus; C. atlanticus; Hemiaulus hauckii; Nitzschia bicapitata; N. sicula; N. braarudii; N. delicatissima; N. marina; Thalassiothrix gibberula; Thalassionema* sp.*; *T. nitzschioides; Mastogloia* (also called *Stigmaphora*) *lanceolata; M. rostrata; Amphora* sp.; *Tropidineis* sp.

Of species commonly found in the eastern region (occurrence in >30% of the samples) only *Chaetoceros coarctatus* and *Pseudoeunotia doliolus* were not among the most abundant to the west (list above). The following species were in general restricted to oligotrophic regions:

Hemiaulus hauckii, Mastogloia (Stigmaphora) rostrata, M. lanceolata, Rhizosolenia calcar-avis, R. styliformis, Asterolampra marylandica, Tropidoneis sp.; *Amphora* sp.

Other species, typified by *Nitzschia bicapitata, N. sicula,* and *Thalassionema nitzschioides* to the west and *Coscinodiscus crenulatus, C. nodulifer, Nitzschia marina* and *Pseudoeunotia doliolus* to the east, occurred mainly in the waters enriched by upwelling. Still other species behaved as if relatively indifferent, being ubiquitous but more abundant in eutrophic waters. Catches with the Hardy plankton recorder (Desrosieres, 1969) illustrated dramatically how the crop of larger phytoplankters decreases proceeding westward from the region of the Galapagos into progressively warmer and less fertile water where the number of species encountered also decreases. The species listed by Desrosieres (1969) are just the larger eutrophic indicator species given by Belyaeva (*Coscinodiscus* sp., *Asteromphalus* sp., *Chaetoceros* sp., *Rhizosolenia bergonii, Pseudoeunotia doliolus, Planktoniella sol,* and *Hemidiscus cuneiformis*). Lapshina (1971) extended observations with net samplings northward towards Baja California. Only 11 species were identified, of which nine are cited by Belyaeva; the other two are known to occur in the region (Smayda, 1966).

Marshall (1972) examined the phytoplankton from about 225 miles west of the Galapagos to a like distance from the Gulf of Guayaquil of the South American coast. He also examined plankton samples taken within the Galapagos Island group. About 58 forms were recorded, of which about 23 were not recorded by Belyaeva (1970, 1971). This reflects two phenomena. First, Marshall sampled by examining the concentrate from a 500 ml sample with the inverted microscope, thus his records of the smaller species are more complete, while rare species could be missed, whether large or small. Secondly, to judge from some species seen by Marshall but not recorded by Belyaeva, the easternmost stations reflected some influence of the cold Peru Current system, which joins the South Equatorial Current. Hasle (1959, 1960) used the inverted microscope to examine concentrated samples from a relatively productive part of the Equatorial Pacific (where populations reached a density of 7×10^4/litre). By far most of the population consisted of small pennate species, largely *Nitzschia delicatissima* and other *Nitzschia* species, a Thalassiothrix (?) species, and another unclassified pennate diatom thought to be a weakly silicified form of *Thalassionema nitzschioides*. In this rather rich tropical oceanic region, 24 of the 56 forms of diatoms seen were pennate ones, conforming to the general opinion that pennate species become important relative to centric ones in proportion to the fertility of the water. Many of these small diatoms are not caught by other methods of sampling. Of the 50-odd species seen, four were reported from the Pacific for the first time and seven were new:

New records: *Thalassionema elegans, Thalassiothrix vanhoffenii, Nitzschia bicapitata, N. sicula.*
New species: *Actinocyclus parvus, Dactyliosolen curvatus, Thalassiothrix gibberula, Nitzschia braarudii, N. longicollum, N. oceanica, N. gaarderi.*

In spite of these new finds, we estimate that 13 of the 33 'most common' species of Belyaeva (1971) were not seen by Hasle. These include *Chaetoceros coarctatus* and *C. crenulatus*, and other large but relatively scarce species easily caught in nets. (What appears to be a remarkable find of one shell of *Thalassiosira antarctica* (Hasle, 1960) is probably to be regarded as a dubious identification; Hasle & Heimdal, 1968.)

Further to the east in the Pacific, cycles of diatom abundance and occurrence in the Gulf of Panama have been followed by Smayda (1963a, 1966), who recorded 175 species (with three new forms and one new species). In this neritic region the population densities of diatoms reached $1 \cdot 5 \times 10^6$/litre at times of upwelling. The communities changed composition considerably, but the important members were generally drawn from these:

Rhizosolenia delicatula, R. stolterfothii, R. fragilissima, Skeletonema tropicum (designated *S. costatum* f. *tropicum*), *Chaetoceros affinis, C. brevis, C. compressus, C. costatus, C. curvisetus, C. didymus, C. laciniosus, C. laevis, C. lorenzianus, Nitzschia delicatissima, N. pacifica* and *pungens, Lauderia annulata, Eucampia cornuta, Bacteriastrum hyalinum, B. elegans, B. varians,* and *Thalassiosira aestivalis.*

Smayda's check-list for the extreme eastern Pacific can be compared with the longer ones of Sournia (1970a) and of Taylor (1966) for the extreme western Indian Ocean; Smayda's list of 175 names contains only 29 species not on Sournia's list and 36 not on Taylor's list.

(g) *Central North Pacific*

Venrick's study (1969, 1971, 1972) in the North Central Pacific along meridian 155°W from coastal waters of the Alaska Peninsula almost to

Hawaii (26°N) included populations from four basically different regions (Fig. 12.4). The three northernmost ones are considered Subarctic Pacific domains. These are: the Coastal (or neritic), lying over the Alaska shelf; the Central Subarctic, associated with the cyclonic Alaskan Gyre; and the Transition Domain, which, containing waters of the warm Kuroshio and the cold Oyashi currents, is part of the large North Pacific anticyclonic gyre. The southernmost region is the Central (North) Pacific Domain. Venrick used the inverted microscope, examining 100 ml subsamples. This technique was adequate in the more eutrophic waters. However, in the oligotrophic Central Pacific Domain, where most species were less abundant than 3 cells/100 ml, the probability of missing a species (in one sampling) is ⩾5%, so that qualitative population sampling was supplemented by use of a 153 μm Phinque net, which yielded some 20 additional species (even though the net sampled only 3% of the crop). Venrick (1969) observed 110 species and 20 varieties.

As a first point, her list contains virtually all the species recorded (often under different names) in the discussion of the Equatorial Pacific region immediately preceding this. In considering the distributions of individual

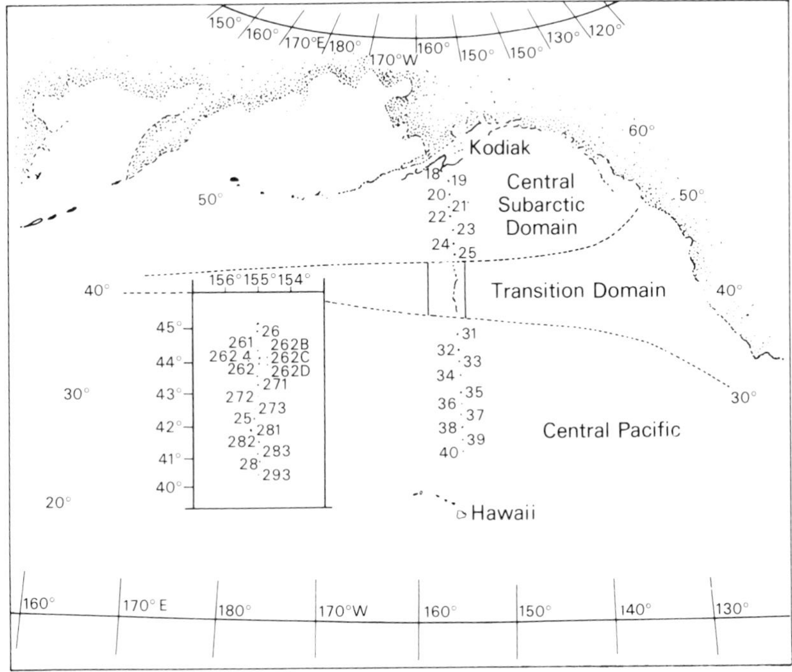

Fig. 12.4. Station positions of expeditions URSA MAJOR Aug–Sept 19yr in relation to the neritic environment (200 m continental shelf contour) and three oceanic environments. From Venrick (1971).

species, Venrick (1969) remarked on the low fidelity of most diatom species to their environment. Isolated and apparently living specimens were encountered surprisingly often at large distances from their known centres of abundance. A cell of one oceanic species was found 600 nautical miles from the nearest record, and one specimen of a benthic diatom (*Actinoptychus undulatus*) was found in the Transition Domain some 3 000 miles downstream from its presumed origin on the coast of Japan. Of the 110 species seen, 14 were called cosmopolitan in the sense that they apparently grew in all four regions of the study. These were:

Cosmopolitan species: *Asteromphalus heptactis, Chaetoceros atlanticus, Thalassiosira excentrica* (ex-*Coscinodiscus excentricus*), *T. excentrica* var.?, *Dactyliosolen mediterraneus, Rhizosolenia alata, R. hebetata* f. *semispina, R. styliformis, Cymbella* sp. ‘a’, *Nitzschia* sp. ‘a’ *(Fragillariopsis), Nitzschia bicapitata/braarudii, Pleurosigma directum, Synedra vaucheria capitellata, Thalassionema nitzschioides.*

Of 65 species found in the Central Pacific, the 40 listed below occurred only there:

Species found only in the Central Pacific: *Actinocyclus ehrenbergii, Asterolampra grevillei, A. marylandica, A. vanheurckii, Asteromphalus arachne, A. roperianus, A.* species ‘a’, *Bacteriastrum biconicum, Chaetoceros dadayi, C. bacteriastroides, C. bacteriastroides* var.?, *C. lorenzianus, C. messanensis, Climacodium biconcavum, C. frauenfeldianum, Coscinodiscus concinnus granii, C. nobilis, C. nodulifer, Ethmodiscus gazellae, E. rex, Gossleriella tropica, Guinardia flaccida, Hemiaulus hauckii, H.* species ‘a’, *Podosira hormoides, Rhizosolenia bergonii, R. calvar-avis, R. castracanei, R. imbricata shrubsolei, R. robusta, Amphiprora* species, *Fragillariopsis (Nitzschia)* species ‘b’, *Licmophora abreviata, Mastogloia capitata, M. rostrata, Nitzschia kolaczekii, N. sicula, N.* species ‘a’, *Synedra* species ‘a’, *Thalassionema baccillaris.*

Three additional species of *Rhizosolenia* occurred regularly, giving this genus the most species of any in the region.

In the Transition Domain there occurred 42 species in all, including cosmopolites and species common to neighbouring domains, but only the following five species occurred only in this Domain: *Hemidiscus cuneiformis, Thalassiosira decipiens* (an unnamed variety), *Thalassiosira subtilis, Actinocyclus* sp. (possibly related to *A. divisus*), an undescribed *Podosira* sp. (At least three of these are ‘endemic’ only in the sense of not being found in the waters immediately to the north and south.) Several other species were also characteristic of the Transition Domain in that their maximum abundances occurred there; flowerings could be in summer, winter, or, in the case of *Rhizosolenia alata,* both.

The Central Subarctic Domain was populated by 24 species, including the 14 cosmopolites. Four species were found only in this domain (though common elsewhere in the oceans); *Achnanthes longipes, Rhizosolenia hebetata* (f. *hiemalis*), *Coscinodiscus marginatus, Actinocyclus* sp. (possibly a form of *A. curvatulus*, which also appears in the literature as *Coscinodiscus curvatulus* and *C. crenulatus*).

The 20 species that occurred only in the Coastal Domain are common neritic forms. All but *Nitzschia heimii* of the following are listed by Cupp (1943) or Gran & Angst (1931):

> **Coastal Domain species:** *Actinoptychus undulatus, Bacteriastrum delicatulum, Biddulphia longicruris, Cerataulina bergonii, Chaetoceros compressus, C. debilis, C. didymus, C. laciniosus, C. radicans, Coscinodiscus angstii, C. wailesii, Ditylum brightwellii, Leptocylindrus danicus, Rhizosolenia fragilissima, R. stolterfothii, Skeletonema costatum, Thalassiosira rotula, Asterionella glacialis, Nitzschia heimii, Pleurosigma normanii.*

N. heimii, once thought restricted to the Antarctic, is known from the North Atlantic also (Hasle, 1965).

Venrick (1969) noted an important point regarding the boundary between the neritic and oceanic (central Subarctic) domains. The 20 species found only in the neritic region (plus two others) dominated its plankton during summer (periods of high population) but comprised not over half the population in winter, when oceanic species such as *Denticula seminae* and *Chaetoceros atlanticus* dominated. The abrupt boundary between the two domains was established by the large gradient in numbers of neritic species, not oceanic ones. Venrick suggests that the neritic species may tend to form rapidly sinking spores under unfavourable growing conditions, thus disappear rapidly when such circumstances occur (as in winter), while the species wholly adapted to pelagic life survive to various extents as neritic and oceanic waters mix, finally outnumbering the depleted neritic species.

The diatom flora of the western North American coast is described in Cupp's (1943) standard work, which serves also as primary source of references. The relation of diatom populations to hydrography in Puget Sound was described by Gran & Thompson (1930). Allen (1945) contributed a seasonal study, using nets, of diatoms off Southern California.

(h) *Gulf of California*

Round (1967) has summarized his own and previous studies in the Gulf of California. To our knowledge only net collections have been examined to date. The flora is a characteristic pantropical one, of mainly oceanic species. The most productive zone (Round's 'zone 3'), which was enriched by offshore upwelling, lies about half way up the Gulf. Its flora was reduced in species compared to that of the oceanic region outside the Gulf, but more abundant in quantity. Large populations occurred there, usually confined locally to a striking degree. In spring the large discoid centric genera *Coscinodiscus, Asteromphalus,* and *Actinoptychus* were dominant in Round's samples. In earlier studies, *Asterionella glacialis* (ex-*japonica*) had been found dominant, but Round did not record the species at all. In the autumn, the large discoid species (and *Pseudoeunotia*) were replaced by more neritic species such as *Skeletonema costatum, Thalassionema nitzschioides,* also by species of *Thalassiothrix, Chaetoceros,* and *Nitzschia seriata*. The neritic zone was less rich than the open waters at all times. The diatom thanatocoenoses, also patchy, occurred in zones agreeing essentially though not exactly with those of the biocoenoses above. The abundant delicate species from the plankton scarcely occurred in most sediments (except possibly in some 'laminated sediments'). As is usual in other regions, the dominant diatoms in the cores are heavy walled species not always abundant above. *Cyclotella striata* and *Nitzschia marina* were common in sediments but not frequent in the plankton—perhaps these are missed by nets and would occur in sedimented samples.

(i) *Peru Current and South Pacific*

Smayda's studies of the rich Gulf of Panama (1963a, 1966) were referred to before (see p. 408). Marshall (1970), using a settling technique and Palmer-Maloney chamber, found 39 diatom species between the Gulf of Panama and coastal Ecuador (8°N to 2°S); virtually all the species are listed by Smayda (1966). Populations did not exceed 7 350/litre. Marshall, citing previous studies of the Peru Current region, did not find certain supposedly characteristic Peru Current species at his stations, which were all in warm water ($T > 25$°C). However, the list of 50-odd species found in Peruvian coastal waters at 12°S–16°S by Strickland *et al.* (1969) was very similar to Marshall's. Notable differences were the absence of *Bacteriastrum* from the Peru Current and of *Actinocyclus, Asterionella,* and *Roperia* from Marshall's list.

Belyaeva (1972), using 140–180 µm mesh nets, studied the distribution of large diatoms (*Asteromphalus, Coscinodiscus, Ethmodiscus,* and *Rhizosolenia*) in the Peru Current region from about 5°S to 30°S. The larger species were restricted to the Equatorial regions, in which upwelling is persistent. She attributes the absence of these species elsewhere to insufficient upward water transport to offset diatom sinking. In any event, their distribution does not correlate with abundance of nutrients, as indicated by phosphate analyses. Semina (1971) observed all net phytoplankton (80 species) over the same area, recognizing five floristic regions.

The annual cycle of phytoplankton (81 species) in Valparaiso Bay (about 33°S) was followed by Avaria (1971). The most important of the 21 diatoms that appeared regularly were *Biddulphia longicruris, Skeletonema costatum, Detonula pumila* (ex-*Schroderella delicatula*; Hasle, 1973a) and *Chaetoceros debilis.* Autumn–winter (cold-preferring) and spring–summer (warm-preferring) communities were recognized. A diatom flora for Conception Bay, Chile (at about 37°S), has been assembled by Rivera (1968); 120 species, many of which are planktonic, are included.

3.4 *Indian Ocean and adjacent Seas*

Source references are the checklists of Wood (1963b), Taylor (1966), Sournia (1970a), Thorrington-Smith (1969, 1970), and Simonsen (1974). Wood's (1963b) rather inaccessible list is by far the largest and gives the known records of occurrence of all species up to that time (with some synonyms), thus imparting an idea of the distribution of species. (It should be noted that Wood included a great number of African freshwater species, mostly named by Cholnoky.) The bibliographies of Taylor (1966), who lists the 238 diatom taxa he saw in the southwest Indian Ocean, and of Sournia (1970a), who cites 345 species seen by himself or others near Madagascar, cover the floristic literature from east African coastal waters to the Bay of Bengal. Thorrington-Smith (1969) identified 136 diatom species in the Agulhas current region (about 31°E, 30°S) below Madagascar, and 143 diatom taxa from *Discovery* stations along the 58° and 67°E meridians from about 16°N to 19°S. Simonsen (1974) recorded 247 taxa from the Northern Indian Ocean and Persian Gulf.

(a) *Biocoenoses and thanatocoenoses*

As to the overall phytoplankton distribution, Kozlova (1971) recognizes five plankton biocoenoses in the Indian Ocean region: Tropical, Subtropical, Temperate, Subantarctic, and Antarctic. (The Indian Ocean proper is generally considered to extend only to the Antarctic Convergence, hence the Antarctic zone would be excluded.) The distribution of siliceous remains of diatoms and silicoflagellates in the surface layers of the sediments (the thanatocoenoses) here (as elsewhere) parallels essentially the biocoenoses above, but shows great selectivity of species (it is a subset of the biocoenoses). The richest deposits lie in Subantarctic waters just South of the Antarctic Convergence, where the flora is not always more abundant than that of other upwelling (or neritic) regions, but contains a larger proportion of diatom species that reach the sediments and preserve well. In the rich coastal zone south of the Antarctic Divergence, the diatoms dissolve, so that the sediments have less than 1% of the concentrations of shells found in the Subantarctic waters. Beneath some Equatorial regions with surface upwelling, there are beds of shells of the giant diatom *Ethmodiscus rex* and of *Thalassiothrix* species. There is a region near the Seychelles where an inexplicably large number of shells of benthic diatoms has been found (Kolbe, 1957, p. 14 *et seq.*). Most benthic species occur near shores of continents or archipelagos.

The areal distribution of living planktonic foraminifera (Bé & Tolderlund, 1971) agrees essentially with that of the living diatoms.

The Indian Ocean differs from the other four major oceans in that diatoms do not constitute the major portion of its phytoplankton biomass overall (Kozlova, 1971)—dinoflagellates generally dominate except in rich neritic waters or in intensive upwelling regions (Sukhanova, 1962).

(b) *Southwest Indian Ocean*

A brief account of regional floras can begin with the accounts of Taylor (1966) and Thorrington-Smith (1969) in the southwest Indian Ocean. The inshore waters have a pantropical or cosmopolitan neritic flora much like that of the Bay of Bengal and the Arabian Sea, as described e.g., by Subramanyan (see Taylor for citation); the species will not be listed. Prominent subtropical oceanic species like *Climacodium frauenfeldianum, C. biconcavum, Eucampia cornuta* and *Ditylum sol* often accompany them. (Other species typically found offshore include *Roperia tesselata, Planktoniella sol,* and *Chaetoceros atlanticus.*) Certain species often considered neritic tended to be found offshore also, but generally when conditions were favourable; these included: *Chaetoceros lorenzianus, C. messanensis, C. peruvianus, Planktoniella sol, Thalassionema nitzschioides, Dactyliosolen mediterraneus,* and *Nitzschia pacifica*. Some important species e.g., *Skeletonema costatum,* never spread offshore. Species that grew in abundance in an upwelling at the southern edge of the continental shelf included *Skeletonema costatum, Asterionella glacialis* (ex-*japonica*), *Chaetoceros lorenzianus, Lauderia annulata, Bacteriastrum minus,* and *Nitzschia pacifica.*

Species that were only found offshore included *Asterolampra marylandica, Gossleriella tropica, Chaetoceros tetrastichon, Hemidiscus cuneiformis, Rhizosolenia hebatata, Mastogloia rostrata* and *M. woodiana*. These species could not successfully enter neritic waters, as could their companion species mentioned before. At times *Thalassiothrix longissima* and *Pseudoeunotia doliolus* entered the area from the colder waters to the south.

(c) *Central regions*

Sukhanova's (1962) phytoplankton survey using nets covered an enormous region, from about 20°N to 20°S over the whole Indian Ocean, and further still to the south in the region west of Australia. Thirty-seven species of diatoms were among the 103 species of phytoplankton found. Two patterns of distribution are illustrated; *Thalassiothrix* species (possibly including *Thalassionema nitzschioides*) were found in varying abundance over all of the region except part of the 'desert' region west of Australia; this region was characterized by *Hemiaulus hauckii* (and possibly *Climacodium frauenfeldianum, Planktoniella sol,* and *Gossleriella tropica*). The diatom species given by Wood (1964) as characteristic of the 'West Australian' and the 'Subequatorial Indian Ocean' communities overlap Sukhanova's listings, and all species mentioned by either author are listed by Sournia (1970a) as found near Madagascar—it is necessary to take synonomy into account. Desrosieres (1965) also studied the eastern Indian Ocean, using the Hardy recorder along a section on 110°E from Java to Freemantle, Western Australia. The track crossed the regions in which Wood (1964) had identified the communities mentioned before; few species listed are in common, but all are characteristic of the kinds of environment traversed. The difference could reflect differences in collecting gear as well as temporal changes. All Desrosieres' species are in Sournia's (1970a) list (*Nitzschia paradoxa* = *Bacillaria paxillifer*).

In the oceanic regions enriched by upwelling, notably in the western Equatorial Indian Ocean at about 55° to 80°E the *Thalassiothrix* sp. and characteristic Indian Ocean dinoflagellates both flourished (Sukhanova, 1962; Desrosieres, 1965). Otherwise, as mentioned before, dinoflagellates dominated.

The mixing of Indian Ocean and Pacific Ocean waters through the Timor, Arafuru, and Coral Seas to the north of Australia is discussed by Wood (1964); diatom communities overlap much in species. Wood felt that *Corethron criophilum, Chaetoceros debilis* and *C. messanensis* were useful indicators of Indian Ocean water flowing eastward.

(d) *Indian waters*

Recent contributions to knowledge of planktonic diatoms around India are by Prasad & Nair (1960), who report 51 species from net tows, and by Durairatnam (1964) who similarly found 33 species. Both papers list largely pantropical and cosmopolitan forms. Prasad & Nair (1960) consider the quantitatively important species to be in *Chaetoceros* (seven species listed as neritic plus *C. peruvianus* and *C. coarctatus*), *Rhizosolenia* (*alata* and *imbricata*) also *Thalassionema nitzschioides, Thalassiothrix frauenfeldii, Bacteriastrum hyalinum* and *Biddulphia sinensis*. In the more open waters, besides *C. coarctatus* and *Biddulphia sinensis, Ditylum brightwellii* and *Hemidiscus hardmanianus (cuneiformis)* were important. Two species considered arctic were also listed: *Fragilaria oceanica* (now *Nitzschia grunowii,* see Hasle, 1972b), and *Nitzschia seriata*. Hasle (1972c) points out that reports of what appear to be *N. seriata* from the tropics probably refer in fact to other very similar members of the subgenus *Pseudonitzschia*.

(e) *Red Sea*

Halim (1969) has reviewed studies of all Red Sea plankton including diatoms; the survey of phytoplankton by Sukhanova (1969) may be added. The Red Sea is unique in its generally high temperature (21·3°C–32°C) and salinity (37–41‰). There is an exchange of surface water with the Indian Ocean over the shallow sill at the Gulf of Aden, with a new inflow of water from about November to March bringing with it the tropical Indian Ocean flora, most of which apparently does not survive over all the Red Sea. However, this season is the time of maximum diatom abundance. The Red Sea flora is impoverished in general, but particularly in diatoms both as to species and numbers. Dinoflagellates generally predominate. Up to Halim's review (1969) some 82 species of diatoms had been recorded (by eight workers, including Halim). The species list (Table 10, Halim, 1969) contains many names not on the list of diatoms found near Madagascar (Sournia, 1970a) but most are synonyms and only about a dozen are not common; most of these dozen species have been found elsewhere in the Indian Ocean, e.g. in the Java Sea (Wood, 1963b). Thus, the diatom flora is not endemic and does represent a reduced Indian Ocean flora of pantropical or cosmopolitan species. Two species not previously seen were recorded by Sukhanova (1969)—*Stephanopyxis palmeriana* and *Gossleriella tropica*. Sukanova, using nets, found 15 species in all, of which the widest spread were *Climacodium frauenfeldianum* and *Planktoniella sol*. Over half the species reported from the Red Sea are in the genera *Chaetoceros, Rhizosolenia,* and *Coscinodiscus*. Species that have been reported to occur in relatively large abundance (for the region) are:

> *Hemidiscus cuneiformis, Striatella delicatula, Thalassiosira monile* (an endemic?), *Rhizosolenia calcar-avis, R. shrubsolei (imbricata), R. alata, Guinardia flaccida, Thalassionema nitzschioides* and *Nitzschia (Bacillaria) paradoxa*.

Asteromphalus sarcophagus Wall. provides an example of growing knowledge of a diatom's distribution. *A. sarcophagus* had been reported only three times (to our knowledge), each time in the Indian Ocean, until it was found in fair abundance and in three morphological forms by Thorrington-Smith (1970), again in the western Indian Ocean. Simonsen (1974) also observed it off Africa and in the Arabian Sea, always rare. Thorrington-Smith (1970) cites a personal communication from G. Fryxell that the species has also been seen in the Gulf of Mexico. Belyaeva (1970) also reported it at a few stations in the equatorial Pacific along meridian 140°W, though not at stations further to the east.

(f) *Persian Gulf*

The Persian Gulf has a reduced Indian Ocean flora, as determined by net samples (Simonsen, 1974). Few species are listed as other than 'rare'; these include *Thalassionema nitzschioides, Bacteriastrum furcatum, Chaetoceros lorenzianus, Rhizosolenia alata* f. *indica,* and a few other species of *Rhizosolenia* and of *Nitzschia.* Hendey (1970) has recorded 205 species of diatoms, mostly littoral ones.

3.5 *Atlantic Ocean and adjacent Seas*

The Atlantic Ocean from about 60°N to the Antarctic Convergence is considered here, higher latitudes having been treated with Polar Seas.

(a) *Biocoenoses and thanatocoenoses*

Diatom thanatocoenoses in the Atlantic have been less studied recently than those of the other major oceans. The thesis by Maynard (1974a), which describes six assemblages associated with major water bodies or current systems, was not available for this review. Kolbe compared the major diatom species in Equatorial cores taken by the 'Albatross' (see Table 12.2 from Kolbe, 1957) with cores from the North Atlantic previously studied by Lohman (Kolbe, 1955). North Atlantic cores had such important cosmopolites as *Thalassiothrix longissima, Coscinodiscus excentricus* (now *Thalassiosira excentrica*), and *C. lineatus* in common with Equatorial cores, but differed strikingly in having species such as *Melosira sulcata, Coscinodiscus radiatus, C. oculus-iridis, C. marginatus* and *Endictya robusta* in abundance while these were scarce or undetected to the south. Thus, the latitudinal variation in deposits seen elsewhere exists also in the Atlantic. Kolbe regarded the first eight species in the Indian Ocean list above, which are ubiquitous, as the 'dominant abyssal assemblage' of equatorial waters. Certain other species were either essentially absent from, or characteristic of, deposits of one ocean or another, though known to be widespread. However, the major differences between equatorial deposits lies in the relative abundances of the important species (see Table 12.2).

Atlantic equatorial deposits are poor in diatom remains, even the richest, occurring off Africa, being poor in comparison with Indo-Pacific oozes. As might be expected, the Indian and Pacific Ocean assemblages are very similar and set apart somewhat from the Atlantic.

In certain Atlantic cores Kolbe (1955) found shells of freshwater diatoms, notably *Melosira granulata,* and also silicified epidermal fragments of grasses and sedges. These were taken as evidence of potamic contribution to the sediments. Aeolian contribution of diatom shells to the sediments was considered significant only in the region of the 'Harmattan' haze, off the West African coast in the Gulf of Guinea (about 15°N to 0°N). Dust in this region (though not elsewhere off Africa) is comprised largely of freshwater diatom shells. *Ethmodiscus rex,* which is important in deposits, has recently been recorded alive in the plankton of the North Atlantic south of Bermuda from about 20°N to 28°N (Swift, 1973).

No large scale survey of diatom biocoenoses over both North and South Atlantic Oceans has been made by a single technique as done in the Pacific and Indian Oceans, though Lohmann (1920) examined centrifuged samples taken between the British Isles and the Brazil Current, noting 54 diatom taxa in all. Phytoplankton in general and diatoms in particular were scarce in the oceanic subtropical and tropical regions, where only species of *Coscinodiscus,* of *Navicula,* and of *Nitzschia seriata*-like forms were significant in numbers. Hentschel (1936) surveyed general plankton biocoenoses over the South Atlantic from about 20°N to 60°S, also using the centrifuge method. Samples were small (not exceeding 540 ml) and diatoms were often identified only to genus (about 58 taxa were recognized). Nevertheless, his charts (Hentschel, 1936, Abb. 15, 16) show the enormous differences in diatom abundance over the region, from about 50 cells/litre in the generally impoverished Central South Atlantic (centered about 20°S, 25°W) to over 10^5 cells/litre in the Antarctic and some African upwelling regions. In the plankton-poor regions diatoms comprised 2·5% or less of the centrifuged plankton, while in the productive regions diatoms were at least 40% and often over 80% of the total. *Nitzschia seriata* and *N. closterium* were the most widespread forms recognized, though *Thalassiosira, Fragilaria, Navicula,* and *Rhizosolenia* were also abundant locally.

Oceanic regions of the North and South Atlantic Oceans will be considered separately. Adjacent seas will be added as appropriate, then coastal regions, and finally the Mediterranean Sea and its neighbours as far as the Aral Sea will be considered.

(b) *North oceanic regions*

Source references for the North Atlantic flora are the works of Hendey (1964) and of Gran (1908), who listed 298 species of planktonic diatoms for which there were records north of about 40°N at the time.

The cruise of the 'Michael Sars' under Murray and Hjort provided the basic survey of diatom distribution over the North Atlantic (Gran, 1912; Gaarder, 1951). Gaarder (1951) recorded 179 taxa in all regions, both neritic and oceanic. The Central Atlantic and Northern Atlantic sections of the cruise are of interest here. The former—roughly from the Canaries (28°N) to Newfoundland (51·5°N)—yielded 113 taxa, chiefly of oceanic species of cosmopolitan or tropical to temperate distribution. The following list contains about 40 species found in some abundance at three stations or more on this section:

Diatoms of Central Atlantic Sections: *Actinocyclus tenuissimus, Asterionella notata, Asterolampra van heurckii, Bacteriastrum* spp., *Cerataulina bergonii, Chaetoceros atlanticus, C. convolutus, C. curvisetus, C. dadayi, C. decipiens, C. lorenzianus, C. messanensis, Cocconeis* sp., *Coscinodiscus divisus, C. lineatus, C. nodulifer, C. radiatus, Dactyliosolen mediterraneus, Eucampia zoodiacus, Gossleriella tropica, Guinardia flaccida, Hemiaulus hauckii, Hemidiscus cuneiformis, Lauderia borealis, Leptocylindrus danicus, Melosira seriata, Planktoniella sol, Rhizosolenia acuminata, R. alata, R. bergonii, R. castracanei, R. cylindrus, R. hebetata, R. hyalina, R. setigera, R. stolterfothii, R. styliformis, Detonula pumila (Schroederella delicatula), Thalassionema nitzschioides, Thalassiosira (Coscinodiscus) excentrica, T. subtilis, Thalassiothrix frauenfeldii, Triceratium antediluvianum,* and *T. formosum.*

Along the northern section (about 48°N) from Newfoundland to Ireland only 59 taxa were recorded in all, with the following the most important:

Diatoms of Northern Sections: *Asteromphalus heptactis, Bacteriastrum* spp., *Chaetoceros borealis, Corethron hystrix, Thalassiosira (Coscinodiscus) excentrica, C. lineatus, C. radiatus, Dactyliosolen mediterraneus, Hemidiscus cuneiformis, Navicula* sp., *Nitzschia seriata, Planktoniella sol, Rhizosolenia alata, R. bergonii, R. hebetata, R. imbricata, Thalassionema nitzschioides,* and *Thalassiosira monile.*

More detail for the Northern regions has been mentioned earlier in connection with studies of the Norwegian Sea, which is often invaded by North Atlantic water. Also, for larger phytoplankters that are caught by the plankton recorder, there are decades of observations in the shipping lanes of the North Atlantic from about 45°N to 64°N near Iceland, and in the North Sea (Edinburgh Oceanographic Laboratory, 1973, and references therein). The distributions of 57 diatom taxa (some species are grouped) as well as of other phytoplankters and of zooplankters are given; organisms are classified as oceanic, neritic, indifferent (panthalassic) or intermediate, the latter presumably indicating preference for the transition zone. Louis & Clarysse (1971) made a detailed seasonal floristic study of phytoplankton collected by filtration (of 25 litres) on a total of 21 voyages between Belgium and Iceland, recording 132 diatoms in all (of 200 phytoplankton species recognized). Eight distributional types of phytoplankton were recognized by seasonal or geographical occurrence and presumably reflect the influence of such factors as light, water column stability, or proximity to the coast.

(c) *Labrador Sea*

Holmes (1956) followed the seasonal cycle of phytoplankton in the southern Labrador Sea. Diatom populations reflected both the mixing of water bodies within the Labrador Sea and admixture of North Atlantic water. The introduction of allochthonous populations was evidenced by the occasional presence of temperate–tropical species (e.g. *Cerataulina bergonii, Chaetoceros pelagicus, Detonula pumila (Schroederella delicatula),* and *Stephanopyxis palmeriana*) as well as by the sizes of the populations, which were sometimes much too large to have been produced under the environmental circumstances obtaining at the time of collection. While boreal and temperate oceanic forms dominated, species generally considered either arctic or temperate–tropical occurred at times. Small species generally dominated, of which *Nitzschia pseudonana* (ex-*Fragilaria nana*) was outstanding at all times. Overall, the diatom flora much resembled that of the Norwegian Sea.

(d) *Central Sargasso Sea*

Studies of phytoplankton seasonality in the north central Sargasso Sea (Riley, 1957) and the western Sargasso Sea (Hulburt *et al.* 1960; Hulburt & Rodman, 1963) have added but a few species to Gaarder's (1951) list—*Asterionella bleakleyi, Chaetoceros pendulus, Hemiaulus membranaceous,* and *Mastogloia (Stigmaphora) rostrata*—though some 50 taxa were recognized. In both regions the diatoms were relatively inconspicuous floral elements except sporadically and to a small degree in winter and during the time of the spring maximum, following stabilization of surface water enriched by mixing. The species reported dominant in the spring 'flowering' by Riley (1957) are *Chaetoceros messanensis, C.* (?) *laciniosus, C.* (?) *lorenzianus, C.* (?) *brevis, Rhizosolenia stolterfothii, R. alata,* and *Nitzschia* sp. Species seen in spring by Hulburt *et al.* (1960) at densities of at least 1 000/litre were *Bacteriastrum delicatulum, B. hyalinum, Cerataulina bergonii, Chaetoceros decipiens, C. laciniosus, Eucampia zoodiacus, Leptocylindrus danicus, Nitzschia delicatissima, Rhizosolenia stolterfothii, Thalassiosira rotula, Skeletonema costatum,* and a small unidentified centric diatom (possible one of the small species from this region later studied in culture; e.g. *Thalassiosira pseudonana* or *Skeletonema menzelii*). The species roster for the region includes not only typical oceanic warm-water forms but temperate oceanic species and warm or temperate neritic species. Some of these were found in patches and presumably were introduced into the region by relatively small scale lateral eddies. Riley (1957) suggested that larger scale water movements may occur that 'constitute a general movement of the surface layer rather than limited penetrations along density surfaces'. These are not easy to detect hydrographically but contain different species. On the other hand, Hulburt & Rodman (1963) found neritic species at all times all the way from New England to Bermuda, albeit in very low numbers offshore. The 'flowerings' are probably the growth to co-dominance of whatever diatoms of autochthonous or allochthonous origin happened to be in the water when it stabilized.

(e) *Eastern North Atlantic and Baltic Sea*

The eastern North Atlantic, from Scandinavia to Gibraltar, surely holds the world's most studied plankton, and its diatom flora is the basis of marine diatom systematics. The monographs of Hustedt (1927–30; 1931–59), Hendey (1964), Cleve-Euler (1951–55), and Gran (1908) as well as check lists mentioned elsewhere in this review serve as basic references. (The new check list for the Oslofjord by Heimdal *et al.* 1973, should be added; it contains 98 diatom taxa.) Most of the literature is available through these, the review of Patrick (1948), or the papers on the Norwegian Sea (treated under Arctic Seas), therefore only a few additional new works are cited here. The detailed study of the North Sea and adjacent waters by the inverted microscope method (Braarud *et al.* 1953) merits special mention for its effort to correlate the regional vegetational types, of which there were 16, with the complex hydrographical conditions of the region. About 131 diatom taxa and about an equal number of other organisms were seen. The plankton recorder studies (Edinburgh Oceanographic Laboratory, 1973) and the study by Louis & Clarysse (1971) cited before also pertain to the North Sea.

The Baltic Sea to the east has a salinity of only 2–8‰ in its main basin, compared to the average salinity of about 32‰ in the North Sea. The phytoplankton is much reduced, and productivity is low, especially in the Gulf of Bothnia to the north, except in regions subject to enrichment from human activities. A survey of the phytoplankton, based largely on the work of Nikolaev, is given by Zenkevitch (1963). Briefly, the spring phytoplankton maximum has an Arctic species component that is replaced largely by blue–green algae during the summer; diatoms are abundant there only in polluted regions. Typically arctic species include *Fragilaria cylindricus, Navicula granii, N. vanhoffenii, Nitzschia frigida, Melosira arctica,* and *Achnanthes taeniata.* Boreal coastal species such as *Skeletonema costatum, Thalassiosira baltica, Nitzschia longissima,* and *Chaetoceros* sp. follow or flourish concurrently in coastal regions. The relative proportions of diatoms and other algae (given as plasma volumes) in the spring plankton of the central and northern regions has been measured by Schnese (1969). The summer plankton is generally dominated by blue–green algae (Zenkevitch, 1963; Niemi, 1971, 1972), but in some places cryptomonads, green algae, or diatoms share significantly in the biomass. Diatoms are sparse in the open waters, where the most abundant species include *Actinocyclus ehrenbergii,*

Coscinodiscus granii, Chaetoceros densus, C. wighami, Rhoicosphenia curvata, Epithemia spp. and *Synedra* spp. Net-caught species (less abundant) include *Melosira moniliformis, Thalassiosira baltica, T. nana, Chaetoceros holsaticus, Cocconeis pediculus, Amphiprora alta* and *Surirella gemma.* In eutrophic regions diatoms are much more abundant and the dominant species are *Melosira italica, Skeletonema costatum, Chaetoceros* spp., *Rhizosolenia minima, Coscinodiscus lacustris, Stephanodiscus hantzschii, Diatoma elongatum, Cyclotella meneghiniana, Asterionella formosa, Epithemia* spp. and *Synedra* spp. Dominant diatoms in the Gulf of Gdansk were about the same (Rumkówna, 1948) but *Chaetoceros ebenii* and *Coscinodiscus oculus-iridis* were included. Niemi *et al.* (1970) included but 52 diatom species in their list of 235 phytoplankters from the south Finnish Coast.

Floristic or seasonal studies along the southern European Atlantic coast have been carried out near Roscoff, in the English Channel (Grall & Jacques, 1964), adjacent to Vigo (Vives & Lopez-Benito, 1958), Cascais (Sousa & Silva, 1949), and Cadiz (Establier & Margalef, 1964) along the Atlantic Iberian Coast. (The adjacent Mediterranean is considered last in this section.)

(f) *Central South Atlantic*

Source reference on diatoms of the central South Atlantic is Hendey's (1937) study of 50 μm mesh net samples taken by the 'Discovery' from about 70°S to the equator. This work is of great biogeographical value because collections from 'Discovery' stations in the eastern Pacific, the western Indian Ocean, and the Southern Ocean were examined by the same authority. A total of 229 diatom species was recorded. In the Atlantic, 23 oceanic to panthalassic and 20 additional neritic species were 'typical of the warm-water flora examined'; of these, only *Fragilaria granulata, Rhizosolenia annulata,* and *Chaetoceros buceros* have not been reported from the North Atlantic (also *Corethron criophilum,* if one distinguishes it from *C. hystrix,* which Hendey, 1964, does not). In the oceanic waters, large discoid genera (*Hemidiscus* and *Asterolampra*) were most common, but *Chaetoceros* and *Bacteriastrum* were scarce. Both Hart (1934) and Hendey (1937) found it impossible to distinguish adequately between tropical and subtropical oceanic floras because diatom populations were too low to permit adequate sampling. According to Hart's sampling, *Asteromphalus heptactis* was the only diatom 'characteristic of subtropical waters', yet this species occurs in the Atlantic from the Arctic to the Antarctic. (Kozlova & Mukhina, 1967, had the same problem with small populations in the tropical Pacific.)

Hendey's (1937) 'species typical of the cold-water flora examined' has some of the species listed before as important in the Southern seas. Besides forms such as *Charcotia bifrons* and *Chuniella oceanica,* there occurred three species of *Chaetoceros (castracanei, chunii,* and *schimperianum),* six of

Rhizosolenia (bidens, chunii, crassa, curva, rhombus, and *truncata)* and no less than 14 of *Coscinodiscus.* The Antarctic Convergence emerges clearly as a distinct boundary floristically as well as otherwise.

(g) *Coastal South Atlantic*

Hart & Currie (1960) is source reference for the southeast Atlantic. They used 50 μm mesh nets to study plankton of both the inshore waters seasonally enriched by upwelling—the Benguela Current proper—and the offshore oceanic waters, which lie in the anticyclonic circulation—the 'south-east trade wind drift'. Of some 208 categories of net plankton, 95 species were diatoms, and these dominated the plankton in general and that of the rich Benguela Current particularly. Dominance was not with the 'monotony' often seen in the Antarctic, where one or two species may comprise most of the catch; here no less than 48 species were among the dominants at one station or another, with *Chaetoceros* and *Rhizosolenia* alone accounting for 28 species. In the coastal enriched water, species of *Chaetoceros, Stephanopyxis turris, Eucampia zoodiacus, Fragilaria karstenii* and *Asterionella glacilais* (ex-*japonica*) dominated, while in oceanic waters the dominants were *Planktoniella sol, Thalassiothrix longissima, Rhizosolenia styliformis* and *Dactyliosolen mediterraneus.* In both domains the flora was strikingly cosmopolitan. Of the 44 most abundant species, all but four are widespread in temperate or tropical neritic or oceanic waters; the four not reported from elsewhere (at the time) are *Chaetoceros strictum, C. tetras, Fragilaria granulata* and *F. karstenii,* all typically 'coastal' species (Hart & Currie, 1960, pp. 263–264). *Chaetoceros tetras* has since been reported from the southern Indian Ocean (Taylor, 1966). Hart & Currie drew up brief 'ecological characterizations' of the more important species based on their own observations of distribution, seasonal dominance, water characteristics, etc., and compared their diagnoses with those of workers observing the same species in other seas. While agreement was generally good, some notable exceptions were observed, having the general trend that certain species elsewhere reported as either neritic or oceanic had panthalassic tendencies here. Species mentioned are *Thalassiosira subtilis, Chaetoceros lorenzianus, C. compressus, C. constrictus, C. curvisetus, Rhizosolenia alata, R. hebetata* f. *semispina, R. imbricata, R. styliformis, Corethron criophilum, Dactyliosolen mediterraneus, Nitzschia delicatissima,* and *N. seriata.* A somewhat different disparity—one of seasonal occurrence—was observed for *Chaetoceros subsecundus.* Both kinds of local differences in 'ecological characterization' have been encountered before in this review.

Wood (1965b) found 90 diatom species in living centrifuged samples from five depths at four stations from 3°N to 7°S along 5°E in the Gulf of Guinea. Virtually all the species were circumtropical forms, having been recorded by himself in the Atlantic from the Florida Straits as well as from the warm Pacific and Indian Ocean waters neighbouring Australia. Chyong (1971) examined settled preserved samples taken along latitudes 11°S and 14°S from about 4°W, in the South Equatorial Current, to the coastal waters. Of 93 species seen, *Hemiaulus hauckii, Rhizosolenia alata, R. calcar-avis, R. hebetata, Corethron hystrix, Thalassiosira (Coscinodiscus) excentrica, Planktoniella sol, Roperia tesselata* and *Nitzschia closterium* were the most important. Curiously, *Hemiaulus hauckii* was defined as neritic rather than oceanic, based on its increased abundance near the coast. Further, Chyong had previously found the neritic *Asterionella glacialis* (ex-*japonica*) to be dominant on sections along 5°S and 8°S; it was not dominant along the more southerly sections nor had it even been reported as present by Wood (1965b). Reyssac (1973) compared net phytoplankters in the upwelling region off Walvis Bay (about 23°S) with those in the Canary Current upwelling at about the same latitude north. Diatoms dominated (58 species were seen off Walvis Bay). Most species were classified as cosmopolitan or temperate; the temperate influence was particularly strong at Walvis Bay, where water temperatures fell as low as 12·3°C. In the warm waters of the Cape Verde Islands, Paredes (1969/70) recognized 127 diatom taxa in netting-concentrated 8 litre samples taken at one neritic and six oceanic stations (15–17°N). The stations lie in two different water bodies, distinguishable by surface temperatures slightly below 27°C to the north and above 27°C to the south. Of the 11 most abundant species, *Nitzschia bicapitata, N. longissima* and *Thalassiosira subtilis* were common to both water bodies. The more southerly region had also *Coscinodiscus crenulatus, Rhabdonema adriaticum, Rhizosolenia calcar-avis, R. hebetata* f. *semispina, R. imbricata* var. *shrubsolei,* and *Streptotheca thamensis,* while the northerly region had only *Cocconeis pelucida* var. *minor* and *Nitzschia closterium.*

The net phytoplankton of coastal and adjacent oceanic waters of the southwest Atlantic has been studied in the region 34°–42°S, 51°–60°W (Frenguelli & Orlando, 1959; Müller-Melchers, 1959) and off Cape Frio, at 23°–24°S (Moreira Filho, 1965). All authors remarked the presence of rare or infrequent species of diverse origins, occurring alone or together in patches. Some of these species are characteristic of tropical waters to the north (e.g. *Biddulphia sinensis* and *Asterolampra marylandica*), while others have their regions of abundance in Subantarctic or even Antarctic waters (e.g. *Corethron criophilum* and *Synedra reinboldii*). Frenguelli & Orlando (1959) list 88 diatom taxa, most of which are cosmopolitan; the following 11 forms were the only ones abundant:

> *Chaetoceros affinis, C. decipiens, Bacteriastrum delicatulum, B. hyalinum, Lauderia annulata, Nitzschia pertenuis, Rhizosolenia alata* (and var. *gracillima*), *R. styliformis, Synedra reinboldii,* and *Thalassionema nitzschioides.*

Equatorial waters along the north coast of Brazil from about Trinidad (about 12°N) to Cape de São Roque are influenced by the Amazon outflow. Müller-Melchers (1957) found the net phytoplankton in the Amazon region reduced in species and numbers compared to the net flora collected to the south (about 30°S). Only *Coscinodiscus concinnus, C. oculus-iridis, Nitzschia pungens* var. *atlantica* and *Ditylum brightwellii* occurred in any abundance, and only the latter was found to the south as well. Wood (1966) examined living specimens by fluorescence microscopy, concentrating samples when the sediment load permitted. He reported 117 diatom taxa, of which about half were categorized as rare or infrequent. Three distinct communities were recognized, including the following diatoms:

North of the Amazon: *Asteromphalus cleveanus, A. elegans, A. flabellatus, Biddulphia aurita, B. sinensis, Cerataulina pelagica, Chaetoceros aequatoriale, C. coarctatus, C. decipiens, Climacodium frauenfeldianum, Hemiaulus indicus, H. sinensis, Lauderia annulata, Pleurosigma distortum, Streptotheca thamensis, Thalassiothrix frauenfeldii, Thalassionema nitzschioides.*

Amazon Estuary species: *Asterionella japonica, Biddulphia aurita, B. sinensis, Chaetoceros didymus, C. paradoxa, C. peruvianus, C. vanheurckii, Coscinodiscus concinnus, Ditylum brightwellii, D. sol, Nitzschia closterium, N. lorenziana, N. seriata, Rhizosolenia stolterfothii, Skeletonema costatum* ('varieties'), and various benthic species.

Southern Region: *Chaetoceros concavicorne, Coscinodiscus africanus, Fragilaria striatula, Triceratium alternans.*

It is obvious that species usually considered either neritic or oceanic are mixed in these communities.

Hulburt & Corwin (1969) examined centrifuged 100 ml samples in connection with enrichment studies made in the rich coastal waters and moderate to barren offshore areas, mostly north of the Amazon. The abundant species were *Skeletonema costatum* (coastal only), *S. tropicum, Thalassionema nitzschioides, Chaetoceros subtilis,* and *Nitzschia seriata; N. delicatissima* was abundant only to 30 miles beyond the 200 m depth, and *Leptocylindrus danicus* and *Hemiaulus hauckii* were moderately abundant both coastally and offshore. *Nitzschia closterium* and the coccolithophore *Coccolithus huxleyi* were ubiquitous but never very numerous. Coastal diatoms were carried offshore by the Amazon effluent and the current system, where they flourished during circumstances that caused upwelling. These normally neritic species are also the ones that increased when offshore surface waters were enriched experimentally, accompanied by *C. huxleyi* and *N. closterium*. The characteristic oceanic phytoplankters did not respond noticeably to experimental enrichment. This accords with Hulburt's previous observations (see references in Hulburt & Corwin, 1969) that natural enrichment of Western Atlantic surface waters by hydrographic events was accompanied by increased abundances of coastal species, mainly diatoms, even in regions far from shore. An extensive study of phytoplankton ecology of coastal and offshore Venezuelan waters from the Orinoco mouth into the Caribbean Sea by Margalef (1965) yielded essentially the same results, at least for the region south of Port-of-Spain. Margalef recorded over 300 phytoplankton species, including 142 diatoms, and remarked that while most of the phytoplankters were widespread tropical species, those dominating in the richer areas were mostly diatom species typical of coastal regions of much more northern latitudes. Margalef used I_2-KI as preservative, and employed the inverted microscope technique; he is the only investigator who recorded chrysophyte and prasinophyte flagellates in his plankton samples. These flagellates sometimes rival dinoflagellates and coccolithophores in numbers, hence may, like them, outnumber diatoms. Presumably diatom gametes would resemble chrysophyte flagellates in samples.

(h) *Caribbean Sea*

Source references for the Caribbean Sea are by Margalef (1965), Hulburt (1968), and Hargraves *et al.* (1970); the latter list 56 benthic and 86 planktonic diatom species caught in nets along the Lesser Antilles. Most species were pantropical or cosmopolitan ones. *Bacteriastrum hyalinum, Hemiaulus hauckii,* and *Chaetoceros coarctatus* were widespread and relatively abundant. Stations that were 'downstream' in the westerly current system passing the island chain generally had more total phytoplankton, more diatoms, and more benthic diatoms, than stations 'upstream'. The Caribbean is interesting because it shows the influences of upwelling, of coastal proximity, and of shallow water, all separately and in the same geographical region. Thus, in the open waters of the Caribbean, total phytoplankton populations were low, seldom exceeding 4×10^3/litre (Margalef, 1965; Hulburt, 1968) and were mainly coccolithophores or naked flagellates as determined from settled preserved samples (Margalef, 1965). The population was larger, up to $1{\cdot}7 \times 10^4$/litre, and contained more diatoms, in an upwelling region beyond the continental shelf off Venezuela (Hulburt, 1968). It increased still further (up to 4×10^4/litre) in diatom-dominated waters over the shelf, showing the influence of land added to that of upwelling. Further to the west, where there is no upwelling and where relatively deep water occurs near shore, an enhanced phytoplankton again showed the stimulation of proximity to land. Far from terrestrial influence, population changes over the Nicaraguan–Honduran shoal demonstrated the influence of shallow water (Hulburt, 1968). The shoal lies above the level of the thermocline, making nutrient addition from below unlikely; the shoal may act like a reef in nutrient-poor ocean water.

As to diatom species, Takano's (1960b) net samples contained 83, 22 species in *Chaetoceros,* 12 in *Rhizosolenia,* and 11 in *Coscinodiscus.* Hulburt's (1968) settled samples give a different picture of relative abundances, as expected. The relatively abundant ubiquitous diatoms were *Nitzschia closterium* and *N. bicapitata,* while the species that flowered locally were *Thalassionema nitzschioides, Nitzschia delicatissima, Bacteriastrum delicatulum, Chaetoceros decipiens, C. diversus, C. laciniosus, C. compressus, Rhizosolenia fragilissima, Thalassiosira nana* (Lohmann) and *Skeletonema costatum.* Additional species, found by Margalef (1965) to exceed 500/litre were *Nitzschia seriata, N. closterium, Leptocylindrus danicus* and *Rhizosolenia delicatula.*

(i) *Gulf of Mexico*

Plankton of the Gulf of Mexico has been examined by sedimentation or centrifugation techniques (Hulburt & Corwin, 1972; Zernova, 1969; these authors refer to the previous literature, which largely deals with coastal plankton; Saunders & Glenn, 1969, has been referred to before, p. 379.) Coastal regions are rich, with populations of diatoms reaching 10^6/litre in places, while total phytoplankters do not exceed 5×10^3/litre over most of the open waters of the Gulf. Coccolithophores, notably *Coccolithus huxleyi* and dinoflagellates, especially *Katodinium rotundatum,* dominate this sparse plankton, and the most abundant diatoms (90–150/litre) were *Hemiaulus membranaceous, Bacteriastrum delicatulum,* and *Nitzschia closterium.* Hulburt & Corwin (1972) found that oceanic stations located in the current entering the Gulf from Yucatan Strait had more abundant phytoplankton than other oceanic stations and many more diatoms; these included *Chaetoceros laciniosus, C. decipiens, Nitzschia delicatissima, Rhizosolenia alata, R. fragilissima, R. stolterfothii, Mastogloia rostrata, Ceraulina bergonii, Thalassionema nitzschioides, Thalassiothrix frauenfeldii,* and especially *Thalassiosira subtilis. T. subtilis* attained a density of about 10^4/litre at depths of 80–100 m but was not detected at shallower depths. This same species was also abundant on the eastern part of Campeche Bank where there is upwelling, and in the area of cyclonic upwelling in western Campeche Bay (Zernova, 1969), far from the stations occupied by Hulburt & Corwin (1972). Zernova (1969) identified 141 planktonic and 15 benthic species of diatoms (not listed in the reference).

(j) *Coastal North Atlantic*

The western Atlantic is of interest because of the proximity of the Gulf Stream, with its essentially oceanic water, to the Continental Shelf. Wood (1968) examined the phytoplankton (by centrifugation and fluorescence counting) just east of the Gulf Stream at stations in or near the deep Tongue of the Ocean, which is mostly surrounded by the shallow Bahama Banks. Waters of the North Equatorial Current mix with those of Caribbean origin and from the Gulf of Mexico, and the diatom flora of 138 species reflected these warm oceanic and coastal origins as well as the influence of the shallow reef waters nearby, manifested by the presence of many benthic species. Coccolithophores dominated in general (38 species, 13 rare) and dinoflagellates (185 species) were next, but when the plankton became more abundant, it was through increases of *Hemiaulus* (four species) or *Rhizosolenia* species.

Further north, there are three recognizable water types. Seaward transects from coastal waters across the Gulf Stream into Sargasso Sea water have been made south of Cape Hatteras at about 30° to 35°N (Hulburt & MacKenzie, 1971; Marshall, 1971; and references in both). Settling techniques were used. Total phytoplankton populations decreased going offshore, from about 10^5/litre to 10^4/litre (usually) in coastal waters—varying seasonally—to about 5×10^3/litre in the Gulf Stream, then remained about the same or decreased to about 200/litre in the Sargasso Sea. Diatoms were the major proportion of the coastal population, single species attaining 10^5/litre; they were perhaps a tenth of the Gulf Stream population (200–800 cells/litre) and were negligible in the Sargasso Sea populations (at the times sampled), not exceeding 120/litre. Marshall (1971) identified 100 diatom species in all, mostly widespread oceanic or coastal forms. Of these, 36 were found in the Sargasso Sea. Eighteen species were found only in coastal waters, three in the Sargasso Sea, and nine only in the Gulf Stream. While some of these nine may be useful indicators of the Gulf Stream, better indicators are the dinoflagellate, silicoflagellate, and coccolithophorid populations, which show much clearer changes. No single diatom species or characteristic community dominated overall at the richer coastal stations. In these, the important species were *Skeletonema costatum, Navicula* sp., *Nitzschia delicatissima, N. seriata, N. closterium, Chaetoceros gracilis, C. didymus, C. teres, Rhizosolenia alata* (and several other species), *Cerataulina bergonii, Leptocylindrus danicus, Thalassiosira gravida,* and *Thalassionema nitzschioides.* A number of species common to the north, especially of *Thalassiosira,* were not reported south of Cape Hatteras, though some of the characteristic Gulf Stream species, such as *Rhizosolenia alata* and *R. calcar-avis* are abundant near Cape Cod in late summer (see Patrick, 1948, p. 488 *et seq.*).

To the north of Cape Hatteras there are four rather than three water types; the so-called slope water lies between the Gulf Stream and the coastal water lying directly over the shelf. There are as yet no definitive studies of the diatom populations of this slope water, though observations of bottom fauna and dinoflagellate cysts suggest that this would be interesting. It is a transitional region.

Numerous studies of coastal phytoplankton have been compiled by Smayda (1973a). Access to the literature for the regions from Long Island to the Gulf of Maine is provided by Riley & Conover (1967), Smayda (1957), Pratt (1959), and Lillick (1940). These supplement the account in Patrick (1948). The similarity of these regions, the Gulf of Maine especially, to European and even other boreal to temperate coastal areas is evident in the seasonal succession and the character of the flora, if not identity of the species (see Lillick, 1940, p. 232). In Long Island Sound (Riley & Conover, 1967; Conover, 1956) some 75 diatom species have been recognized. Overall these comprised the dominant organisms, though flagellates have not been adequately studied and they outnumber diatoms during the summer. Twenty-four diatoms were dominants (over 5% of the population) at some time or place. These are shown below; the asterisk indicates that the species was also a dominant in Narragansett Bay, which is discussed next.

Dominant diatoms in Long Island Sound: *Skeletonema costatum**, *Thalassiosira nordenskioldii**, *Chaetoceros compressus, C. radians* and *tortissimus, C. affinis, C. decipiens, C. danicus, C. didymus, Leptocylindrus danicus**, *Rhizosolenia delicatula, Thalassionema nitzschioides, Asterionella glacialis (japonica)**, *Lauderia borealis, Detonula pumila (Schroederella delicatula), Thalassiosira rotula**, *T. gravida**, *Cerataulina pelagica, Guinardia flaccida, Rhizosolenia fragilissima**, *Corethron criophilum, Coscinodiscus perforatus cellulosa, C. radiatus,* and *Paralia sulcata.*

Skeletonema costatum contributed by far most to the crop, attaining $3{\cdot}5 \times 10^7$/litre at one time, when all diatoms totalled 4×10^7/litre. The plant crop per unit area rivalled that of the Gulf of Maine, which has a deeper water column.

About 71 diatom taxa have been recognized in Narragansett Bay (Smayda, 1957; Pratt, 1959). To the list of important species (designated by asterisks in the list preceding) should be added *Nitzschia seriata, Rhizosolenia fragilissima, Detonula confervacea* (ex-*cystifera*), *Chaetoceros curvisetus* and *Thalassiosira nana* (not identified by electron microscopy to our knowledge). All are boreal-neritic species. *Skeletonema costatum* and *Thalassiosira nordenskioldii* were again important, the former during spring blooms attaining densities of 10^7/litre and constituting as much as 80% of the population.

Lillick (1940) recognized 131 diatom taxa in the Gulf of Maine and provides an extensive seasonal and regional study. The sparse winter flora was dominated (among the diatoms) by *Coscinodiscus excentricus* (now *Thalassiosira excentrica*), *C. centralis,* and *Thalassionema nitzschioides;* 14 other boreal diatoms were recorded. In spring the expected great flowering of

Thalassiosira species occurred, followed by the dominance of some 18 species of *Chaetoceros*. During the *Chaetoceros* dominance the arctic species *Fragilaria oceanica, Navicula vanhoeffenii,* and *Achnanthes taeniata* were seen. *Skeletonema costatum,* a summer form in the Gulf of Maine, never exceeded 7×10^5/litre. Saifullah & Steven (1974) observed an annual cycle of phytoplankton (65 diatom species) in St. Margaret's Bay, northern Gulf of Maine.

Brunel (1962) has compiled a phytoplankton flora of the Baie des Chaleurs in the Gulf of St. Lawrence; it contains 74 diatom taxa.

(k) *Mediterranean Sea*

The manual of Mediterranean plankton by Trégouboff & Rose (1957), which lists 121 diatom species, also gives characteristic habitats or areas of occurrence for most diatoms in this Sea, largely based on earlier work by Pavillard. Reviews of Mediterranean phytoplankton by F. Bernard and of zooplankton by M. Bernard (1967) stress the hydrographical heterogeneity of the Mediterranean. In brief, the several regions of the western basin are influenced to varying extents by inflowing Atlantic surface water, which is somewhat richer in plankton (and phosphorus) than the higher salinity Mediterranean water proper. Atlantic influence is detected as far east as Sicily, at places even close to shore, particularly along the Algerian coastline. Thus a moderate to very poor oceanic phytoplankton is usual, dominated by coccolithophores except near the large European rivers, where there is a coastal flora containing at least 200 diatom species. The eastern basin is oligotrophic save near the Nile delta, and locally near shore. The Adriatic Sea is divided, the northern region, which becomes cold in winter, has a population like that of northern European seas, while the sourthern portion shows the influence of the oceanic eastern basin. F. Bernard (1967) should be consulted for the references.

Recent studies in the western Mediterranean include observations of net plankton in the open Ligurian Sea (Léger, 1973a,b), revealing about 30 oceanic species, and include a study made from a buoy laboratory off Nice (at 42°47′N, 7°29′W) by the inverted microscope method supplemented by net samples (Léger, 1972, and earlier studies in this same series, cited therein). In this oceanic region, in which 83 diatom species were seen, neritic species such as *Skeletonema costatum* rarely occurred, even during periods of maximum populations (which were only of the order of $1{\cdot}5 \times 10^4$ cells/litre of any one species). The ubiquitous *Nitzschia seriata* and *Thalassiothrix frauenfeldii, Bacteriastrum delicatulum* and *Nitzschia delicatissima* occurred occasionally in relatively large abundance at one depth or another, apparently controlled largely by light and turbulence. (Nannoflagellates, however, tended to be uniformly distributed.) Net samples also yielded the much less abundant but characteristic large or colonial species such as *Rhizosolenia calcar-avis* or *Chaetoceros decipiens*. The populations observed near the Italian coast by Bernhard & Rampi (1967) attained at least 10 times the densities seen by Léger and contained *Asterionella japonica, Chaetoceros affinis, C. decipiens, Cerataulina bergonii, Dactyliosolen mediterraneus, Rhizosolenia stolterfothii* and *Thalassionema nitzschioides* as additional dominants. *Leptocylindrus danicus* became very abundant, but at certain times and depths only. *Skeletonema costatum,* though recorded, was never abundant; on the other hand, in the coastal water at Banyuls in the Golfe du Lion, to the west, this species was the dominant phytoplankter (Jacques, 1968). It was accompanied by several species listed above, and by *Chaetoceros insignis* and *Nitzschia pungens*. This is the richest area of the western Mediterranean; *S. costatum* alone attained 5×10^6 cells/litre at one time. As expected, diatoms dominate over dinoflagellates here in number, and the two groups are almost equal in numbers of species (88 diatoms, 104 peridinians; dinoflagellates usually outnumber diatoms by almost two to one). Jacques (1968) points out that the plankton off the Spanish coast, studied by Margalef and co-workers (see Jacques, 1968 for references) shows much less coastal influence; populations are about an order of magnitude lower, dinoflagellates dominate, and several diatoms dominant at Banyuls are not important off Spain. The coastal water along the Algerian shore is of Atlantic oceanic origin and no neritic influence is seen. The diatoms found in net tows near shore were typical oceanic ones (Sequin, 1968). Note, however, that neritic species such as *Skeletonema costatum* occur in abundance in enclosed areas (F. Bernard, 1967).

To the review of Adriatic phytoplankton by Bernard (1967) should be added the recent inverted microscope studies of Pucher-Petrović (1966, 1969), dealing with the open waters of the central Adriatic and the eastern shore, which shows relatively little neritic influence (unlike the Italian shore, which has a more abundant plankton poorer in species; see Pucher-Petrović, 1966 for references). The study by Marchesoni (1954) is a convenient source reference for the diatom flora of the northern Adriatic, which shows not only the north temperate neritic aspect referred to before, but the presence of Mediterranean oceanic forms. There were 219 diatom species (vs. 71 peridinians and 12 silicoflagellates and coccolithophores) within and just outside the Lagoon of Venice. Only 67 of the diatoms were considered pelagic species, and of these, nine were considered oceanic. The species that dominated during the spring maxima (January–May) were *Bacteriastrum hyalinum, Chaetoceros rostratus, C. curvisetus, C. diversus, Skeletonema costatum,* and several species of *Rhizosolenia* (Marchesoni, 1954). The eastern Adriatic nearshore coastal flora is generally similar, but *Thalassiosira decipiens* is also recorded there (Pucher-Petrović, 1966). In the pelagic central Adriatic, which is even less fertile than the adjacent Mediterranean, the most abundant diatoms (of 90 species) were certain of the permanent residents:

Rhizosolenia calcar-avis, R. alata, Bacteriastrum delicatulum, Chaetoceros decipiens, Hemiaulus hauckii, Thalassiothrix longissima, Thalassiosira (ex-*Coscinodiscus*) *excentrica, Asterolampra marylandica* and *Thalassiothrix frauenfeldii.* More abundant when present, but only occurring sporadically, were several usually coastal species including members of *Chaetoceros* and *Rhizosolenia,* also *Leptocylindrus danicus, Guinardia flaccida,* and *Cerataulina bergonii* (Pucher-Petrović, 1969).

There are still few studies in the eastern basin of the Mediterranean. Denisenko (1964) compared abundances of 48 diatom species in samples from the Black Sea to the Adriatic. Ignatiades (1969) followed succession in the net phytoplankton of Saronicos Bay, southern Aegean Sea, where the plankton was numerically dominated by a *Chaetoceros–Rhizosolenia* association (33 of 58 diatom species belonged to these genera). Aleem & Dowidar (1967) observed the seasonal phytoplankton composition and productivity along the Egyptian Coast, in the region enriched by the late summer Nile floods. The ensuing blooms (reaching $9{\cdot}4 \times 10^6$ cells/litre) consisted of 99% diatoms with about a dozen species (of some 60) dominating markedly at certain times or places. Species that at one time or another constituted over 10% of the population included *Chaetoceros socialis, C. curvisetus, C. costatus, Cerataulina bergonii, Skeletonema costatum, Hemiaulus sinensis, Asterionella glacialis, Thalassionema nitzschioides.* The result of restriction of the Nile outflow has not been reported to our knowledge.

(l) *Black Sea and Azov Sea*

An account of the biota, including the phytoplankton, of the Black, Azov, Caspian, and Aral Seas is included in the treatise by Zenkevitch (1963). The biota reflects the long sequence of geological events and climatic changes leading to the present state of partial or total isolation from the world ocean with the attendant lowering of salinity and alteration of the ionic composition of the water. The relict 'caspian' fauna is best preserved in the Caspian Sea (Zenkevitch, 1963, p. 354); this fauna is present in the Black and Azov Seas only in the least saline regions, where numbers of freshwater or mixohaline algae also occur. Zenkevitch (1963, p. 564, p. 587) cites Makarova's point that endemism among diatoms is very low even in the Caspian Sea, where endemism among animals is high. The phytoplankton is much like that of northern Europe, though reduced in numbers of species. Most of the *Chaetoceros* species found in the Black Sea are relict boreal neritic species still found elsewhere (Mikhailova, 1964). Some of these are extremely abundant at times in the Black Sea at a salinity (about $18^0/_{00}$) considerably lower than that at which they are usually found elsewhere. Evidence for the existence of races is discussed earlier in this review in connection with the concept of species among diatoms.

Proshkina-Lavrenko (1955, 1963) has monographed the diatoms of the Black and Azov Seas, and Proshkina-Lavrenko & Makarova (1968) treat the diatoms of the Caspian Sea. This review deals briefly with other literature concerning these seas.

For the Black Sea, Zenkevitch (1963), quoting work of Morozova-Wodjanitzkaja and Belogorskaya, lists *Skeletonema costatum, Chaetoceros radians, Cerataulina bergonii, Leptocylindrus danicus, Thalassionema nitzschioides, Rhizosolenia calcar-avis,* and *R. fragilissima* as the most abundant of 150 diatom species found there. The average salinity of surface waters is 17–18‰; that of the adjacent Mediterranean, in which these species are also abundant in places, exceeds 35‰.

The flora of the Romanian coastal region now numbers 320 algal species, bringing the total for the Black Sea to 771 species or varieties in 157 genera (Bodeanu, 1969). (Ivanov 1965 had tallied 695 species in 149 genera for the Black Sea as a whole; his list was not available for the review.) In Romanian waters alone Bodeanu (1969) reports 216 diatom species found in the plankton, though many of these are primarily benthic or freshwater species, as were 36 of the 37 species described as new to the Black Sea. On the other hand, *Thalassiothrix mediterranea,* the 37th, is planktonic and may be a newcomer rather than just a new record. (This species is usually scarce.) The most abundant diatoms, e.g., *Skeletonema costatum,* attained densities to 10^7 cells/litre in places; to the list of important species should be added *Chaetoceros socialis, Nitzschia delicatissima, Cyclotella caspia,* and *Thalassiosira parva.* Quantitative information on abundance and distribution over the Romanian shelf is given by Bodeanu (1969) and Skolka (1969; 146 diatom species in settled samples) and for the Northwest Caspian by Ivanov (1960; 157 diatom species), and Makarova (1961; 68 diatoms in net samples).

Maynard (1974b) found unexpectedly few diatom shells in Black Sea deposits and saw only 34 taxa. She suggested that conditions in the sediments bring about solution of even those shells that survive sinking to the bottom.

The Sea of Azov has salinities varying locally from 1‰ to about 17‰ where it joins the Black Sea. Diatoms (41 species) comprised but 21·8% of the phytoplankton species though they formed 55% of the biomass overall and over 98% of it during blooms. Diatoms attaining greatest abundance were *Skeletonema costatum, Thalassiosira nana, Coscinodiscus biconicus, Leptocylindrus danicus, Rhizosolenia calcar-avis, R. radiatus, Chaetoceros subtilis, Biddulphia mobiliensis, Ditylum brightwellii* and *Thalassionema nitzschioides* (Zenkevitch, 1963, quoting Usachev and Pitzik). The biomass in the Sea of Azov is stated to be enormous; a biomass of 270 mg/litre consisted of almost pure *Rhizosolenia calcar-avis* (Zenkevitch, 1963, p. 485).

Changes in the Azov Sea biota due to recent diversion of the River Don waters for irrigation are not documented to our knowledge, but the works of Proshkina-Lavrenko (1963) and Vodyanitskii (1965) were not seen for this review.

(m) *Caspian Sea*

The landlocked Caspian Sea has a salinity of about 12·8‰. The monograph on its plankton by Proshkina-Lavrenko & Makarova (1968) was not available for our review, which summarizes briefly the chapter by Zenkevitch (1963) and the data of Makarova (1961) for the northern Caspian Sea. At least 68 species of diatoms have been found in the plankton, and these constitute somewhat over a third of the total flora. Diatoms, blue–green algae, green algae, and the dinoflagellates *Exuviaella, Prorocentrum,* and *Gonyaulax* share importance in the biomass. Diatoms abundant in various locations include *Skeletonema costatum, Actinocyclus ehrenbergii,* about a dozen species of *Chaetoceros, Coscinodiscus biconicus,* several species of *Thalassiosira* (some apparently endemic), *Melosira* (both freshwater and brackish species), the freshwater *Asterionella formosa* and the 'immigrant' *Rhizosolenia calcar-avis,* which was just detected in the Caspian Sea in 1934 and rapidly became a dominant. During 1955–1959 *R. calcar-avis* became less abundant and the blue–green alga *Aphanizomenon flos-aquae* more so in the Central Caspian Sea (Makarova, 1961).

(n) *Aral Sea*

The Aral Sea is also landlocked, has a salinity of about 10‰, and is even higher in Mg and Ca than the Caspian. It has a relatively poor biota in species and abundance both, and marks the most eastward extension of the Mediterranean fauna (Zenkevitch, 1963). Only 39 algal species have been observed in the plankton (see Zenkevitch, 1963) of which 18 were diatoms. *Actinocyclus ehrenbergii* var. *crassa* was the chief form, other relatively abundant ones being *Chaetoceros wighamii, C. subtilis, Coscinodiscus granii* var. *aralensis, Skeletonema costatum,* and *Thalassiosira decipiens.* We are unaware of any new studies in this sea.

3.6 *Summary and conclusions*

1 Despite difficulties in sampling and identification of species, it seems established that the major climatic zones and hydrographical features such as gyres or currents are inhabited by characteristic groups of species (which may contain cosmopolitan forms). The broad patterns of distribution of living diatoms (the biocoenoses) are reflected in the thanatocoenoses below, which have fewer species present. Diatom biocoenoses resemble those of other groups of planktonic organisms, as far as can be told.

2 Latitudinally, temperature is the overriding environmental factor. Species (and races) adapted to different temperature regimes have been found.

3 The important differences between the physiological requirements of neritic and oceanic species remain unknown, though the influence of higher levels of organic chelators in neritic waters is suspected.

4 Transition zones between water bodies remain problematical. Whether these zones invariably have a characteristic flora is not clear, and, in the affirmative cases, the adaptations of the flora remain unexplained.

5 The problem of speciation in many important planktonic forms remains; small centric diatoms as well as small pennate species may be evolving explosively. Breeding experiments are urgently needed to supplement the morphological, physiological, and enzymatic studies now possible. Regions such as the Black and Caspian Sea must be inhabited by physiologically adapted races of species found elsewhere.

6 The physiology of diatoms should be investigated under conditions of low temperature and high pressure such as exist at great depths in the sea. Survival of diatoms at such depths may be much longer than suspected with significant ecological consequences regarding dispersal and 'seeding'.

4 SEASONAL SUCCESSION OF MARINE PLANKTONIC DIATOMS

The seasonal succession of marine phytoplankton is an extremely complex ecological phenomenon. Summaries of the multitude of often interrelated physical and chemical factors that have been proposed to account for the seasonal variation of phytoplankton are given in Margalef (1958, 1967a, 1967b), Braarud (1962), Raymont (1963), Smayda (1963b, 1973a), Fogg (1965), Lund (1965, 1966), Wimpenny (1966a), Hutchinson (1967), Riley (1971) and Parsons & Takahashi (1973a). Lund's summary is presented in Table 12.3. Because the seasonal succession of planktonic diatoms involves such a large number of interrelated biotic and abiotic factors few studies of seasonal succession are particularly comprehensive. In most cases changes in the abundance of individual species are reported and attempts are made to relate the observed changes to variations in a few selected nutrients in addition to light, temperature and stratification. Only on rare occasions have either laboratory or *in situ* experiments been performed to test hypotheses generated from the field observations. Progress in understanding the seasonal succession of planktonic diatoms has been hampered to some extent in recent years by the whole-community approach to phytoplankton ecology. Attention has been focused on such community properties as primary productivity, biomass, chlorophyll *a* content, and particle concentration and not on the intricate biological properties of the organisms which actually make up the community (Smayda, 1973b).

Table 12.3. List of factors that have been proposed to account for the distribution and seasonal succession of phytoplankton. From Lund, 1966.

I Physical

1 Temperature. Range within which growth is possible; effect on rate of growth, on nutrient demands and on enzymatic processes; thermal stratification.

2 Light. Length and brightness of day; spectral composition; light saturation; inhibitory or lethal intensities; IR absorption.

3 Water movements. Horizontal and vertical transport into and out of an area or depth zone; invasions; eddy diffusion.

4 Density distribution; effects of salinity, temperature, metabolism or gas production, in relation to the sinking or rising rate of organisms.

II Chemical

1 Inorganic substances. Iron, nitrogen compound; phosphates; silicon; sulphides; trace elements; oxygen; ionic ratios and salinity; redox-potentials; pH.

2 Organic substances. Vitamins (B_{12} and thiamine), acids (glycolic, glutamic), chelates, unknown or imperfectly known compounds such as 'humus', natural chelates and most extracellular compounds.

III Biological

1 Inhibitory or stimulatory substances; through the activities of previous populations or the organism's own extracellular products (e.g. lag phases).

2 'Clock' mechanisms.

3 Life histories, changes in the response of an alga to a given environment in relation to: growth; sexual recombinations; morphology and habit; resting stages.

4 Symbiosis *sensu lato* (a living together). Bacteria on algal cells or in their mucilage.

5 Grazing; qualitative and quantitative effects.

6 Parasitism.

7 Rate of sinking, variations in weight of organisms in water.

IV Human

1 Sampling and experimental techniques. Imperfections in sampling; identification of species; area sampled and frequency of sampling; uses of bacterized or axenic cultures, or enrichment experiments; effect of sterilization or filtration or of the walls of a vessel on the medium (e.g. adsorption of phosphates); size, shape and location of experimental containers.

2 Interpretation of the data obtained.

All species of planktonic diatoms are very different from one another. The patterns of seasonal succession are a product of these differences. Braarud (1962) lists the following autecological characteristics which may differ from species to species: temperature tolerance range and temperature-growth curve, salinity tolerance range and salinity-growth curve, light-growth curve, nutrient requirements and tolerance range (for inorganic and organic compounds), motility and flotation properties, life-cycle features (resting stages, modes of reproduction, etc.), growth rate range, and competitive characteristics (including external metabolites). The suitability of diatoms as prey is another important characteristic.

4.1 *Three stages of succession*

We have modified Margalef's (1958, 1962, 1967a) analysis of the three stages of succession in a number of minor ways in order to relate the stages he recognizes to research published in recent years.

Stage I begins with the enrichment of the euphotic zone. Enrichment can be caused by upwelling, strong mixing, or runoff from land and the accumulation of nutrients during the winter. Release from zooplankton grazing pressure has also been suggested as a possible factor triggering the first stage of succession (Pratt, 1965; Martin, 1965). Diatoms that bloom during this stage are small and have, on the average, high surface:volume ratios (~ 1 in terms of $\mu m^2/\mu m^3$) and consequently usually have high maximum growth rates. Division rates are often in excess of one division per day and total community densities between 10^6 and 10^7 cells/litre are not uncommon. Population regulation at this stage may often be due to dilution of populations by diffusion, or dispersal of cells around patches of high productivity. Mucilaginous materials characteristically envelop clusters of cells and few species with rigid appendages are observed. In very cold waters in the northern hemisphere (1°–5°C) *Chaetoceros debilis, Thalassiosira antarctica, Thalassiosira gravida* and *Thalassiosira nordenskioldii* often initiate the first stage of succession. In warmer waters, or as the temperature of the water increases, *Skeletonema costatum, Chaetoceros affinis, C. compressus, C. radians, C. socialis, C. tortissimus, Nitzschia kerguelensis* (ex-*Fragilaria antarctica*), *Nitzschia pseudonana* (ex-*Fragilaria nana*), *Leptocylindus danicus, Thalassionema nitzschioides*, and *Porosira glacialis* are commonly reported.

Stage II is usually characterized by the appearance of a large number of medium sized species of *Chaetoceros*. Surface:volume ratios are often between 0·2 and 0·5 (in $\mu m^2/\mu m^3$, excluding appendages). The bulk of genera of planktonic diatoms are present at this time and species diversity is relatively high. Population densities are generally lower than during the first period of succession but densities from 2×10^4 to 10^5 cells/litre are observed. The *Chaetoceros* species which are abundant during this stage often have long stiff bristles and form long chains. In addition to a large number of *Chaetoceros* species various species of *Bacteriastrum, Corethron, Nitzschia* and *Rhizosolenia* are also found.

Stage III contains only the few species of diatoms that are able to grow slowly under nutrient poor conditions. Most of the diatoms that thrive during the second stage of succession form resting spores and settle rapidly once nutrients are virtually exhausted in the euphotic zone and stratified 'oceanic' conditions prevail. The diatoms characterizing the third stage of succession are generally present during the second stage. They become dominant in the third period by virtue of persistence. Their surface : volume ratio is generally very low and total population densities are often considerably less than 10^4 cells/litre. Many of the diatoms found in nutrient poor waters form symbiotic associations with a blue–green alga which is suspected of fixing elemental nitrogen, or with various protozoans. The third stage of succession is characterized by *Hemiaulus hauckii, Rhizosolenia alata, R. calcar-avis, R. hebetata, Mastogloia rostrata* and a few *Chaetoceros* species.

Recent information suggests that very small diatoms ($<10\ \mu m$) may also be present in oligotrophic waters but their identification and importance are virtually unknown (see Malone, 1971; Sheldon *et al.* 1973). Nutrient enrichment experiments using surface water from the Sargasso Sea reveal the presence of *Chaetoceros simplex* and other small diatoms that are undetectable in unenriched samples (Menzel, Hulburt & Ryther, 1963; Hargraves & Guillard, 1974). Enrichment techniques will probably be required to study the distribution and taxonomy of these minute algae.

The pattern of seasonal succession often reflects the geographical distribution of planktonic diatoms. Diatom species that flower early in succession are characteristically abundant in eutrophic estuarine and coastal waters while those that develop later are often common in oligotrophic environments.

In many regions of the ocean, Margalef's three stages of succession in modified form are repeated over and over again. Diatom successions are most rapid in mid- and high-latitudes and least pronounced in tropical regions. The rate of succession is often determined by the loss of diatoms from the euphotic zone through diffusion, sinking and grazing and the rate of nutrient regeneration. Increases in mixing and grazing can push succession back towards the first stage (Margalef, 1967b). Grazing, which regenerates nutrients, has a tendency to prolong diatom successions (Corner & Davies, 1971; Margalef, 1967b). In general, succession is most rapid in the absence of strong mixing and heavy grazing pressure.

Considerable controversy has developed over which ecological factors are most important in determining the course of succession. The major argument has concerned the roles played by ambient nutrient concentration and grazing in the development and dissipation of phytoplankton blooms. Harvey *et al.* (1935) were among the first to suggest that zooplankton grazing was probably chiefly responsible for the decline of phytoplankton populations. This point of view has received particularly strong support in the papers of Cushing (1959, 1964). Fleming (1939) used a simple modification of the exponential growth equation to examine the effect of varying death rates (owing primarily to zooplankton grazing) on the growth and decline of diatom populations. The basic equation is:

$$\frac{dN}{dt} = rN$$

where N is the population size and r, the growth rate, is equal to the birth rate (division rate) minus the death rate. Fleming (1939) and Cushing (1958, 1959) have employed this equation and appropriate data from the literature to demonstrate that slight increases in diatom mortality caused by grazing can rapidly terminate a diatom flowering. In support of his argument that grazing is of particular importance in determining the fate of phytoplankton populations Cushing (1964) points out that control of the algal outburst in the North Sea, Atlantic Ocean in 1954 could be attributed solely to grazing and that the effect of nutrients was irrelevant. Steemann Nielsen (1972) has suggested that at least 75% of the daily primary production (on a worldwide basis) is consumed by zooplankton.

Both nutrients and grazing are of particular importance in determining the fate of phytoplankton blooms and the course of successions (see Riley, 1963b). In Trondheimsfjord in Norway two spring phytoplankton blooms occur (Sakshaug & Myklestad, 1973). The first flowering begins in March as light intensities increase, and the diatoms grow nearly exponentially until the nutrients accumulated over the winter are exhausted. Diatoms collected near the end of the augmentation had biochemical properties similar to those observed in old diatom cultures. Throughout the period of rapid growth grazers were virtually absent, and at the end many of the diatom species formed resting spores and settled out. The first diatom bloom had all of the characteristics of a typical laboratory batch culture. In this case nutrient depletion was probably responsible for its decline. The second spring flowering occurred in May and June. The phytoplankton remained in good physiological condition and Sakshaug and Myklestad consider this second bloom analogous to a continuous culture. A continuous supply of nutrients was carried by rivers flowing into the fjord and a portion of the population was continually transported out to sea. This second bloom was often suppressed by overgrazing. These data indicate that, depending on the circumstances, nutrient supply, grazing and diffusion play important roles in diatom successions.

Recent investigations in Trondheimsfjord (Sakshaug, 1972; Haug *et al.* 1973; Jensen & Sakshaug, 1973; Sakshaug & Myklestad, 1973) and on Narragansett Bay in the United States (Smayda, 1957, 1973a, 1973b; Pratt, 1959, 1965, 1966; Martin, 1965, 1968, 1970) are probably the most comprehensive studies of diatom succession undertaken to date. In Narragansett Bay the population growth of the dominant diatom *Skeletonema costatum* is regulated by temperature, light, nutrients, grazing, competition and hydrographic disturbances (Smayda, 1973b). At various times certain factors seem more important than others in limiting population growth. Temperature limits the growth of *Skeletonema* in January and February. The growth rate of *Skeletonema* increases as temperatures progressively rise from February onwards. Nutrient enrichment experiments indicated that both nitrogen and silicon were limiting to growth at one time or another from mid-March through June. Smayda suggests that when diatom growth occurred, it usually considerably exceeded the grazing rate, but when growth was poor, *Skeletonema* populations declined owing to a combination of natural mortality and grazing. The effects of light and species interactions on the growth of *Skeletonema* in Narragansett Bay has also been studied but the results are not conclusive. Smayda (1973b) suggests that a seasonal succession of primary limiting factors occurs, but poorly understood factor interactions complicate the situation.

5 ECOLOGY OF DIATOMS THAT ARE CHARACTERISTIC OF OLIGOTROPHIC WATERS

There are relatively few species of planktonic diatoms whose distribution and abundance are in any sense of the word predictable. A few species of mid- and low-latitude oceanic diatoms, however, regularly occur in biologically barren regions characterized by sparse phytoplankton standing crops and extremely low rates of primary production. *Hemiaulus hauckii, Mastogloia (Stigmaphora) rostrata* and a number of *Rhizosolenia* species are usually the dominant diatoms when low nutrient conditions prevail. In tropical regions these diatoms are usually components of a very diverse community characterized by *Trichodesmium (Oscillatoria) thiebautii* and a large number of different species of coccolithophores and dinoflagellates. These species associations have been observed in the Central North Pacific, the Sargasso Sea (Atlantic Ocean), the Indian Ocean and the Mediterranean Sea. In the Central North Pacific, *Hemiaulus hauckii* and *Mastogloia rostrata* are often the dominant diatoms (Venrick, 1971, 1974; Gilmartin & Revelante, 1974). *Rhizosolenia* species belonging to the taxonomically difficult *R. styliformis–R. hebetata* complex are also present though rarely numerically dominant (Venrick, 1974). In the Sargasso Sea, *Hemiaulus hauckii* and *Mastogloia (Stigmaphora) rostrata* are the most conspicuous diatoms after the formation of the summer thermocline in April or May (Hulburt *et al.* 1960; Hulburt, 1966). Based on his own research in the Mediterranean Sea, Margalef (1962, 1967a) has generalized that only a few species of diatoms such as *Hemiaulus hauckii* and *Rhizosolenia calcar-avis* can persist in stratified nutrient poor water. Similar observations have been made by Ignatiades (1969) who examined the seasonal succession of net phytoplankton in Saronicos Bay, Aegean Sea. She found that *Hemiaulus hauckii, Rhizosolenia alata* and *Chaetoceros decipiens* were the only species to remain in the plankton after the spring diatom bloom.

Very little is known about the autecology of most of the diatoms found in oligotrophic oceans, but they seem to differ from diatoms that predominate in nutrient rich waters in that they often form biological associations with a number of other organisms. In recent years considerable attention has been focused on the association between *Richelia intracellularis* Schmidt, a heterocystous cyanophyte, and individuals of the genus *Rhizosolenia.* A number of species of *Rhizosolenia* contain *Richelia* as an endophyte. *Richelia* has also been reported in association with several species of *Chaetoceros, Bacteriastrum* and *Hemiaulus* (Sournia, 1970b; Mague, Weare & Holm-Hansen, 1974; Venrick, 1974). *Richelia* has a heterocyst and there is strong circumstantial evidence that this blue–green alga can fix elemental nitrogen (Mague, Weare & Holm-Hansen, 1974). In the North Pacific Central Gyre, blooms of *Richelia*-containing species of *Rhizosolenia (R. cylindrus* and *R. hebetata semispina)* have been studied on a number of occasions. *Hemiaulus hauckii* and *Mastogloia rostrata* show strong increases in abundance when *Richelia* 'blooms' occur. During nonbloom conditions populations of *Hemiaulus hauckii* rarely exceed 120/litre, while under bloom conditions populations as large as $2{\cdot}5 \times 10^4$/litre have been observed (Venrick, 1974). Venrick has hypothesized that phytoplankton blooms in these waters are stimulated and sustained by an increase in the available nitrogenous nutrients resulting from the nitrogen fixation of *Richelia.*

Numerous examples of the association of oceanic diatoms with other plants and animals are known (see Cupp, 1943). Oceanic diatoms, for example, have been reported to attach to animals. *Hemiaulus hauckii* and *Chaetoceros peruvianus* are often associated with *Tintinnus lusus-undae,* a tintinnid (Protozoa) and *Chaetoceros tetrastichon* and *C. dadayi* live attached to *Tintinnus inquilinus* (Cupp, 1943).

One, as yet unanswered, question is repeatedly asked by investigators who have studied the distribution and abundance of planktonic diatoms in both eutrophic and oligotrophic regions of the ocean. That question is 'why are relatively large diatoms such as *Hemiaulus hauckii* and *Rhizosolenia alata* able to persist often in extremely low numbers (50 cells/litre) in extremely oligotrophic regions and why are these species often present though in proportionally low numbers even in more eutrophic regions?' In one form or another this question has been asked by Allen (1945, p. 332), Halldal (1953, pp. 39–40) and Hulburt (1962, 1966). *Hemiaulus hauckii* and *Rhizosolenia alata* are often referred to as ubiquitous species because they occur throughout the year under either oligotrophic or eutrophic conditions. They are rarely dominant except when nutrients are extremely scarce. As relatively large diatoms with low surface to volume ratios their maximum growth rates are presumably fairly low. The maximum growth rate of *Rhizosolenia alata* has been estimated on at least two occasions to be approximately one division/day (Gran, 1933; Lanskaya, in Riley, 1963a, p. 77), and to our knowledge the growth rate of *Hemiaulus hauckii* has never been determined. A number of investigators have suggested that large diatoms with low maximum growth rates are able to persist in oligotrophic regions because their nutrient requirements per unit time are lower (Gran & Braarud, 1935; Halldal, 1953; Lund, 1964; Moss, 1973). The argument is difficult to understand because even small eutrophic species with high maximum growth rates would grow slowly under conditions of nutrient-limited growth.

Another extremely puzzling aspect of the ecology of *Hemiaulus hauckii* and *Rhizosolenia alata* is that they are sometimes capable of blooming and remaining abundant when there is a sudden increase of nutrients in an essentially oligotrophic environment. *Hemiaulus hauckii* flowers ($2{\cdot}5 \times 10^4$/litre) when *Richelia* 'blooms' in the Central North Pacific Gyre and *Rhizosolenia alata* blooms in the Gulf of Maine when mixing on Georges Bank brings nutrients into circulation (Gran, 1933; Gran & Braarud, 1935). On the other hand, neither *Hemiaulus hauckii* nor *Rhizosolenia alata* increase in numbers during the spring diatom bloom in the Sargasso Sea (Hulburt *et al.* 1960). Are these diatoms only capable of rapid increases in numbers when diatoms with higher maximum growth rates are either not present or in some way inhibited? The possibility that neritic diatoms are sometimes inhibited by 'antimetabolites' has been explored by Johnston (1963a,b). It is also conceivable that the nature and abundance of grazers may determine whether or not these diatoms can bloom. Sheldon *et al.* (1973) have estimated that a 10% decrease in grazing pressure for only a few days in the Sargasso Sea would permit a diatom flowering equal in magnitude to the annual spring outburst (assuming adequate nutrients were available). A sudden increase in nutrients in an oceanic region dominated by microzooplankton might also favour the flowering of large-celled diatoms. This would happen if the microzooplankton suppressed the growth of the nannoplankton and there was not sufficient time for the development of macrozooplankton populations large enough to check a bloom of the large-celled diatoms.

6 IMPORTANCE OF PROXIMATE AND ULTIMATE FACTORS IN ECOLOGICAL SUCCESSIONS

Approaches to the study of the succession of planktonic diatoms have changed considerably over the past 75 years. Early investigations were primarily floristic though some attention was paid to various relatively easily measurable environmental factors. In the 1940s Riley and his associates (Riley, 1947; Riley *et al.* 1949) developed what is now known as the 'systems approach' to study the ecology of marine phytoplankton communities. This largely deterministic approach to mathematical modelling has been very successful in relating successional changes in productivity to a host of physical, chemical and biological factors. On the other hand, mathematical modelling has provided few insights into the distribution and succession of individual species of planktonic diatoms. In the mid-1950s and early 1960s many phytoplankton ecologists, following Gran, Allen and Nelson, and others (see Introduction to this chapter) were convinced that culture studies of individual species of planktonic diatoms would reveal the chemical and physical factors primarily responsible for successional changes observed in nature. Culture studies have greatly increased our understanding of the physiology of particular clones of individual species of diatoms under laboratory conditions, but results obtained from culture studies are often difficult to interpret in relation to field observations.

The study of the proximate (i.e. non-evolutionary) factors which interact to control the growth rate of planktonic diatoms has progressed rapidly since Dugdale (1967) suggested that the Michaelis-Menten or Monod equations could be used to model nutrient-limited growth and the kinetics of nutrient uptake. Examples of this approach can be found in papers of Eppley *et al.* (1969), Uhlmann (1971), Paasche (1973a), Parsons & Takahashi (1973b), Droop (1973), Goldman & Carpenter (1974) and O'Brien (1974). In general, these are multiplicative models which contain separate terms (often in the basic Monod form) which model the effect of various parameters (e.g. nutrients, light, temperature, etc.) on the growth rate of particular species of algae. It has been suggested that given sufficient data multiplicative models of the Monod type could eventually be used to predict successional changes in species composition in nature (Dugdale, 1967; Eppley *et al.* 1969; Parsons & Takahashi, 1973b; Goldman & Carpenter, 1974). To date deterministic models of this sort have met with limited success in predicting the pattern of seasonal succession. Even if ample data were available it seems improbable that models based almost exclusively on proximate physical and chemical factors will ever be sufficiently comprehensive to model phytoplankton succession predictably. The effect of proximate factors on algal growth rates can be used to define the dimensions of the fundamental niche (*sensu* Hutchinson, 1967) of particular diatom species, but they are of little use in determining the dimensions of the realized niche because its dimensions are determined by a large number of complex biological factors and their proximate and ultimate effects (e.g. competition, predation, sinking, life history strategy, etc.). This problem is clearly illustrated by the fact that temperature optima of particular species of planktonic diatoms in laboratory cultures often do not correspond with ambient temperatures recorded during periods of maximum abundance in nature (Braarud, 1961). For example, *Thalassiosira nordenskioldii* is usually an early spring dominant when water temperatures are approximately 2° to 3°C. As the water warms this diatom is usually replaced in dominance by *Skeletonema costatum* or some other species. *T. nordenskioldii* is rarely particularly abundant when ambient water temperatures reach 10°C. These field observations contrast markedly with data from culture studies which indicate that *T. nordenskioldii* obtains its maximum growth rate between 10° to 11°C (see data and references in Durbin, 1974). The difference may stem at least in part from the provision of high nutrient levels in the experiments showing the dependence of growth rate on the environmental variables, temperature, light, and salinity. To our knowledge the only study of any growth responses of a diatom to such variables at both low and high nutrient levels is that of Maddux & Jones (1964), carried out in continuous (turbidostat) culture. Their higher levels of N and P (10 mM and 0·52 mM respectively) are typical of nutrient solutions, while their lower levels of N and P (8·9 μM and 0·42 μM) correspond to moderate natural levels. The optimum light intensity for growth of *Cylindrotheca fusiformis* (ex-*Nitzschia closterium*; Reimann & Lewin, 1964) was higher in the solution of higher nutrient levels. The experiment involving the temperature–nutrient interaction was complicated by a difference in light intensity also; adjusting for this by the previous result shows that at high nutrient levels the temperature optimum shifted to higher values. Maddux & Jones (1964) suggested that the anomalies observed by previous workers might be explained as due to the use of unnaturally high nutrient concentrations, provided the species they studied were typical.

Culture studies on the influence of day length, light intensity and temperature on the growth of the early spring dominants in northern coastal waters have revealed that these diatoms have higher growth rates at lower light levels and lower temperatures than the diatoms which replace them at higher temperatures and light intensities. *Thalassiosira nordenskioldii* and *Detonula confervacea* appear to have higher growth rates at lower temperatures and light levels than *Skeletonema costatum* which replaces them in dominance as growth conditions improve (Durbin, 1974; Holt & Smayda, 1974; Conover, 1956). These experimental studies are important because they provide one of the few well documented examples of the influence of proximate factors in seasonal succession.

The role of ambient nutrient concentrations in determining the pattern of seasonal succession is unclear. Dugdale (1967) suggested that changing ambient nutrient concentrations might play an important role in the phytoplankton succession if some species were found to have higher growth rates (or velocities of nutrient uptake) than others, at lower ambient nutrient concentrations. Experimental studies, on a limited number of planktonic diatoms, have shown that Michaelis-Menten half-saturation constants (K_s) for nitrate uptake velocity, for example, are different for all species investigated thus far. Even though species specific differences occur in K_S values for nutrient uptake and growth, the available data indicate that K_S values may be primarily related to cell size. Eppley *et al.* (1969) and Parsons & Takahashi (1973b, 1974) have suggested that small diatoms, in general, have low K_S values for nitrate uptake and large diatoms have high K_S values. The investigations of Finenko & Krupatkina-Akinina (1974) can be interpreted to show that a similar relationship between K_S value and cell size also exists for phosphorus-limited growth. A few general trends can be seen in most if not all of the data on K_S values for growth and uptake collected to date. These trends were pointed out by Eppley *et al.* (1969) and we have modified them only slightly. The following trends are observed: (1) Diatoms with a high K_S for one nutrient usually have a high K_S for other nutrients as well. (2) Large-celled species tend to show high K_S values and small-celled species tend to show low K_S values. (3) Oceanic species (or clones) have lower K_S values than estuarine species (or clones) of similar cell size (see Carpenter & Guillard, 1971; Guillard *et al.* 1973). (4) Fast-growing diatoms tend to have lower K_S values than slow growers. In general, diatom growth rates decrease with increasing cell size (Williams, 1964; Eppley & Sloan, 1966; Findley, 1972; Paasche, 1973b). We are not sure if these general trends which seem to relate K_S values to cell size can be used to evaluate the role of ambient nutrient concentrations in determining the course of seasonal succession in nature because K_S values also depend on temperature and nutrient stores within cells. Diatom growth rates in nature are also directly affected in a multiplicative way by other proximate factors such as light and temperature (see Eppley *et al.* 1969; Droop, 1973; Parsons & Takahashi, 1973b; Goldman & Carpenter, 1974). In spite of these complicating factors one might expect to observe a general relationship between ambient nutrient concentrations and diatom cell sizes as ambient nutrient concentrations decrease during seasonal succession. If ambient nutrient concentrations play a primary role in determining the pattern of seasonal succession, small-celled diatoms with low K_S values should replace large-celled diatoms with high K_S values as nutrient levels decline. But, the pattern of succession generally observed in nature indicates that small-celled diatoms predominate early in the successional sequence when nutrients are high. Large-celled diatoms are characteristic of the final stages of succession when ambient nutrient concentrations are extremely low. These observations, though far from conclusive, suggest that understanding the factors that determine diatom cell size may provide major clues to ecological processes underlying the commonly recognized patterns of seasonal succession.

It is possible that one reason that the autoecology of planktonic diatoms is only partially understood is that phytoplankton ecology has, in general, been approached from a purely mechanistic point of view. We have attempted to explain the succession of planktonic diatoms almost entirely in terms of chemical and physical factors that are governed by the basic laws of physics and chemistry. At the same time we have largely ignored the laws of evolutionary change. Williams (1966) states that 'the laws of physical science plus natural selection can furnish a complete explanation for any biological phenomenon, these principles can explain adaptation in general and in the abstract and any particular example of an adaptation'.

Emlen (1973) gives an example of these contrasting approaches to ecology which we can paraphrase in terms of the ecology of planktonic diatoms. He suggests that to a biological oceanographer a diatom forms a resting spore, perhaps, because of salinity, temperature or nutrient changes in the environment. An evolutionist on the other hand would suggest that a particular diatom forms a resting spore because in the past those ancestors which possessed genetic material predilecting them to form resting spores under appropriate conditions more often survived to pass on that genetic material than individuals that possessed other genes. In order to obtain a greater understanding of the ecology of planktonic diatoms it seems clear that we will have to pay closer attention to individual adaptations to the environment. It will not be enough to examine individual adaptation almost exclusively from either the mechanistic (proximate) or evolutionary (ultimate) points of view (see Emlen, 1973; Pianka, 1974).

7 EVOLUTIONARY ECOLOGY OF PLANKTONIC DIATOMS

Comprehensive examination of the biological properties of planktonic diatoms from an evolutionary point of view has never been attempted. The biological properties of planktonic diatoms represent an adapted complex which includes adaptations at the biochemical, physiological, morphological, behavioural and life history levels. In recent years it has become increasingly evident that understanding the ecology of planktonic diatoms will require considerable knowledge of the selective pressures which mould and maintain the biological properties of a particular diatom species. We have little understanding of the adaptive significance of cell size, cell morphology, coloniality, maximum growth rate, cyclomorphosis, sinking rate, resting spore formation, differences in the parameters of nutrient-limited growth and life history. It is clear that patterns of seasonal succession will only be understood once the changing selective pressures of continually changing marine environments are identified. The phenotypic attributes of any particular diatom represent a series of selective compromises to a large number of environmental variables. Therefore identification of the primary selective pressures in any given environment may prove extremely difficult.

Attempts to explain the distribution and succession of marine phytoplankton in terms of a few proximate factors will probably not prove to be particularly successful. It is certainly not possible, at present, to explain the three stages of succession recognized by Margalef (see above) in terms of varying ambient nutrient concentrations. Margalef (1958, 1962, 1967b) views the ecological succession of marine phytoplankton as commencing with a developmental stage (or immature stage) and proceeding until a mature stage is finally reached. In his analysis succession is related to the environment. By the process of succession he suggests that the community becomes more precisely adjusted to the environment. This point of view is open to criticism because the organisms involved are probably precisely adjusted to environmental conditions at all stages in a succession. Organisms replace one another as a succession proceeds because every species is adapted to a particular set of biotic and abiotic factors. Correlations between changing environmental conditions and the biological properties of dominant organisms at each stage along a succession are the foundation upon which a growing (though controversial) school of evolutionary ecology is based. This school has used a wide variety of terms to classify the organisms which occur early in a succession (immature stage). Among these are fugitive species, *r*-selected species, colonist species, weedy species and opportunistic species (see Grassle & Grassle, 1974). The terms *K*-selected or equilibrium are often used to describe organisms that are dominant in the later (mature) stages of a succession.

Tables summarizing the biological properties of organisms and environmental conditions at early and late stages in a succession have been constructed by Odum (1969) and Pianka (1970, 1974). They suggest that a continuum exists between the biological characteristics of species occurring early and late in an ecological succession. In Table 12.4 we present a modified version of the summary table of Pianka (1970). Because there are significant differences between planktonic diatoms and higher plants and animals we have excluded irrelevant correlations from our table. Even though many ecologists are highly critical of this form of ecological analysis (see Ricklefs, 1973; Wilbur, Tinkel & Collins, 1974) it may afford useful insights into the ecology of planktonic diatoms. In addition it should be possible to discuss phytoplankton ecology in terms of many of the concepts that are currently exciting to terrestrial ecologists.

Correlations of the type we have presented in Table 12.4 are interpreted by many ecologists as providing a link between ecology in general and the selective process of evolution. Because organisms are adapted to the environment in which they survive and reproduce, natural selection is believed to favour organisms with contrasting biological properties under different environmental conditions. Variable environments with plentiful resources usually appear to selectively favour smaller organisms with high maximum

Table 12.4. Some correlates of *r*- and *K*-selection (from Pianka, 1970). Diatom species abundant early in succession appear to be more *r*-selected while species common later in succession are more *K*-selected.

	r-*selection*	K-*selection*
Climate or Hydrography	Variable and/or unpredictable; uncertain	Fairly constant and/or predictable; more certain
Mortality	Often catastrophic, non-directed, density-independent	More directed, density-dependent
Population size	Variable in time, non-equilibrium; usually well below carrying capacity of environment; unsaturated communities or portions thereof; ecologic vacuums; recolonization each year	Fairly constant in time, equilibrium; at or near carrying capacity of the environment; saturated communities; no recolonization necessary
Intra- and inter-specific competition	Variable, often lax	Usually keen
Selection favours	**1** High maximal rate of increase, r_{max} **2** Small body size	**1** Lower maximal rate of increase r_{max} **2** Larger body size
Leads to	Productivity	Efficiency

growth rates while stable environments with scarce resources selectively favour organisms that delegate a higher proportion of their metabolic resources to processes not directly related to reproduction. Perhaps the major criticism of this approach to evolutionary ecology is that it is extremely difficult to determine which selective pressure or combination of selective pressures has been of primary importance for the evolution of the biological properties of organisms characteristically found in different environments. The terms *r*-selection and *K*-selection imply that competition for resources has provided the primary selective pressure. Other ecologists have suggested that these terms are potentially misleading because predation, for example, could selectively favour the evolution of similar biological properties (see Schmalhausen, 1949, Ch. 4; Wilbur, Tinkle & Collins, 1974).

Phytoplankton ecologists have been involved in similar arguments. Schütt (1892) suggested that differences in the size and shape of diatom cells and colonies are mainly the result of the evolution of adaptations related to keeping diatoms afloat. This point of view has been challenged by Munk & Riley (1952) who argue convincingly that diatom speciation is the product of evolutionary responses to the problems of flotation, nutrient absorption and predation. Their point of view is supported by Beklemishev (1959) who suggests that the evolution of coloniality in diatoms is primarily the result of selective predation. Smayda (1970) has questioned the belief that flotation and predation are particularly important problems. He suggests that the various morphological adaptations of diatoms are related to nutrient assimilation and this is aided by adaptations which permit these organisms to sink or rise and rotate. The observation that small-celled diatoms usually occur in eutrophic environments and large-celled diatoms predominate in oligotrophic environments (see above) suggests that size-selective predation is an extremely important selective pressure. If competition for resources were the primary selection pressure all planktonic diatoms should be $<20\ \mu m$ in diameter and small-celled diatoms with low K_S values should be dominant in nutrient-poor waters. Munk & Riley (1952) have calculated that diatoms $<20\ \mu m$ in diameter should not have significant problems with either flotation or nutrient absorption. Predation, on the other hand, is a major problem. Some diatoms surmount herbivory by having a rapid growth rate. Others escape by becoming too large to ingest. This can be accomplished by forming chains, star-shaped colonies, large agglomerations of cells or by growing spines or increasing cell volume (see Munk & Riley, 1952; Beklemishev, 1959). There are numerous marine predators that consume diatoms, and antipredation devices that work well against one type of predator may not work well against another type, but it seems reasonable to believe that predation has been an extremely important selective pressure over evolutionary time. It seems probable that size-selective predation as the major selective pressure moulds and maintains many of the adaptations often attributed to other selective pressures. If large cell size is an adaptation maintained by predation, secondary adaptations related to flotation and nutrient absorption should also be favoured because large cell size hinders flotation and nutrient absorption (Munk & Riley, 1952).

8 SIZE SELECTIVE PREDATION

At present the literature concerning size selective predation by copepods on diatoms is contradictory and confusing. Marshall (1973) came to this conclusion in her comprehensive review on the respiration and feeding of marine copepods. Some investigators (Hargrave & Geen, 1970; Martin, 1970) have demonstrated experimentally that copepods selectively feed on smaller diatoms while other workers have reached the opposite conclusion (Mullin, 1963; Richman & Rogers, 1969). Poulet (1974) suggests that copepods graze the size fraction of a natural population that contains the highest concentration of particles (in terms of ppm). He considers particle size and particle number to be of lesser importance.

The selective grazing of marine diatoms by zooplankton other than copepods has rarely been studied. Wimpenny (1973) investigated the effect of grazing by *Artemia salina* (a brine shrimp) on the size distributions of *Skeletonema costatum* and *Ditylum brightwelli* in laboratory cultures. He observed that *Artemia salina* appears to selectively consume the smaller size classes of these diatoms. Wimpenny (1966b, 1973) believes that the size distributions of large diatoms observed in nature are associated with the effects of grazing.

9 SUMMARY AND CONCLUSIONS

1 All species of planktonic diatoms are ecologically very different from one another. The patterns of seasonal succession are a product of these differences.

2 Three stages of succession are commonly observed. The first stage is characterized by small diatoms (e.g. *Skeletonema costatum* and *Chaetoceros socialis*). Larger diatoms mainly of the genus *Chaetoceros* are dominant during the second stage. The final stage is comprised of large diatoms, which become dominant by virtue of persistence when stratified 'oceanic' conditions prevail (e.g. *Hemiaulus hauckii, Rhizosolenia calcar-avis* and *Mastogloia (Stigmaphora) rostrata*).

3 Nutrients and grazing are of particular importance in determining the fate of phytoplankton blooms and the course of diatom successions.

4 Oligotrophic oceanic regions are characterized by relatively large diatoms which often form symbiotic associations with blue–green algae suspected of fixing elemental nitrogen or with various protozoans. The ecology of these diatoms is poorly understood.

5 The effect of proximate (i.e. non-evolutionary) factors on algal growth rates can be used to define the dimensions of the fundamental niche of particular diatom species but they are of little use in determining the dimensions of the realized niche. Complex biological factors and their proximate and ultimate effects determine the dimensions of the realized niche.

6 The role of ambient nutrient concentrations in determining the pattern of seasonal succession is unclear. If nutrient concentrations in the environment play a primary role, small-celled diatoms with low K_S values should replace large-celled diatoms with high K_S values as nutrients decline. They apparently do not. This suggests that understanding the factors that determine cell size in planktonic diatoms may provide major clues to the ecological processes underlying the commonly recognized patterns of succession.

7 Correlations between changing environmental conditions and the biological properties of the dominant organisms suggest that diatoms occurring early in succession appear to invest more energy in growth than those species common towards the end of succession which invest more energy in adaptations not directly related to growth (e.g. antipredation mechanisms etc.). Size selective predation rather than competition for resources may provide the major selective pressure which moulds and maintains the biological properties of planktonic diatoms.

8 If size selective predation has been a major selective pressure over evolutionary time, diatoms may have evolved in such a way as to surmount herbivory by having a rapid growth rate or by becoming too large to ingest.

9 Understanding the process of seasonal succession will require a thorough knowledge of both the proximate (mechanistic) and ultimate (evolutionary) factors which affect planktonic diatoms in marine environments.

ACKNOWLEDGMENTS

This review was supported by National Science Foundation Grants 33288 (to R.R.L.G.) and GB-41315 (to P.K.). Ideas concerning the application of the concepts of *r*- and *K*-selection to phytoplankton ecology were developed in conversation with Robert E. Hecky. Discussions with Karen G. Porter and David S. Wethey increased our understanding of the importance of grazing. Susan S. Kilham, David S. Wethey, and G. David Titman critically read portions of the manuscript. We gratefully acknowledge the help of Mary Hedberg and the staff of the library of the Marine Biological Laboratory, Woods Hole, with problems of literature search. The National Oceanographic Data Centre kindly provided a great number of citations through its Biological Information Retrieval Service. We thank Alicja Mann for translations from Slavic languages, and our special thanks go to Vicki L. McAlister, who not only assisted in literature search and manuscript preparation, but organized and checked the difficult bibliographic entries.

Contribution 3652 from the Woods Hole Oceanographic Institution.

10 REFERENCES

AIKAWA H. (1936) On the diatom communities in the waters surrounding Japan. *Rec. oceanogr. Wks Japan* **8,** 1–159.

ALEEM A.A. & DOWIDAR N. (1967) Phytoplankton production in relation to nutrients along the Egyptian Mediterranean Coast. *Stud. trop. Oceanogr. Miami* **5,** 305–23.

ALLEN M.B. (1971) High-latitude phytoplankton. *Ann. Rev. Ecol. Syst.* **2,** 261–76.

ALLEN T.F.H. & SKAGEN S. (1973) Multivariate geometry as an approach to algal community analysis. *Br. phycol. J.* **8,** 267–87.

ALLEN W.E. (1945) Seasonal occurrence of marine plankton diatoms off Southern California in 1938. *Bull. Scripps Instn Oceanogr. tech. Ser.* **5,** 293–333.

ALLEN W.E. & CUPP E.E. (1935) Planktonic diatoms of the Java Sea. *Annls. Jard. bot. Buitenz.* **44,** 101–74.

ARCTIC BIBLIOGRAPHY Vol. III, Index (1953). U.S. Gov't. Printing Office, Washington, D.C. 3 329 p.

AVARIA S. (1971) Variaciones mensuales del fitoplancton de la Bahia de Valparaiso, entre Julio de 1963 y Julio de 1966. *Revta Biol. mar.* **14,** 15–43.

BAINBRIDGE R. (1957) The size, shape and density of marine phytoplankton concentrations. *Biol. Rev.* **32,** 91–115.

BALECH E., EL SAYED S.Z., HASLE G., NEUSHUL M. & ZANEYELD J.S. (1968) Primary productivity and benthic marine algae of the Antarctic and Subantarctic. American Geographical Society of New York. *Antarctic Map Folio Series,* Folio 10.

BÉ A.W.H. & TOLDERLUND D.S. (1971) Distribution and ecology of living planktonic Foraminifera in surface waters of the Atlantic and Indian Oceans. In *Micropaleontology of Oceans* (eds. B.M. Funnell & W.R. Riedel) pp. 105–49. Cambridge Univ. Press, Cambridge.

BEKLEMISHEV C.W. (1959) Sur la colonialité des diatomées planctoniques. *Int. Revue ges. Hydrobiol.* **44,** 11–26.

BEKLEMISHEV K.V. (1964) Concerning the phytogeographic division of the Antarctic pelagic region. *Inf. Byull. sov. antarkt. Eksped.* **II,** 272–5.

BELYAEVA T.V. (1970) Taxonomy and distribution patterns of plankton diatoms in the equatorial Pacific (in Russian, English summary). *Okeanologiya* **10,** 132–9; Engl. trans. (1970) *Oceanology* **10,** 101–7.

BELYAEVA T.V. (1971) Quantitative distribution of planktonic diatoms in the western tropical Pacific (in Russian, English summary). *Okeanologiya* **11,** 687–94; Engl. trans. (1971) *Oceanology* **11,** 578–85.

BELYAEVA T.V. (1972) Distribution of large diatoms in the southeastern Pacific (in Russian, English summary). *Okeanologiya* **12,** 475–84; Engl. trans. (1972) *Oceanology* **12,** 400–7.

BERNARD F. (1967) Research on phytoplankton and pelagic protozoa in the Mediterranean Sea from 1953 to 1966. *Oceanogr. mar. Biol. Ann. Rev.* **5,** 205–29.

BERNARD M. (1967) Recent advances in research on the zooplankton of the Mediterranean Sea. *Oceanogr. mar. Biol. Ann. Rev.* **5,** 231–55.

BERNHARD M. & RAMPI L. (1965) Horizontal microdistribution of marine phytoplankton in the Ligurian Sea. *Bot. Gothoburg.* **III,** 13–24.

BERNHARD M. & RAMPI L. (1967) The annual cycle of the 'Utermöhl-phytoplankton' in the Ligurian Sea in 1959 and 1962. *Pubbl. Staz. zool. Napoli* **35,** 137–69.

BLASCO D. (1971) Composición y distribución del fitoplancton en la región del afloramiento de las costas peruanas. *Investigación pesq.* **35,** 61–112.

BODEANU N. (1969) Cercetări asupra fitoplanctonului din zona de mică adincime de la litoralul Românesc al Mării Negre. *Ecologie mariná* (Bucharest) **3,** 65–147.

BRAARUD T. (1935) The ØST Expedition to the Denmark Strait, 1929. II. The phytoplankton and its conditions of growth. *Hvalråd. Skr.* **10,** 173 p.

BRAARUD T. (1961) Cultivation of marine organisms as a means of understanding environmental influences on populations. In *Oceanography* (ed. M. Sears) pp. 271–98. Am. Ass. Adv. Sci. No. 67, Washington.

BRAARUD T. (1962) Species distribution in marine phytoplankton. *J. oceanogr. Soc. Japan,* 20th Anniversary Volume, 628–49.

BRAARUD T., GAARDER K. & G. JUL (1953) The phytoplankton of the North Sea and adjacent waters in May 1948. *Rapp. P.-v. Réun. Cons. perm. int. Explor. Mer.* **133,** 1–87.

BRUNEL J. (1962) Le phytoplancton de la Baie des Chaleurs. *Contr. Inst. bot. Univ. Montréal* **77,** 365 p.

BUNT J.S. (1963) Diatoms of Antarctic Sea ice as agents of primary production. *Nature* **199,** 1 255–7.

BUNT J.S. (1967) Some characteristics of microalgae isolated from Antarctic Sea ice. *Antarct. Res. Ser.* **11,** 1–14.

BUNT J.S. & LEE C.C. (1972) Data on the composition and dark survival of four sea-ice microalgae. *Limnol. Oceanogr.* **17,** 458–61.

BUNT J.S. & WOOD E.J.F. (1963) Microalgae and Antarctic sea-ice. *Nature* **199,** 1 254–5.

BUNT J.S., OWENS O. v. H. & HOCH G. (1966) Exploratory studies on the physiology and ecology of a psychrophilic marine diatom. *J. Phycol.* **2,** 96–100.

BURKHOLDER P.R. & MANDELLI E.F. (1965) Productivity of microalgae in Antarctic Sea ice. *Science* **149,** 872–3.

BURSA A.S. (1961a) Phytoplankton of the 'Calanus' Expeditions in Hudson Bay, 1953 and 1954. *J. Fish. Res. Bd. Can.* **18,** 51–83.

BURSA A.S. (1961b) The annual oceanographic cycle at Igloolik in the Canadian Arctic. II. The phytoplankton. *J. Fish. Res. Bd Can.* **18,** 563–615.

BURSA A.S. (1963) Phytoplankton in coastal waters of the Arctic Ocean at Point Barrow, Alaska. *Arctic* **16,** 239–62.

CARPENTER E.J. & GUILLARD R.R.L. (1971) Intraspecific differences in nitrate half-saturation constants for three species of marine phytoplankton. *Ecology* **52,** 183–5.

CASSIE V. (1961) Marine phytoplankton in New Zealand waters. *Botanica mar.* **2** (Suppl.), 54 pp.

CHOE S. (1966) Phytoplankton studies in Korean waters. I. Phytoplankton survey of the surface in the Korea Strait in summer of 1965 (in Korean, English abstract). *J. oceanol. Soc. Korea* **1,** 14–21.

CHOE S. (1967) Phytoplankton studies in Korean waters. II. Phytoplankton in the coastal waters of Korea (in Korean, English abstract). *J. oceanol. Soc. Korea* **2,** 1–12.

CHYONG NGOK AN (1971) Phytoplankton of the Atlantic Ocean south of the Guinean Gulf in the sections along 11° and 14° south latitudes (in Russian, English summary). *Okeanologiya* **11,** 1 082–9; Engl. trans. (1971) *Oceanology* **11,** 896–901.

CLARKE G.L. & BUMPUS D.F. (1950) The plankton sampler – an instrument for quantitative plankton investigation. *Spec. Publs limnol. oceanogr. Soc. Am.* **5,** 1–8.

CLEVE P.T. (1897) *Phytoplankton of the Atlantic and its tributaries.* P.T. Cleve, Uppsala. 27 pp.

CLEVE EULER A. (1951–55) Die Diatomeen von Schweden und Finnland. *K. svenska Vetensk Akad. Hadl.,* Ser. 4. I. 2, 1, 1–163, 1951. V. 3, 3, 1–153, 1952. II. 4, 1, 1–158, 1953a. III. 4, 5, 1–255, 1953b. IV. 5, 4, 1–232, 1955.

CONOVER S.A.M. (1956) Oceanography of Long Island Sound, 1952–54. IV. Phytoplankton. *Bull. Bingham oceanogr. Coll.* **15,** 62–112.

CORNER E.D.S. & DAVIES A.G. (1971) Plankton as a factor in the nitrogen and phosphorus cycles in the sea. *Adv. mar. Biol.* **9,** 101–204.

CROSBY L.H. & WOOD E.J.F. (1958) Studies on Australian and New Zealand diatoms. I. Planktonic and allied species. *Trans. R. Soc. N.Z.* **85,** 483–530.

CUPP E.E. (1943) Marine plankton diatoms of the west coast of North America. *Bull. Scripps. Instn. Oceanogr. tech. Ser.* **5,** 1–238.

CUSHING D.H. (1958) The effect of grazing in reducing the primary production: a review. *Rapp. P.-v. Réun. Cons. perm. int. Explor. Mer.* **144,** 149–54.

CUSHING D.H. (1959) The seasonal variation in oceanic production as a problem in population dynamics. *J. Cons. perm. int. Explor. Mer.* **24,** 455–64.

CUSHING D.H. (1964) The work of grazing in the sea. In *Grazing in Terrestrial and Marine Environments* (ed. D.J. Crisp) pp. 207–25. Blackwell Scientific Publications, Oxford.

DANDONNEAU Y. (1971) Étude du phytoplancton sur le plateau continental de côte d'Ivoire. I. Groupes d'espèces associées. *Cah. ORSTOM-Oceanog.* **9,** 247–65.

DAKIN W.J. & COLEFAX A.N. (1940) The plankton of the Australian Coastal waters off New South Wales. *Publs Univ. Sydney Dep. Zool. Monog.* **1,** 221 p.

DEACON G.E.R. (1963) The southern ocean. In *The Sea,* Vol. 2 (ed. M.N. Hill) pp. 281–96. Interscience Publishers, New York.

DENISENKO V.V. (1964) On the phytoplankton of the Adriatic, Ionian, Aegean and Black Seas in August 1958 (in Russian). *Trudý sevastopol. biol. Sta.* **17,** 13–20.

DESROSIERES R. (1965) Observations sur le phytoplancton superficiel de l'ocean Indien Oriental. *Cah. ORSTOM – Oceanogr.* **3,** 31–7.

DESROSIERES R. (1969) Surface macrophytoplankton of the Pacific Ocean along the equator. *Limnol. Oceanogr..* **14,** 626–32.

DREBES G. (1967 *Bacteriastrum solitarium* Mangin, a stage in the life history of the centric diatom *Bacteriastrum hyalinum. Mar. Biol.* **1,** 40–2.

DROOP M.R. (1973) Some thoughts on nutrient limitation in algae. *J. Phycol.* **9,** 264–72.

DUGDALE R.C. (1967) Nutrient limitation in the sea: dynamics, identification, and significance. *Limnol. Oceanogr.* **12,** 685–95.

DURAIRATNAM M. (1964) Some planktonic diatoms from the Indian Ocean. *Bull. Fish. Res. Stn Ceylon* **17,** 159–68.

DURBIN E.G. (1974) Studies on the autecology of the marine diatom *Thalassiosira nordenskiökdii* Cleve. I. The influence of daylength, light intensity, and temperature on growth. *J. Phycol.* **10,** 220–25.

EDINBURGH, OCEANOGRAPHIC LABORATORY (1973) Continuous plankton records: a plankton atlas of the North Atlantic and the North Sea. *Bull. mar. Ecol.* **7,** 1–174.

EKMAN S. (1953) *Zoogeography of the Sea.* 417 pp. Sidgwick and Jackson.

EL-SAYED S.Z. (1971) Observations on phytoplankton bloom in the Weddell Sea. *Antarct. Res. Ser.* **17,** 301–12.

EMLEN J.M. (1973) *Ecology: An Evolutionary Approach.* 493 pp. Addison-Wesley, Reading.

EPPLEY R.W. & SLOAN P.R. (1966) Growth rates of marine phytoplankton: correlation with light absorption by cell chlorophyll a. *Physiologia Pl.* **19,** 47–59.

EPPLEY R.W., ROGERS J.N. & MCCARTHY J.J. (1969) Half-saturation 'constants' for uptake of nitrate and ammonium by marine phytoplankton. *Limnol. Oceanogr.* **14,** 912–20.

ESTABLIER R. & MARGALEF R. (1964) Fitoplancton e hidrografiá de las costas de Cádiz (Barbate), de junio de 1961 a agosto de 1962. *Investigación pesq.* **25,** 5–31.

FEDOROV V.D. (1969) The correlation between biomass of a species and maximum population number in a phytoplankton community (in Russian). *Dokl. Akad. Nauk SSSR* **188,** 694–7; Engl. trans. (1969) *Dokl.* (Proc.) *Acad. Sci. U.S.S.R.* (Biol. Sci. section) **188,** 649–51.

FINDLAY I.W.O. (1972) Effect of external factors and cell size on the cell division rate of a marine diatom, *Coscinodiscus pavillardii* Fott. *Int. Revue ges. Hydrobiol.* **57,** 523–33.

FINENKO Z.Z. & KRUPATKINA-AKININA D.K. (1974) Effect of inorganic phosphorus on the growth rate of diatoms. *Mar. Biol.* **26,** 193–201.

FISHER N.S., GRAHAM L.B., CARPENTER E.J. & WURSTER C.F. (1973) Geographic differences in phytoplankton sensitivity to PCBs. *Nature* **241,** 548–9.

FLEMING R.H. (1939) The control of diatom populations by grazing. *J. Cons. perm. int. Explor. Mer* **14,** 210–27.

FOGG G.E. (1965) *Algal Cultures and Phytoplankton Ecology*. University of Wisconsin, Madison. 126 pp.

FRENGUELLI J. (1960) Diatomeas y silicoflagelados recogidas en Tierra Adélia durante las Expediciones Polares Francesas de Paul-Emile VICTOR (1950–52). *Revue Algol.* N.S. **5,** 3–48.

FRENGUELLI J. & ORLANDO H.A. (1958) Diatomeas y silicoflagelados del sector Antartico Sudamericano. *Publnes Inst. antart. argent.* **5,** 191 pp.

FRENGUELLI J. & ORLANDO H.A. (1959) Operacion MERLOZA. Diatomeas y silicoflagelados del plancton del 'VI Crucero'. *Servicio Hidrogr. Naval., Argentina* H. **619,** 5–62.

FRYXELL G.A. & HASLE G.R. (1971) *Corethron criophilum* Castracane; its distribution and structure. *Antarct. Res. Ser.* **17,** 335–46.

FRYXELL G.A. & HASLE G.R. (1972) *Thalassiosira eccentrica* (Ehrenb.) Cleve, *T. symmetrica* sp. nov., and some related centric diatoms. *J. Phycol.* **8,** 297–317.

GAARDER K.R. (1951) Bacillariophyceae from the 'Michael Sars' North Atlantic Deep-Sea Expedition 1910. *Rep. scient. Results Michael Sars N. Atlant. Deep-Sea Exped.* **2,** 1–36.

GILMARTIN M. & REVELANTE N. (1974) The 'island mass' effect on the phytoplankton and primary production of the Hawaiian Islands. *J. exp. mar. Biol. Ecol.* **16,** 181–204.

GOLDMAN J.C. & CARPENTER E.J. (1974) A kinetic approach to the effect of temperature on algal growth. *Limnol. Oceanogr.* **19,** 756–66.

GRALL J.R. & JACQUES G. (1964) Etude dynamique et variations saisonnieres du plancton de la region de Roscoff. *Cah. Biol. mar.* **5,** 423–55.

GRAN H.H. (1904a) Diatomaceae from the ice floes and plankton of the Arctic Ocean. *Scient. Results Norw. N. polar Exped.* IV (XI). 74 pp.

GRAN H.H. (1904b) Die Diatomeen der arktischen Meere. I. Teil: Die Diatomeen des Planktons. *Fauna arct.* **3,** 511–54.

GRAN H.H. (1908) Diatomeen. In *Nordisches Plankton, Botanischer Teil* (eds. K. Brandt & C. Apstein) Chap. XIX pp. 1–146. Lipsius and Tischer, Kiel and Leipzig.

GRAN H.H. (1912) Pelagic plant life. In *The Depths of the Ocean* (eds. J. Murray & J. Hjort) pp. 307–86. Macmillan, London.

GRAN H.H. (1933) Studies on the biology and chemistry of the Gulf of Maine, II. Distribution of phytoplankton in August, 1932. *Biol. Bull.* **64,** 159–82.

GRAN H.H. & ANGST E.C. (1931) Plankton diatoms of Puget Sound. *Publs. Puget Sound mar. biol. Stn.* **7,** 417–519.

GRAN H.H. & BRAARUD T. (1935) A quantitative study of the phytoplankton in the Bay of Fundy and the Gulf of Maine (including observations on the hydrography, chemistry and turbidity). *J. Biol. Bd Can.* **1,** 279–476.

GRAN H.H. & THOMPSON G.T. (1930) Diatoms and the physical and chemical conditions of water of the San Juan archipelago. *Publs Puget Sound mar. biol. Stn* **7,** 169–204.

GRANT B.R. & KERR J.D. (1970) Phytoplankton numbers and species at Port Hacking Station and their relationship to the physical environment. *Aust. J. mar. Freshwat. Res.* **21,** 34–45.

GRASSLE J.F. & GRASSLE J.P. (1974) Opportunistic life histories and genetic systems in marine benthic polychaetes. *J. mar. Res.* **32,** 253–84.

GUILLARD R.R.L. (1962) Salt and osmotic balance. In *Physiology and Biochemistry of Algae* (ed. R.A. Lewin) pp. 529–40. Academic Press, New York.

GUILLARD R.R.L. (1973) Division rates. In *Handbook of Phycological Methods – Culture Methods and Growth Measurements* (ed. J.R. Stein) pp. 289–311. Cambridge Univ. Press, Cambridge.

GUILLARD R.R.L. & Hellebust J.A. (1971). Growth and the production of extracellular substances by two strains of *Phaeocystis poucheti. J. Phycol.* **7,** 330–8.

GUILLARD R.R.L. & MYKLESTAD S. (1970) Osmotic and ionic requirements of the marine diatom *Cyclotella nana. Helgoländer wiss. Meeresunters.* **20,** 104–10.

GUILLARD R.R.L. & RYTHER J.H. (1962) Studies of marine planktonic diatoms. I. *Cyclotella nana* Hustedt, and *Detonula confervacea* (Cleve) *Gran. Can. J. Microbiol.* **8,** 229–39.

GUILLARD R.R.L., CARPENTER E.J. & REIMANN B.E.F. (1974) *Skeletonema menzelii* sp. nov., a new diatom from the western Atlantic Ocean. *Phycologia* **13,** 131–8.

GUILLARD R.R.L., KILHAM P. & JACKSON T.A. (1973) Kinetics of silicon-limited growth in the marine diatom *Thalassiosira pseudonana* Hasle and Heimdal (=*Cyclotella nana* Hustedt). *J. Phycol.* **9,** 233–7.

HALIM Y. (1969) Plankton of the Red Sea. *Oceanogr. mar. Biol. Ann. Rev.* **7,** 231–75.

HALLDAL P. (1953) Phytoplankton investigations from weather ship M in the Norweigian Sea, 1948–49. *Hvalråd. Skr.* **38,** 3–91.

HARGRAVE B.T. & GEEN G.H. (1970) Effects of copepod grazing on two natural phytoplankton populations. *J. Fish. Res. Bd Can.* **27,** 1 395–403.

HARGRAVES P.E. & GUILLARD R.R.L. (1974) Structural and physiological observations on some small marine diatoms. *Phycologia* **13,** 163–72.

HARGRAVES P.E., BRODY R.W. & BURKHOLDER P.R. (1970) A study of phytoplankton in the lesser Antilles region. *Bull. mar. Sci.* **20,** 331–49.

HART T.J. (1934) On the phytoplankton of the south-west Atlantic and the Bellingshausen Sea. *'Discovery' Rep.* **8,** 1–268.

HART T.J. (1942) Phytoplankton periodicity in Antarctic surface waters. *'Discovery' Rep.* **21,** 261–356.

HART T.J. & CURRIE R.I. (1960) The Benguela Current. *'Discovery' Rep.* **31,** 123–298.

HARVEY H.W., COOPER L.H.N., LEBOUR M.V. & RUSSEL F.S. (1935) Plankton production and its control. *J. mar. biol. Ass. U.K.* **20,** 407–41.

HASLE G.R. (1959) A quantitative study of phytoplankton from the equatorial Pacific. *Deep Sea Res.* **6,** 38–55.

HASLE G.R. (1960) Phytoplankton and ciliate species from the tropical Pacific. *Skr. norske Vidensk-Akad.* No. **2,** 50 pp.

HASLE G.R. (1965) *Nitzschia* and *Fragilariopsis* species studied in the light and electron microscopes. II. The group Pseudonitzschia. *Skr. norske Vidensk-Akad.* No. **18,** 45 pp.

HASLE G.R. (1968a) The valve processes of the centric diatom genus *Thalassiosira. Norw. J. Bot.* **15,** 193–201.

HASLE G.R. (1968b) Observations on the marine diatom *Fragilariopsis kerguelensis* (O'Meara) Hust. in the scanning electron microscope. *Norw. J. Bot.* **15,** 205–8.

HASLE G.R. (1969) An analysis of the phytoplankton of the Pacific Southern Ocean: abundance, composition and distribution during the Brategg Expedition, 1947–48. *Hvalråd Skr.* **52,** 6–128.

HASLE G.R. (1971) *Nitzschia pungiformis* (Bacillariophyceae) a new species of the *Nitzschia seriata* group. *Norw. J. Bot.* **18,** 139–44.

HASLE G.R. (1972a) *Thalassiosira subtilis* (Bacillariophyceae) and two allied species. *Norw. J. Bot.* **19,** 111–37.

HASLE G.R. (1972b) *Fragilariopsis* Hustedt as a section of the genus *Nitzschia Hassall. Nova Hedwigia, Beih.* **39,** 111–9.

HASLE G.R. (1972c) The distribution of *Nitzschia seriata* Cleve and allied species. *Nova Hedwigia, Beih.* **39,** 171–90.

HASLE G.R. (1972d) The inclusion of *Coscinosira* Gran (Bacillariophyceae) in *Thalassiosira* Cleve. *Taxon* **21,** 543–4.

HASLE G.R. (1973a) Some marine plankton genera of the diatom family *Thalassiosiraceae. Nova Hedwigia, Beih.* **45,** 1–49.

HASLE G.R. (1973b) Morphology and taxonomy of *Skeletonema costatum* (Bacillariophyceae). *Norw. J. Bot.* **20,** 109–37.

HASLE G.R. & HEIMDAL B.R. (1968) Morphology and distribution of the marine diatom *Thalassiosira antarctica* Comber. *Jl R. microsc. Soc.* **88,** 357–69.

HASLE G.R., HEIMDAL B.R. & FRYXELL G.A. (1971) Morphologic variability in fasciculated diatoms as exemplified by *Thalassiosira tumida* (Janisch) Hasle, comb. nov. *Antarct. Res. Ser.* **17,** 313–33.

HAUG A., MYKLESTAD S. & SAKSHAUG E. (1973) Studies on the phytoplankton ecology of the Trondheimsfjord. I. The chemical composition of phytoplankton populations. *J. exp. mar. Biol. Ecol.* **11,** 15–26.

HEIDEN H. & KOLBE R.W. (1928) Die marinen Diatomeen der Deutschen Südpolar-Expedition, 1901–3. *Dt. Südpol.-Exped.* **8,** 447–715.

HEIMDAL B.R., HASLE G.R. & THRONDSEN J. (1973). An annotated check-list of plankton algae from Oslofjord, Norway (1951–72). *Norw. J. Bot.* **20,** 13–9.

HELMCKE J.G. (1961) Versuch einer Gestaltsanalyse an Diatomeenschalen. *Rec. Adv. Bot.* (IX Int. Bot. Congr. 1959) **1,** 216–21.

HENDEY N.I. (1937) The plankton diatoms of the Southern Seas. *'Discovery' Rep.* **16,** 151–364.

HENDEY N.I. (1964) *An Introductory Account of the Smaller Algae of British Coastal Waters. V. Bacillariophyceae (Diatoms).* 318 pp. HMSO, London.

HENDEY N.I. (1970) Some littoral diatoms of Kuwait. *Nova Hedwigia, Beih.* **31,** 107–67.

HENDEY N.I. (1974) A revised check-list of British marine diatoms. *J. mar. biol. Ass. U.K.* **54,** 277–300.

HENTSCHEL E. (1936) Allgemeine Biologie des Südatlantischen Ozeans. *Dt. Atlant. Exp. 'Meteor' 1925–1927,* **11,** 343 pp.

HOLMES R.W. (1956) The annual cycle of phytoplankton in the Labrador Sea, 1950–51. *Bull. Bingham oceanogr. Coll.* **16,** 1–74.

HOLMES R.W. (1967) Auxospore formation in two marine clones of the diatom genus *Coscinodiscus. Am. J. Bot.* **54,** 163–8.

HOLMES R.W. & REIMANN B.E.F. (1966) Variation in valve morphology during the life cycle of the marine diatom *Coscinodiscus concinnus. Phycologia* **5,** 233–44.

HOLT M.G. & SMAYDA T.J. (1974) The effect of daylength and light intensity on the growth rate of the marine diatom *Detonula confervacea* (Cleve) Gran. *J. Phycol.* **10,** 231–7.

HORNER R. & ALEXANDER V. (1972) Algal populations in arctic sea ice: an investigation of heterotrophy. *Limnol. Oceanogr.* **17,** 454–7.

HULBURT E.M. (1962) Phytoplankton in the southwestern Sargasso Sea and North Equatorial Current, February 1961. *Limnol. Oceanogr.* **7,** 307–15.

HULBURT E.M. (1966) The distribution of phytoplankton, and its relationship to hydrography, between southern New England and Venezuela. *J. mar. Res.* **24,** 67–81.

HULBURT E.M. (1968) Phytoplankton observations in the western Caribbean Sea. *Bull. mar. Sci.* **8,** 388–99.

HULBURT E.M. & CORWIN N. (1969) Influence of the Amazon River outflow on the ecology of the western tropical Atlantic. III. The planktonic flora between the Amazon River and the Windward Islands. *J. mar. Res.* **27,** 55–72.

HULBURT E.M. & CORWIN N. (1972) A note on the phytoplankton distribution in the offshore water of the eastern and central Gulf of Mexico. *Caribb. J. Sci.* **12,** 29–38.

HULBERT E.M. & GUILLARD R.R.L. (1968) The relationship of the distribution of the diatom *Skeletonema tropicum* to temperature. *Ecology* **49,** 337–9.

HULBERT E.M. & MACKENZIE R.S. (1971) Distribution of phytoplankton species at the western margin of the north Atlantic Ocean. *Bull. mar. Sci.* **21,** 603–12.

HULBERT E.M. & RODMAN J. (1963) Distribution of phytoplankton species with respect to salinity between the coast of southern New England and Bermuda. *Limnol. Oceanogr.* **8,** 263–9.

HULBERT E.M., RYTHER J.H. & GUILLARD R.R.L. (1960) The phytoplankton of the Sargasso Sea off Bermuda. *J. Cons. perm. int. Explor. Mer.,* **25,** 115–27.

HUSTEDT F. (1927–30) Die Kieselalgen Deutschlands, Österreichs und der Schweiz mit Berücksichtigung der übrigen Länder Europas sowie der angrenzenden Meeresgebiete. In *Kryptogamen – Flora* **7,** part I (ed. L. Rabenhorst) pp. 1–272 (1927); pp. 273–464 (1928); pp. 465–608 (1929); pp. 609–784 (1930); pp. 785–925 (1930). Leipzig.

HUSTEDT F. (1931–59) Die Kieselalgen Deutschlands, Österreichs und der Schweiz mit Berücksichtigung der übrigen Länder Europas sowie der angrenzenden Meeresgebiete. In *Kryptogamen – Flora 7,* part II (ed. L. Rabenhorst) pp. 1–176 (1931); pp. 177–320 (1932); pp. 321–432 (1933); pp. 433–576 (1933); pp. 577–736 (1937); pp. 737–845 (1959). Leipzig.

HUSTEDT F. (1958) Diatomeen aus der Antarktis und dem Südatlantik. *Dt. Antarkt. Exped. 1938–1939* **2,** 103–91.

HUTCHINSON G.E. (1967) *A Treatise on Limnology.* Vol. II. 1 115 pp. John Wiley & Sons, New York.

IBANEZ F. (1972) Interprétation de données écologiques par l'analyse des composantes principales: Ecologie planctonique de la Mer du Nord. *J. Cons. perm. int. Explor. Mer* **34,** 323–40.

IGNATIADES L. (1969) Annual cycle, species diversity and succession of phytoplankton in lower Saronicos Bay, Aegean Sea. *Mar. Biol.* **3,** 196–200.

IVANOV A.I. (1960) Characteristics of qualitativive composition and quantitative distribution of phytoplankton in the NW Black Sea (In Russian). *Trudÿ vses. gidrobiol. Obshch.* **10,** 182–96.

IVANOV A.I. (1965) Qualitative classification of the composition of phytoplankton in the Black Sea. In *Plankton Studies in the Black and Avoz Seas* (ed. V.A. Vodyanitskii) pp. 17–35 (in Russian). Naukova Dumka, Kiev.

JACQUES G. (1968) Aspects quantitatifs de phytoplancton de Banyuls-sur-Mer (Golfe du Lion). III. Diatomées et dinoflagellés de juin 1965 a juin 1968. *Vie Milieu* (B) **20,** 91–126.

JANNASCH H.W. & WIRSEN C.O. (1973) Deep-sea microorganisms: *in situ* response to nutrient enrichment. *Science* **180,** 641–3.

JENSEN A. & RYSTAD B. (1974) Heavy metal tolerance of marine phytoplankton. I. The tolerance of three algal species to zinc in coastal sea water. *J. exp. mar. Biol. Ecol.* **15,** 145–57.

JENSEN A. & SAKSHAUG E. (1973) Studies on the phytoplankton ecology of the Trondheimsfjord. II. Chloroplast pigments in relation to abundance and physiological state of phytoplankton. *J. exp. mar. Biol. Ecol.* **11,** 137–55.

JOHNSON M.W. (1956) The plankton of the Beaufort and Chukchi Sea areas of the Arctic and its relation to hydrography. *Arctic* **1,** 32 pp.

JOHNSTON R. (1963a) Antimetabolites as an aid to the study of phytoplankton nutrition. *J. mar. biol. Ass. U.K.* **43,** 409–25.

JOHNSTON R. (1963b) Sea water, the natural medium of phytoplankton. I. General features. *J. mar. biol. Ass. U.K.* **43,** 427–56.

JOUSÉ A.P., KOZLOVA O.G. & MUKHINA V.V. (1971) Distribution of diatoms in the surface layer of sediment from the Pacific Ocean. In *Micropaleontology of Oceans* (eds. B.M. Funnel & W.R. Riedel) pp. 263–69. Cambridge Univ. Press, Cambridge.

KANAYA T. & KOIZUMI I. (1966) Interpretation of diatom thanatocoenoses from the north Pacific applied to a study of core V20–130. (Studies of a deep-sea core V20–130 Part IV). *Sci. Rep. Res. Insts. Tôhoku Univ., Second Series (Geol.)* **37,** 89–130.

KAROHJI K. (1972) Regional distribution of phytoplankton in the Bering Sea and western and northern subarctic regions of the North Pacific Ocean in summer. In *Biological Oceanography of the Northern North Pacific Ocean* (ed. A.Y. Takenouti) pp. 99–115. Idemitsu Shoten, Japan.

KARSTEN G. (1905) Das Phytoplankton des Antarktischen Meeres nach dem Material der deutschen Tiefsee-Expedition 1898–1899. *Dt. Tiefsee-Exped. 1898–9* **2(2),** 1–136.

KAWARADA Y. (1960) A contribution of microplankton observations to the hydrography of the northern North Pacific and adjacent seas. III. Plankton diatoms of the western Okhotsk Sea in the period from June to August 1957. *Mem. mar. Obs. Kobe* **14,** 74–80.

KAWARADA Y., KITOU M., FURUHASHI K. & SANO A. (1968a) Plankton in the western North Pacific in the winter of 1967 (CSK). *Oceanogrl. Mag.* **20,** 9–29.

KAWARADA Y., KITOU M., FURUHASHI K. & SANO A. (1968b) Distribution of plankton in the waters neighboring Japan in 1966 (CSK). *Oceanogrl. Mag.* **20,** 187–212.

KIMBALL J.F. JR., CORCORAN E.F. & WOOD E.J.F. (1963) Chlorophyll-containing microorganisms in the aphotic zone of the oceans. *Bull. mar. Sci.* **13,** 574–7.

KISSEKEW I.A. (1959) The composition of the phytoplankton of sea waters if South Sakhalin and South Kuril Islands. *Issled. dal'nevost. Morei SSSR* **6,** 162–72.

KISSELEW J. (1932) Beiträge zur Mikrofiora des süd-östlichen Teiles des Laptev Meers (in Russian). Trudȳ arkt. nauchno-issled Inst. **37.** (After Zenkevich.)

KOLBE R.W. (1955) Diatoms from equatorial Atlantic cores. *Rep. Swed. deep Sea Exped. 1947–1948* **7,** 151–85.

KOLBE R.W. (1957) Diatoms from equatorial Indian Ocean cores. *Rep. Swed. deep Sea Exped. 1947–1948* **9,** 1–50.

KONOPLYA L.A. (1973) Vertical distribution of phytoplankton in the Karelian coastal waters of the White Sea (in Russian, English summary). *Okeanologiya* **13,** 314–320; Engl. trans. (1973) *Oceanology* **13,** 256–61.

KONOVALOVA G.A. (1972) Seasonal characteristics of phytoplankton in Amur Bay, Sea of Japan (in Russian, English summary). *Okeanologiya* **12,** 123–8; Engl. trans. (1972) *Oceanology* **12,** 100–5.

KÖRNER H. (1970) Morphologie und Taxonomie der Diatomeengattung *Asterionella*. *Nova Hedwigia* **20,** 557–724.

KOZLOVA O.G. (1964) *Diatoms of the Indian and Pacific sectors of the Antarctic*. Translated by Israel Program for Scientific Translations, 1966, 191 pp. U.S. Dept. of Comm., Clearinghouse for Fed. Sci. Tech. Inf., Springfield, Va. (Russian original published by the Acad. Sci. U.S.S.R., Inst. Oceanol.)

KOZLOVA O.G. (1971) The main features of diatom and silicoflagellate distribution in the Indian Ocean. In *Micropaleontology of Oceans* (eds. B.M. Funnell & W.R. Riedel) pp. 271–5. Cambridge Univ. Press, Cambridge.

KOZLOVA O.G. & MUKHINA V.V. (1967) Diatoms and silicoflagellates in suspension and floor sediments of the Pacific Ocean. *Int. Geol. Rev.* **9,** 1 322–42.

LAPSHINA V.I. (1971) Plankton characteristic in the eastern tropical and equatorial zone of the eastern Pacific Ocean (in Russian). *Izv. tikhookean. nauchno-issled. Inst. rȳb. Khoz. Okeanogr.* **79,** 100–26.

LAURITIS J.A., HEMMINGSEN B.B. & VOLCANI B.E. (1967) Propagation of *Hantzschia* sp. Grunow daughter cells by *Nitzschia alba* Lewin and Lewin. *J. Phycol.* **3,** 236–7.

LEBOUR M.V. (1930) *The Planktonic Diatoms of Northern Seas*. 244 pp. The Ray Society, London.

LÉGER G. (1972) Les populations phytoplanctoniques au point Ø=42°47′N, G=7°29′E Greenwich. F. Discussion générale. *Bull. Inst. océanogr. Monaco* **70,** (1420) 1–11.

LÉGER G. (1973a) Diatomées et Dinoflagellés de la Mer Ligure. Systématique et distribution en juillet 1963. *Bull. Inst. océanogr. Monaco* **71,** (1425) 1–36.

LÉGER G. (1973b) Diatomées et Dinoflagellés de la côte Est de Corse. Systématique et distribution en juillet 1964. *Bull Inst. océanogr. Monaco* **71,** (1426) 1–31.

LEWIN J.C. (1974) Blooms of surf-zone diatoms along the coast of the Olympic Peninsula, Washington, III. Changes in the species composition of the blooms since 1925. *Nova Hedwigia, Beih.* **45,** 251–6.

LEWIN J.C. & GUILLARD R.R. L. (1963) Diatoms. *A. Rev. Microbiol.* **17,** 373–414.

LILLICK L.C. (1940) Phytoplankton and planktonic protozoa of the offshore waters of the Gulf of Maine. Part II. Qualitative composition of the planktonic flora. *Trans. Am. phil. Soc. n.s.* **31,** 193–237.

LOHMANN H. (1920) Die Bevölkerung des Ozeans mit Plankton nach den Ergebnissen der Zentrifugenfänge während der Ausreise der ‘Deutschland’ 1911. – Zugleich ein Beitrag zur Biologie des Atlantischen Ozeans. *Arch. Biontol.* **IV(3),** 1–617.

LOUIS A. & CLARYSSE R. (1971) Contribution á la connaissance du phytoplancton de l’Atlantic Nord-Est et de la mer du Nord. *Biol. Jahrb.* **39,** 261–337.

LUND J.W.G. (1964) Primary production and periodicity of phytoplankton *Verh. int. Verein. theor. angew. Limnol* **15,** 37–56.

LUND J.W.G. (1965) The ecology of the freshwater phytoplankton. *Biol. Rev.* **40,** 231–93.

LUND J.W.G. (1966) Summation. In *Marine Biology 2* (ed. C.H. Oppenheimer) pp. 227–49. The New York Academy of Sciences, New York.

LUND J.W.G., KIPLING C. & LECREN E.D. (1958) The inverted microscope method of estimating algal numbers and the statistical basis of estimations by counting. *Hydrobiologia* **2,** 143–70.

MADDUX W.S. & JONES R.F. (1964) Some interactions of temperature, light intensity, and nutrient concentrations during the continuous culture of *Nitzschia closterium* and *Tetraselmis* sp. *Limnol. Oceanogr.* **9,** 79–86.

MAGUE T.H., WEARE N.M. & HOLM-HANSEN O. (1974) Nitrogen fixation in the North Pacific Ocean. *Mar. Biol.* **24,** 109–19.

MAKAROVA I.V. (1961) The phytoplankton of the north Caspian Sea (in Russian). *Bot. Zh.* **46,** 1 669–78.

MALONE T.C. (1971) The relative importance of nannoplankton and netplankton as primary producers in tropical and neritic phytoplankton communities. *Limnol. Oceanogr.* **16,** 633–9.

MALONE T.C., GARSIDE C., ANDERSON R. ROELS O.A. (1973) The possible occurrence of photosynthetic microorganisms in deep-sea sediments of the North Atlantic. *J. Phycol.* **9,** 482–8.

MANGIN L. (1915) Phytoplancton de l'Antarctique. *Deux. Exp. Ant. Française (1908–10),* 95 pp.

MANGUIN E. (1960) Les diatomées de la terre Adélie. Campagne du 'Commandant Charcot' 1949–1950. *Annls Sci. nat. (Bot.) 12e série,* **1,** 223–363.

MANN A. (1907) Report on the diatoms of the *Albatross* voyages in the Pacific Ocean, 1888–1904. *Contr. U.S. natn. Herb.* **10,** 221–419.

MANN A. (1925) Marine diatoms of the Philippines Islands. *Bull. U.S. natn. Mus. No. 100,* **6,** 1–182.

MARCHESONI V. (1954) Il trofismo della Laguna Veneta e la vivificazione marina. III. Ricerche sulle variazioni quantitative del fitoplancton. *Archo Oceanogr. Limnol.* **9,** 153–285.

MARGALEF R. (1958) Temporal succession and spatial heterogeneity in phytoplankton. In *Perspectives in Marine Biology* (ed. A.A. Buzzati-Traverso) pp. 323–49. University of California Press, Berkeley.

MARGALEF R. (1962) Succession in marine populations. *Advg. Front. Pl. Sci.* **2,** 137–88.

MARGALEF R. (1965) Composicion y distribution del fitoplancton. *Mems Soc. Cienc. nat. 'La Salle'* **25,** 141–205.

MARGALEF R. (1967a) The food web in the pelagic environment. *Helgoländer wiss. Meeresunters.* **15,** 548–59.

MARGALEF R. (1967b) Some concepts relative to the organization of plankton. *Oceanogr. mar. Biol. Ann. Rev.* **5,** 257–89.

MARGALEF R. & GONZÁLAS-BERNÁLDEZ F. (1969) Grupos de especies asociadas en el fitoplancton del mar Caribe (NE de Venezuela). *Investigación pesq.* **33,** 287–312.

MARSHALL H.G. (1970) Phytoplankton in tropical surface waters between the coast of Ecuador and the Gulf of Panama. *J. Wash. Acad. Sci.* **60,** 18–21.

MARSHALL H.G. (1971) Composition of phytoplankton off the southeastern coast of the United States. *Bull. mar. Sci.* **21,** 806–25.

MARSHALL H.G. (1972) Phytoplankton composition in the southeastern Pacific between Ecuador and Galapagos Islands (Archipelago de Colon). *Proc. biol. Soc. Wash.* **85,** 1–37.

MARSHALL S.M. (1973) Respiration and feeding in copepods. *Adv. mar. Biol.* **11,** 57–120.

MARTIN J.H. (1965) Phytoplankton–zooplankton relationships in Narragansett Bay. *Limnol. Oceanogr.* **10,** 185–91.

MARTIN J.H. (1968) Phytoplankton–zooplanklton relationships in Narragansett Bay. III. Seasonal changes in zooplankton excretion rates in relation to phytoplankton abundance. *Limnol. Oceanogr.* **13,** 63–71.

MARTIN J.H. (1970) Phytoplankton–zooplankton relationships in Narragansett Bay. IV. The seasonal importance of grazing. *Limnol. Oceanogr.* **15,** 413–8.

MARUMO R. (1954) Diatom plankton in the south of Cape Shionomisaki in 1953. *Oceanogrl. Mag.* **6,** 145–52.

MARUMO R. (1956) Diatom communities in Bering Sea and its neighbouring waters in the summer of 1954. *Oceanogrl. Mag.* **8**, 69–73.

MAYNARD N.G. (1974a) The distribution of diatoms in the surface sediments of the Atlantic Ocean and their relationship to the biological and physical oceanography of the overlying waters. Ph. D. Thesis. Univ. Miami (Diss. Abstr. Dec., 1974).

MAYNARD N.G. (1974b) Diatoms in Pleistocene deep Black Sea sediments. In *The Black Sea – Geology, Chemistry and Biology* (eds. E.T. Degens & D.A. Ross) pp. 389–95. Mem. 20 Am. Assoc. Petrol. Geologists, Tulsa, Oklahoma.

MCALICE B.J. (1970) Observations on the small-scale distribution of estuarine phytoplankton. *Mar. Biol.* **7**, 100–11.

MCGOWAN J.A. (1971) Oceanic biogeography of the Pacific. In *The Micropaleontology of Oceans* (eds. B.M. Funnell & W.R. Riedel) pp. 3–74. Cambridge Univ. Press, Cambridge.

MCGOWAN J.A. (1974) The nature of oceanic ecosystems. In *The Biology of the Pacific Ocean* (ed. C.B. Miller) pp. 9–28. Oregon State Univ. Press, Corvallis.

MEGURO H., ITO K. & FUKUSHIMA H. (1967) Ice flora (bottom type): a mechanism of primary production in polar seas and the growth of diatoms in sea ice. *Arctic* **20**, 114–33.

MENZEL D.W., HULBURT E.M. & RYTHER J.H. (1963) The effects of enriching Sargasso Sea water on the production and species composition of the phytoplankton. *Deep Sea Res.* **10**, 209–19.

MEUNIER A. (1910) *Microplankton des Mers de Barents et de Kara. Campagne Arctique de 1907 Duc d'Orleans.* Imprimerie Scientifique, Bruxelles. 355 pp.

MIKHAILOVA N.F. (1964) On the spreading of the habitat into the Black Sea of the species of the genus *Chaetoceros* of northern seas and their biogeography (in Russian). *Trudȳ sevastopol'. biol. Sta.* **17**, 231–48.

MOREIRA FILHO H. (1965) Contribuiçao ao estudo das diatomáceas da região de Cabo Frio (Estado do Rio de Janeiro-Brasil). *Anais. Acad. bras. Cienc.* **37** (Suppl.), 231–8.

MOSS B. (1973) The influence of environmental factors on the distribution of freshwater algae: An experimental study. III. Effects of temperature, vitamin requirements and inorganic nitrogen compounds on growth. *J. Ecol.* **61**, 179–211.

MOTODA S. & MARUMO R. (1963) Plankton of the Kuroshio water. *Proc. Symp. Kuroshio,* 40–61.

MÜLLER-MELCHERS F.C. (1957) Plankton diatoms of the 'Toko-maru' voyage (Brazil Coast). *Bolm Inst. Oceangr., S. Paulo* **8**, 111–26.

MÜLLER-MELCHERS F.C. (1959) Plankton diatoms of the southern Atlantic Argentina and Uruguay coast. *Comun. bot. Mus. Hist. nat. Montev.* **3**, 1–45.

MULLIN M.M. (1963) Some factors affecting the feeding of marine copepods of the genus *Calanus*. *Limnol. Oceanogr.* **8**, 239–50.

MUNK W.H. & RILEY G.A. (1952) Absorption of nutrients by aquatic plants. *J. mar. Res.* **11**, 215–40.

MURPHY L.S. & GUILLARD R.R.L. (1976) Biochemical taxonomy of marine phytoplankton by enzyme electrophoresis. I. The centric diatoms *Thalassiosira pseudonana* Hasle and Heimdal, and *Thalassiosira fluviatilis* Hustedt. *J. Phycol.* **12**, 9–13.

MURRAY J. (1895) A summary of the scientific results. Second part. 1608 pp. *Rep. scient. Results Voyage Challenger 1873–6.*

NEL E.A. (1968) Diatoms in the aphotic zone of the southwest Indian Ocean. *Fish. Bull. Misc. Contrib. Oceanogr. Fish. Biol. S. Afri.* **5**, 11–31.

NIEMI A. (1971) Late summer phytoplankton of Kimito Archipelago (SW coast of Finland). *Merentutkimuslait. Julk./HavsforskInst. Skr.* No. **233,** 3–17.

NIEMI A. (1972) Observations on phytoplankton in eutrophied and noneutrophied archipelago waters of the southern coast of Finland. *Memo. Soc. Fauna Flora fenn.* **48,** 63–74.

NIEMI A., SKUJA H. & WILLÉN T. (1970) Phytoplankton from the Pojoviken-Tvärminne area, S. coast of Finland. *Memo. Soc. Fauna Flora fenn.* **46,** 14–28.

O'BRIEN W.J. (1974) The dynamics of nutrient limitation of phytoplankton algae: A model reconsidered. *Ecology* **55,** 135–41.

ODUM E.P. (1969) The strategy of ecosystem development. *Science* **164,** 262–70.

OHWADA M. (1972) Vertical distribution of diatoms in the Sea of Japan. In *Biological Oceanography of the Northern North Pacific Ocean* (ed. A.Y. Takenouti) pp. 145–64. Idemitsu Shoten, Japan.

OKADA H. & HONJO S. (1973) The distribution of oceanic coccolithophorids in the Pacific. *Deep Sea Res.* **20,** 355–74.

PAASCHE E. (1960) Phytoplankton distribution in the Norwegian Sea in June, 1954, related to hydrography and compared with primary production data. *FiskDir. Skr. serie Havundersøkelser.* **12,** 77 pp.

PAASCHE E. (1961) Notes on phytoplankton from the Norwegian Sea. *Botanica mar.* **2,** 197–210.

PAASCHE E. (1973a) Silicon and the ecology of marine plankton diatoms. I. *Thalassiosira pseudonana (Cyclotella nana)* grown in a chemostat with silicate as limiting nutrient. *Mar. Biol.* **19,** 117–26.

PAASCHE E. (1973b) The influence of cell size on growth rate, silica content, and some other properties of four marine diatom species. *Norw. J. Bot.* **20,** 197–204.

PAASCHE E. & ROM A.-M. (1962) On the phytoplankton vegetation of the Norwegian Sea in May 1958. *Norw. J. Bot.* **9,** 33–60.

PAREDES J.F. (1969/1970) Subsídios para o conhecimento do Plancton marinho de Cabo Verde. I. Diatomáceas, Silicoflagelados e Dinoflagelados. *Mems Inst. Invest. cient. Moçamb. Ser. A* **10,** 3–107.

PARSONS T.R. & TAKAHASHI M. (1973a) *Biological Oceanographic Processes.* 186 pp. Pergamon Press, Oxford.

PARSONS T.R. & TAKAHASHI M. (1973b) Environmental control of phytoplankton cell size. *Limnol. Oceanogr.* **18,** 511–5.

PARSONS T.R. & TAKAHASHI M. (1974) A rebuttal to the comment by Hecky and Kilham. *Limnol. Oceanogr.* **19,** 366–8.

PATRICK R. (1948) Factors effecting the distribution of diatoms. *Bot. Rev.* **14,** 473–524.

PHIFER L.D. (1934) The occurrence and distribution of plankton diatoms in the Bering Sea and Bering Strait, July 26–August 24, 1934. *Rep. Oceanogr. cruise U.S. Coast Guard Cutter Chelan 1934,* Part II (A), 1–44.

PIANKA E.R. (1970) On r- and K-selection. *Am. Nat.* **104,** 592–7.

PIANKA E.R. (1974) *Evolutionary Ecology.* 356 pp. Harper and Row, New York.

PLATT T. (1972) Local phytoplankton abundance and turbulence. *Deep Sea Res.* **19,** 183–7.

POULET S.A. (1974) Seasonal grazing of *Pseudocalanus* on particles. *Mar. Biol.* **25,** 109–23.

PRASAD R.R. & NAIR P.V.R. (1960) Observations on the distribution and occurrence of diatoms in the inshore waters of the Gulf of Mannar and Palk Bay. *Indian J. Fish.* **7,** 49–68.

PRATT D.M. (1959) The phytoplankton of Narragansett Bay. *Limnol. Oceanogr.* **4,** 425–40.

PRATT D.M. (1965) The winter–spring diatom flowering in Narragansett Bay. *Limnol. Oceanogr.* **10,** 173–84.

PRATT D.M. (1966) Competition between *Skeletonema costatum* and *Olisthodiscus luteus* in Narragansett Bay and in culture. *Limnol. Oceanogr.* **11,** 447–55.

PROASHKINA-LAVRENKO A.I. (1955) *Diatom Plankton of the Black Sea* (in Russian) 222 pp. Moscow.

PROSHKINA-LAVRENKO A.I. (1961) Concerning variability of some Black Sea diatoms (in Russian). *Bot. Zh.* **46,** 1 794–7.

PROSHKINA-LEVRENKO A.I. (1963) *Diatom Plankton of the Sea of Azov* (in Russian). Moscow, 190 pp.

PROSHKINA-LAVRENKO A.I. & MAKAROVA I.V. (1968) *Plankton of the Caspian Sea* (in Russian) 191 pp. Leningrad.

PUCHER-PETKOVIĆ T. (1966) Vegetation des diatomees pelagiques de l'Adriatique moyenne. *Acta adriat.* **13,** 1–97.

PUCHER-PETKOVIĆ T. (1969) Oceanographic conditions in the middle Adriatic area. IV. Seasonal and spatial distribution of diatomaceous populations. *Thalassia jugosl.* **5,** 267–75.

RAMSFJELL E. (1960) Phytoplankton distribution in the Norwegian Sea in June 1952 and 1953. *FiskDir. Skr. Serie Havundersøkelser.* **12,** 112 pp.

RAYMONT J.E.G. (1963) *Plankton and Productivity in the Oceans.* 660 pp. Pergamon Press, Oxford.

REIMANN B.E.F. & LEWIN J.C. (1964) The diatom genus *Cylindrotheca* Rabenhorst. *J. R. microsc. Soc.* **83,** 283–96.

REYSSAC J. (1973) Diatomées et dinoflagellés au large de Walvis Bay (Sud-Ouest africain). *Bull. Inst. fondam. Afrique noire, Ser. A.* **35,** 273–97.

REYSSAC J. & ROUX M. (1972) Communautés phytoplanctoniques dans les eaux de Cote d'Ivoire. Groupes de'espèces associées. *Mar. Biol.* **13,** 14–33.

RICHMAN S. & ROGERS J.N. (1969) The feeding of *Calanus helgolandius* on synchronously growing populations of the marine diatom *Ditylum brightwellii. Limnol. Oceanogr.* **14,** 701–9.

RICKLEFS R.E. (1973) *Ecology.* 861 pp. Chiron Press, Newton.

RILEY G.A. (1947) Factors controlling phytoplankton populations on Georges Bank. *J. mar. Res.* **6,** 54–73.

RILEY G.A. (1957) Phytoplankton of the north central Sargasso Sea. *Limnol. Oceanogr.* **2,** 252–70.

RILEY G.A., ed. (1963a) *Marine Biology I.* 286 pp. Amer. Inst. Biol. Sci., Washington.

RILEY G.A. (1963b) Theory of food-chain relations in the ocean. In *The Sea,* Vol. 2 (ed. M.N. Hill) pp. 438–63. John Wiley & Sons, New York.

RILEY G.A. & CONOVER S.A.M. (1967) Aspects of oceanography of Long Island Sound. I. Phytoplankton of Long Island Sound 1954–55. *Bull. Bingham oceanogr. Coll.* **19,** 5–34.

RILEY G.A., STOMMEL H. & BUMPUS D.F. (1949) Quantitative ecology of the plankton of the western North Atlantic. *Bull. Bingham oceanogr. Collect.* **12,** 1–169.

RILEY J.P. (1971) Primary and secondary production in the marine environment. In *Introduction to Marine Chemistry* (ed. J.P. Riley & R. Chester) pp. 219–82. Academic Press, New York.

RIVERA R.P. (1968) Sinopsis de las diatomeas de la Bahia de Concepcion, Chile. *Gayana (Botanica)* **18,** 111 pp.

ROBINSON G.A. (1961) Continuous plankton records: Contribution towards a plankton atlas of the north-eastern Atlantic and the North Sea. Part I: Phytoplankton. *Bull. mar. Ecol.* **5,** 81–9.

ROBINSON G.A. & COLBOURN D.J. (1970) Continuous plankton records: Further studies on the distribution of *Rhizosolenia styliformis* Brightwell. *Bull. mar. Ecol.* **6,** 303–31.

ROUND F.E. (1967) The phytoplankton of the Gulf of California. Part I. Its composition, distribution and contribution to the sediments. *J. exp. mar. Biol. Ecol.* **1,** 76–9.

ROUND F.E. (1970) The delineation of the genera *Cyclotella* and *Stephanodiscus* by light microscopy, transmission and reflecting electron microscopy. *Nova Hedwigia* **31,** 591–601.

ROUND F.E. (1972) Some observations on colonies and ultrastructure of the frustule of *Coenobiodiscus muriformis* and its transfer to *Planktoniella. J. Phycol.* **8,** 222–31.

RUMKÓWNA A. (1948) List of the phytoplankton species occurring in the superficial water layers in the Gulf of Gdansk (in Polish). *Biul. morsk. Inst. ryb. Gdyni* **4,** 139–41.

RYTHER J.H. & HULBURT E.M. (1960) On winter mixing and the vertical distribution of phytoplankton. *Limnol. Oceanogr.* **5,** 337–8.

SAIFULLAH S.M. & STEVEN D.M. (1974) The phytoplankton of St. Margaret's Bay. *Botanica mar.* **17,** 107–12.

SAKSHAUG E. (1972) Phytoplankton investigations in Trondheimsfjord, 1963–66. *K. norske Vidensk. Selsk. Skr.* **1,** 1–56.

SAKSHAUG E. & MYKLESTAD S. (1973) Studies on the phytoplankton ecology of the Trondheimsfjord. III. Dynamics of the phytoplankton blooms in relation to environmental factors, bioassay experiments and parameters for the physiological state of the population. *J. exp. mar. Biol. Ecol.* **11,** 157–88.

SAUNDERS R.P. & GLEN D.A. (1969) Diatoms. *Mem. Hourglass Cruises* I, Part III, 1–119. Fla. Dept. Nat. Resources Mar. Res. Lab., St. Petersburg.

SCHMALHAUSEN I.I. (1949) *Factors of Evolution: The Theory of Stabilizing Selection.* 327 pp. The Blakiston Company, Philadelphia.

SCHNESE W. (1969) Untersuchungen über die Produktivität der Ostsee. II. Das Phytoplankton in der mittleren Ostsee und in der Bottensee im April/Mai 1967. *Beitr. Meeresk.* **26,** 11–20.

SCHÜTT F. (1892) Das Pflanzenleben der Hochsee. *Ergebn. Atlant. Ozean Planktonexped. Humboldt-stift.* **1A,** 243–314.

SEATON D.D. (1970) Reproduction in *Rhizosolenia hebetata* and its linkage with *Rhizosolenia styliformis. J. mar. biol. Ass. U.K.* **50,** 97–106.

SEGUIN G. (1968) Le plancton de la côte nord de la Tunisie (note préliminaire). *Pelagos* **9,** 73–84.

SEIDENFADEN G. (1947) Marine phytoplankton. In *Botany of the Canadian Eastern Arctic.* Part II, Thallophyta and Bryophyta (ed N. Polunin) pp. 138–233. *Nat. Museum of Canada Bull.* **97** (Biol. Series No. 26).

SEMINA H.J. (1971) Distribution of plankton in the south-eastern Pacific (in Russian, English Summary). *Trudy̆ Inst. Okeanol.* **89,** 43–59.

SHELDON R.W., SUTCLIFFE W.H., Jr. & PRAKASH A. (1973) The production of particles in the surface waters of the ocean with particular reference to the Sargasso Sea. *Limnol. Oceanogr.* **18,** 719–33.

SHIRSHOV P. (1936) The plankton as an indicator of the ice regime of the sea. Scientific results of the expedition on the ice-breaker *Krassin* in 1935 (After Zenkevich).

SIMONSEN R. (1974) The diatom plankton of the Indian Ocean expedition of RV 'Meteor' 1964–1965. *'Meteor' Forschungsergebnisse* Reihe D No. **19,** 1–61. Gebrüder Borntraeger, Berlin.

SIMONSEN R. & KANAYA T. (1961) Notes on the marine species of the diatom genus *Denticula* Kütz. *Int. Revue ges. Hydrobiol.* **46,** 498–513.

SKOLKA V.H. (1969) Dinamica fitoplanctonului din zona de larg a platformei continentale Romånesti a Mării Negre in Anii 1964–1967. *Ecologie Mariná* (Bucharest) **3,** 159–226.

SMAYDA T.J. (1957) Phytoplankton studies in lower Narragansett Bay. *Limnol. Oceanogr.* **2,** 342–59.

SMAYDA T.J. (1958) Biogeographical studies of marine phytoplankton. *Oikos* **9,** 158–91.

SMAYDA T.J. (1963a) A quantitative analysis of the phytoplankton of the Gulf of Panama. I. Results of the regional phytoplankton surveys during July and November, 1957 and March 1958. *Bull. inter.-Am. trop. Tuna Commn.* **7,** 193–253.

SMAYDA T.J. (1963b) Succession of phytoplankton and the ocean as an holocoenotic environment. In *Symposium on Marine Microbiology* (ed. C.H. Oppenheimer) pp. 260–74. Charles C. Thomas, Springfield.

SMAYDA T.J. (1965) A quantitative analysis of the phytoplankton of the Gulf of Panama. II. On the relationship between C^{14} assimilation and the diatom standing crop. *Bull. inter-Am. trop. Tuna Commn* **9,** 466–531.

SMAYDA T.J. (1966) A quantitative analysis of the phytoplankton of the Gulf of Panama. III. General ecological conditions and the phytoplankton dynamics at 80° 45′N, 79° 23′W from November 1954 to May 1957. *Bull. inter-Am. trop. Tuna Commn* **11,** 355–612.

SMAYDA T.J. (1970) The suspension and sinking of phytoplankton in the sea. *Oceanogr. mar. Biol. Ann. Rev.* **8,** 353–414.

SMAYDA T.J. (1971) Normal and accelerated sinking of phytoplankton in the sea. *Marine Geol.* **11,** 105–22.

SMAYDA T.J. (1973a) A survey of phytoplankton dynamics in coastal waters from Cape Hatteras to Nantucket. In *Coastal and Offshore Environmental Inventory, Cape Hatteras to Nantucket Shoals* (Marine Publication Series No. 2) pp. 3-1—3-100. Univ. of Rhode Island.

SMAYDA T.J. (1973b) The growth of *Skeletonema costatum* during a winter–spring bloom in Narragansett Bay. *Norw. J. Bot.* **20,** 219–47.

SMIRNOVA L.I. (1959) Phytoplankton in the Okhotsk Sea and the Kuriles (in Russian). *Trudӯ Inst. Okeanol.* **30,** 3–51.

SOURNIA A. (1970a) A checklist of planktonic diatoms and dinoflagellates from the Mozambique channel. *Bull. mar. Sci.* **20,** 678–96.

SOURNIA A. (1970b) Les cyanophycées dans le plancton marin. *Année biol.* (Ser. 4), 63–76.

SOUSA E SILVA E. (1949) Diatomáceas e Dinoflagelados de Baía de Casais. *Port. Acta biol. Série B,* Vol: Julio Henriques, 300–83.

STEBBINS G.L. (1950) *Variation and Evolution in Plants.* Columbia University Press, New York. 643 pp.

STEEMANN NIELSEN E. (1972) The rate of primary production and the size of the standing stock of zooplankton in the oceans. *Int. Revue ges. Hydrobiol.* **57,** 513–6.

STEYAERT J. (1973a) Distribution of plankton diatoms along an African–Antarctic transect. *Investigación pesq.* **37,** 295–328.

STEYAERT J. (1973b) Difference in diatom abundance between the two summer periods of 1965 and 1967 in Antarctic inshore waters (Breid Bay). *Investigación pesq.* **37,** 517–32.

STEYAERT J. (1974) Distribution of some selected diatom species during the Belgo-Dutch Antarctic expedition of 1964–65 and 1966–67. *Investigación pesq.* **38,** 259–87.

STRICKLAND J.D.H. (1965) A comparison of profiles of nutrient and chlorophyll concentrations taken from discrete depths and by continuous recording. *Limnol. Oceanogr.* **13,** 388–91.

STRICKLAND J.D.H., EPPLEY R.W. & ROJAS DE MENDIOLA B. (1969) Phytoplankton populations, nutrients and photosynthesis in Peruvian coastal waters. *Bol. Inst. Mar. Peru* **2,** 4–45.

SUKHANOVA I.N. (1962) On the specific composition and distribution of the phytoplankton in the northern Indian Ocean (in Russian, English summary). *Trudȳ Inst. Okeanol.* **58,** 27–39.

SUKHANOVA I.N. (1969) Some data on the phytoplankton of the Red Sea and the western Gulf of Aden (in Russian, English summary). *Okeanologiya* **9,** 295–300; Engl. trans. (1969) *Oceanology* **9,** 243–7.

SVERDRUP H.U., JOHNSON M.W. & FLEMING R.H. (1942) *The Oceans, their Physics, Chemistry, and General Biology.* 1087 pp. Prentice-Hall, New York.

SWIFT E. (1973) The marine diatom *Ethmodiscus rex:* its morphology and occurrence in the plankton of the Sargasso Sea. *J. Phycol.* **9,** 456–60.

TAKANO H. (1960a) Diatoms in pearl shell fishing grounds in the Arafura Sea. *Bull. Tokai reg. Fish. Res. Lab.* **27,** 1–6.

TAKANO H. (1960b) Plankton diatoms in the eastern Caribbean Sea. *J. oceanogr. Soc. Japan* **16,** 180–4.

TAKANO H. (1972) Remarks on the morphology of the diatom *Rhizosolenia hebetata* forma *semispina* occurring in cold and warm waters. In *Biological Oeanography of the Northern North Pacific Ocean* (ed. A.Y. Takenouti) pp. 165–72. Idemitsu Shoten, Tokyo.

TAYLOR F.J.R. (1966) Phytoplankton of the south western Indian Ocean. *Nova Hedwigia* **12,** 433–76.

TAYLOR F.J.R. (1970) A preliminary annotated check list of diatoms from New Zealand coastal waters. *Trans. R. Soc. N.Z., Biol. Sci.* **12,** 153–74.

THORRINGTON-SMITH M. (1969) Phytoplankton studies in the Agulhas current region off the Natal coast. *Invest. Rep. Oceanogr. Res. Inst. S. Afr. Ass. mar. Biol.* **23,** 1–20.

THORRINGTON-SMITH M. (1970) Some new and little known planktonic diatoms from the west Indian Ocean. *Nova Hedwigia, Beih.* **31,** 815–35.

THORRINGTON-SMITH M. (1971) West Indian Ocean phytoplankton, a numerical investigation of phytohydrographic regions and their characteristic phytoplankton associations. *Mar. Biol.* **9,** 115–37.

TRÉGOUBOFF G. & ROSE M. (1957) *Manuel de planctonologie Méditerranéenne. Tome 1.* 587 pp. Centre National de la Recherche Scientifique, Paris.

UHLMANN D. (1971) Influence of dilution, sinking and grazing rate on phytoplankton populations of hyperfertilized ponds and microecosystems. *Mitt. int. Verein. theor. angew. Limnol.* **19,** 100–24.

UHM K.B. & YOO K.I. (1967) Diatoms in the Korea Strait. *Rep. Inst. mar. Biol., Seoul Univ.* **1,** 1–6

USATCHEV P. (1947) General characteristics of the phytoplankton of the seas of the USSR (in Russian). *Recent. Adv. Biol.* **23** (After Zenkevich).

UTERMÖHL H. (1958) Zur Vervollkommnung der quantitativen Phytoplankton-Methodik. *Mitt. int. Verein. Limnol.* **9,** 38 pp.

VALENTINE J.W. (1973) *Evolutionary Paleoecology of the Marine Biosphere.* 472 pp. Prentice-Hall, New Jersey.

VAN DER SPOEL S., HALLEGRAEFF G.M. & VAN SOEST R.W.M. (1973) Notes on variation of diatoms and silicoflagellates in the south Atlantic Ocean. *Neth. J. Sea Res.* **6,** 518–41.

VAN DER WERFF A. (1954) Diatoms in plankton samples of the William Barendsz-Expedition 1947. *Hydrobiologia* **6,** 331–2.

VAN LANDINGHAM S.L. (1967) *Catalogue of the Fossil and Recent Genera and Species of Diatoms and their Synonyms.* (A revision of F.W. Mills', 'An Index to the Genera and Species of the Diatomaceae and their Synonyms'). Part I. *Acanthoceras* through *Bacillaria,* 493 + xi p. Verlag J. Cramer, Lehre, Germany.

VAN LANDINGHAM S.L. (1968) *Catalogue of the Fossil and Recent Genera and Species of Diatoms and their Synonyms.* (A revision of F.W. Mills', 'An Index to the Genera and Species of the Diatomaceae and their Synonyms'.) Part II. *Bacteriastrum* through *Coscinodiscus,* pp. 494–1 086, + v-vii p. Verlag J. Cramer, Lehre, Germany.

VAN LANDINGHAM S.L. (1969) *Catalogue of the Fossil and Recent Genera and Species of Diatoms and their Synonyms.* (A revision of F.W. Mills', 'An Index to the Genera and Species of the Diatomacae and their Synonyms'.) Part III. *Coscinophaena* through *Fibula,* pp. 1087–756, + viii-x p. Verlag J. Cramer, Lehre, Germany.

VAN LANDINGHAM S.L. (1971) *Catalogue of the Fossil and Recent Genera and Species of Diatoms and their Synonyms.* (A revision of F.W. Mills', 'An Index to the Genera and Species of the Diatomaceae and their Synonyms'.) Part IV. *Fragilaria* through *Naunema,* pp. 1 757–2 385, + xi-xiv p. Verlag J. Cramer, Lehre, Germany.

VAN LANDINGHAM S.L. (1975) *Catalogue of the Fossil and Recent Genera and Species of Diatoms and their Synonyms.* (A revision of F.W. Mills', 'An Index to the Genera and Species of the Diatomaceae and their Synonyms'.) Part V. *Navicula,* p. 2 386–963. Verlag J. Cramer, Lehre, Germany.

VENRICK E.L. (1969) The distribution and ecology of oceanic diatoms in the North Pacific. Ph.D. Thesis, Univ. California, San Diego. 684 pp. Univ. Microfilms. Ann. Arbor, Mich. (Diss. Abstr. **30,** 2330–B).

VENRICK E.L. (1971) Recurrent groups of diatom species in the North Pacific. *Ecology* **52,** 614–25.

VENRICK E.L. (1972) Small-scale distributions of oceanic diatoms. *Fish. Bull. U.S. Nat. Mar. Fish. Serv. NOAA* **70,** 363–72.

VENRICK E.L. (1974) The distribution and significance of *Richelia intracellularis* Schmidt in the North Pacific Central Gyre. *Limnol. Oceanogr.* **19,** 437–45.

VIVES F. & LÓPEZ-BENITO M. (1958) El fitoplancton de la Riá de Vigo y su relación con los factores tĕrmicos y energéticos. *Investigación pesq.* **13,** 87–125.

VODYANITSKII V.A. (ed.) (1965) *Plankton Investigations in the Black and Azov Seas* (in Russian). 'Naukova Dumka', Kiev.

WILBUR H.M., TINKLE D.W. & COLLINS J.P. (1974) Environmental certainty, trophic level, and resource availability in life history evolution. *Am. Nat.* **108,** 805–17.

WILLIAMS G.C. (1966) *Adaptation and Natural Selection; a Critique of some Current Evolutionary Thought.* 307 pp. Princeton University Press, Princeton.

WILLIAMS R.B. (1964) Division rates of salt marsh diatoms in relation to salinity and cell size. *Ecology* **45,** 877–80.

WIMPENNY R.S. (1966a) *The Plankton of the Sea.* 426 pp. Faber and Faber, London.

WIMPENNY R.S. (1966b) The size of diatoms. IV. The cell diameter in *Rhizosolenia styliformis* var. *oceanica. J. mar. biol. Ass. U.K.* **46,** 541–6.

WIMPENNY R.S. (1973) The size of diatoms. V. The effect of animal grazing *J. mar. biol. Ass. U.K.* **53,** 957–74.

WOMERSLEY H.B.S (1959) The marine algae of Australia. *Bot. Rev.* **25,** 545–614.

WOOD E.J.F (1959) An unusual diatom from the Antarctic. *Nature* **184,** 1 962–3.

WOOD E.J.F. (1960) Antarctic phytoplankton studies. *Proc. Linn. Soc. N.S.W.,* **85,** 215–29.

WOOD E.J.F. (1961) Studies on Australian and New Zealand diatoms. V. The Rawson collection of recent diatoms. *Trans. R. Soc. N.Z.* **88,** 699–712.

WOOD E.J.F. (1963a) Studies on Australian and New Zealand diatoms. VI. Tropical and subtropical species. *Trans. R. Soc. N.Z. (Botany)* **2,** 189–218.

WOOD E.J.F. (1963b) Checklist of diatoms recorded from the Indian Ocean. *Rep. Div. Fish. Oceanogr. C.S.I.R.O. Aust.* **36,** 1–304.

WOOD E.J.F. (1963c) A study of the diatom flora of fresh sediments of the south Texas bays and adjacent waters. *Publs Inst. mar. Sci. Univ. Tex.* **9,** 237–310.

WOOD E.J.F. (1964) Studies on microbial ecology of the Australasian region. I. Relation of oceanic species of diatoms and dinoflagellates to hydrology. *Nova Hedwigia* **8,** 5–19.

WOOD E.J.F (1965a) A note on diatoms occurring in Milford Sound. *Bull. N.Z. Dep. scient. ind. Res.* **157,** 197.

WOOD E.J.F. (1965b) Protoplankton of the Benguela-Guinea current region. *Bull. mar. Sci.* **15,** 475–9.

WOOD E.J.F. (1966) A phytoplankton study of the Amazon region. *Bull. mar. Sci.* **16,** 102–23.

WOOD E.J.F. (1967) *Microbiology of Oceans and Estuaries.* 319 pp. Elsevier Publishing Co., Amsterdam.

WOOD E.J.F. (1968) Studies of phytoplankton ecology in tropical and subtropical environments of the Atlantic Ocean. Part 3. Phytoplankton communities in the Providence Channels and the Tongue of the Ocean. *Bull. mar. Sci.* **18,** 481–543.

WOOD E.J.F. & CROSBY L.H. (1959) Studies on Australian and New Zealand diatoms. III. Descriptions of further discoid species. *Trans. R. Soc. N.Z.* **87,** 211–9.

ZENKEVICH L.A. (1963) *Biologiya morei SSSR.* Moscow. Izd. Akad. Nauk SSSR. 793 pp. Engl. trans. (Zenkevitch 1963) *Biology of the Seas of the USSR.* 955 pp. Interscience Pub., New York.

ZERNOVA V.V. (1969) The horizontal distribution of phytoplankton in the Gulf of Mexico (in Russian, English summary). *Okeanologiya* **9,** 695–706; Engl. trans. (1969) *Oceanology* **9,** 565–75.

ORGANISM INDEX

Genera and species of diatoms and other organisms are also included in this index. Higher taxonomic categories such as families and orders are to be found in the Subject Index. A genus, e.g. *Achnanthes*, may include also unidentified species (*Achnanthes* sp.).

SUBJECT INDEX